D0910161

Chris Rauwendaal

Polymer Extrusion

SPE Books from Hanser Publishers

Chris Rauwendaal

Polymer Extrusion

Third, revised edition

Hanser Publishers, Munich Vienna New York

Hanser/Gardner Publications, Inc., Cincinnati

Distributed in the USA and in Canada by
Hanser/Gardner Publications, Inc.
6600 Clough Pike, Cincinnati, Ohio 45244-4090, USA
Fax: +1(513) 527-8950

Distributed in all other countries by
Carl Hanser Verlag
Postfach 860420, 81631 München, Germany
Fax: +49(89)981264

The use of general descriptive names, trademarks, etc, in this publication, even if the former are not especially identified, is not to be taken as a sign that such names, as understood by the Trade Marks and Merchandise Marks Act, may accordingly be used freely by anyone.

While the advice and information in this book are believed to be true and accurdate at the date of going to press, neither the authors nor the editors nor the publisher can accept any legal responsibility for any errors or omissions that may be made. The publisher makes no warranty, express or implied, with respect to the material contained herein.

Library of Congress Cataloging-in-Publication Data

Rauwendaal, Chris.
    Polymer extrusion / Chris Rauwendaal. – – 3rd. rev. ed.
    p. cm.
    Includes bibliographical references and indexes.
    ISBN 1-56990-140-6
    1. Plastic – – Extrusion.   I. Title.
  TP 1175.E9R37   1994
  668.4´13 – – dc20   94-29397

Die Deutsche Bibliothek – CIP-Einheitsaufnahme

Rauwendaal, Chris:
Polymer extrusion / Chris Rauwendaal. – 3., rev. ed. – Munich ;
Vienna ; New York : Hanser, 1994
    ISBN 3-446-17960-7 (Munich...)
    ISBN 1-56990-140-6 (New York)

Copyright © Carl Hanser Verlag, Munich Vienna New York 1994
Printed in Germany by Druckerei Appl, Wemding

To Sietske, Randy-Roel and Lisette

FOREWORD

The Society of Plastics Engineers is particularly pleased to sponsor and endorse this new "Polymer Extrusion" technical volume. The author, Chris Rauwendaal, has long been associated with the SPE Educational Seminar Program as one of its premier instructors. Through the medium of his two seminars on basic and advanced aspects of extrusion, he has been instrumental in providing a deeper insight to over a thousand engineers at these continuing education programs. In addition, this new volume itself fills a much needed void in the literature.

SPE, through its Technical Volumes Committee, has long sponsored books on various aspects of plastics and polymers. Its involvement has ranged from identification of needed volumes to recruitment of authors. An ever-present ingredient, however, has been review of the final manuscript to insure accuracy of the technical content.

This technical competence pervades all SPE activities, not only in publication of books but also in other activities such as technical conferences and educational programs. In addition, the Society publishes four periodicals - *Plastics Engineering, Polymer Engineering and Science, Journal of Vinyl Technology* and *Polymer Composites* - as well as conference proceedings and other selected publications, all of which are subject to the same rigorous technical review procedure.

The resource of some 26,000 practicing plastics engineers has made SPE the largest organization of its type in plastics worldwide. Further information is available from the Society at 14 Fairfield Drive, Brookfield Center, Connecticut 06805.

Robert D. Forger

Executive Director
Society of Plastics Engineers

Technical Volumes Committee

Donald V. Rosato, Chairman
Lee L. Blyler, Jr.
Jerome L. Dunn
Tom W. Haas

Timothy Lim
Eldridge M. Mount, III
James L. Throne
Sheldon M. Wecker

PREFACE

This book is primarily the result of the author's involvement in teaching seminars and classes on extrusion. Frequently, attendees would request an expanded text covering the seminar contents more completely. Considering that the available handout material was already about 70 pages, it became obvious that an extended, more detailed writeup could easily evolve into a full-fledged book. The original goal was to keep the book to about 300 pages. However, it soon became obvious that a complete text on extrusion would require more pages. Thus, the book has grown to a size of about twice the original goal. In future seminars, the author can expect complaints about the seminar not completely covering the text; however, the opposite complaint should be largely eliminated. An important reason for writing this book is to report on the many new developments that have occurred in the field of extrusion since about 1970. Several of these have impacted and will impact the state of the art in extrusion quite dramatically.

This book is intended primarily for practicing polymer process engineers and chemists. Being a practicing polymer process engineer myself and having dealt with many others, I am familiar with the needs and requirements of this group. The text is intended to be used in intensive short courses on extrusion. The material in the book can also provide a basis for a one-semester course on extrusion in a polymer processing engineering program. Hopefully, polymer processing will soon be more widely recognized as an important engineering discipline and will be taught in a number of colleges and universities throughout the United States.

The major aim of this book is to bridge the gap between theory and practice. At this point in time, the extrusion theory is reasonably well developed; however, application of the theory into practice leaves much to be desired. The emphasis of this book is on demonstrating how the extrusion theory can be applied to actual extrusion problems such as screw design, die design, troubleshooting, etc. Screw design, in particular, is an area where misconceptions still abound. A major chapter is devoted to a logical, step-by-step analysis of the screw design and optimization process. A sincere attempt is made to keep the mathematical complexity to a minimum. On the other hand, care is taken not to introduce too many simplifying assumptions in order to maintain predictive capability and accuracy. The objective is to obtain useful and sufficiently accurate results in a short time and with little effort.

The author would like to thank Dr. E. Immergut for his encouragement to start this project. He also wants to express his appreciation to Raychem Corporation for their support in his educational activities in the field of extrusion and in writing this book. The author wishes to thank his colleagues in Raychem Corporation for their contributions to the experimental and theoretical work described in portions of this text. In particular, he would like to thank Mr. P. Turk, Mrs. K. Erbes-Mrsny, Mr. F. Fernandez, Mr. S. Basu, and Mr. W. C. Johnson. The author gratefully acknowledges the thorough review of the manuscript by Dr. H. E. H. Meijer and his many helpful comments. He also wants to thank Mrs. K. Erbes-Mrsny for proofreading the manuscript and improving the grammar and punctuation quite substantially. Further, the author is very grateful to Ms. S. Christensen, who did an absolutely outstanding job on word processing the entire manuscript and who never lost her patience, even after the tenth revision. Above all, the author wants to thank his wife for her support and patience and his children for putting up with his frequent absence and absent-mindedness in the course of writing this book.

Menlo Park, California, March 1985
Chris Rauwendaal

CONTENTS

1 INTRODUCTION

1.1 BASIC PROCESS

The extruder is indisputably the most important piece of machinery in the polymer processing industry. To extrude means to push or to force out. Material is extruded when it is pushed through an opening. When toothpaste is squeezed out of a tube, it is extruded. The part of the machine containing the opening through which the material is forced is referred to as the extruder die. As material passes through the die, the material acquires the shape of the die opening. This shape generally changes to some extent as the material exits from the die. The extruded product is referred to as the extrudate.

Many different materials are formed through an extrusion process: metals, clays, ceramics, foodstuffs, etc. The food industry, in particular, makes frequent use of extruders to make, for instance, noodles, sausages, snacks, cereal, and numerous other items. In this book, the materials that will be treated are confined to polymers or plastics.

Polymers can be divided in three main groups: thermoplastics, thermosets, and elastomers. Thermoplastic materials soften when they are heated and solidify when they are cooled. If the extrudate does not meet the specifications, the material can generally be reground and recycled. Thus, the basic chemical nature of a thermoplastic usually does not change significantly as a result of the extrusion process. Thermosets undergo a crosslinking reaction when the temperature is raised above a certain temperature. This crosslinking bonds the polymer molecules together to form a three-dimensional network. This network remains intact when the temperature is reduced again. Crosslinking causes an irreversible change in the material. Therefore, thermosetting materials cannot be recycled as thermoplastic materials. Elastomers or rubbers are materials capable of very large deformations with the material behaving in a largely elastic manner. This means that when the deforming force is removed, the material completely, or almost completely, recovers. The emphasis of this book will be on thermoplastics; only a relatively small amount of attention will be paid to thermosets and elastomers.

Materials can be extruded in the molten state or in the solid state. Polymers are generally extruded in the molten state; however, some applications involve solid state extrusion of polymers. If the polymer is fed to the extruder in the solid state and the material is melted as it is conveyed by the extruder screw from the feed port to the die, the process is called plasticating extrusion. In this case, the extruder performs an additional function, namely melting, besides the regular extrusion function. Sometimes the extruder is fed with molten polymer; this is called melt fed extrusion. In melt fed extrusion, the extruder acts purely as a pump, developing the pressure necessary to force the polymer melt through the die.

There are two basic types of extruders: continuous and discontinous or batch type extruders. Continuous extruders are capable of developing a steady, continuous flow of material, whereas batch extruders operate in a cyclic fashion. Continuous extruders utilize a rotating member for transport of the material. Batch extruders generally have a reciprocating member to cause transport of the material.

1.2   SCOPE OF THE BOOK

This book will primarily deal with plasticating extrusion in a continuous fashion. Chapters 2, 3 and 4 deal with a description of extrusion machinery. Chapter 5 will briefly review the fundamental principles that will be used in the analysis of the extrusion process. Chapter 6 deals with the polymer properties important in the extrusion process. This is a very important chapter, because one cannot understand the extrusion process if one does not know the special characteristics of the material to be extruded. Understanding just the extrusion machinery is not enough. Rheological and thermal properties of a polymer determine, to a large extent, the characteristics of the extrusion process. The process engineer, therefore, should be a mechanical or chemical engineer on the one hand, and a rheologist on the other hand. Since most engineers have little or no formal training in polymer rheology, the subject is covered in Chapter 6, inasmuch as it is necessary in analysis of the extrusion process.

Chapter 7 covers the actual analysis of the extrusion process. The process is analyzed in discrete functional zones with particular emphasis on developing a quantitative understanding of the mechanisms operational in each zone.

The theory developed in Chapter 7 is applied in Chapter 8 to the design of extruder screws and in Chapter 9 to the design of extruder dies. Chapter 10 is devoted to twin screw extruders. Twin screw extruders have become an increasingly important branch of the extrusion industry. It is felt, therefore, that any book on extrusion cannot ignore this type of extruder. Troubleshooting extruders is covered in Chapter 11. This is perhaps the most critical and most important function performed in industrial operation of extruders. Extrusion problems causing downtime or off-spec product can become very costly in a short period of time: losses can often exceed the purchase price of the entire machine in a few hours or a few days. Thus, it is very important that the process engineer can troubleshoot quickly and accurately. This requires a solid understanding of the operational principles of extruders and the basic mechanisms behind them. Therefore, the chapter on troubleshooting is a practical application of the functional process analysis developed in Chapter 7.

The book is, thus, divided in three main parts. The first part deals with the hardware and/or mechanical aspects of extruders: this is covered in Chapters 2, 3 and 4. The second part deals with the process analysis: this is covered in Chapters 5, 6 and 7. The third part deals with practical applications of the extrusion theory: this is covered in Chapters 8, 9, 10 and 11. Parts I and II can be appreciated by themselves; Part III, however, cannot be fully appreciated without studying Part II.

1.3   GENERAL LITERATURE SURVEY

Considering there are already several books on extrusion of polymers, the question can be asked why the need for another book on extrusion. The three most comprehensive books written on extrusion are the books by Bernhardt (1), Schenkel (2), and Tadmor (5). The book by Bernhardt is on polymer processing, but has a very good chapter on extrusion. It is well written and describes the extrusion theory and practical applications to screw and die design. Because of the age of the book, however, the extrusion theory is incomplete in that it does not cover plasticating – this theory was developed later by Tadmor (5) – and devolatilization theory.

The book by Schenkel (2) is a translation of the German text "Kunststoff Extruder-Technik" (3), which is an extension of the original book "Schneckenpressen fuer Kunststoffe" (4). This book is very complete; dealing with flow properties of polymers, extrusion theory, and design of extrusion equipment. The emphasis of Schenkel's book is on the mechanical or machinery aspects of extruders. As in Bernhardt's book, the book by Schenkel suffers from the fact that it was written a long time ago, in the early sixties. Thus, the extrusion theory is incomplete and new trends in extrusion machinery are absent. The book by Tadmor (5) is probably the most complete theoretical treatise on extrusion. From a theoretical point of view, it is an outstanding book. The book is almost completely devoted to a detailed engineering analysis of the extrusion process. Consequently, relatively little attention is paid to extrusion machinery and practical applications to screw and die design. In order to fully appreciate this book, the reader should possess a significant degree of mathematical dexterity. The book is ideally suited for those who want to delve into detailed mathematical analysis and computer simulation of the extrusion process. The book is less suited for those who need to design extruder screws and/or extruder dies, or those who need to solve practical extrusion problems.

A more recent book on extrusion is "Plastics Extrusion Technology" edited by Hensen et al. (44). This is quite a comprehensive book dealing in detail with a number of extrusion processes, such as compounding, pipe extrusion, profile extrusion, etc. The book gives good information on the various extrusion operations with detailed information on downstream equipment. This book is an English version of the two-volume German original. "Kunststoff-Extrusionstechnik I und II" (45, 46). Volume I, Grundlagen (Fundamentals), covers the fundamental aspects such as rheology, thermodynamics, fluid flow analysis, single- and twin-screw extruders, die design, heating and cooling, etc. Volume II, Extrusionsanlagen, covers various extrusion lines; this is the volume that is translated into English.

Another important addition to the extrusion literature is the book on twin screw extrusion by White (47). This book has an excellent coverage of the historical developments of various twin screw extruders. It also discusses in some detail recent experimental work and developments in twin screw theory. This book covers intermeshing and non-intermeshing extruders, both co- and counter-rotating extruders. One extrusion process that has been gaining interest and is being used more widely is reactive extrusion. A good book on this subject was edited by Xanthos (48); it covers applications, review, and engineering fundamentals of reactive extrusion.

Statistical process control in extrusion is covered by a book by Rauwendaal (49). Mixing in extrusion processes is covered by a book edited by Rauwendaal (50). It covers basic aspects of mixing and mixing in various extrusion machinery, such as single screw extruders, twin screw extruders, reciprocating screw extruders, internal mixers, and corotating disk processors. Another book on mixing is the book edited by Manas-Zloczower and Tadmor (51). This book covers four major sections, mixing mechanisms and theory, modeling and flow visualization, material considerations and mixing practices.

In order not to duplicate other texts on extrusion, this book will emphasize new trends and developments in extrusion machinery. The extrusion theory will be covered as completely as possible; however, the mathematical complexity will be kept to a minimum. This is done to enhance the ease of applying the theory to practical cases and to make the book accessible to a larger number of people. A significant amount of attention will be paid to practical application of the extrusion theory to screw and die design and solving extrusion problems. After all, practical applications are most important to practicing polymer process engineers or chemists.

Various other books on extrusion have been written (6–23). These books generally cover only specific segments of extrusion technology: e.g., screw design, twin screw extrusion, etc., or provide only introductory material. Thus, it is felt that a comprehensive book on polymer extrusion with recent information on machinery, theory and applications does fulfill a need. There are several books on the more general subject of polymer processing (24–36), in addition to the book by Bernhardt (7) already mentioned. Many of these give a good review of the field of polymer processing, some emphasizing the general principles involved in process analysis (24–33) or machine design (34), while others concentrate more on a description of the process machinery and products (35, 36). Since the field of polymer processing is tremendously large, books on the subject inevitably cover extrusion in less detail than possible in a book devoted exclusively to extrusion.

1.4   HISTORY OF POLYMER EXTRUSION

The first machine for extrusion of thermoplastic materials was built around 1935 by Paul Troester in Germany (37). Before this time, extruders were primarily used for extrusion of rubber. These earlier rubber extruders were steam-heated ram extruders and screw extruders, with the latter having very short length to diameter (L/D) ratios, about 3 to 5. After 1935 extruders evolved to electrically heated screw extruders with increased length. Around this time, the basic principle of twin screw extruders for thermoplastics was conceived in Italy by Roberto Colombo of LMP. He was working with Carlo Pasquetti on mixing cellulose acetate. Colombo developed an intermeshing corotating twin screw extruder. He obtained patents in many different countries and several companies acquired the rights to use these patents. Pasquetti followed a different concept and developed and patented the intermeshing counterrotating twin screw extruder.

The first detailed analyses of the extrusion process were concerned with the melt conveying or pumping process. The earliest publication was an anonymous article (38), which is often erroneously credited to Rowell and Finlayson, who wrote an article of the same title in the same journal six years later (39). Around 1950, scientific studies of the extrusion process started appearing with increased frequency. In the mid-fifties, the first quantitative study on solids conveying was published by Darnell and Mol (40). An important conference in the development of extrusion theories was the 122 nd ACS meeting in 1953. At this symposium, members of the Polychemicals Department of E. I. du Pont de Nemours & Co. presented the latest developments in extrusion theory (41). These members, Carley, Strub, Mallouk, McKelvey, and Jepson, were honored in 1983 by the SPE Extrusion Division for original development of extrusion theories. In the mid-sixties, the first quantitative study on melting was published by Tadmor (42), based on earlier qualitative studies by Maddock (43). Thus, it was not until about 1965 that the entire extrusion process, from the feed hopper to the die, could be described quantitatively. The theoretical work since this time has concentrated to a large extent on generalizing and extending the extrusion theory and the development of numerical techniques and computer methods to solve equations that can no longer be solved by analytical methods.

As a result, there has been a shift in the affiliation of the workers involved in scientific extrusion studies. While the early work was done mostly by investigators in the polymer industry, later workers, to a large extent, have been academicians. This has created somewhat of a gap between extrusion theoreticians and practicing extrusion technologists. This is aggrevated by the fact that some workers are so concerned about the scientific pureness of the work, which is in itself commendable, that it becomes increasingly unappealing to the industrial process engineer who wants to apply the work. One of the objectives of this book is to bridge this gap between theory and practice by demonstrating, in detail, how the theory can be applied and by analyzing the limitations of the theory.

Another interesting development in practical extrusion technology has been the concept of feed-controlled extrusion. In this type of extrusion, the performance is determined by the solids conveying zone of the extruder. By the use of grooves in the first portion (close to the feed port) of the extruder barrel, the solids conveying zone is capable of developing very high pressures and quite positive conveying characteristics (i.e., throughput independent of pressure). In this case, the diehead pressure does not need to be developed in the pumping or melt conveying zone; it is developed in the solids conveying zone. The feed zone overrides both the plasticating and the melt conveying (pumping) zone.

Thus, the extruder becomes solids-conveying or feed-controlled. This concept has now become an accepted standard in Western Europe, particularly in West Germany. It is estimated that more than 5,000 extruders have been equipped with grooved barrel sections in West Germany.

In the U.S., this concept has met with a substantial amount of reluctance and skepticism. For the longest time, the few proponents of the feed-controlled extrusion were heavily outnumbered by the many opponents. As a consequence, only a small fraction of the extruders in the U.S. have been equipped with grooved barrel sections. However, there does seem to be a trend towards increasing acceptance of the new concept.

REFERENCES – CHAPTER 1

1. E. C. Bernhardt, Ed., "Processing of Thermoplastic Materials", Reinhold, N. Y. (1959).
2. G. Schenkel, "Plastics Extrusion Technology and Theory", Illiffe Books Ltd., London (1966), published in the U.S.A. by American Elsevier, N.Y. (1966).
3. G. Schenkel, "Kunststoff Extruder-Technik", Carl Hanser Verlag, Munich (1963).
4. G. Schenkel, "Schneckenpressen fuer Kunststoffe", Carl Hanser Verlag, Munich (1959).
5. Z. Tadmor and I. Klein, "Engineering Principles of Plasticating Extrusion", Van Nostrand Reinhold, N. Y. (1970).
6. H. R. Simonds, A. J. Weith, and W. Schack, "Extrusion of Rubber, Plastics and Metals", Reinhold, N. Y. (1952).
7. E. G. Fisher, "Extrusion of Plastics", Illiffe Books Ltd., London (1954).
8. H. R. Jacobi, "Grundlagen der Extrudertechnik", Carl Hanser Verlag, Munich (1960).
9. W. Mink, "Grundzuege der Extrudertechnik", Rudolf Zechner Verlag, Speyer am Rhein (1963).
10. A. L. Griff, "Plastics Extrusion Technology", Reinhold, N. Y. (1968).
11. R. T. Fenner, "Extruder Screw Design", Illiffe Books, Ltd., London (1970).
12. N. M. Bikales, Ed., "Extrusion and Other Plastics Operations", Wiley, N. Y. (1971).
13. P. N. Richardson, "Introduction to Extrusion", Society of Plastics Engineers, Inc. (1974).
14. L. P. B. M. Janssen, "Twin Screw Extrusion", Elsevier, Amsterdam (1978).
15. J. A. Brydson and D. G. Peacock, "Principles of Plastics Extrusion", Applied Science Publishers Ltd., London (1973).
16. F. G. Martelli, "Twin Screw Extrusion, A Basic Understanding", Van Nostrand Reinhold, N. Y. (1983).
17. "Kunststoff-Verarbeitung im Gespraech, 2 Extrusion", BASF, Ludwigshafen (1971)
18. "Der Extruder als Plastifiziereinheit", VDI-Verlag, Duesseldorf (1977).

19. S. Levy., "Plastics Extrusion Technology Handbook", Industrial Press Inc., N. Y. (1981).
20. H. Potente, "Auslegen von Schneckenmaschinen-Baureihen, Modellgesetze und ihre Anwendung", Carl Hanser Verlag, Munich (1981).
21. H. Herrmann, "Schneckenmaschinen in der Verfahrenstechnik", Springer-Verlag, Berlin (1972)
22. W. Dalhoff, "Systematische Extruder-Konstruktion", Krausskopf-Verlag, Mainz (1974).
23. E. Harms, "Kautschuk-Extruder, Aufbau und Einsatz aus verfahrenstechnischer Sicht", Krausskopf-Verlag Mainz, Bd. 2, Buchreihe Kunststofftechnik (1974).
24. J. M. McKelvey, "Polymer Processing", Wiley, N. Y. (1962).
25. R. M. Ogorkiewicz, "Thermoplastics: Effects of Processing", Illiffe Books Ltd., London (1969).
26. J. R. A. Pearson, "Mechanical Principles of Polymer Melt Processing", Pergamon, Oxford (1966)
27. S. Middleman, "The Flow of High Polymers", Interscience, (1968).
28. R. V. Torner, "Grundprozesse der Verarbeitung von Polymeren", VEB Deutscher Verlag fuer Grundstoffindustrie, Leipzig (1973).
29. S. Middleman, "Fundamentals of Polymer Processing", McGraw-Hill, N. Y. (1977).
30. H. L. Williams, "Polymer Engineering", Elsevier, Amsterdam (1975).
31. J. L. Throne, "Plastics Process Engineering", Marcel Dekker, Inc., N. Y. (1979).
32. Z. Tadmor and C. Gogos, "Principles of Polymer Processing", Wiley, N. Y. (1979).
33. R. T. Fenner, "Principles of Polymer Processing", MacMillan Press Ltd., London (1979).
34. N. S. Rao, "Designing Machines and Dies for Polymer Processing with Computer Programs", Carl Hanser Verlag, Munich (1981).
35. J. Frados, Ed., "Plastics Engineering Handbook", Van Nostrand Reinhold, N. Y. (1976).
36. S. S. Schwartz and S. H. Goodman, "Plastics Materials and Processes", Van Nostrand Reinhold, N. Y. (1982).
37. M. Kaufman, Plastics & Polymers, 37, 243 (1969).
38. N. N., Engineering, 114, 606 (1922).
39. H. S. Rowell and D. Finlayson, Engineering, 126, 249–250, 385–387, 678 (1928).
40. W. H. Darnell and E. A. J. Mol, SPE Journal, 12, 20 (1956).
41. J. F. Carley, R. A. Strub, R. S. Mallouk, J. M. McKelvey, and C. H. Jepson, 122 nd Meeting of the American Chemical Society, Atlantic City, N. J. (1953). The seven papers were published in Ind. Eng. Chem., 45, 970–992 (1953).
42. Z. Tadmor, Polym. Eng. Sci., 6, 3, 1 (1966).
43. B. H. Maddock, S. P. E. Journal, 15, 383 (1959).
44. F. Hensen, W. Knappe, and H. Potente, Editors, "Plastics Extrusion Technology," Carl Hanser Verlag, Munich (1988).
45. F. Hensen, W. Knappe, and H. Potente, Editors, "Handbuch der Kunststoff-Extrusionstechnik I Grundlagen," Carl Hanser Verlag, Munich (1989).
46. F. Hensen, W. Knappe, and H. Potente, Editors, "Handbuch der Kunststoff-Extrusionstechnik II Extrusionsanlagen," Carl Hanser Verlag, Munich (1989).
47. J. L. White, "Twin Screw Extrusion," Carl Hanser Verlag, Munich (1991).
48. M. Xanthos, "Reactive Extrusion," Carl Hanser Verlag, Munich (1992).
49. C. Rauwendaal, "Statistical Process Control in Extrusion," Carl Hanser Verlag, Munich (1993).
50. C. Rauwendaal, Ed., "Mixing in Polymer Processing," Marcel Dekker, New York (1991).
51. I. Manas-Zloczower and Z. Tadmor, Editors, "Mixing and Compounding – Theory and Practice," Carl Hanser Verlag, Munich (1994).

PART I

EXTRUSION MACHINERY

2 DIFFERENT TYPES OF EXTRUDERS

Extruders in the polymer industry come in many different designs. The main distinction between the various extruders is their mode of operation: continuous or discontinuous. The latter type extruder delivers polymer in an intermittent fashion and, therefore, is ideally suited for batch type processes, such as injection molding and blow molding. As mentioned earlier, continuous extruders have a rotating member, whereas batch extruders have a reciprocating member. A classification of the various extruders is shown in Table 2-1.

TABLE 2-1. CLASSIFICATION OF POLYMER EXTRUDERS

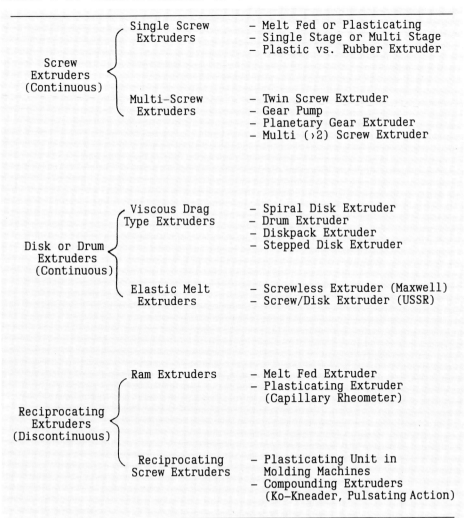

Screw Extruders (Continuous)	Single Screw Extruders	– Melt Fed or Plasticating – Single Stage or Multi Stage – Plastic vs. Rubber Extruder
	Multi-Screw Extruders	– Twin Screw Extruder – Gear Pump – Planetary Gear Extruder – Multi (>2) Screw Extruder
Disk or Drum Extruders (Continuous)	Viscous Drag Type Extruders	– Spiral Disk Extruder – Drum Extruder – Diskpack Extruder – Stepped Disk Extruder
	Elastic Melt Extruders	– Screwless Extruder (Maxwell) – Screw/Disk Extruder (USSR)
Reciprocating Extruders (Discontinuous)	Ram Extruders	– Melt Fed Extruder – Plasticating Extruder (Capillary Rheometer)
	Reciprocating Screw Extruders	– Plasticating Unit in Molding Machines – Compounding Extruders (Ko-Kneader, Pulsating Action)

2.1   THE SINGLE SCREW EXTRUDER

Screw extruders are divided into single screw and multi screw extruders. The single screw extruder is the most important type of extruder used in the polymer industry. Its key advantages are relatively low cost, straightforward design, ruggedness and reliability, and favorable performance/cost ratio. A detailed description of the hardware components of a single screw extruder is given in Chapter 3.

The extruder screw of a conventional plasticating extruder has three geometrically different sections, see Figure 2-1.

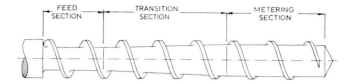

Figure 2 – 1. Geometry Conventional Extruder Screw

This geometry is also referred to as a "single stage." The single stage refers to the fact that the screw has only one compression section, even though the screw has three distinct geometrical sections! The first section (closest to the feed opening) generally has deep flights. The material in this section will be mostly in the solid state. This section is referred to as the feed section of the screw. The last section (closest to the die) usually has shallow flights. The material in this section will be mostly in the molten state. This screw section is referred to as the metering section or pump section. The third screw section connects the feed section and the metering section. This section is called the transition section or compression section. In most cases, the depth of the screw channel (or the height of the screw flight) reduces in a linear fashion, going from the feed section towards the metering section, thus causing a compression of the material in the screw channel. Later, it will be shown that this compression, in many cases, is essential to the proper functioning of the extruder.

The extruder is usually designated by the diameter of the extruder barrel. In the U.S., the standard extruder sizes are $\frac{3}{4}$, 1, $1\frac{1}{2}$, 2, $2\frac{1}{2}$, $3\frac{1}{2}$, $4\frac{1}{2}$, 6, 8, 10, 12, 14, 16, 18, 20 and 24 inches.

Obviously, the very large machines are much less common than the smaller extruders. Some machines go up in size as large as 35 inches. These machines are used in specialty operations, such as melt removal directly from a polymerization reactor. In Europe, the standard extruder sizes are 20, 25, 30, 35, 40, 50, 60, 90, 120, 150, 200, 250, 300, 350, 400, 450, 500, and 600 millimeters. Most extruders range in size from 1 to 6 inches or from 25 to 150 mm. An additional designation often used is the length of the extruder, generally expressed as length to diameter (L/D) ratio. Typical L/D ratios range from 20 to 30, with 24 being very common. Extruders used for extraction of volatiles (vented extruders, see Section 2.1.2) can have an L/D ratio as high as 35 or 40 and sometimes even higher.

2.1.1   BASIC OPERATION

The basic operation of a single screw extruder is rather straightforward. Material enters from the feed hopper. Generally, the feed material flows by gravity from the feed hopper down into the extruder barrel. Some materials do not flow easily in dry form and special measures have to be taken to prevent hang-up (bridging) of the material in the feed hopper.

As the material falls down in the extruder barrel, it is situated in the annular space between the extruder screw and barrel, and further bounded by the passive and active flanks of the screw flight: the screw channel. The barrel is stationary and the screw is rotating. As a result, frictional forces will act on the material, both on the barrel as well as on the screw surface. These frictional forces are responsible for the forward transport of the material, at least as long as the material is in the solid state (below its melting point).

As the material moves forward, it will heat up as a result of frictional heat generation and heat conducted from the barrel heaters. When the temperature of the material exceeds the melting point, a melt film will form at the barrel surface. This is where the plasticating zone starts. It should be noted that this point generally does not coincide with the start of the compression section. The boundaries of the functional zones will depend on polymer properties, machine geometry, and operating conditions. Thus, they can change as operating conditions change. However, the geometrical sections of the screw are fixed by the design and will not change with operating conditions. As the material moves forward, the amount of solid material at each location will reduce as a result of melting. When all solid polymer has disappeared, the end of the plasticating zone has been reached and the melt conveying zone starts. In the melt conveying zone, the polymer melt is simply pumped to the die.

As the polymer flows through the die, it adopts the shape of the flow channel of the die. Thus, as the polymer leaves the die, its shape will more or less correspond to the crosssectional shape of the final portion of the die flow channel. Since the die exerts a resistance to flow, a pressure is required to force the material through the die. This is generally referred to as the diehead pressure. The diehead pressure is determined by the shape of the die (particularly the flow channel), the temperature of the polymer melt, the flow rate through the die, and the rheological properties of the polymer melt. It is important to understand that the diehead pressure is caused by the die, and not by the extruder! The extruder simply has to generate sufficient pressure to force the material through the die. If the polymer, the throughput, the die, and the temperatures in the die are the same, then it does not make any difference whether the extruder is a gear pump, a single screw extruder, a twin screw extruder, etc.; the diehead pressure will be the same. Thus, the diehead pressure is caused by the die, by the flow process taking place in the die flow channel. This is an important point to remember.

2.1.2   VENTED EXTRUDERS

Vented extruders are significantly different from nonvented extruders in design and in functional capabilities. A vented extruder is equipped with one or more openings (vent ports) in the extruder barrel, through which volatiles can escape. Thus, the vented extruder can extract volatiles from the polymer in a continuous fashion. This devolatilization adds a functional capability not present in non-vented extruders. Instead of the extraction of volatiles, one can use the vent port to add certain components to the polymer, such as additives, fillers, reactive components, etc. This clearly adds to the versatility of vented extruders, with the additional benefit that the extruder can be operated as a conventional non-vented extruder by simply plugging the vent port and, possibly, changing the screw geometry.

A schematic picture of a vented extruder is shown in Figure 2-2.

The design of the extruder screw is very critical to the proper functioning of the vented extruder. One of the main problems that vented extruders are plagued with is vent flow. This is a situation where not only the volatiles are escaping through the vent port, but also some amount of polymer. Thus, the extruder screw has to be designed in such a way that there

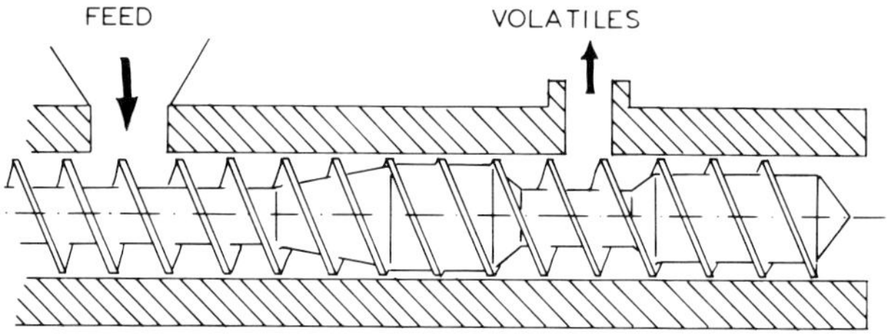

Figure 2-2. Schematic Vented Extruder

will be no positive pressure in the polymer under the vent port (extraction section). This has led to the development of the two-stage extruder screw, especially designed for devolatilizing extrusion. Two-stage extruder screws have two compression sections separated by a decompression/extraction section. It is somewhat like two single-stage extruder screws coupled in series along one shaft. The details of the design of two-stage extruder screws will be covered in Chapter 8. Vented extruders are used for the removal of monomers and oligomers, reaction products, moisture, solvents, etc.

The devolatilization capability of single screw extruders of conventional design is limited compared to twin screw extruders. Twin screw extruders can handle solvent contents of 50 percent and higher, using a multiple stage extraction system, and solvent content of up to 15 percent using single stage extraction. Single screw vented extruders of conventional design usually cannot handle more than 5 percent volatiles; this would require multiple vent ports. With a single vent port, a single screw vented extruder of conventional design can generally reduce the level of volatiles only a fraction of one percent, depending, of course, on the polymer/solvent system.

Because of the limited devolatilization capacity of single screw extruders of conventional design, they are sometimes equipped with two or more vent ports. A drawback of such a design is that the length of the extruder can become a problem. Some of these extruders have a L/D ratio of 40 to 50! This creates a problem in handling the screw, for instance when the screw is pulled, and increases the chance of mechanical problems in the extruder (deflection, buckling, etc.). If substantial amounts of volatiles need to be removed, a twin screw extruder may be more cost-effective than a single screw extruder. However, some vented single screw extruders of more modern design have substantially improved devolatilization capability and deserve equal consideration, see Section 8.5.2.

2.1.3   RUBBER EXTRUDERS

Extruders for processing elastomers have been around longer than any other type of extruder. Industrial machines for rubber extrusion were built as early as the second half of the nineteenth century. Some of the early extruder manufacturers were John Royle in the U.S. and Francis Shaw in England. One of the major rubber extruder manufacturers in Germany was Paul Troester; in fact, it still is a major producer of extruders. Despite the fact that rubber extruders have been around for more than a century, there is a surprising lack of literature on the subject of rubber extrusion. Some of the handbooks on rubber (1–5) discuss rubber extrusion, but in most cases the information is very meager and of limited usefulness. Harms' book on rubber extruders (13) appears to be the only book devoted exclusively to rubber extrusion. The few publications on rubber extrusion stand in sharp contrast to the abundance of books and articles on plastic extrusion. Considering the commercial significance of rubber extrusion, this is a surprising situation.

The first rubber extruders were built for hot feed extrusion. These machines are fed with warm material from a mill or other mixing device. Around 1950, machines were developed for cold feed extrusion. The advantages of cold feed extruders are thought to be:

— Less capital equipment cost
— Better control of stock temperature
— Reduced labor cost
— Capable of handling a wider variety of compounds

However, there is no general agreement on this issue. As a result, hot feed rubber extruders are still in use today.

Cold feed rubber extruders, nowadays, do not differ too much from thermoplastic extruders. Some of the differences are: i) reduced length, ii) heating and cooling, iii) feed section, and iv) screw design. There are several reasons for the reduced length. The viscosity of rubbers is generally very high compared to most thermoplastics; about an order of magnitude higher (5). Consequently, there is a substantial amount of heat generated in the extrusion process. The reduced length keeps the temperature build-up within limits. The specific energy requirement for rubbers is generally low, partly because they are usually extruded at relatively low temperatures (from 20 to 120° C). This is another reason for the short extruder length. The length of the rubber extruder will depend on whether it is a cold or hot feed extruder. Hot feed rubber extruders are usually very short, about 5 D (D = diameter). Cold feed extruders range from 15 D to 20 D. Vented cold feed extruders may be even longer than 20 D.

Rubber extruders used to be heated quite frequently with steam because of the relatively low extrusion temperatures. Today, many rubber extruders are heated like thermoplastic extruders with electrical heater bands clamped around the barrel. Oil heating is also used on rubber extruders and the circulating oil system can also be used to cool the rubber. Many rubber extruders use water cooling because it allows effective heat transfer.

The feed section of the rubber extruders has to be designed specifically to the feed stock characteristics of the material. The extruder may be fed with either strips, chunks or pellets. If the extruder is fed from an internal mixer (e.g., Banbury, Shaw, etc.), a power-operated ram can be used to force the rubber compound into the extruder. The feed opening can be undercut to improve the intake capability of the extruder. This can be useful, because the rubber feed stock at times comes in relatively large particles of irregular shape. When the material is supplied in the form of a strip, the feed opening is often equipped with a driven roll parallel to the screw to give a "roller feed." Material can also be supplied in powder form. It has been shown that satisfactory extrusion is possible if the powder is consolidated by "pill-making" techniques. Powdered rubber technology is discussed in detail in (6).

The rubber extrusion technology appears to be considerably behind the plastics extrusion technology. Kennaway, in one of the few articles on rubber extrusion (8), contributes this situation to two factors. The first is the frequent tendency of rubber process personnel to solve extrusion problems by changing the formulation of the compound. The second is the widespread notion that the extrusion behavior of rubbers is substantially different from plastics, because rubbers crosslink and plastics generally do not. This is a misconception, however, because extrusion characteristics of rubber and plastics are actually not substantially different (9).

When the rubber is slippery, as in dewatering rubber extruders, the feed section of the barrel is grooved to prevent slipping along the barrel surface or the barrel I. D. may be fitted with pins. This significantly improves the conveying action of the extruder. The same technique has been applied to thermoplastic extrusion, as discussed in Section 1.4 and Section 7.2.2.2.

The extruder screw for rubber often has constant depth and variable decreasing pitch (VDP); many rubber screws use a double-flighted design, see Figure 2-3. Screws for thermoplastics usually have a decreasing depth and constant pitch, see Figure 2-1.

Figure 2-3. Typical Screw Geometry for Rubber

Another difference with the rubber extruder screw is that the channel depth is usually considerably larger than with a plastic extruder screw. The larger depth is to reduce the shearing of the rubber and the resulting viscous heat generation. There is a large variety of rubber extruder screws, as is the case with plastic extruder screws. Figure 2-4 shows the "Plasti-screw" manufactured by NRM.

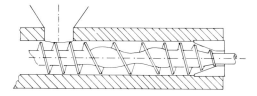

Figure 2 – 4. NRM Plastiscrew

Figure 2-5 shows the Pirelli rubber extruder screw. This design uses a feed section of large diameter, reducing quickly to the much smaller diameter of the pumping section. The conical feed section uses a large clearance between screw flight and barrel wall. This causes a large amount of leakage over the flight and improves the batch-mixing capability of the extruder. Figure 2-6 shows the EVK screw by Werner & Pfleiderer (14).

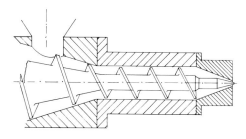

Figure 2 – 5. Pirelli Rubber Extruder Screw

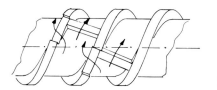

Figure 2 – 6. EVK Screw

This design features cross-channel barriers with preceeding undercuts in the flights to provide a change in flow direction and increased shearing as the material flows over the barrier or the undercut in the flight. A rather unusual design is the Transfermix (10–13) extruder/mixer which has been used for compounding rubber formulations. This machine features helical channels in both the screw and barrel, see Figure 2-7.

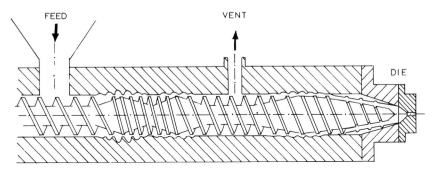

Figure 2 – 7.   Transfermix

By a varying root diameter of the screw, the material is forced in the flow channel of the barrel. A reduction of the depth of the barrel channel forces the material back into the screw channel. This frequent reorientation provides good mixing. However, the machine is difficult to manufacture and expensive to repair in case of damage.

A fairly recent development is the QSM extruder (15–20, 7). QSM stands for the German words "Quer Strom Misch", meaning cross-flow mixing. This extruder has adjustable pins in the extruder barrel that protrude all the way into the screw channel, see Figure 2-8.

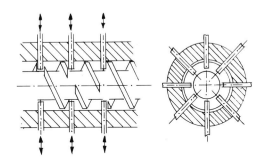

Figure 2 – 8.   QSM Extruder (Pin Barrel Extruder)

The screw flight has slots at the various pin locations. The advantages of this extruder in rubber applications are good mixing capability with a low stock temperature increase and low specific energy consumption. This extruder was developed by Harms in Germany and is manufactured and sold by Troester and other companies. Even though the QSM extruder has become popular in the rubber industry, its applications clearly extend beyond just rubber extrusion. In thermoplastic extrusion, its most obvious application would be in high viscosity, thermally less stable resins: PVC could possibly be a candidate, although dead spots may create problems with degradation. However, it can probably be applied wherever good mixing and good temperature control are required.

Another typical rubber extrusion piece of hardware is the roller die. A schematic representation is shown in Figure 2-9.

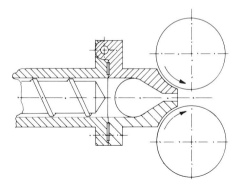

Figure 2 – 9. Roller Die

The roller die (B. F. Goodrich, 1933) is a combination of a standard sheet die and a calender. It allows high throughput by reducing the diehead pressure, it reduces air entrapment and provides good gauge control.

2.2   THE MULTI SCREW EXTRUDER

2.2.1   THE TWIN SCREW EXTRUDER

A twin screw extruder is a machine with two Archimedian screws. Admittedly, this is a very general definition. However, as soon as the definition is made more specific, one limits it to a specific class of twin screw extruders. There is a tremendous variety of twin screw extruders, with vast differences in design, principle of operation, and field of application. It is, therefore, difficult to make general comments about twin screw extruders. The differences between the various twin screw extruders are much larger than the differences between single screw extruders. This is to be expected, since the twin screw construction substantially increases the number of design variables, such as direction of rotation, degree of intermeshing, etc. A classification of twin screw extruders is shown in Table 2-2. This classification is primarily based on the geometrical configuration of the twin screw extruder. Some twin screw extruders function much in the same fashion as single screw extruders. Other twin screw extruders operate quite differently from single screw extruders and are used in very different applications. The design of the various twin screw extruders with their operational and functional aspects will be covered in more detail in Chapter 10.

TABLE 2-2. CLASSIFICATION OF TWIN SCREW EXTRUDERS

Intermeshing Extruders	Corotating Extruders	– Low Speed Extruders (profile extrusion) – High Speed Extruders (compounding, devolatilization)
	Counterrotating Extruders	– Conical Extruders (profile extrusion) – Cylindrical Extruders (profile extrusion)
Nonintermeshing Extruders	Counterrotating Extruders	– Equal Screw Length – Unequal Screw Length
	Corotating Extruders	– Not used in practice
	Coaxial Extruders	– Inner Melt Transport Forward – Inner Melt Transport Rearward – Inner Solids Transport Rearward – Inner Plasticating, Rearward Transport

2.2.2   THE MULTI SCREW EXTRUDER WITH MORE THAN TWO SCREWS

There are several types of extruders that incorporate more than two screws. One relatively well-known example is the planetary roller extruder, see Figure 2-10.

This extruder looks similar to a single screw extruder. The feed section is, in fact, the same as on a standard single screw extruder. However, the mixing section of the extruder looks considerably different. In the planetary roller section of the extruder, six or more planetary screws, evenly spaced, revolve around the circumference of the main screw. In the planetary screw section, the main screw is referred to as the sun screw. The planetary screws intermesh with the sun screw and the barrel. The planetary barrel section, therefore, must have helical grooves corresponding to the helical flights on the planetary screws. This planetary barrel section is generally a separate barrel section with a flange-type connection to the feed barrel section.

In the first part of the machine, before the planetary screws, the material moves forward as in a regular single screw extruder. As the material reaches the planetary section, being at this point largely plasticated, it is exposed to intensive mixing by the rolling action between the planetary screws, the sun screw and the barrel. The helical design of the barrel, sun screw and planetary screws result in a large surface area relative to the barrel length. The small clearance between the planetary screws and the mating surfaces, about 1/4 mm, allows thin layers of compound to be exposed to large surface areas, resulting in effective devolatiliza-

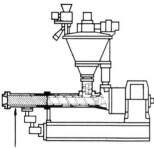

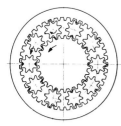

planetary gear section

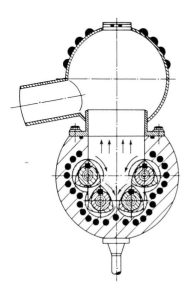

Figure 2 – 10.   Planetary Roller Extruder

detail planetary gear section

tion, heat exchange and temperature control. Thus, heat-sensitive compounds can be processed with a minimum of degradation. For this reason, the planetary gear extruder is frequently used for extrusion/compounding of PVC formulations, both rigid and plasticized
(21,22). Planetary roller sections are also used as add-ons to regular extruders to improve
mixing performance (97,98). Another multi-screw extruder is the four-screw extruder,
shown in Figure 2-11.

Figure 2 – 11.   Four – Screw Extruder
(Courtesy Werner & Pfleiderer Corporation)

This machine is used primarily for devolatilization of solvents from 40 percent to as low as 0.3 percent (23). Flash devolatilization occurs in a flash dome attached to the barrel. The polymer solution is delivered under pressure and at temperatures above the boiling point of the solvent. The solution is then expanded through a nozzle into the flash dome. The foamy material resulting from the flash devolatilization is then transported away by the four screws. In many cases, downstream vent sections will be incorporated to further reduce the solvent level.

2.2.3   THE GEAR PUMP EXTRUDER

Gear pumps are used in some extrusion operations at the end of a plasticating extruder, either single screw or twin screw (99–106). Strictly speaking, the gear pump is a closely intermeshing counterrotating twin screw extruder. However, since gear pumps are solely used to generate pressure, they are generally not referred to as an extruder even though the gear pump is an extruder. One of the main advantages of the gear pump is its good pressure-generating capability and its ability to maintain a relatively constant outlet pressure even if the inlet pressure fluctuates considerably. Some fluctuation in the outlet pressure will result from the intermeshing of the gear teeth. This fluctuation can be reduced by a helical orientation of the gear teeth instead of an axial orientation.

Gear pumps are sometimes referred to as positive displacement devices. This is not completely correct because there must be mechanical clearances between the gears and the housing, which causes leakage. Therefore, the gear pump output is dependent on pressure, although the pressure sensitivity will generally be less than that of a single screw extruder. The actual pressure sensitivity will be determined by the design clearances, the polymer melt viscosity, and the rotational speed of the gears. A good method to obtain constant throughput is to maintain a constant pressure differential across the pump. This can be done by a relatively simple pressure feedback control on the extruder feeding into the gear pump (102). The non-zero clearances in the gear pump will cause a transformation of mechanical energy into heat by viscous heat generation, see Section 5.3.4. Thus, the energy efficiency of actual gear pumps is considerably below 100 percent; the pumping efficiency generally ranges from 15 to 35 percent. The other 65 to 85 percent goes into mechanical losses and viscous heat generation. Mechanical losses usually range from about 20 to 40 percent and viscous heating from about 40 to 50 percent. As a result, the polymer melt going through the gear pump will experience a considerable temperature rise, typically 5 to 10° C. However, in some cases the temperature rise can be as much as 20 to 30° C. Since the gear pump has limited energy efficiency, the combination extruder-gear pump is not necessarily more energy-efficient than the extruder without the gear pump. Only if the extruder feeding into the gear pump is very inefficient in its pressure development will the addition of a gear pump allow a reduction in energy consumption. This could be the case with co-rotating twin screw extruders or single screw extruders with inefficient screw design.

The mixing capacity of gear pumps is very limited. This was clearly demonstrated by Kramer (106) by comparing melt temperature fluctuation before and after the gear pump, which showed no distinguishable improvement in melt temperature uniformity. Gear pumps are often added to extruders with unacceptable output fluctuations. In many cases this constitutes treating the symptoms but not curing the actual problem. Most single screw extruders, if properly designed, can maintain their output to within ± 1 percent. If the output fluctuation is considerably larger than 1 percent, there is probably something wrong with the machine; very often incorrect screw design. In these cases, solving the actual problem will generally be more efficient than adding a gear pump. For an efficient extruder-gear pump system, the extruder screw has to be modified to reduce the pressure-generating capacity of the screw.

Gear pumps can be used advantageously:

a) On extruders with poor pressure-generating capability (e. g., corotating twin screws, multi−stage vented extruders, etc.)

b) When output stability is required of better than 1 percent, i. e., in close tolerance extrusion (e. g., fiber spinning, cable extrusion, medical tubing, coextrusion, etc.).

Gear pumps can cause problems when:

a) The polymer contains abrasive components; because of the small clearances, the gear pump is very susceptible to wear.

b) When the polymer is susceptible to degradation; gear pumps are not self − cleaning and combined with the exposure to high temperatures this will result in degraded product.

2.3   DISK EXTRUDERS

There are a number of extruders that do not utilize an Archimedean screw for transport of the material, but still fall in the class of continuous extruders. Sometimes these machines are referred to as screwless extruders. These machines employ some kind of disk or drum to extrude the material. One can classify the disk extruders according to their conveying mechanism (see Table 2-1). Most of the disk extruders are based on viscous drag transport. One special disk extruder utilizes the elasticity of polymer melts to convey the material and to develop the necessary diehead pressure.

Disk extruders have been around for a long time, at least since 1950. However, at this point in time, the industrial significance of disk extruders is still relatively small compared to screw extruders.

2.3.1   VISCOUS DRAG DISK EXTRUDERS

2.3.1.1   STEPPED DISK EXTRUDER

One of the first disk extruders was developed by Westover at Bell Telephone Laboratories: it is often referred to as a stepped disk extruder or slider pad extruder (24). A schematic picture of the extruder is shown in Figure 2-12.

The heart of the machine is the stepped disk positioned a small distance from a flat disk. When one of the disks is rotated with a polymer melt in the axial gap, a pressure build-up will occur at the transition of one gap size to another, smaller gap size, see Figure 2-13.

If exit channels are incorporated into the stepped disk, the polymer can be extruded in a continuous fashion. The design of this extruder is based on Rayleigh's (25) analysis of hydrodynamic lubrication in various geometries. He concluded that the parallel stepped pad

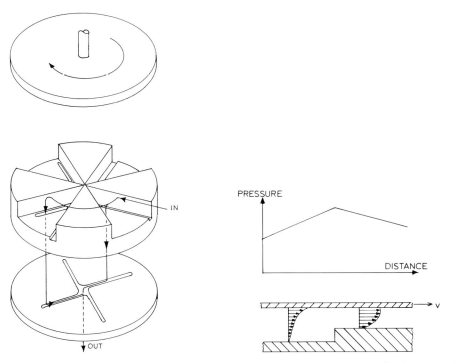

Figure 2–12.   Stepped Disk Extruder   Figure 2–13.   Pressure Generation in Step Region

was capable of supporting the greatest load. The stepped disk extruder has also been de-signed in a different configuration using a gradual change in gap size. This extruder, thus, has a wedge-shaped disk with a gradual increase in pressure with radial distance.

A practical disadvantage of the stepped disk extruder is the fact that the machine is difficult to clean because of the intricate design of the flow channels in the stepped disk.

2.3.1.2   DRUM EXTRUDER

Another rather old concept is the drum extruder. A schematic picture of a machine manu-factured by Schmid & Kocher in Switzerland is shown in Figure 2-14.

Material is fed by a feed hopper into an annular space between rotor and barrel. By the ro-tational motion of the rotor, the material is carried along the circumference of the barrel. Just before the material reaches the feed hopper it encounters a wiper bar. This wiper bar scrapes the polymer from the rotor and deflects the polymer flow into a channel that leads to the extruder die. Several patents (26, 27) were issued on this design; however, these pat-ents have long since expired.

A very similar extruder was developed by Askco Engineering and Cosden Oil & Chemical in a joint venture. Two patents have been issued on this design (28,29), even though the con-cept is almost identical to the Schmid & Kocher design, see Figure 2-15.

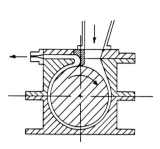

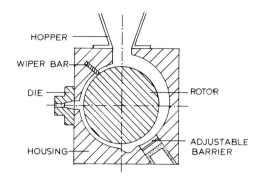

Figure 2 – 14. Drum Extruder
(Schmid & Kocher)

Figure 2 – 15. Drum Extruder (Askco/Cosden)

One special feature of this design is the local gap adjustment capability by means of a choker bar, similar to the gap adjustment in a flat sheet die, see Section 9-2. The choker bar in this drum extruder is activated by adjustable hydraulic oil pressure.

2.3.1.3   SPIRAL DISK EXTRUDER

The spiral disk extruder is another type of disk extruder that has been known for many years. Several patented designs were described by Schenkel (Chapter 1, reference 3) some twenty years ago. Similar to the stepped disk extruder, the development of the spiral disk extruder is closely connected to spiral groove bearings. It has long been known that spiral groove bearings are capable of supporting substantial loads. Ingen Housz (30) has analyzed the melt conveying in a spiral disk extruder with logarithmic grooves in the disk, based on Newtonian flow behavior of the polymer melt. In terms of melt conveying capability, the spiral disk extruder seems comparable to the screw extruder; however, the solids conveying capability is questionable.

2.3.1.4   DISKPACK EXTRUDER

A recent development in disk extruders has been the diskpack extruder. Tadmor originated the idea of the diskpack machine, which is covered under several patents (31–33). The development of the machine was undertaken by the Farrel Machinery Group of Emhart Corporation in cooperation with Tadmor (34–39). The basic concept of the machine is shown in Figure 2-16.

Material drops in the axial gap between relatively thin disks mounted on a rotating shaft. The material will move with the disks almost a full turn, then it meets a channel block. The channel block closes off the space between the disks and deflects the polymer flow to either an outlet channel or to a transfer channel in the barrel. The shape of the disks can be optimized for specific functions: solids conveying, melting, devolatilization, melt conveying, and mixing. A detailed functional analysis can be found in Tadmor's book on polymer processing (Chapter 1, reference 32).

It is claimed that this machine will be able to perform all basic polymer processing operations with efficiency equaling or surpassing existing machinery. Clearly, if the claim is true and can be delivered for a competitive price, the diskpack extruder will become an impor-

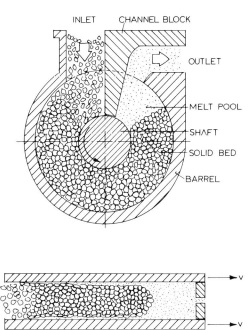

Figure 2 – 16.   Diskpack Extruder

tant machine in the extrusion industry. The first diskpack machines were delivered to the industry in 1982, still under a joint development type of agreement. It is, therefore, at the time of this writing, too early to establish the actual commercial significance of this machine. In any case, from a technical point of view, it is reassuring that the development and commercialization of an extruder based on a new concept is still possible today. In most of the publications, the diskpack machine is referred to as a polymer processor. This term is probably used to indicate that this machine can do more than just extruding, although the diskpack is, of course, an extruder.

The diskpack machine incorporates some of the features of the drum extruder and single screw extruder. One can think of the diskpack as a single screw extruder using a screw with zero helix angle and very deep flights. Forward axial transport can only take place by transport channels in the barrel, with the material forced into those channels by a restrictor bar, as with the drum extruder. The use of restrictor bars (channel block) and transfer channels in the housing make it considerably more complex than the barrel of a single screw extruder.

One of the advantages of the diskpack is that mixing blocks and spreading dams can be incorporated into the machine as shown in Figure 2-17 a.

Mixing blocks of various shapes can be positioned externally into the processing chamber. This is similar to the QSM extruders, with the one difference being that in the QSM extruder the screw flight has to be interrupted to avoid contact. This is not necessary in the diskpack because the flights have zero helix angle; in other words, they run perpendicular to the rotor axis. By the number of blocks, the geometry of the block, and the clearance between block and disk, one can tailor the mixing capability to the particular application. The use of spreading dams allows the generation of a thin film with large surface area; this results in effective devolatilization capability. The diskpack has inherently higher pressure-generat-

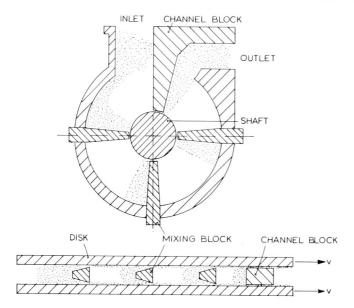

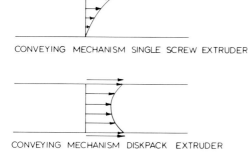

Figure 2 – 17 a. Mixing
Blocks in Diskpack

ing capability than the single screw extruder. This is because the diskpack has two dragging
surfaces, while the single screw extruder has only one, see Figure 2-17 b.

At the same net flow rate, equal viscosity, equal plate velocity, and optimum plate separation, the theoretically maximum pressure gradient of the diskpack is eight times higher than
the single screw extruder (34). Thus, high pressures can be generated over a short distance,
which allows a more compact machine design. The energy efficiency of the pressure generation is also better than the single screw extruder; the maximum pumping efficiency approaches 100 percent, see derivation in Appendix 2-1. For the single screw extruder, the
maximum pumping efficiency is only 33 percent as discussed in Section 7.4.1.3, Power Consumption in Melt Conveying. The energy efficiency of pressure generation is the ratio of the
theoretical energy requirement (flow rate times pressure rise) to the actual energy requirement (wall velocity times shear stress integrated over wall surface). It should be noted that
the two dragging plates pumping mechanism exists only in tangential direction. The pumping in axial (forward) direction occurs only in the transfer channel in the housing. This forward pumping occurs by the one dragging plate mechanism as in the single screw extruder.
The maximum efficiency for forward pumping, therefore, is only 33 percent. The overall

Figure 2 – 17 b. Comparison Conveying
Mechanism, Diskpack to Single Screw Extruder

pumping efficiency will be somewhere between 33 percent and 100 percent if the power consumption in the disk and channel block clearance is neglected.

Studies on melting in the diskpack were reported by Valsamis et al (96). It was found that two types of melting mechanism can take place in the diskpack: the drag melt removal (DMR) mechanism and the dissipative melt mixing (DMM) mechanism. The DMR melting mechanism is the predominant mechanism in single screw extruders; this is discussed in detail in Section 7.3, Plasticating. In the DMR melting mechanism, the solids and melt coexist as two largely continuous and separate phases. In the DMM melting mechanism, the solids are dispersed in the melt; there is no continuous solid bed. It was found that the DMM mechanism can be induced by promoting back leakage of the polymer melt past the channel block. This can be controlled by varying the clearance between the channel block and the disks. The advantage of the DMM melting mechanism is that the melting rate can be substantially higher than with the DMR mechanism, reportedly by as much as a factor of 3 (96).

Of all the elementary polymer processing functions, the solids conveying in a diskpack extruder has not been discussed to any extent in the open technical literature. Considering that the solids conveying mechanism is a frictional drag mechanism (as in single screw extruders) and not a positive displacement type of transport (as in intermeshing counterrotating twin screw extruders, see Sections 10.2 and 10.4), it can be expected that the diskpack will have solids conveying limitations similar to those of single screw extruders, see Section 7.2.2, Drag Induced Solids Conveying. This means that powders, blends of powders and pellets, slippery materials, etc., are likely to encounter solids conveying problems in a diskpack extruder unless special measures are taken to enhance the solids conveying capability (e.g., crammer feeder, grooves in the disks, etc.).

Because of the more complex machine geometry, the cost per unit throughput of the diskpack is likely to be higher than the conventional single screw extruder. Therefore, the diskpack will probably not compete directly with single screw extruders.

The probable area of applications for the diskpack will be specialty polymer processing operations, such as polymerization, post-reactor processing (devolatilization), continuous compounding, etc. Thus, it can be expected that the diskpack will probably compete mostly with twin screw extruders. Presently, twin screw extruders are usually the first choice when it comes to specialty polymer processing operations.

2.3.2   THE ELASTIC MELT EXTRUDER

The elastic melt extruder was developed in the late fifties by Maxwell and Scalora (40, 41). The extruder makes use of the viscoelastic, in particular the elastic, properties of polymer melts. When a viscoelastic fluid is exposed to a shearing deformation, normal stresses will develop in the fluid that are not equal in all directions, as opposed to a purely viscous fluid. In the elastic melt extruder, the polymer is sheared between two plates, one stationary and one rotating, see Figure 2-18.

As the polymer is sheared, normal stresses will generate a centripetal pumping action. Thus, the polymer can be extruded through the central opening in the stationary plate in a continuous fashion. Because normal stresses generate the pumping action, this machine is sometimes referred to as a normal stress extruder.

This extruder is quite interesting from a rheological point of view, since it is probably the only extruder that utilizes the elasticity of the melt for its conveying. Thus, several detailed experimental and theoretical studies have been devoted to the elastic melt extruder (42-47).

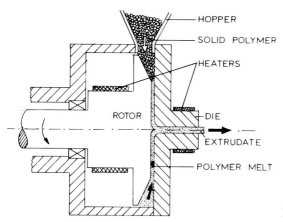

HOPPER

SOLID POLYMER

HEATERS

ROTOR

DIE

EXTRUDATE

POLYMER MELT

Figure 2 – 18. The Elastic Melt Extruder

The detailed study by Fritz (44) concluded that transport by normal stresses only could be as much as two orders of magnitude lower than a corresponding system with forced feed. In addition, substantial temperature gradients developed in the polymer, causing substantial degradation in high molecular weight polyolefins. The scant market acceptance of the elastic melt extruder would tend to confirm Fritz's conclusions.

Several modifications have been proposed to improve the performance of the elastic melt extruder. Fritz (43) suggested incorporation of spiral grooves to improve the pressure generating capability, essentially combining the elastic melt extruder and the spiral disk extruder into one machine. In Russia (47), several modifications were made to the design of the elastic melt extruder. One of those combined a screw extruder with the elastic melt extruder to eliminate the feeding and plasticating problem. Despite all of these activities, the elastic melt extruder has not been able to acquire a position of importance in the extrusion industry.

2.3.3   OVERVIEW OF DISK EXTRUDERS

Many attempts have been made in the past to come up with a continuous plasticating extruder of simple design that could perform better than the single screw extruder. It seems fair to say that, at this point in time, no disk extruder has been able to meet this goal. The simple disk extruders do not perform nearly as well as the single screw extruder. The more complex disk extruder, such as the diskpack, can possibly outperform the single screw extruder. However, this is at the expense of design simplicity, thus increasing the cost of the machine. Disk extruders, therefore, have not been able to seriously challenge the position of the single screw extruder. This is not to say, however, that it is not possible for this to happen some time in the future. But, considering the long dominance of the single screw extruder, it is not probable that a new disk extruder will come along that can challenge the single screw extruder.

2.4   RAM EXTRUDERS

Ram or plunger extruders are simple in design, rugged, and discontinuous in their mode of operation. Ram extruders are essentially positive displacement devices and are able to generate very high pressures. Because of the intermittent operation of ram extruders, they are ideally suited for cyclic processes, such as injection molding and blow molding. In fact, the early molding machines were almost exclusively equipped with ram extruders to supply the polymer melt to the mold. Certain limitations of the ram extruder have caused a switch to reciprocating screw extruders or combinations of the two. The two main limitations are: i) limited melting capacity, and ii) poor temperature uniformity of the polymer melt.

Presently, ram extruders are used in relatively small shot size molding machines and certain specialty operations where use is made of the positive displacement characteristics and the outstanding pressure generation capability. There are basically two types of ram extruders: single ram extruders and multi ram extruders.

2.4.1   SINGLE RAM EXTRUDERS

The single ram extruder is used in small general purpose molding machines, but also in some special polymer processing operations. One such operation is extrusion of intractible polymers, such as ultrahigh molecular weight polyethylene (UHMWPE) or polytetrafluoroethylene (PTFE). These polymers are not considered to be melt processable on conventional melt processing equipment. Teledynamik (48) has built a ram injection molding machine under a license from Th. Engel who developed the prototype machine. This machine is used to mold UHMWPE under very high pressures. The machine uses a reciprocating plunger that densifies the cold incoming material with a pressure up to 300 MPa (about 44,000 psi). The frequency of the ram can be adjusted continuously with a maximum of 250 strokes/minute. The densified material is forced through heated channels into the heated cylinder where final melting takes place. The material is then injected into a mold by a telescoping injection ram. Pressures up to 100 MPa (about 14,500 psi) occur during the injection of the polymer into the mold.

Another application is the extrusion of PTFE, with again the primary ingredients for successful extrusion being very high pressure. Granular PTFE can be extruded at slow rates in a ram extruder (49–51). The powder is compacted by the ram, forced into a die where the material is heated above the melting point and shaped into the desired form. PTFE is often processed as a PTFE paste (52, 53). This is small particle size PTFE powder (about 0.2 mm) mixed with a processing aid such as naphta. PTFE paste can be extruded at room temperature or slightly above room temperature. After extrusion, the processing aid is removed by heating the extrudate above its volatilization temperature. The extruded PTFE paste product may be sintered if the application requires a more fully coalesced product.

2.4.1.1   SOLID STATE EXTRUSION

An extrusion technique that has slowly been gaining popularity is solid state extrusion. The polymer is forced through a die while it is below its melting point. This causes substantial deformation of the polymer in the die, but since the polymer is in the solid state, a very effective molecular orientation takes place. This orientation is much more effective than that

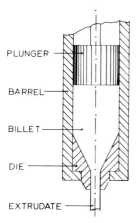

PLUNGER

BARREL

BILLET

DIE

EXTRUDATE

Figure 2-19. Direct Solid State Extrusion

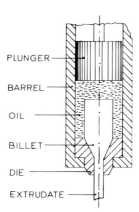

PLUNGER

BARREL

OIL

BILLET

DIE

EXTRUDATE

Figure 2-20. Hydrostatic Solid State Extrusion

which occurs in conventional melt processing. As a result, extraordinary mechanical properties can be obtained.

Solid state extrusion is a technique borrowed from the metal industry, where solid state extrusion has been used commercially since the late 1940's. Bridgman (54) was one of the first to do a systematic study on the effect of pressure on the mechanical properties of metals. He also studied polymers and found that the glass transition temperature was raised by the application of pressure.

There are two methods of solid state extrusion; one is direct solid state extrusion, the other is hydrostatic extrusion. In direct solid state extrusion, a preformed solid rod of material (a billet) is in direct contact with the plunger and the walls of the extrusion die, see Figure 2-19.

The material is extruded as the ram is pushed towards the die. In hydrostatic extrusion, the pressure required for extrusion is transmitted from the plunger to the billet through a lubricating liquid, usually castor oil. The billet must be shaped to fit the die to prevent loss of fluid. The hydrostatic fluid reduces the friction, thereby reducing the extrusion pressure, see Figure 2-20.

In hydrostatic extrusion, the pressure-generating device for the fluid does not necessarily have to be in close proximity to the forming part of the machine. The fluid can be supplied to the extrusion device by high pressure tubes.

Judging from publications in the open literature, most of the work on solid state extrusion of polymers is done at universities and research institutes. It is possible, of course, that some companies are working on solid state extrusion but are keeping the information proprietary. A major research effort in solid state extrusion is being made at the University of Amherst, Massachusetts (50-64), University of Leeds, England (65-72), Fyushu University, Fukuoka, Japan (73-76), Research Institute for Polymers and Textiles, Yokohama, Japan (77-79), Battelle, Columbus, Ohio (80-83), and Rutgers University, New Brunswick, New Jersey (84-86). As mentioned earlier, publications from other sources are considerably less plentiful (87-90). Efforts have also been made to achieve the same high degree of orientation in a more or less conventional extrusion process by special die design and temperature control in the die region (91).

TABLE 2-3. COMPARISON OF MECHANICAL PROPERTIES

Material	Tensile Modulus (Pa)	Tensile Strength (Pa)	Elongation %	Density (g/cc)
Carbon Steel:				
Annealed SAE 1020	21E10	4.1E8	35	7.86
W – 200°F SAE 1020	21E10	7.2E8	6	7.86
Stainless Steel:				
Annealed 304	20E10	5.9E8	50	7.92
Aluminum 1100-0	7E10	0.9E8	45	2.71
HDPE Solid State Extr.	7E10	4.8E8	3	0.97
HDPE Melt Extruded	1E10	0.3E8	20-1000	0.96

Table 2-3 shows a comparison of mechanical properties between steel, aluminum, solid state extruded HDPE, and HDPE extruded by conventional means.

Table 2-3 clearly indicates that the mechanical properties of solid state extruded HDPE are much superior to the melt extruded HDPE. In fact, the tensile strength of solid state extruded HDPE is about the same as carbon steel! There are some other interesting benefits associated with solid state extrusion of polymers. There is essentially no die swell at high extrusion ratios. (Extrusion ratio is the ratio of the area in the cylinder to the area in the die.) Thus, the dimensions of the extrudate closely conform to those of the die exit. The surface of the extrudate produced by hydrostatic extrusion has a lower coefficient of friction than that of the unoriented polymer. Above a certain extrusion ratio (about ten), polyethylene and polypropylene become transparent. Further, solid state extruded polymers maintain their tensile properties at elevated temperatures. Polyethylene maintains its modulus up to 120°C when it is extruded in the solid state at a high extrusion ratio. The thermal conductivity in the extrusion direction is much higher than that of the unoriented polymer, as much as 25 times higher. The melting point of the solid state extruded polymer increases with the amount of orientation. The melting point of HDPE can be shifted to as high as 140°C.

Solid state extrusion has also been practiced with coextrusion of different polymers (93). Despite the large amount of research on solid state extrusion and the outstanding mechanical properties that can be obtained, there does not seem to be much interest in the polymer industry. The main drawbacks, of course, are that solid state extrusion is basically a discontinuous process, it cannot be done on conventional polymer processing equipment, and very high pressures are required to achieve solid state extrusion. Also, one should keep in mind that very good mechanical properties can be obtained by taking a profile (fiber, film, tube, etc.) produced by conventional, continuous extrusion and exposing it to controlled deformation at a temperature below the melting point. This is a well established technique in many extrusion operations (fiber spinning, film extrusion, etc.), and it can be done at a high rate. Thus, this method of producing extrudates with very good mechanical properties is likely to be more cost-effective than solid state extrusion.

2.4.2 MULTI RAM EXTRUDER

As mentioned before, the main disadvantage of ram extruders is their intermittent operation. Several attempts have been made to overcome this problem by designing multi-ram extruders that work together in such a way as to produce a continuous flow of material.

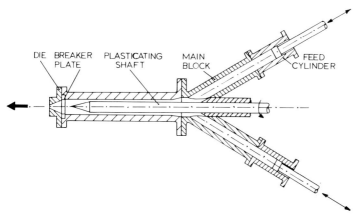

Figure 2 – 21. Twin Ram Extruder

Westover (94) designed a continuous ram extruder that combined four plunger-cylinders. Two plunger-cylinders were used for plasticating and two for pumping. An intricate shuttle valve connecting all the plunger cylinders provides continuous extrusion.

Another attempt to develop a continuous ram extruder was made by Yi and Fenner (95). They designed a twin ram extruder with the cylinders in a V-configuration, see Figure 2-21.

The two rams discharge into a common barrel in which a plasticating shaft is rotating. Thus, solids conveying occurs in the two separate cylinders and plasticating and melt conveying occurs in the annular region between the barrel and plasticating shaft. The machine is able to extrude; however, the throughput uniformity is poor. The performance could probably have been improved if the plasticating shaft had been provided with a helical channel. But, of course, then the machine would have become a screw extruder with a ram-assisted feed. This only goes to demonstrate that it is far from easy to improve upon the simple screw extruder.

APPENDIX   2 – 1

PUMPING EFFICIENCY IN DISKPACK EXTRUDER

The velocity profile between the moving walls is a parabolic function when the fluid is a Newtonian fluid, see Section 6.2.1 and Figure 2-22.

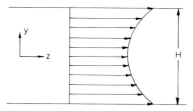

Figure 2-22. Velocity Profiles Between Moving Walls at Various Pressure Gradients

The velocity profile for a Newtonian fluid is:

$$v(y) = v - \left(1 - \frac{4y^2}{H^2}\right) \frac{H^2 \Delta P}{8 \mu \Delta L} \tag{1}$$

where v is the velocity of the plates, H the distance between the plates, and μ the viscosity of the fluid. The flow rate can be found by integrating v(y) over the width and depth of the channel; this yields:

$$\dot{V} = vWH - \frac{H^3 W \Delta P}{12 \mu \Delta L} \tag{2}$$

The first term on the right-hand side of the equal sign is the drag flow term. The second term is the pressure flow term. The ratio of pressure flow to drag flow is termed the throttle ratio, r_d (see Section 7.4.1.3):

$$r_d = \frac{H^2 \Delta P}{12 v \mu \Delta L} \tag{3}$$

The flow rate can now be written as:

$$\dot{V} = vWH \ (1 - r_d) \tag{4}$$

The shear stress at the wall is obtained from:

$$\tau = \mu \left. \frac{dv}{dy} \right|_{\frac{1}{2}H} = \frac{H \Delta P}{2 \Delta L} \tag{5}$$

The power consumption in the channel is:

$$Z_{ch} = 2 \tau W \Delta L v = vHW \Delta P \tag{6}$$

The energy efficiency for pressure generation is:

$$\varepsilon = \frac{\dot{V} \Delta P}{Z_{ch}} = 1 - r_d \tag{7}$$

Equation 7 is not valid for $r_d = 0$ because in this case both numerator and denominator become zero. From equation 7 it can be seen that when the machine is operated at low r_d values, the pumping efficiency can come close to 100 percent. This is considerably better than the single screw extruder where the optimum pumping efficiency is 33 percent at a throttle ratio value of 0.33 ($r_d = 1/3$).

REFERENCES - CHAPTER 2

1. J. Le Bras, "Rubber, Fundamentals of its Science and Technology", Chemical Publ. Co., NY (1957).
2. W. S. Penn, "Synthetic Rubber Technology, Volume I", MacLaren & Sons, Ltd., London (1960).
3. W. J. S. Naunton, "The Applied Science of Rubber", Edward Arnold Ltd., London (1961).
4. C. M. Blow, "Rubber Technology and Manufacture", Butterworth & Co. Ltd., London (1971).
5. F. R. Eirich, ed., "Science and Technology of Rubber", Academic Press, NY (1978).
6. C. W. Evans, "Powdered and Particulate Rubber Technology", Applied Science Publ. Ltd., London (1978).
7. G. Targiel et al., 10. IKV-Kolloquium, March 12-14, 45 (1980).
8. A. Kennaway, Kautschuk und Gummi, Kunststoffe 17, 378-391 (1964).
9. G. Menges and J. P. Lehnen, Plastverarbeiter 20, 1, 31-39 (1969).
10. C. M. Parshall and A. J. Saulino, Rubber World, 2, 5, 78-83 (1967).
11. S. E. Perlberg, Rubber World, 2, 6, 71-76 (1967).
12. H. G. Gohlisch, Gummi. Asbest. Kunststoffe, 25, 9, 834-835 (1972).
13. E. Harms, "Kautschuk-Extruder, Aufbau und Einsatz aus Verfahrenstechnischer Sicht", Krausskopf-Verlag Mainz, Bd. 2, Buchreihe Kunststofftechnik (1974).
14. G. Schwarz, Eur. Rubber Journal, Sept., 28-32 (1977).
15. G. Menges and E. G. Harms, Kautschuk & Gummi, Kunststoffe 25, 10, 469-475 (1972).
16. G. Menges and E. G. Harms, Kautschuk & Gummi, Kunststoffe 27, 5, 187-193 (1974).
17. E. G. Harms, Elastomerics, 109, 6, 33-39 (1977).
18. E. G. Harms, Eur. Rubber Journal, 6, 23 (1978).
19. E. G. Harms, Kunststoffe, 69, 1, 32-33 (1979).
20. E. G. Harms, Dissertation RWTH Aachen, Germany (1981).
21. S. H. Collins, Plastics Compounding, Nov./Dec., 29 (1982).
22. D. Anders, Kunststoffe, 69, 194-198 (1979).
23. D. Gras and K. Eise, SPE Tech. Papers (ANTEC), 21, 386 (1975).
24. R. F. Westover, SPE Journal, 18, 12, 1473 (1962).
25. Lord Raleigh, Philosophical Magazine, 35, 1-12 (1918).
26. German Patent: DRP 1,129,681.
27. British Patent: BP 759,354.
28. U. S. Patent: 3,880,564.
29. U. S. Patent: 4,012,477.
30. J. F. Ingen Housz, Plastverarbeiter, 10, 1 (1975).
31. U. S. Patent: 4,142,805.
32. U. S. Patent: 4,194,841.
33. U. S. Patent: 4,213,709.
34. Z. Tadmor, P. Hold, and L. Valsamis, SPE Tech. Papers (ANTEC), 25, 193 (1979).
35. P. Hold, Z. Tadmor, and L. Valsamis, SPE Tech. Papers (ANTEC), 25, 205 (1979).
36. Z. Tadmor, P. Hold, and L. Valsamis, Plastics Engineering, Nov., 20-25 (1979).
37. Z. Tadmor, P. Hold, and L. Valsamis, Plastics Engineering, Dec., 30-38 (1979).
38. Z. Tadmor et al, The Diskpack Plastics Processor, Farrel Publication, Jan. (1982).
39. L. Valsamis, AIChE Meeting, Washington, D. C., Oct. (1983).
40. B. Maxwell and A. J. Scalora, Modern Plastics, 37, 107, Oct. (1959).
41. U. S. Patent: 3,046,603.
42. L. L. Blyler, Ph. D. thesis, Princeton University, N. J. (1966).
43. H. G. Fritz, Kunststofftechnik, 6, 430 (1968).
44. H. G. Fritz, Ph. D. thesis, Stuttgart University, Germany (1971).

45. C. W. Macosko and J. M. Starita, SPE Journal 27, 30 (1971).
46. P. A. Good, A. J. Schwartz, and C. W. Macosko, AIChE Journal, 20, 1, 67 (1974).
47. V. L. Kocherov, Y. L. Lukach, E. A. Sporyagin, and G. V. Vinogradov, Polym. Eng. Sci., 13, 194 (1973).
48. J. Berzen and G. Braun, Kunststoffe, 69, 2, 62–66 (1979).
49. R. S. Porter et al, J. Polym. Sci., 17, 485–488 (1979).
50. C. A. Sperati, Modern Plastics Encyclopedia, McGraw-Hill, NY (1983).
51. S. S. Schwartz and S. H. Goodman, see Chapter 1, reference 36.
52. G. R. Snelling and J. F. Lontz, J. Appl. Polym. Sci., 3, 9, 257–265 (1960).
53. D. C. F. Couzens, Plastics and Rubber Processing, March, 45–48 (1976).
54. P. W. Bridgman, "Studies in Large Plastic Flow and Fracture", McGraw-Hill, NY (1952).
55. H. L. D. Pugh, "The Mechanical Behavior of Materials Under Pressure", Elsevier, Amsterdam (1970).
56. H. L. D. Pugh and A. H. Low, J. Inst. Metals, 93, 201 (1964/65).
57. F. Slack, Mach. Design, Oct. 7, 61–64 (1982).
58. J. H. Southern and R. S. Porter, J. Appl. Polym. Sci., 14, 2305 (1970).
59. J. H. Southern and R. S. Porter, J. Macromol. Sci. Phys., 3–4, 541 (1970).
60. J. H. Southern, N. E. Weeks, and R. S. Porter, Macromol. Chem., 162, 19 (1972).
61. N. J. Capiati and R. S. Porter, J. Polym. Sci., Polym. Phys. Ed., 13, 1177 (1975).
62. R. S. Porter, J. H. Southern, and N. E. Weeks, Polym. Eng. Sci., 15, 213 (1975).
63. A. E. Zachariades, E. S. Sherman, and R. S. Porter, J. Polym. Sci. Polym. Lett. Ed., 17, 255 (1979).
64. A. E. Zachariades and R. S. Porter, J. Polym. Sci. Polym. Lett. Ed., 17, 277 (1979).
65. B. Parsons, D. Bretherton, and B. N. Cole, in: Advances in MTDR, 11th Int. Conf. Proc., S. A. Tobias and F. Koeningsberger, eds, Pergamon Press, London, Vol. B, 1049 (1971).
66. G. Capaccio and I. M. Ward, Polymer 15, 233 (1974).
67. A. G. Gibson, I. M. Ward, B. N. Cole and B. Parsons, J. Mater. Sci., 9, 1193–1196 (1974).
68. A. G. Gibson and I. M. Ward, J. Appl. Polym. Sci. Polym. Phys. Ed., 16, 2015–2030 (1978).
69. P. S. Hope and B. Parsons, Polym. Eng. Sci., 20, 589–600 (1980).
70. P. S. Hope, I. M. Ward, and A. G. Gibson, J. Polym. Sci. Polym. Phys. Edn., 18, 1242–1256 (1980).
71. P. S. Hope, A. G. Gibson, B. Parsons, and I. M. Ward, Polym. Eng. Sci, 20, 54–55 (1980).
72. B. Parsons and I. M. Ward, Plast. Rubber Proc. Appl., 2, 3, 215–224 (1982).
73. K. Imada, T. Yamamoto, K. Shigematsu, and M. Takayanagi, J. Mater. Sci., 6, 537–546 (1971).
74. K. Nakamura, K. Imada, and M. Takayanagi, Int. J. Polym. Mater., 2, 71 (1972).
75. K. Imada and M. Takayanagi, Int. J. Polym. Mater., 2, 89 (1973).
76. K. Nakamura, K. Imada, and M. Takayanagi, Int. J. Polym. Mater., 3, 23 (1974).
77. K. Nakayama and H. Kanetsuna, J. Mater. Sci., 10, 1105 (1975).
78. K. Nakayama and H. Kanetsuna, J. Mater. Sci., 12, 1477 (1977).
79. K. Nakayama and H. Kanetsuna, J. Appl. Polym. Sci., 23, 2543–2554 (1979).
80. D. M. Bigg, Polym. Eng. Sci., 16, 725 (1976).
81. D. M. Bigg, M. M. Epstein, R. J. Fiorentino, and E. G. Smith, Polym. Eng. Sci., 18, 908 (1978).
82. D. M. Bigg and M. M. Epstein, "Science and Technology of Polymer Processing", N. S. Suh and N. Sung, Eds., 897, MIT Press (1979).
83. D. M. Bigg, M. M. Epstein, R. J. Fiorentino, and E. G. Smith, J. Appl. Polym. Sci., 26, 395–409 (1981).
84. K. D. Pae and D. R. Mears, J. Polym. Sci., B-6, 269 (1968).
85. K. D. Pae, D. R. Mears, and J. A. Sauer, J. Polym. Sci. Polym. Lett. Ed.., 6, 773 (1968).

86. D. R. Mears, K. P. Pae, and J. A. Sauer, J. Appl. Phys., 40, 11, 4229–4237 (1969).
87. L. A. Davis and C. A. Pampillo, J. Appl. Phys., 42, 12, 4659–4666 (1971).
88. A. Buckley and H. A. Long, Polym. Eng. Sci., 9, 2, 115–120 (1969).
89. L. A. Davis, Polym. Eng. Sci., 14, 9, 641–645 (1974).
90. R. K. Okine and N. P. Suh, Polym. Eng. Sci., 22, 5, 269–279 (1982).
91. J. R. Collier, T. Y. T. Tam, J. Newcome, and N. Dinos, Polym. Eng. Sci., 16, 204–211 (1976).
92. J. H. Faupel and F. E. Fisher, "Engineering Design", Wiley, NY (1981).
93. A. E. Zachariades, R. Ball, and R. S. Porter, J. Mater. Sci., 13, 2671–2675 (1978).
94. R. F. Westover, Modern Plastics, March (1963).
95. B. Yi and R. T. Fenner, Plastics and Polymers, Dec., 224–228 (1975).
96. A. Mekkaoui and L. N. Valsamis, Polym. Eng. Sci., 24, 1260–1269 (1984).
97. H. Rust, Kunststoffe, 73, 342–346 (1983).
98. J. Huszman, Kunststoffe, 73, 347–348 (1983).
99. J. M. McKelvey, U. Maire, and F. Haupt, Chem. Eng., Sept. 27, 94–102 (1976).
100. K. Schneider, Kunststoffe, 68, 201–206 (1978).
101. W. T. Rice, Plastic Technology, 87–91, Febr. (1980).
102. Harrel Corp., "Melt Pump Systems for Extruders", Product Description TDS-264 (1982).
103. J. M. McKelvey and W. T. Rice, Chem. Eng., 90, 2, 89–94 (1983).
104. K. Kapfer, K. Eise, and H. Herrmann, SPE ANTEC, Chicago, 161–163 (1983).
105. C. L. Woodworth, SPE ANTEC, New Orleans, 122–126 (1984).
106. W. A. Kramer, SPE ANTEC, Washington, DC, 23–29 (1985).

3   EXTRUDER HARDWARE

In this chapter, the hardware components of a typical single screw extruder will be described. Each major component will be discussed with respect to its major function, the possible design alternatives, and how important the component is to the proper functioning of the extruder.

3.1   EXTRUDER DRIVE

The extruder drive has to turn the extruder screw at the desired speed. It should be able to maintain a constant screw speed because fluctuations in screw speed will result in throughput fluctuations which, in turn, will cause fluctuations in the dimensions of the extrudate. Thus, constancy of speed is a very important requirement for an extruder drive. The drive also has to be able to supply the required amount of torque to the shank of the extruder screw. A third requirement for most extruder drives is the ability to vary the speed over a relatively wide range. In most cases, one would desire a screw speed that is continuously adjustable from almost zero to maximum screw speed. Over the years, various drive systems have been employed on extruders. The main drive systems are: AC motor drive systems, DC motor drive systems, and hydraulic drives.

3.1.1   AC MOTOR DRIVE SYSTEM

The two AC drive systems used on extruders are the adjustable transmission ratio drive and the adjustable frequency drive. The adjustable transmission ratio drive can be either a mechanical adjustable speed drive or an electric friction clutch drive.

3.1.1.1   MECHANICAL ADJUSTABLE SPEED DRIVE

There are four basic types of mechanical adjustable speed (MAS) drives: belt, chain, wooden block, and traction type. The latter two are not used on extruders because they are limited to low input speeds and easily damaged by shock loads (1,2).

Belt drives use adjustable sheaves. The axial distance between the sheaves can be varied; this changes the effective pitch at which the belt contacts the sheave. This, in turn, changes the transmission ratio. The speed is usually varied by a vernier screw mechanism, which is hand cranked or activated electrically. Belt drives are used up to 100 hp. The largest speed ratio is about 10:1, and a maximum speed typically 4000 rpm. Belt drives have a reasonable efficiency, tolerate shock loads, and provide optimum smoothness in a mechanical drive. Disadvantages are heat generation, possibility of slippage, and relatively poor speed control. In addition, belt drives are subject to wear and, thus, are maintenance-intensive; belts generally have to be replaced every 2000 hours.

Chain drives come in two different designs. One design uses a chain where each link is composed of a number of laminated carriers through which a stack of hardened steel slats slide. The slats fit in the grooves of conical, movable sheaves. The other design uses a conventional sprocket chain, but uses extended pins to contact the sheaves. Chain drives are more durable than belts and can transmit higher torques. Speed control is better than belt drives, about 1 percent of setting. Chain drives are more compact than belt drives and can operate at higher temperatures. On the negative side, chain drives are about twice as expensive as belt drives, they provide little shock load protection, they are suited only to relatively low speed operation, and its speed ratio is about half that of belt drives. The efficiency of both the chain and belt drive is about 90 percent.

Mechanical adjustable speed drives, nowadays, are rarely used in extruders because they are maintenance intensive, have limited speed control, limited speed ratio, and their power efficiency is not very good.

3.1.1.2   ELECTRIC FRICTION CLUTCH DRIVE

In the electric friction clutch drive, there is no direct mechanical connection between input and output shaft, eliminating mechanical friction and wear. Electrical forces are used to engage the input and output shaft. The three main types are hysteresis, eddy-current, and magnetic particle clutches. In the extrusion industry, the eddy-current drive has been widely applied in the past. The majority of the older extruders were equipped with eddy-current drives.

The popularity of this drive was, and still is to a large extent, due to the simplicity of the drive. In simple terms, the eddy-current drive consists of a fixed speed AC motor driving a steel drum, see Figure 3-1.

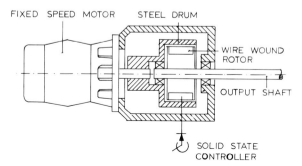

Figure 3 – 1. Eddy – Current Clutch

Inside this drum, a wire wound rotor is positioned, with a small annular gap between rotor and drum. When a low level current is applied to the rotor, it is dragged by the rotation of the drum at a somewhat lower rotational speed. When the voltage to the rotor is reduced, the slippage between the rotor and drum will increase. Thus, reduced voltage reduces the rotor speed, since the speed of the drum is constant. By controlling the voltage to the rotor, the rotor speed can be varied or it can be maintained at a steady speed under varying load.

Typical operating characteristics for eddy-current drives are (3):

- 30:1 speed range at constant torque.
- Intermittent torque to 200 percent of rated torque.
- Speed regulation: 0.5 percent of maximum speed.

- Drift: 0.05 percent of maximum speed per degree C.
- Ability to deliver rated torque at stall conditions.

The efficiency of the eddy-current drive is proportional to the difference between input and output speed. Thus, when an extruder is operated at low speeds for extended periods of time, the eddy-current drive would not be a good candidate from an energy consumption point of view. It is possible to reduce this problem by using a two-speed AC motor to drive the eddy-current clutch (3,4).

3.1.1.3   ADJUSTABLE FREQUENCY DRIVE

Adjustable frequency drives use an AC squirrel cage induction motor connected to a solid state power supply capable of providing an adjustable frequency to the AC motor. The AC squirrel cage induction motor has several advantages: low price, simplicity, ruggedness, no commutators and brushes, low maintenance, and compact construction. The cost of the adjustable frequency drive is mostly determined by the solid state power supply. The power supply converts AC power to DC power. It then inverts the DC energy so as to provide the required frequency and voltage for the AC motor. Thus, full power is handled through two sets of solid state devices, unlike the single conversion used in the DC-thyristor system. Therefore, the cost of an adjustable frequency drive is higher than a comparable DC drive, even though the DC motor itself will be more expensive than the AC motor.

A common power supply is the six-step variable voltage inverter, see Figure 3-2.

Incoming three-phase AC power is rectified and smoothed to generate a variable voltage DC supply. This DC voltage is then switched among the three output phases to generate a step waveform that approximates a sinusoidal waveform. The switching is done by six SCR's (silicon-controlled rectifiers), which are sequentially fired at the proper frequency by a solid-state control circuit.

The ratio of voltage to frequency must be held constant to maintain a constant torque capability as the motor speed is varied. Almost any speed/torque characteristic can be obtained by varying the voltage to frequency ratio. Because of limitations of the SCR cells, the maximum rating of adjustable frequency drives is presently around 300 hp. As better SCR cells are developed, this maximum rating is likely to increase.

Other power supplies are the pulse-width modulated power supply and the current-source adjustable frequency power supply. Power transistors solve many of the problems associated with SCR's. A transistor is not fixed like an SCR, but it conducts when a current is applied to the base and stops conducting when the current is removed. Size is the main limitation on transistorized variable frequency drives. The present limit is about 10 hp.

The efficiency of the six-step variable frequency power supply is quite high, despite the poor approximation to a sinusoidal output. The jagged output waveform causes losses when the output is fed to a motor. The core heating losses as a result of this non-ideal sinus-

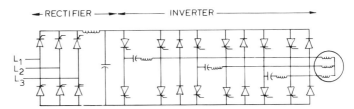

Figure 3-2. Six-Step
Variable Voltage Inverter

oidal motor input can increase by as much as 25 percent. These losses can be reduced by more closely approximating the sinusoidal waveform. This can be done by using 12 SCR's to generate a 12-step waveform. In this case, the core heating losses may be increased by only 10 percent. It is to be noted, however, that the total motor losses may be only 5 percent of full power. Thus, the overall improvement in efficiency may be quite small.

The reliability of the variable frequency power supply is generally good. However, failure is unpredictable and little can be done in the way of preventive maintenance. When failure does occur, the average plant electrician is often unable to repair the electronic system. This situation is a major impediment to the acceptance of variable frequency drives. Another disadvantage of the variable frequency AC drive is the fact that it does not handle high start-up torques as well as a DC drive. Therefore, the drive may have to be oversized considerably in order to handle these temporary high load situations.

3.1.2   DC MOTOR DRIVE SYSTEM

Some early DC extruder drives used fixed-speed AC motors to drive DC generators that produced the variable voltage for the DC motor. Nowadays, the DC motor drives usually operate from a solid state power supply, since this power supply is generally more cost-effective than the motor generator set. The DC motor drive remains much simpler and cheaper than the variable frequency drive, even when the higher cost of the DC motor is included. The smaller number of solid state devices tends to give the DC drive a better reliability than the variable frequency drive. Brushes and commutator maintenance is the principle drawback to the use of DC motors. If the drive has to be explosion-proof, the additional expense associated with this option may be quite large for a DC drive; more so than with a variable frequency AC drive or hydraulic drive.

The DC drive can provide a speed range of up to 100:1. The DC motor can handle either a constant torque or constant load, and in some cases both (with field weakening). Any overload capacity presented to the motor must be provided in the sizing of the solid state power supply. Generally, drives are provided with an overload capacity of 150 percent for one minute. The DC motor can be readily reversed by reversing the armature of the motor. For rapid stopping, resistors can be connected across the armature by a contactor, thereby providing dynamic braking at relatively low cost. DC motors respond quickly to changes in control signal due to their high ratio of torque to inertia.

The DC voltage from the solid state power supply generally has a rather poor form factor. The magnitude of the form factor is dependent on the configuration of the rectifier circuitry. The poorer the form factor, the higher the ripple current in the DC motor. This increases motor heating and reduces the power efficiency. Several three-phase rectifier circuits are available for the AC line power into DC. Most drives over 5 hp use three-phase full wave circuitry. Figure 3–3 shows a three-phase half-controlled full wave rectifier.

This circuit uses only three thyristors and four diodes. The drawback of this rectifier circuit is its high ripple current; typical form factor is 1.05 with the ripple current frequency 180 Hz.

Another popular rectifier circuit is the full controlled three-phase full wave rectifier, see Figure 3–4.

This circuit is more expensive because six thyristors are used. However, the form factor is much better, about 1.01, and the ripple current 360 Hz. The higher frquency makes it easier to filter the ripple current. The half-controlled three-phase bridge rectifier circuit may require armature current smoothing reactors to reduce the ripple current. Another problem

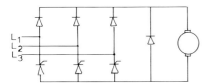

Figure 3 3. Three-Phase Half-Controlled Full
Wave Rectifier

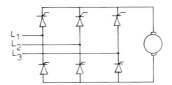

Figure 3-4. Full controlled Three-Phase
Full Wave Rectifier

associated with the nonuniform DC input to the motor is the commutation. The motor must commutate under a relatively high degree of leakage reactance.

A potential safety hazard is the fact that with armature current feedback, the armature current is connected to the operator control circuit and potentiometers may be operated at high potentials (500 V). This problem can be eliminated through the use of isolation transformers or DC to DC chopper circuits.

3.1.3   HYDRAULIC DRIVE SYSTEM

A hydraulic drive generally consists of a constant speed AC motor driving a hydraulic pump, which, in turn, drives a hydraulic motor and, of course, the associated controls. The entire package is often referred to as a hydrostatic drive. Some of the advantages of a hydrostatic drive are stepless adjustment of speed, torque, and power; smooth and controllable acceleration; ability to be stalled without damage; and easy controllability.

Over the years, considerable improvements have been made to pumps and motors, resulting in improved stability, controllability, efficiency, reduced noise, and reduced cost. Hydrostatic drives are presently used in many demanding applications. At least three types of output performance are commonly available. Variable power, variable torque transmissions are based on a variable displacement pump supplying a variable displacement motor. These transmissions provide a combination of constant torque and constant power. These units are the most adjustable, most flexible, and most expensive. Constant torque, variable power transmissions are based on a variable displacement pump supplying fluid to a fixed displacement motor under constant load. Speed is controlled by varying pump delivery. This is generally considered the best general purpose drive, with wide speed ranges, up to 40:1, and simple controls. Constant power, variable torque transmissions are based on a variable displacement pump with a power limiter, driving a fixed displacement motor. The main strength of this transmission is its efficiency; however, the speed range is usually limited to 4:1.

Hydrostatic transmissions are available where the pump and motor are combined in a single rigid unit, usually referred to as close-coupled transmissions. This produces a very compact drive that can be encased in a sealed housing to protect it from its environment. By eliminating external plumbing, close coupling reduces noise, vibration, flow losses, and leakage.

Hydraulic drives can be controlled quite well. Advances in control sophistication have helped broaden the field in which hydrostatic drives are applied. Fast-acting pressure compensators reduce heat generation, eliminate the need for crossport relief valves, and simplify other control circuits. Load sensing controls increase operating efficiency by reducing unnecessary loads on the pump. Brake and bypass circuits eliminate the need for external ac-

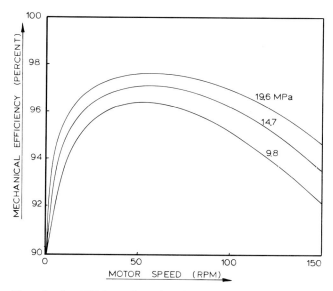

Figure 3 – 5 a.   Efficiency Curve for High Torque, Low Speed Hydraulic Motor

tuation mechanical brakes. Power limiters eliminate prime-mover stalls by limiting the transmission output to a maximum value. And speed controls hold the output speed at a constant value, regardless of the prime-mover speed.

The overall efficiency of hydrostatic drives can be quite good. For a well-designed system, the overall efficiency can be as high as 70 percent over a reasonably wide range. The use of accumulators and logic-type valves (cartridge) is beneficial to the efficiency of the drive, particularly when they are connected to programmable controls. Typical efficiency curves for a high torque, low speed hydraulic motor (6) are shown in Figure 3-5 a.

It is seen that an excellent efficiency can be obtained in the range from zero to 150 rpm, which is the typical operating range for extruders. It can be further noticed that higher operating pressures increase the mechanical efficiency. The drawback of a higher working pressure is more strain on the seals, couplings, and other components; this can increase the cost of the drive.

One significant advantage of hydrostatic drives for extruders is the elimination of the need for a transmission between the hydraulic motor and extruder screw. Since low speed hydraulic motors are readily available, one can do away with the bulky, expensive gear box found on most extruders. Therefore, when comparing the cost of a hydrostatic drive to a DC drive, one should include the cost of the gear box in the cost of the DC drive. Another advantage is the fact that other components of the machine or auxiliary equipment can be operated hydraulically from the same hydraulic power supply. This is particularly advantageous for screen changers where the screen is changed by a hydraulic cylinder or in molding machines where the mold is opened and closed hydraulically. For these reasons, hydraulic drives have become almost standard for reciprocating extruders used in injection molding and blow molding.

It should be realized, however, that the operation of the reciprocating extruder as a plasticating unit of a molding machine is considerably different from a conventional rotating extruder. The screw rotation is stopped abruptly at the end of the plasticating cycle, then the screw is moved forward and remains in the forward position for some time. Then the screw starts rotating again and as material accumulates at the tip of the screw, the screw moves backward until enough material has accumulated. Then the cycle repeats itself. Thus, in this operation there is a frequent stop and go motion. The hydraulic drive is ideally suited for this type of operation. The operation of a conventional extruder is much more continuous and, therefore, its drive requirements are different from a reciprocating extruder.

Despite the attractive features of hydrostatic drives, they are rarely used on regular (non-reciprocating) extruders. The reason for this situation is not obvious since the hydrostatic drive is in many respects competitive with, for instance, the SCR DC drive and in some respects better; e.g., no need for a gear box. A possible reason is that the hydrostatic drive is still regarded with suspicion by many people. This is because early hydraulic drives were not very reliable and accurate. However, this situation has changed dramatically over the years, but the hydrostatic drive still seems to suffer from its early unfavorable reputation. Very few U.S. companies supply hydraulic drives for extruders: Feed Screws Division of New Castles Industries and Wilmington Plastics Machinery. It is claimed (5) that hydraulic drives are less expensive than DC drives on smaller extruders, up to about 90 mm.

3.1.4   COMPARISON VARIOUS DRIVE SYSTEMS

Tables 3–1 and 3–2 compare the various adjustable speed drives with respect to speed range and control, power efficiency, and cost. From this comparison, several interesting points become clear. The efficiency of the eddy-current drive is the worst; only if the eddy-current drive is operated at full speed is the efficiency halfway decent. However, at lower speeds the eddy-current drive becomes very inefficient. In most cases, the efficiency increases with the horsepower of the drive. While the SCR DC and variable frequency AC have roughly the same line-to-shaft efficiency for small horsepower drives, the SCR DC becomes considerably more efficient at larger horsepower drives. High horsepower DC drives achieve a line-to-shaft efficiency as high as 90 percent, which is better than the other drives.

However, the line-to-shaft efficiency does not tell the full story. The power factor is also an important factor to consider. The power factor of the DC drive reduces slightly with load, but it drops drastically with speed. This becomes more severe as the horsepower increases. The power factor of the variable frequency AC drive is higher than the power factor of the DC drive and it is much less affected by speed. In fact, for small horsepower variable frequency AC drives, the power factor is essentially independent of load and speed. The effect of the power factor on overall energy cost cannot be simply calculated by multiplying the line-to-shaft efficiency with the power factor. The cost of electric power generally depends both on the actual power (KW) and the apparent power (KVA). The power factor is the ratio of actual power to apparent power. Most utilities incorporate a certain cost penalty if the power factor is considerably below one for extended periods of time. The actual cost penalty will differ from one utility to another. Also, to determine the actual expenses for electric power for a certain production facility, an energy survey must be made of the entire facility. It is possible that the extruder drive has a small effect on the overall power factor. In that case, the power factor of the extruder drive is of little concern. If the overall power factor is strongly affected by the extruder drive(s), the power factor of the electric motor of the extruder drive may be a serious concern. In this case, an energy management system should be used to keep the overall power factor as high as possible.

With DC electric motors, the power factor correction is sometimes used to improve the power factor of the drive (30). This is done by incorporating capacitive components into the circuit. Capacitors produce leading reactive power whereas the phase controlled rectifiers

TABLE 3–1. COMPARISON OF FOUR ADJUSTABLE SPEED DRIVES (1/4 – 20 HP) (Courtesy Louis Allis)

	SCR DC	Var. Freq. AC Static Inverter	Eddy Current Air Cooled	Mechanical Var. Pitch Pulley
0.25-5 hp range				
Speed range (max)	30:1	30:1	35:1	10:1
Speed regulation %	2.0 – 0.1	2.0 – 0.1	0.5 – 0.1	3.0 – 0.5
Power factor				
Full load, full speed	78	95	76	76
Full load, half speed	78	95	76	76
Half load, full speed	65	95	50	50
Half load, half speed	65	95	50	50
Line-to-shaft efficiency				
Full load, full speed	78	79	71	70
Full load, half speed	75	76	36	70
Half load, full speed	76	79	67	65
Half load, half speed	72	76	33	65
Cost comparison				
Initial cost	0.8	1.0	1.0	0.7
Installation	1.0	1.0	1.0	0.9
Mech. maintenance	Low	Very Low	Low	Medium
Elect. maintenance	Very Low	Low	Very Low	None
7.5-20 hp range				
Speed range (max)	30:1	30:1	35:1	8:1
Speed regulation %	2.0 – 0.1	2.0 – 0.1	0.5 – 0.1	3.0 – 0.5
Power factor				
Full load, full speed	88	95	85	85
Full load, half speed	61	95	85	85
Half load, full speed	86	95	70	70
Half load, half speed	57	95	70	70
Line-to-shaft efficiency				
Full load, full speed	87	79	77	83
Full load, half speed	84	76	39	83
Half load, full speed	85	79	77	85
Half load, half speed	81	76	39	85
Cost comparison				
Initial cost	0.5	1.0	0.4	0.3
Installation	1.0	1.0	1.0	0.9
Mech. maintenance	Low	Very Low	Low	Medium
Elect. maintenance	Very Low	Low	Very Low	None

TABLE 3-2. COMPARISON OF ADJUSTABLE SPEED DRIVES (25 – 4000 HP) (Courtesy Louis Allis)

	SCR DC	Var. Freq. AC Static Inverter	Eddy Current Liquid Cooled	Eddy Current Air Cooled	Mechanical Var. Pitch Pulley
25-150 hp range					
Speed range (max)	30:1	30:1	17:1	17:1	6:1
Speed regulation, %	2:0 – 0.1	2.0 – 0.1	0.5 – 0.1	0.5 – 0.1	3.0 – 0.5
Power factor					
Full load, full speed	88	90	86	86	85
Full load, half speed	61	64	86	86	85
Half load, full speed	86	81	74	74	70
Half load, half speed	57	55	74	74	70
Line-to-shaft efficiency					
Full load, full speed	89	83	83	83	87
Full load, half speed	86	80	41	41	87
Half load, full speed	87	83	84	84	88
Half load, half speed	84	80	42	42	88
Cost comparison					
Initial cost	0.5	1.0	0.4	0.4	0.3
Installation	0.9	0.9	1.0	0.9	0.8
Mech. maintenance	Low	Very Low	Medium	Low	Medium
Elect. maintenance	Very Low	Low	Very Low	Very Low	None
200-4000 hp range					
Speed range (max)	30:1	30:1	17:1		
Speed regulation %	2:0 – 0.1	2.0 – 0.1	0.5 – 0.1		
Power factor					
Full load, full speed	78	91	89		
Full load, half speed	41	65	89		
Half load, full speed	75	82	84		
Half load, half speed	38	56	84		
Line-to-shaft efficiency					
Full load, full speed	91	83	85		
Full load, half speed	89	81	42		
Half load, full speed	89	83	84		
Half load, half speed	87	81	42		
Cost comparison					
Initial cost	0.4	1.0	0.3		
Installation	0.8	0.8	1.0		
Mech. maintenance	Low	Very Low	Medium		
Elect. maintenance	Very Low	Low	Very Low		

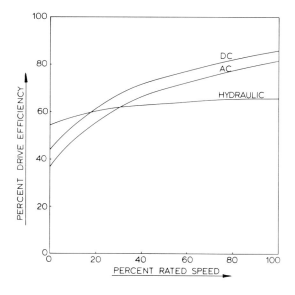

Figure 3-5b. Comparison Drive Efficiency

produce lagging reactive power. Thus, by adding appropriately sized capacitors, the power factor of the drive can be improved.

The line-to-shaft efficiency of DC motors is around 0.85 and about 0.80 for variable frequency AC motors. Considering that a typical two-stage gear box has an efficiency of about 0.95, the overall efficiency for a DC drive is about 0.80 and about 0.75 for a variable frequency AC drive. The overall efficiency of a well-designed hydrostatic drive is around 0.70. Thus, the hydrostatic drive compares reasonably well with the AC and DC drives with regards to overall efficiency. A comparison of the drive efficiency of the hydraulic drive, the DC motor drive, and the adjustable frequency drive is shown in Figure 3-5b (31). The drive efficiency is plotted versus percent rated speed. The drive efficiency of the DC motor drive and the adjustable frequency drive increases with rated speed, whereas the drive efficiency of the hydraulic drive is relatively independent of rated speed. The efficiency of the DC motor drive is better than any other drive in the range of 20 to 100 percent rated speed; below 20 percent rated speed the hydraulic drive is more efficient.

The mechanical drive has a reasonable overall efficiency at full load and full speed. The advantage of this drive is that the efficiency does not change with speed. Thus, at low speeds this drive can actually be more efficient than the DC or AC drive. A drawback of the mechanical drive is the higher maintenance requirement.

Another consideration in choosing a drive is the speed drift. The eddy current drive has a speed drift of about 0.4 percent per degree centigrade. The adjustable frequency AC drive has a speed drift of 0.05 percent or better. With a DC (SCR) drive using armature voltage control, the speed can vary 10 percent for the initial warm-up period of about 15 to 30 minutes. After the warm-up period, the drift will be about 1 percent. The speed drift can be reduced by using tachometer feedback and regulated field; this reduces the drift of the DC drive to about 0.25 percent.

Cost is obviously also a factor of importance. In order to evaluate the cost factor requests for quotations were sent to several manufacturers of SCR DC drives, variable frequency AC drives, hydrostatic drives, and eddy-current AC drives. Quotes were requested for a 100 hp drive capable of a 30:1 speed control to be used in an extrusion application. The cost of the

TABLE 3-3. VARIOUS U. S. SUPPLIERS OF VARIABLE SPEED DRIVES

Supplier	SCR DC	Var. Freq. AC	Eddy-Current	Mechanical
Allen-Bradley Corp. Cedarburg, WI	1-2000 hp	1-40 hp		
Burton Industries, Inc. Pawtacket, RI	20-1000			
Cleveland Machine Control Cleveland, OH	1-500			
Contraves Goerz Corp., Motion Control Division Pittsburg, PA	5-500			
Eaton Corp., Kenosha, WI		1-350	0.25-18,000	
Eaton Corp., Specific Industry Control Div. Milwaukee, WI	5-600			
Electric Regular Norwalk, CT	0.25-1000			
Black Clawson Corp., Electro-Flyte Div., Fulton, NY	40-1500			
General Electric Co., Drive Systems Dept., Salem, VA		1500-3000		
General Electric Co., Speed Variator Products Erie, PA	0.17-3500	1-800		
Litton Industries, Louis Allen Division New Berlin, WI	0.25-4000	1-2000	1-2250	1-50
Parametries, Orange, CT		1-30		
Reliance Electric Cleveland, OH	0.25-6000	0.25-500		0.25-50
Robicon Corp., Barber-Colman Pittsburg, PA	5-1750	5-1500		
Subina Electric & Engineering Anaheim, CA	0.25-5000	0.17-40		
Emerson Electric WER International Division Grand Island, NY	0.25-2000			
Westinghouse Electric Corp. Buffalo, NY	5-10,000	0.25-30,000		

SCR DC, eddy-current AC, and the hydrostatic drives was quite close, within 20 percent. The average price for the 100 hp drive, including controls, was around $15,000, early 1983 prices. The variable frequency AC drive was considerably more expensive, running around $35,000. Considering that the hydrostatic drive does not require a gear box, this type of drive appears attractive from a cost point of view.

Finally, Table 3-3 lists various U.S. suppliers of variable speed drives, categorized by the types of drive and the horsepower range.

3.1.5   REDUCER

With AC or DC drives, a reducer is generally required to match the low speed of the screw to the high speed of the drive. The typical reduction ratio ranges from 15:1 to 20:1. The type of reducer most frequently used is the spur gear reducer, often in a two-step configuration; i.e., two sets of intermeshing gears. A popular type of spur gear is the herringbone gear because the V-shaped tooth design practically eliminates axial loads on the gears. The efficiency of these gears is high, about 98 percent at full load and 96 percent at low load.

Some gear boxes are equipped with a quick-change gear provision, which allows one to change gear ratio rather quickly and easily. This feature can add significantly to the flexibility and versatility of the extruder. Of course, one has to make sure that the quick-change gear unit is designed in such a way that it does not significantly affect the transmission efficiency of the reducer.

Worm reduction gears have been used on rare occasions. Their advantage is low cost and compactness, but the efficiency is rather poor, between 90 and 75 percent. Some extruders do not have a direct connection between the drive and reducer, but employ either a chain or a belt transmission. This type of setup allows a relatively simple change in overall reduction ratio by changing the sprocket or sheave diameter. An advantage of the belt drive is the fact that it provides protection against excessive torque. A distinct disadvantage is increased power consumption, as much as 5 to 10 percent. Another drawback is the fact that the chain or belt transmission is less reliable than the spur gear reducer and requires more maintenance.

3.1.6   CONSTANT TORQUE CHARACTERISTICS

Most extruder drives have a so-called "constant torque" characteristic. This means that the maximum torque obtainable from the drive, for all practical purposes, remains constant over the range of screw speed. The torque-speed characteristic can be used to determine the power-speed characteristic by using the well-known relationship between torque and power:

$$P = CTN \qquad\qquad (3-1)$$

where T is torque, P power, N screw speed or motor speed, and C a constant ($C = 2\pi/60 \simeq 0.01$ when N is expressed in revolutions per minute).

Thus, if torque is constant with speed, then it follows directly from expression 3-1 that the power is directly proportional to speed, see Figure 3-6.

This means that the maximum power of the drive can only be utilized if the motor is running at full speed. Whenever the extruder output is power-limited, it is good practice to make sure the motor is running at full speed. If it is not, then a simple gear change can often alleviate the problem.

It is generally very expensive to solve a power limitation problem by installing a more powerful motor. The reason is that most gear boxes are matched in terms of power rating to the motor driving the gear box. Thus, if the motor power is increased significantly, quite often the gear box has to be replaced concurrently. This makes for a very expensive replacement. In fact, it could be more cost-effective to purchase an entirely new extruder. Mechanical power consumption is, to a very large extent, determined by the design of the extruder screw. There are many changes that one can make to the screw design that will reduce the power consumption of the drive (see Chapter 8).

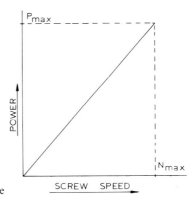

Figure 3-6. Power-Speed Curve With Constant Torque Drive

3.2   THRUST BEARING ASSEMBLY

The thrust bearing assembly is usually located at the point where the screw shank connects with the output shaft of the drive, which is generally the output shaft of the gear box. Thrust bearing capability is required because the extruder generally develops substantial diehead pressure in the polymer melt. This diehead pressure is necessary to push the polymer melt through the die at the desired rate. However, since action = reaction, this pressure will also act on the extruder screw and force it towards the feed end of the extruder. Therefore, the thrust bearing capability has to be available to take up the axial forces acting on the screw. Clearly, the load on the thrust bearing is directly determined by the diehead pressure. The actual force on the screw is obtained by multiplying the diehead pressure with the cross-sectional area of the screw. Thus, when the size of the extruder increases, the load on the thrust bearing will increase with the diameter squared. In scale-up, therefore, the thrust bearing capacity has to increase at least as fast. A 150-mm (6-inch) extruder running with a die head pressure of 35 MPa (about 5000 psi) will experience an axial thrust of about 620 kN (about 140,000 lbf). This illustrates that significant forces are acting on the screw and proper design and dimensioning of the thrust bearing is critical to the troublefree operation of the extruder.

Figure 3-7 shows a typical thrust bearing arrangement for a single screw extruder.

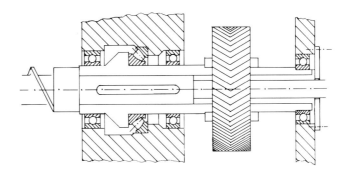

Figure 3-7. Thrust Bearing
Assembly for a Single Screw
Extruder

The screw shank is usually keyed or splined and fits into a driving sleeve in the bearing housing. Thrust bearings are designed to last a certain number of revolutions at a certain thrust load. Under normal operating conditions and reasonable diehead pressures (0–35 MPa), the thrust bearing will generally last as long as the life of the extruder. However, if the extruder operates with sharp fluctuations in diehead pressure and/or if the diehead pressure is unusually high (40–70 MPa), the life expectancy of the thrust bearing can reduce dramatically, particularly when the extruder screw runs at high speed.

The statistical rated life of a bearing is generally predicted using the following formula:

$$L_{10} = \left(\frac{C}{P}\right)^{K} \times 10^{6} \quad [\text{rev}] \qquad (3\text{-}2)$$

where L_{10} = rated life in revolutions
        C = basic load rating
        P = equivalent radial load
        K = constant, 3 for ball bearings,
           and 10/3 for roller bearings

It is important to notice that increased load, i.e., increased diehead pressure, reduces the bearing life by a power of three or more! Also, the life L will be reached more quickly when the extruder runs at high speed. The predicted life, L_y, expressed in years is obtained from the following expression:

$$L_y = \frac{1.9}{N} \left(\frac{C}{P}\right)^{K} \qquad (3\text{-}3)$$

where L_y is predicted life in years and N the screw speed in revolutions per minute. This is based on 24 hours per day, 365 days per year operation. Thus, the expected lifetime of the thrust bearing is inversely proportional to the screw speed. Sharp fluctuations to the thrust load can further reduce the thrust bearing life. The effect of load fluctuation is usually assessed by means of load factors used in equations 3–2 and 3–3. The handbook of the particular thrust bearing manufacturer or the extruder manufacturer should be consulted for the proper values of these load factors.

Extruder manufacturers often give the rated life of the thrust bearing as a B-10 life. This is expressed in hours at a particular diehead pressure, 35 MPa (5,000 psi), and screw speed, 100 revolutions/minute. The B-10 life represents the life in hours at a constant speed that 90 percent of an apparently identical group of bearings will complete or exceed before the first evidence of fatigue develops; i.e., 10 out of 100 bearings will fail before rated life. The B-10 life at normal operating load should be at least 100,000 hours in order to get an useful life of more than 10 years out of the thrust bearing.

The predicted B-10 life at any diehead pressure and/or screw speed can be found by the following relationship:

$$B\text{-}10 \; (P,N) = B\text{-}10_{Std} \; \times \; \frac{100}{N} \; \times \; \left(\frac{35}{P}\right)^{K} \qquad (3\text{-}4)$$

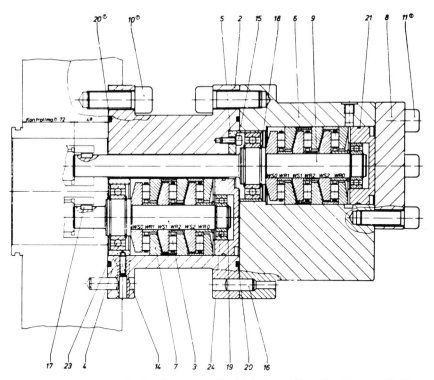

Figure 3-8. Thrust Bearing Assembly of a Counterrotating Twin Screw Extruder (Courtesy Reifen-hauser – Nabco, Inc.)

where P is diehead pressure in MPa, N screw speed in rpm, constant K is 3 for ball bearings and 10/3 for roller bearings, and B-10$_{Std}$ is the B-10 life at P = 35 MPa and N = 100 rpm.

In single screw extruders, the design of the thrust bearing assembly is relatively easy since the diameter of the bearings can be increased without much of a problem in order to obtain the required load carrying capability. This situation is entirely different in twin screw extruders because of the close proximity of the two screws. This severe space limitation makes the proper design of the thrust bearing assembly in twin screw extruders considerably more difficult than in single screw extruders. Older twin screw extruders were often limited in diehead pressure capability exactly because of this thrust bearing problem. A number of newer twin screw extruders have solved this problem sufficiently and can generally withstand almost the same diehead pressures as single screw extruders, although the rated life for the thrust bearings of a twin screw extruder is generally lower.

Figure 3-8 shows an example of a thrust bearing assembly of a counterrotating twin screw extruder.

The thrust bearing assembly consists of four or five roller bearings in tandem arrangement with a special pressure balancing system. Figure 3-9 shows a picture of the thrust bearing assembly of a Bausano twin screw extruder.

Advantages and disadvantages of certain types of thrust bearings are listed in Table 3-4.

Fluid film thrust bearings have been applied to extruders on a few occasions. Their load carrying capability at low speed is generally poor and a loss of fluid film would have disas-

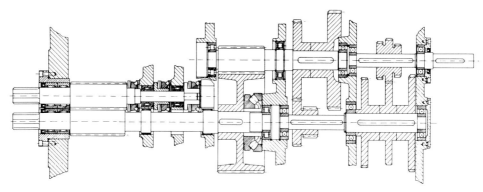

Figure 3–9. Thrust Bearing Assembly of a Bausano Twin Screw Extruder (Courtesy Urimplex-Bausano s.r.l.)

TABLE 3–4. COMPARISION OF VARIOUS TYPES OF THRUST BEARINGS

Type	Advantages	Disadvantages
TVB	High speed capability	Low thrust capacity
TVL	High speed capability Radial load capability	Lower thrust capability
TP/TPS	Higher capacity Minimum cost Static misalignment tolerability (TPS)	Not true rolling contact More heat generation
TSR	Dynamic misalignment tolerability Almost true rolling contact Radial load capability	High cost/capacity ratio Difficult to lubricate
TTHD	True rolling contact Very high capacity Low flange loading	Difficult to lubricate Minimum sizes available Special alignment considerations
TTVF	True rolling contact	Higher flange loading
TTVS	Greater size flexibility Least cost per L–10 life Easy to lubricate Static misalignment tolerability	

Type description:

TVB	Ball thrust bearing
TVL	Angular contact ball thrust bearing
TP	Cylindrical roller thrust bearing
TPS	TP-aligning
TSR	Spherical roller thrust bearing
TTHD	Tapered roller thrust bearing
TTVF	Tapered roller thrust bearing, V-flat
TTVS	TTVF-aligning

trous results. If a hydraulic drive is used to turn the screw, application of hydraulic thrust bearings may deserve some consideration. Reference 28 describes an extruder with hydraulic drive that incorporates a patented thrust bearing assembly with hydraulic axial screw adjustment. By measuring the pressure of the hydrostatic chamber of the thrust bearing, the pressure in the polymer melt at the end of the screw can be determined.

3.3   BARREL AND FEED THROAT

The extruder barrel is the cylinder that surrounds the extruder screw. The feed throat is the section of the extruder where material is introduced into the screw channel; it fits around the first few flights of the extruder screw. Some extruders do not have a separate feed throat unit; in these machines, the feed throat is an integral part of the extruder barrel. However, there are some drawbacks to this type of design. The feed throat casting is generally water-cooled. This is done to prevent an early temperature rise of the polymer. If the polymer temperature rises too high it may stick to the surface of the feed opening, causing a restriction to the flow into the extruder. Polymer sticking to the screw surface also causes a solids conveying problem, because the polymer particles adhering to the screw will not move forward and restrict the forward movement of the other polymer particles.

At the point where the feed throat casting connects with the barrel, a thermal barrier should be incorporated to prevent heat from the barrel from escaping to the feed throat unit. In a barrel with an integral feed opening this is not possible. Therefore, heat losses will be greater and there is also the chance of overheating of the feed throat.

The geometry of the feed port should be such that the material can flow into the extruder with minimum restriction. Cross-sections of various feed port designs are shown in Figure 3–10.

Figure 3–10a shows the usual feed port design. Figure 3–10b shows an undercut feed port as is often used on melt fed extruders. The danger with this design is the wedging section between screw and feed opening. If the melt is relatively stiff and/or highly elastic, significant lateral forces will act on the screw. These forces can be high enough to deflect the screw and force it against the barrel surface. This, of course, will lead to severe wear if the contact pressure is sufficiently high. This problem is more severe when this geometry is used for feeding solid polymer powder or pellets. This geometry, therefore, should only be used for feeding molten polymers. A better geometry would be the one shown in Figure 3–10c. It has an undercut to improve the intake capability, but the pronounced wedge is eliminated by a flat section oriented more or less in radial direction.

The shape of the inlet opening is usually circular or square. The smoothest transition from feed hopper to feed throat will occur if the cross-sectional shape of the hopper is the same as the shape of the feed opening. Thus, a circular hopper should feed into a circular feed port. A study done by Miller (7) on various feed port openings did not reveal a noticeable

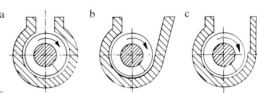

Figure 3–10. Different Feed Port Geometries

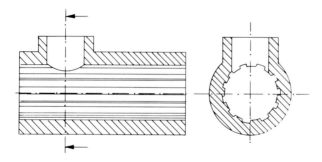

Figure 3-11. Grooved Feed
Throat Section

advantage of increasing the length of the opening beyond one diameter. This conclusion was based on solids conveying with open discharge; i. e., little or no pressure development. This is not typical for solids conveying in a real extruder and, therefore, this conclusion should be treated with caution.

Extruders equipped with grooved barrel sections often have a specially designed feed throat section to accommodate this grooved section. A schematic feed throat section is shown in Figure 3-11, where the effective length of the grooves may range from three to five diameters.

The depth of the grooves varies with axial distance. The depth is maximum at the start of the grooved bushing and reduces to zero where the grooved section meets with the smooth extruder barrel.

Several important requirements in this feed section design are i) very good cooling capability, ii) good thermal barrier between feed section and barrel, and iii) large pressure capability. The requirement of good cooling is due to the large amount of frictional heat generated in the grooved barrel section. Heating of the polymer must be avoided as much as possible in order to maximize the efficiency of the grooved barrel section. If the heat is not carried away quickly enough, the polymer will soften or even melt. This will severely diminish the effectiveness of the grooved barrel section. The requirements of a good thermal barrier between the grooved feed section and extruder barrel is to minimize the heat flow from the barrel to the feed section, so as to maximize the cooling capacity of the feed throat section. Very large pressures can be generated in the grooved feed section, from 100 to 300 MPa (about 15,000 to 45,000 psi). The feed section should be designed with the ability to withstand pressures of this magnitude, otherwise spectacular modes of failure may occur.

The stresses between the polymer and the grooves can be very high. As a result, wear can be a significant problem, particularly when the polymer contains abrasive components. The splines in the grooved bushing, therefore, are generally made out of highly wear-resistant material (see Section 11.2.1 on wear).

The extruder barrel is simply a flanged cylinder. It has to withstand relatively high pressures, as high as 70 MPa (10,000 psi), and should possess good structural rigidity to minimize sagging or deflection. Many extruder barrels are made with a wear-resistant inner surface to increase the service life. The two most common techniques are nitriding and bimetallic alloying.

Nitriding can be done by ion-nitriding (by glow discharge or plasma) or by conventional nitriding techniques (gas nitriding or liquid bath nitriding). It is generally recognized that ion-nitriding yields superior results. In ion-nitriding, the barrel is first hardened and tempered to achieve the desired core properties. The barrel is then placed in a vacuum chamber and connected in a high voltage DC circuit with the barrel as the cathode and the vacuum chamber as the anode. The chamber is evacuated and nitrogen bearing process gas is introduced. A potential difference of about 400 to 1000 volts is applied between vacuum chamber and barrel. This causes the nitrogen molecules to ionize and the nitrogen ions collect on the barrel surface. The dissipation of the kinetic energy of the ions heats the barrel surface to the nitriding temperature. The nitrogen ions combine with the surface constituents to form the nitrides that impart hardness to the surface. The surface layer consists of a compound zone and a diffusion zone. The compound zone is usually about 5 to 8 microns thick (0.2–0.3 milli-inch); it can be tailored to be either wear-resistant or corrosion-resistant. The total nitriding depth is about 0.4 mm (16 milli-inch).

Bimetallic barrels are made by centrifugally casting a bimetallic alloy onto the inside of the barrel. The melting point of the bimetallic alloy is considerably lower than the melting point of the barrel material. The barrel is charged with the alloy, capped, and heated while rotating slowly. When the proper temperature is reached the barrel is rotated at a very high speed (27), forcing the molten alloy to form a uniform layer with a strong bond with the barrel. The last finishing step is honing to form a smooth surface. The depth of the bimetallic liner is usually about 1.5 to 2.0 mm (60 to 80 milli-inch) with a uniform consistency throughout the depth of the liner.

Comparative wear tests (27) indicate that the wear performance of bimetallic barrel liners is better than nitrided barrel surfaces, with a predicted improvement in the barrel life of about four to eight times the life of a nitrided barrel. An additional drawback of the nitrided surface is that the hard compound zone is quite thin. Once the compound zone is worn away, wear will increase more rapidly because the diffusion zone is not as hard and wear-resistant. This becomes more severe as the diffusion zone wears away more deeply.

3.4   FEED HOPPER

The feed hopper feeds the granular material to the extruder. In most cases, the material will flow by gravity, unaided, from the feed hopper into the extruder. Unfortunately, this is not possible with all materials. Some bulk materials have very poor flow characteristics and additional devices may be required to ensure steady flow into the extruder. Sometimes this can be a vibrating pad attached to the hopper to dislodge any bridges as soon as they form. In some cases, stirrers are used in the feed hopper to mix the material (and prevent segregation) and/or to wipe material from the hopper wall, if the bulk material tends to stick to the wall.

Crammer feeders are used for bulk materials that are very difficult to handle. Other materials, particularly those with low bulk density, tend to entrap air. If the air cannot escape through the feed hopper, it will be carried with the polymer and eventually appear at the die exit. In most cases, this will cause surface imperfections of the extrudate. In some cases, it causes small explosions when the air escapes from the die.

One method to overcome this air entrapment problem is to use a vacuum feed hopper. In principle, this is quite simple; however, in practice, a vacuum feed hopper is not a trivial matter. The first problem is how to load the hopper without losing vacuum. This has led to the development of double hopper vacuum systems, where material is loaded into a top hopper and the air is removed before the material is dumped in the main hopper, see Figure 3–12.

A second critical point is the rear vacuum seal around the screw shank. This seal is exposed to sometimes gritty materials. Leakage of air at this point can cause fluidization of the mate-

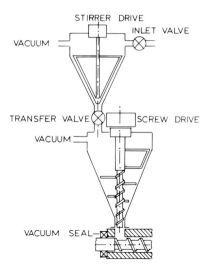

Figure 3-12. Double Feed Hopper Vacuum System

rial and adversely affect solids conveying in the extruder. Another method to avoid air entrapment is to use a two-stage extruder screw with a vent port in the barrel to extract air and any other volatiles that might be present in the polymer.

An important bulk material property with respect to the design of the hopper is the angle of internal friction (see Section 6.1). The angle of the side wall of the hopper to horizontal should be larger than the angle of internal friction, as a rule of thumb. If the bulk material has a very large angle of internal friction, it will bridge in essentially any hopper. In this case, force feeding may be the only way out.

3.5   EXTRUDER SCREW

The extruder screw is the heart of the machine. Everything revolves around the extruder screw, literally and figuratively! The rotation of the screw causes forward transport, contributes to a large extent to the heating of the polymer, and causes homogenization of the material.

In simple terms, the screw is a cylindrical rod of varying diameter with a helical flight(s) wrapped around it. The outside diameter of the screw, from flight tip to flight tip, is constant on most extruders. The clearance between screw and barrel is usually small. Generally, the ratio of radial clearance to screw diameter is around 0.001, with a range of about 0.0005 to 0.0020.

The details of screw design will be discussed in Chapter 8. In the U.S. a very common screw material is 4140 steel, which is a medium carbon, relatively low-cost material. A table of common screw materials used in polymer extrusion with their chemical composition is shown in Table 3-5. The next table, 3-6, shows some physical properties and cost comparison data. The selection of the proper screw base material and hardfacing material will be discussed in detail in Section 11.2.1.4, Solution to Wear Problems.

TABLE 3-5. COMPOSITION OF VARIOUS MATERIALS USED IN EXTRUDER SCREWS

	C	Si	Mn	P	S	Cr	Mo	Ni	V	Al	Cu	W	Co	Fe
Low Carbon Steel														
8620 21NiCrMo2	.21	.30	.80	.035	.035	.50	.20	.55						97.37
Medium Carbon Steel														
4140 HT 42CrMo4	.42	.30	.80	.035	.035	1.05	.23							97.13
Nitralloy 41CrALMoZ 135M	.41	.60		.025	.025	1.60	.35			1.10				95.89
Stainless Steel														
17-4 PH	.04	1.00	.40			16.50		4.80			4.00			73.26
304 X5CrNi189	.07	1.00	2.00	.045	.030	18.50		9.20						69.16
316 X5CrNi189	.07	1.00	2.00	.045	.030	17.50	2.25	12.00						65.10
Tool Steel														
H-13 X40CrMoV51	.40	1.00	.35			5.35	1.35		1.00					90.55
D-2 X155CrMo121	1.50	.25	.30			12.00	.80		.60					84.55
D-7	2.35	.40	.40			12.50	.95		4.00					79.40
Nickel Base														
Hastelloy C-276	.02	.05	1.00	.030	.030	15.50	16.00	55.50	.35			4.00	2.5	5.00
Duranickel	3.00	1.00	.50		.010			94.90						.60
Hardfacing Material														
Stellite 6	1.00	1.25				28.00						4.00	65.75	
Stellite 12	1.25					29.00						8.00	61.75	

TABLE 3-6. PROPERTIES OF VARIOUS SCREW MATERIALS

	Ultimate Tensile Strength After HT (MPa)	Max. Surface Hardness After HT (Rockwell C)	Finished Screw Cost Ratio*	Used With Hardfacing	Used With Chrome
8620	900	60	1.5	No	Yes
4140 HT	2000	55-60	1.0	Yes	Yes
Nitralloy 135M	1400	60-74	1.2	Yes	Not rec.
17-4 PH	1400	65	2.0	Yes	No
304			1.5	Yes	No
316			1.5	Yes	No
H-13	1800	60-74	1.7	No	Yes
D-2	1650		1.7	No	Yes
D-7	1650		3.0	No	Yes
Hastelloy C-276			3.0	Yes	No
Duranickel 301	1100		3.0	Yes	No

* Relative to 4140 HT

3.6   DIE ASSEMBLY

In many extruders, a breaker plate is incorporated between the barrel and die assembly. The breaker plate is a thick metal disk with many, closely spaced parallel holes, parallel to the screw axis. There are two main reasons for using a breaker plate. One reason is to arrest the spiraling motion of the polymer melt and to force the polymer melt to flow in a straight line fashion. Without a breaker plate, the spiraling motion could extend to the die exit and cause extrudate distortion. Another reason is to put screens in front of the breaker plate; the breaker plate then acts as a support for the screens. Screens are generally used for filtering contaminants out of the polymer. Sometimes screens are used for the sole purpose of raising the diehead pressure in order to improve the mixing efficiency of the extruder. This situation often indicates the use of an improper screw design. Another function of the breaker plate is improved heat transfer between the metal and the polymer melt. The reduced heat transfer distances in the breaker plate can improve the thermal homogeneity of the polymer melt.

If the exit opening of the extruder barrel does not match up with the entry opening of the die, an adaptor is used between barrel and die. Dies specifically designed for a certain extruder will usually not require an adaptor. However, since there is little standardization in extruder design and die design, the use of adaptors is quite common.

The die is one of the most critical parts of the extruder. It is here that the forming of the polymer takes place. The rest of the extruder has basically only one task: to deliver the polymer melt to the die at the required pressure and consistency. Thus, the die forming function is a very important part of the entire extrusion process.

Analysis of flow in extrusion dies is very difficult because of the nature of the polymer melt. Die design, therefore, is to a large extent still an empirical science. Flow behavior in flow channels will be discussed in 7.5 and die design will be discussed in detail in Chapter 9.

3.6.1   SCREENS AND SCREEN CHANGERS

The screens before the breaker plate are generally incorporated to filter out contaminants. The coarsest screen (lowest mesh number) is usually placed against the breaker plate for support, with successively finer screens placed against it. A typical screen pack is one 100-mesh screen followed by one 60-mesh and one 30-mesh screen, with the 30-mesh placed against the breaker plate. Some extrusion operations use as many as twenty 325-mesh screens backed up by coarser screens, as reported by Flathers (10).

There are three important types of metallic filter medium: wire mesh, sintered powder, and sintered fiber. Wire mesh comes in a square weave or a Dutch twill (woven in parallel diagonal lines). The different filter media do not perform equally with respect to their ability to hold contaminant, capture gels, etc. (11,12). A relative performance comparison is shown in Table 3-7.

The commonly used square weave wire mesh has poor filtering performance; the only redeeming quality is good permeability. It is clear, therefore, that if filtering is really important, another filter media should be employed. Metal fibers stand out in ability to capture gels and hold contaminants. Gel problems are particularly severe in small gauge extrusion such as low denier fibers, thin films, etc. It is particularly in these applications that metal fiber filters have been applied. If the polymer is heavily contaminated, the screen will clog rather quickly. If the screens have to be replaced frequently, an automatic screen changer is often employed. In these devices, the pressure drop across the screens is monitored continuously. If the pressure drop exceeds a certain value, a hydraulic piston moves the breaker

TABLE 3-7. PERFORMANCE COMPARISON OF DIFFERENT FILTER MEDIA

	Wire Mesh Square Weave	Wire Mesh Dutch Twill	Sintered Powder	Metal Fibers
Gel Capture	Poor	Fair	Good	Very Good
Contaminant Capacity	Fair	Good	Fair	Very Good
Permeability	Very Good	Poor	Fair	Good

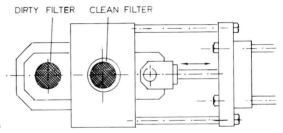

Figure 3-13. Slide Plate Screen Changers

plate with screen pack out of the way and at the same time, a breaker plate with fresh screens is moved in position. These units are referred to as slide-plate screen changers, see Figure 3-13.

With some screen changers, the screen change operation can be performed without having to shut down the extruder. The old screens can be removed and new screens put in place and the screen changer is ready for a new cycle. In operations where the polymer contains a high level of contaminants, screen changes may have to be made as often as every 5 to 10 minutes. Usually, however, the cycle time is measured in hours and not in minutes.

The breaker plate allows only a limited surface area to be used for filtering. If a substantial amount of filtering is required, downstream filtering units can be used. These devices use filter elements with large surface area; a substantial amount of contaminants can be filtered out before the filter clogs up. Various companies manufacture these filtration systems. The filter elements come in various forms: plain cylinders, pleated cylinders, and leaf discs. Many of these devices can change over from one filter to another without disrupting the flow; in other words, without having to stop the extruder.

Another type of screen is the "autoscreen" system. This consists of a continuous steel gauze which moves very slowly across the melt stream in a continuous fashion (see Figure 3-14).

The movement occurs by the pressure drop over the screen; the higher the pressure drop, the more lateral force will be exerted on the screen. In other units, the movement of the screen occurs by a motorized screen takeup. The seal is established by solidified or partially solidified polymer. The autoscreen allows a small amount of polymer to escape with the screen in a controlled fashion to carry the contaminants out and to provide a seal. Keeping a good seal requires close temperature control. Reduction of temperature can cause hang-up of the screen and increased temperatures can cause substantial leakage.

A listing of some U.S. suppliers of various screen changers and continuous filtration systems is shown in Table 3-8. In a few applications, a sandbed has been used as a filter medium (29). This is used in the manufacture of very thin and high quality film.

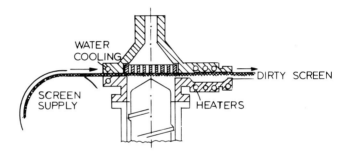

Figure 3-14. Autoscreen

TABLE 3-8. SOME U.S. SUPPLIERS OF POLYMER FILTRATION
           EQUIPMENT

	Slide Plate Non-Continuous	Screen Changers Continuous	Continuous Filtration Systems
Beringer Co. Inc.	X	X	
Berlyn Corp.	X	X	
Gala Industries	X		
Gloucester Engineering Inc.	X		
Heston Products, Inc.	X		
Sterling Extruder Corp.	X		
Sudbury Co. Inc.	X		
High Technology Corp.		X	
Thermoplas Machinery Inc.	X	X	
Welding Engineers		X	
American Barmag			X
Fluid Dynamics			X
SNC Co.			X
Michigan Dynamics			X

Several attempts have been made to model the flow through porous media and to predict the pressure drop as a function of flow rate and polymer flow properties. A number of interesting articles (13–26) are listed in the references.

3.7   HEATING AND COOLING SYSTEMS

Heating of the extruder is required for bringing the machine up to the proper temperature for startup and for maintaining the desired temperature under normal operations. There are three methods of heating extruders: electric heating, fluid heating, and steam heating. Electric heating is the most common type of heating in extruders.

3.7.1   ELECTRIC HEATING

Electric heating has significant advantages over fluid and steam heating. It can cover a much larger temperature range, it is clean, easy to maintain, low cost, efficient, etc. For these reasons, electric heating has displaced fluid and steam heating in most applications. The electrical heaters are normally placed along the extruder barrel grouped in zones. Small extruders usually have two to four zones, while larger extruders have five to ten zones. In most cases, each zone is controlled independently, so that a temperature profile can be maintained along the extruder. This can be a flat profile, increasing profile, decreasing profile, and combinations thereof; all dependent on the particular polymer and operation.

3.7.1.1   RESISTANCE HEATING

The most common barrel heaters are electric resistance heaters. This is based on the principle that if a current is passed through a conductor, a certain amount of heat is generated, depending on the resistance of the conductor and the current passed through it. The amount of heat generated is:

$$\dot{Q}_C = I^2 R = VI = \frac{V^2}{R} \tag{3-5}$$

where I is the current, R the resistance, and V voltage. This expression is valid for direct (DC) as well as single phase alternating current (AC), provided the current and voltage are expressed as root-mean-square (rms) values and the circuit is purely resistive (phase difference zero). With three phase circuits, the heat generation is:

$$\dot{Q}_C = 3VI \tag{3-5a}$$

Early band heaters used a resistance wire insulated with mica strips and encased in flexible sheet steel covers. These heaters are compact and low cost, but they are also fragile, not very reliable, and have limited power density. The maximum loading of these heaters is about $50\,kW/m^2$ ($30\ W/inch^2$) and maximum temperature about $500°C$. Newer types of mica heaters reportedly can handle power densities up to $165\ kW/m^2$ ($100\ W/inch^2$). The efficiency of the heater and its life are to a large extent determined by the goodness of the contact between the heater and the barrel over the entire contact area. Improper contact will cause local overheating and this will result in reduced heater life or even premature burnout of the heater element. Special pastes are commercially available to improve the heat transfer between heater and barrel.

Ceramic bandheaters generally last much longer than the mica insulated heaters and they can withstand higher power densities, up to $160\,kW/m^2$ ($100\ W/inch^2$) or higher, and block temperatures to $750°C$. The disadvantages of heaters with ceramic insulation are that they are not flexible and tend to be bulky. However, some ceramic band heaters have a thinline design with minimal space requirements. They usually come in halves that have to be bolted together around the extruder barrel.

Another type of heater is the "cast-in" heater. In this heater, the heating elements are cast in semi-circular or flat aluminum blocks. The heat transfer in this heater is very good. This heater is reliable and gives good service life. Cast aluminum heaters have a maximum watt density of about $55\,kW/m^2$ ($35\ watts/inch^2$) with a maximum operating temperature of about $400°C$. Bronze castings can increase the power density to about $80\,kW/m^2$ ($50\ watts/inch^2$) and a maximum operating temperature of about $550°C$.

3.7.1.2   INDUCTION HEATING

In induction heating, an alternating electric current is passed through a primary coil that surrounds the extruder barrel. The alternating current causes an alternating magnetic field of the same frequency. This magnetic field induces an electromotive force in the barrel, causing eddy-currents. The I^2R losses of the circulating current are responsible for the heating effect.

The depth of heating reduces with frequency. At normal frequencies of 50 or 60 Hz, the depth is approximately 25 mm. This is similar to the thickness of a typical extruder barrel. The advantage of this system, therefore, is much reduced temperature gradients in the extruder barrel because the heat is generated quite evenly throughout the depth of the barrel – as opposed to resistance-type barrel heaters.

Another advantage of inductive heating is reduced time lag in power input changes. Local overheating because of poor contact does not occur. Power consumption is low because of efficient heating and reduced heat losses, in spite of the fact that the power factor is lower than that of resistance heaters. It is possible to have a cooling system directly on the barrel surface, allowing accurate temperature control with fast response. A major drawback of induction heating is its high cost.

3.7.2   FLUID HEATING

Fluid heating allows even temperatures over the entire heat transfer area, avoiding local overheating. If the same heat transfer fluid is used for cooling, an even reduction in temperatures can be achieved. The maximum operating temperature of most fluids is relatively low, generally below 250° C. A few fluids can operate at high temperature; however, they often produce toxic vapors – this constitutes a considerable safety hazard. Fluid heating systems require considerable space, and installation and operating expenses are high. Another drawback with fluid heating is that if several zones need to be maintained at different temperatures, several independent fluid heating systems are required. This becomes rather expensive, bulky, and ineffective.

Steam heating is rarely used on extruders anymore, although most of the very early extruders were heated this way, particularly rubber extruders. Steam is a good heat transfer fluid because of its high specific heat. However, it is difficult to increase the temperature to sufficiently high temperatures (200 °C and over) as required in polymer extrusion. This requires very high steam pressures; most extrusion plants, nowadays, are not equipped with proper steam generating facilities to do this. Additional problems are bulkiness, chance of leakage, corrosion, heat losses, etc.

3.7.3   EXTRUDER COOLING

Extruder cooling, in most extrusion operations, is a necessary evil. In all cases, cooling should be minimized as much as possible; preferably, it should be eliminated altogether. The reason is that any amount of extruder cooling reduces the energy efficiency of the process, because cooling translates directly into lost energy. Heating of the extruder generally reduces the motor power consumption and, thus, contributes to the overall power requirement of the process. However, cooling does not contribute to the overall power requirement and the energy extracted by cooling is wasted.

If an extrusion process requires a substantial amount of cooling, this is usually a strong indication of improper process design. This could mean improper screw design, excessive length to diameter ratio, or incorrect choice of extruder; e.g., single screw versus twin screw extruder.

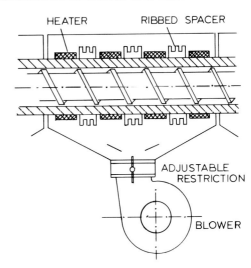

Figure 3-15. Extruder Cooling by Forced Air

The extrusion process is generally designed such that the majority of the total energy re-quirement is supplied by the extruder drive. The rotation of the screw causes frictional and viscous heating of the polymer, which constitutes a transformation of mechanical energy from the drive into thermal energy to raise the temperature of the polymer. The mechanical energy generally contributes 70 to 80 percent of the total energy. This means that the barrel heaters contribute only 20 to 30 percent, discounting any losses.

If the majority of the energy is supplied by the screw, there is a reasonable chance that local internal heat generation in the polymer is higher than required to maintain the desired pro-cess temperature. Thus, some form of cooling is usually required. Many extruders use forced-air cooling by blowers mounted underneath the extruder barrel, see Figure 3-15.

The external surface of the heaters or the spacers between the heaters is often made with cooling ribs to increase the heat transfer area and, thus, the cooling efficiency. Small extrud-ers can often do without forced-air cooling because their barrel surface area is quite large compared to the channel volume, providing a relatively large amount of radiative heat losses.

Some extruders operate without any forced cooling or heating. This is the so-called "auto-genous" extrusion operation, not to be confused with adiabatic operation. An autogenous process is a process where the heat required is supplied entirely by the conversion of me-chanical energy into thermal energy. However, heat losses can occur in an autogenous proc-ess. An adiabatic process is one where there is absolutely no exchange of heat with the sur-roundings. Clearly, an autogenous extrusion operation can never be truly adiabatic, only by approximation.

In practice, autogenous extrusion does not occur often because it requires a delicate bal-ance between polymer properties, machine design, and operating conditions. A change in any of these factors will generally cause a departure from autogenous conditions. The closer one operates to autogenous conditions, the more likely it is that cooling will be required. Given the large differences in thermal and rheological properties of various polymers, it is difficult to design an extruder that can operate in an autogenous fashion with several differ-ent polymers. Therefore, most extruders are designed to have a reasonable amount of ener-gy input from external barrel heaters.

On the other hand, the energy input from the barrel heaters should not be too large. The problem with heating from external heating is that this is associated with relatively large temperature gradients. In materials with low thermal conductivity, large temperature gradients are required to heat up the material by external heating at a reasonable rate. Since polymers have a low thermal conductivity, raising the polymer temperature by external heating is a slow process and involves large temperature gradients. Thus, locally high temperatures will occur at the metal/polymer interface. The combination of high temperatures and long heating times makes for a high chance of degradation. The heating by viscous heat generation is much more favorable in this respect, because the polymer is heated relatively uniformly throughout its mass. Thus, one would generally want the mechanical energy input to be more than 50 percent of the total energy requirement, but less than about 90 percent.

Air cooling is a fairly gentle type of cooling, because the heat transfer rates are relatively small. This is not good if intensive cooling is required. On the other hand, it has an advantage in that when the air cooling is turned on, the change in temperature occurs gradually. With water cooling, a rapid and steep change in temperature will occur as soon as the water cooling is activated. From a control point of view, the latter situation can be more difficult to handle.

When substantial cooling is required, fluid cooling is used, with water being the most common heat transfer medium. It was mentioned under 3.3 that grooved barrel sections require intense cooling to be effective. In most cases, water is used to cool the grooved barrel section, just as it is used to cool the feed throat casting. One of the complications with water cooling is that evaporation can occur if the water temperature exceeds the boiling temperature. This is an effective way to extract heat, but causes a sudden increase in cooling rate. From a control point of view, this constitutes a non-linear effect and it is more difficult to properly control the extruder temperature if such sudden, non-linear effects occur. Thus, water cooling may place much higher demands on the temperature control system as compared to air cooling. The cooling efficiency of air can be increased by wetting the air; however, this requires cooling channels made out of corrosion-resistant material. This technique is used in a patented vapor cooling system (8,9). The latent heat of a vapor which circulates around the extruder barrel is extracted by a water cooling system that surrounds a condensing chamber located away from the barrel. A schematic of this vapor cooling system is shown in Figure 3–16.

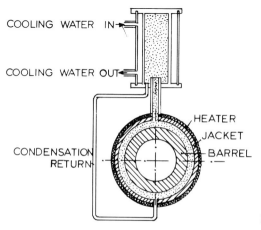

Figure 3–16. Extruder Vapor Cooling System

This type of vapor cooling is claimed to have a smooth operating characteristic and good temperature control.

Oil or air cooled extruders can use stepless cooling control, using proportional valves and positioning motors. These systems are relatively expensive, but they are reliable and require little maintenance. With water cooling, the cooling power is generally controlled by energizing a solenoid valve. For low temperatures (no flashing), usually a constant cycle rate is used with variable pulse width. The pulse width varies in proportion to the cooling power required. At high temperatures, where water is flashed to steam, more intensive cooling is possible. In these cases, a different cooling control can be used, known as a constant pulse width system. When cooling is required, the solenoid is energized by a pulse signal of predetermined length. The frequency of the pulse is varied in proportion to the cooling power required.

Finally, it should always be remembered that cooling is a waste of energy and should be minimized as much as possible.

3.7.4   SCREW HEATING AND COOLING

Thus far, the discussion has been focused on barrel heating and cooling. It is important to realize, though, that the barrel-polymer interface constitutes only about 50 percent of the total polymer-metal interface. Thus, with only barrel heating and/or cooling, only about 50 percent of the total surface area available for heat transfer is being utilized. The screw surface, therefore, constitutes a very important heat transfer surface. Many extruders do not use screw cooling or heating; they run with a so-called "neutral screw." If the external heating or cooling requirements are minor, then screw heating or cooling is generally not necessary. However, if the external heating or cooling requirements are substantial, then screw heating or cooling can become very important, sometimes a necessity.

It is obvious that heating or cooling of the screw is slightly more difficult than barrel heating or cooling, because the screw is in motion. This means that one has to use rotary unions, slip rings, or other devices to transfer energy in or out of the extruder screw. These devices, however, have become rather standard and are commonly available. Water cooling can be done even without the use of a rotary union. This involves running some copper tubing down into the bore of the extruder screw and connecting a water supply to the copper tube. The water will flow towards the end of the screw through the tube and will then flow back in the annular space between the tube and the bore of the screw. As the water reaches the shank end of the screw, it will simply drain away. This is a crude but effective type of screw cooling, see Figure 3–17.

Screw cooling is generally arranged such that the fluid, water or oil, enters the screw via a rotary union, flowing into a pipe in the bore of the screw. The screw is cooled most at the discharge end of the pipe, further cooling occurs upstream as the fluid flows back to the rotary union. The point of maximum cooling can be adjusted by changing the location of the end of the pipe.

Sometimes the end of the pipe in the screw bore is made with an axially adjustable seal. This allows cooling of a selected section of the screw. Other adjustments that can be made are

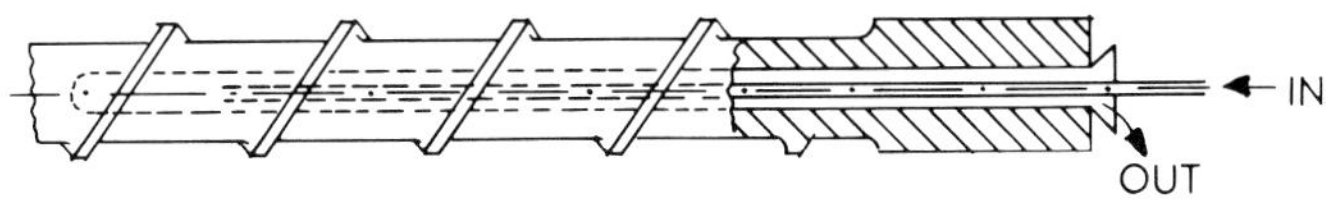

Figure 3–17. Simple Screw
Cooling System

the flow rate and inlet temperature of the cooling medium. Thus, a considerable degree of flexibility in cooling conditions is possible. This is why screw cooling or heating adds, to a significant extent, to the controllability of the process. In fact, some extrusion operations are simply impossible without the use of screw cooling. The two fluids most often used for cooling are water and oil.

In some cases, screw cooling is used to improve the pressure generating capability of the screw. The screw is cooled all the way into the metering section. The colder screw surface freezes the polymer close to the screw or, at least, significantly increases the viscosity of the polymer melt close to the screw surface. This reduces the effective channel depth of the extruder screw and can result in improved pressure generating capacity if the screw was cut too deep to begin with. Therefore, if screw cooling is required to obtain sufficient pressure at the die, this is a strong indication that the design of the extruder screw is incorrect. Instead of cooling the screw, a better solution would be to modify the screw design; specifically, to reduce the channel depth in the metering section. Determination of the optimum channel depth for pressure generation will be discussed in Chapter 8.

An interesting concept of screw cooling is the application of a heat pipe in the extruder screw. The heat pipe extends from the feed section to the metering section. In most cases, the temperature of the metering section will be considerably higher than the temperature of the feed section. This temperature difference will set up a heat flow in the heat pipe trying to diminish this temperature difference. Thus, the metering section will be cooled and the feed section will be heated. The advantage of this system is that it is self-contained, sealed, and heat losses from cooling are very small. A disadvantage is the fact that no external means of control is possible.

Screw heating is sometimes done with cartridge heaters located in the bore of the extruder screw. Power is supplied to the heater by slip rings on the shank of the screw. If the heater is axially adjustable, then the location of heating can be changed as desired. With this type of heating, one has to be careful to establish good contact between heater and screw. This almost requires installation of the heater in a preheated screw. If the heater is installed in a cold screw, good contact between heater and screw can be lost when the screw heats up. The requirement for tight contact between screw and heater would make it difficult to have an axially adjustable heater. In this case, one would have to use a heat transfer paste between screw and heater that allows good heat transfer but still a reasonable axial movement of the heater.

REFERENCES – CHAPTER 3

 1. K. Rape, Power Transmission Design, 8, 36–38 (1982).
 2. N. N., Modern Materials Handling, Oct. 6, 66–69 (1982).
 3. L. J. McCullough, Tech. Papers IEEE Meeting, 755–757 (1978).
 4. N. N., Generation Planbook, 144–147 (1982).
 5. S. Collings, Plastics Machinery & Equipment, Sept., 26–29 (1982).
 6. Machine Design, Fluid Power Reference Issue (1982).
 7. R. L. Miller, SPE Journal, Nov., 1183–1188 (1964).
 8. U. S. Patent 2,796,632.
 9. W. H. Willert, SPE-Journal, 13, 6, 122–123 (1957).
10. N. T. Flathers, et al., Int. Plast. Eng., 1, 256 (1961).
11. H. M. Kennard, Plast. Eng., 30, 12, 59 (1974).
12. J. S. Singleton, paper presented at Filtration Society Conference, London, Sept. (1973).

13. W. C. Smith, Ph. D. thesis, University of Colorado (1974).
14. T. J. Sadowski and R. B. Bird, Trans. Soc. Rheol., 9, 2, 243 (1965).
15. R. J. Marshall and A. B. Metzner, Ind. Eng. Chem. Fund, 6, 393 (1967).
16. R. H. Christopher and S. Middleman, Ind. Eng. Chem. Fund, 4, 422 (1965).
17. D. F. James and D. R. McLaren, J. Fluid Mech., 70, 733 (1975).
18. R. E. Sheffield and A. B. Metzner, AIChE. J., 22, 736 (1976).
19. G. Laufer, C. Gutfinger, and N. Abuaf, Ind. Eng. Chem. Fundamentals, 15, 77 (1976).
20. E. H. Wissler, Ind. Eng. Chem. Fund, 10, 411 (1971).
21. Z. Kamblowski and M. D. Ziubinski, Rheol. Acta., 17, 176 (1978).
22. "Filtration of Polymer Melts", VDI publication, Duesseldorf, Germany (1981).
23. J. A. Deiber and W. R. Schowalter, AIChE J., 27, 6, 912 (1981).
24. M. L. Booy, Polym. Eng. Sci., 22, 14, 895 (1982).
25. S. H. Collins, Plast. Compounding, March/April, 57–70 (1982).
26. D. S. Done and D. G. Baird, Techn. Papers 40th ANTEC, 454–457 (1982).
27. K. O'Brien, Plast. Technology, Feb., 73–74 (1982).
28. N. N., Int. Plast. Eng., 2, 92–95 (1962).
29. A. Bres, VDI-Nachrichten, 19, 42, 14 (1965).
30. R. G. Schieman, Reliance Electric Publication, D-7115, 1, 7 (1983).
31. C. J. Ceroke, Chemical Engineering, Nov 12, 133–134 (1984).

4   INSTRUMENTATION AND CONTROL

4.1   INSTRUMENTATION REQUIREMENTS

From a hardware point of view, extruder instrumentation is one of the most critical components of the entire machine. An important reason for this is that the internal workings of the extruder are totally obscured by the barrel and the die. In many cases, the only visual observation that can be made is of the extrudate leaving the die. When a problem is noticed in the extrudate, it is difficult to determine the source and location of the problem. Instrumentation is required to find out what is happening inside the extruder.

Good instrumentation enables a continuous monitoring of the "vital signs" of the extruder. These vital signs are pressure, temperature, power, and speed. These important process parameters need to be measured for process control, but they are also of vital importance in troubleshooting. Troubleshooting is only possible with good instrumentation; visual observation of the extrudate is definitely not enough to determine the cause of a problem. A medical doctor does not diagnose a patient by just looking at the patient. For an accurate diagnosis, he will at least measure the vital signs of the patient: blood pressure, pulse, temperature, etc. By the same token, a process engineer needs to know pressure, screw speed, and temperature in order to properly diagnose an extrusion problem. A complete set of instrumentation should include:

1. Diehead pressure before and after screen pack
2. Rotational speed of the screw
3. Temperature of polymer melt at the die
4. Temperatures along barrel and die
5. Cooling rate at each heat zone
6. Power consumption of each heat zone
7. Power consumption of the drive

This is a minimum requirement. In many cases additional measurements are required; e. g., in a vented extrusion operation the vacuum of the vent port should be monitored continuously, in some cases, one may want to measure polymer melt temperature at various locations in the die to determine the melt temperature distribution (there is not just one melt temperature!), etc. Incomplete instrumentation can severely hamper quick and accurate troubleshooting; in fact, it can turn troubleshooting from a logical step-by-step process into a guessing game. Thus, it can be days, weeks, or even months before a problem is located and solved. When an extrusion problem results in off-quality product or downtime, it is very important to find the cause of the problem quickly because these can be very costly problems. In some instances, a downtime of just one day is more expensive than an entire new extruder! Thus, it is easy to see that trying to save money on instrumentation is like being penny-wise and pound-foolish.

Good instrumentation allows problems to be detected early before becoming more severe and causing substantial damage to the extruder hardware or to the extrudate quality. It also allows process characterization for process development and optimization. It is further important for production control and recordkeeping, and it provides a means of interfacing the extruder to a computer.

4.2   PRESSURE MEASUREMENT

In pressure measurements, it is important to determine the absolute level of pressure, but it is equally important to determine fluctuations in pressure with time. Fluctuations in die-head pressure cause variations in the flow rate through the die. At constant take-up speed, this will cause fluctuations in the dimensions of the extrudate. Thus, the pressure fluctuation correlates closely with fluctuation in extrudate dimensions. Since high frequency (cycle time < 1 second) pressure fluctuations are quite common in extrusion, a fast response measurement is important.

Pressure measurement on early extruders was done with grease-filled Bourdon gauges. The reliability of these gauges was not very good. At high temperatures, the grease tends to leak out; this causes inaccurate readings and product contamination. Polymer can hang up in the grease cavity; with time, it can form a hard plug, again causing inaccurate readings. The temperature dependence of the Bourdon gauge pressure measurement is also quite high.

Newer pressure transducers often use bonded wire strain gauges to measure the pressure. These units have a fast response and good resolution. Since the strain gauge cannot be exposed to high temperatures, it is placed away from the heated polymer/ barrel environment. Therefore, a mechanical or hydraulic coupling is used to transmit the deflection of the diaphragm to the strain gauge. Another technique was developed at the Centraal Laboratorium TNO Delft, The Netherlands. In this transducer, the deflection of the diaphragm is transformed to a rotational movement, which is virtually independent of temperature. The angular movement is measured indirectly.

Another transducer uses compressed air, see Figure 4-1. The regulated air pressure is controlled such that there is a force balance between the measuring diaphragm and the balancing diaphragm. This is a purely pneumatic system.

Another transducer uses a measuring element with an unbonded wire strain gauge. The wires are part of a full bridge circuit. The measuring element is located directly behind the diaphragm and the deflection of the diaphragm is transferred to the measuring element with a short rod. Temperature changes at the diaphragm affect all four arms of the bridge circuit and, therefore, have little effect on the measurement. Some pressure transducers are avail-

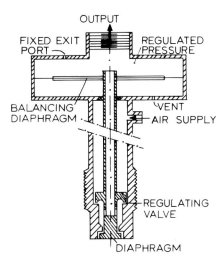

Figure 4-1. Pneumatic Pressure Transducer

able with an internal thermocouple to provide temperature measurement capacity in addition to the pressure measurement capability.

Other pressure transducers use a piezoelectric element. A piezoelectric material has the ability to transform a very small mechanical deformation (input signal) into an electric output signal (voltage or current) without any external electric power supply. Quartz pressure transducers have been developed to measure large pressures in high temperature polymer melts. The absence of a membrane allows a very robust construction. This has led to widespread use in injection molding. One of the main advantages of piezo pressure transducers is their outstanding dynamic response. The natural frequency of the piezo transducers is above 40 kHz. This allows accurate pressure measurements at a frequency up in the low kHz range. This is about three orders of magnitude better than the capillary type strain gauge pressure transducer. One of the main drawbacks of piezoelectric pressure transducers is that they cannot measure steady pressure accurately because the signal decays. Therefore, piezoelectric pressure transducers are limited to applications when the pressure changes with time over relatively short time frames (of the order of a few seconds or less).

4.2.1   MECHANICAL CONSIDERATIONS

Most diaphragms are made of stainless steel, 17–4 PH or Type 304. Type 17–4 stainless steel has a higher tensile strength and, under the same conditions, will be stressed at a lower percentage of yield, thus improving fatigue life. Another important design aspect is the connection between the diaphragm and the main body. This is generally done by welding; two welding configurations are used, see Figure 4–2.

In the "stress plane" weld configuration, the area of the diaphragm under maximum stress coincides with the weld. This can create problems because the weld is different from the other parts of the diaphragm. The weld and the area right around it will be more susceptible to stress corrosion and brittleness problems. This is particularly true of Type 304 stainless steel because of embrittlement from carbide precipitation at the weld zone.

The integral weld places the weld zone in a nonstressed location. This configuration is typical of the force-rod design and can readily be recognized by the lack of weld bead at the circumference of the diaphragm. Some newer pressure transducers do not employ the welded diaphragm construction but use a nonwelded, one-piece diaphragm and tip assembly.

If the pressure transducer is exposed to highly corrosive substances, the diaphragm can be made out of a highly corrosion-resistant material such as Hastelloy. This improvement in corrosion resistance is generally at the expense of nonlinearity and hysteresis. These specifications can increase by as much as 0.5 percent. If the transducer is exposed to highly abrasive components, a very thin wear-resistant coating can be applied to the diaphragm. One

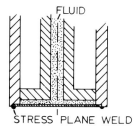

STRESS PLANE WELD

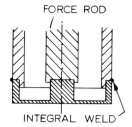

INTEGRAL WELD

Figure 4–2. Two Weld Configurations

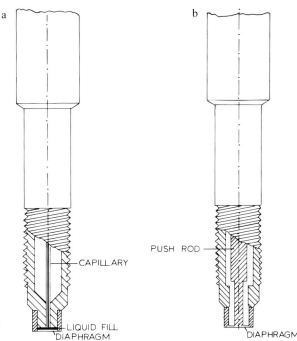

Figure 4–3. Two Types of Pressure
Transducers

manufacturer uses an electrolyzing process to deposit a thin coating on the diaphragm similar to hard chrome, but more wear-resistant.

The transducers with hydraulic coupling between the diaphragm and the strain gauge often use mercury, see Figure 4–3a. The amount of mercury is very small. The channel between the diaphragm and strain gauge is a small capillary. This transducer, therefore, is sometimes referred to as a capillary type transducer. Mercury is used because of its low thermal expansion and high boiling point. The effect of temperature on the pressure measurement is rather small. The capillary-type transducer offers the advantage of a uniform liquid support behind the diaphragm. A potential hazard is rupture of the diaphragm. This would release mercury and would contaminate the extrudate and the production area.

Transducers with a mechanical coupling between diaphragm and strain gauge often use a force rod, see Figure 4–3b.

The force rod design offers the dependability of a direct mechanical link between the diaphragm and the strain gauge. The force rod construction does not result in a uniform loading of the diaphragm.

Piezoelectric pressure transducers can be designed in a very compact package. The required deflection to obtain a pressure reading can be very small because of the high sensitivity of the piezoelectric element. The deflection is generally measured in microns (μm); full pressure can be reached with a deflection of less than 10 μm (1,2). The temperature range is primarily determined by the insulation. Transducers with ceramic insulation can be exposed to temperatures as high as 350°C. The more common PTFE insulation allows temperatures up to 240°C. The linearity of these transducers is better than 1 percent. The sensitivity is in the range of a few pC/bar at pressures of up to 7500 bar (about 750 MPa or 110,000 psi).

4.2.2   SPECIFICATIONS

Specifications on pressure transducers from different manufacturers can vary significantly. It is important to understand how and to what extent certain specifications affect the accuracy of the measurement. Some of the more important specifications will, therefore, be reviewed.

An ideal transducer would have an exactly linear relationship between pressure and output voltage. In reality, there will always be some deviation from the ideal linear relationship; this is referred to as nonlinearity. A "best straight line" (BSL) is fitted to the nonlinear curve. The deviation from BSL is quoted in the specifications and expressed as a percent of full scale (FS). The nonlinear calibration curve is determined in ascending direction, i. e., with the pressure going from zero to full rating.

The pressure measured in the ascending mode will be slightly different from the pressure measured in the descending mode, see Figure 4–4.

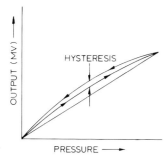

Figure 4–4. Hysteresis in a Pressure Transducer

This difference is termed hysteresis; it is expressed as a percent of full scale. The nonlinearity and hysteresis errors can be reduced by using a 75 to 80 percent shunt resistor to calibrate the output indicator (3). This procedure essentially reduces the nonlinearity error at the 75-percent reading to zero by impressing an additional voltage on the indicator that raises the nonlinearity and hysteresis curves. The maximum deviation now occurs at full scale and maximum precision occurs at mid-range, where the transducer is most likely to be used in normal operating conditions.

Repeatability is a measure of the ability of a transducer to reproduce output readings when measuring pressure consecutively and in the same direction. It specifies the maximum deviation obtained by comparing output readings for the three ascending-descending full scale loadings.

Temperature changes will cause variations in pressure reading. Thermal shift specifications indicate the maximum deviations expected due to temperature changes from room temperature to the specified limits of the operating range both for zero and for up-scale conditions. Deviations are expressed as a percent of full-scale rating for each degree of temperature above room temperature.

When a transducer is being selected, one should consider the accuracy of both the transducer itself and the readout equipment. The meter readout should be readable to an accuracy at least as good as the accuracy of the transducer.

In Germany, the VDMA (Verein Deutsche Maschinenbau Anstalten) has issued a publication (VDMA 24456) describing in detail the various aspects of pressure transducers. This

TABLE 4-1. DYNAMIC DATA OF VARIOUS PRESSURE TRANSDUCERS
(Courtesy Reference 4)

Company	Model	Range [bar]	Natural Frequency [sec⁻¹]	Damping [sec⁻¹]	Limiting Frequency [Hz]
Dr. Staiger, Mohilo & Co.		200	353	49	17
Dynisco	PT 422A	350	182	23	9
Dynisco	PT 420/12	350	628	69	30
Brosa		500	Aperiodic Damping		90
Rosemount	1401 A1	105	Aperiodic Damping		1

publication also describes various test setups that can be used to test pressure transducers; both static and dynamic testing are discussed.

The dynamic behavior of pressure transducers is of particular interest in the analysis of extrusion instabilities. The procedure for dynamic testing described in VDMA 24456 was used to test various commercial pressure transducers. Puetz (4) has published some dynamic test data for different pressure transducers, see Table 4-1.

Dynamic testing of pressure transducers can be done by measuring the response to a pulse input. If the pulse occurs over a small time period, the system will respond with a damped oscillation. A typical response is shown in Figure 4-5.

The response can be described by a function of form:

$$f(t) = A\exp(-\delta t)\sin(\omega_d t + \varphi) \tag{4-1}$$

where: A = amplitude of oscillation
δ = damping constant
ω_d = natural frequency
of the system

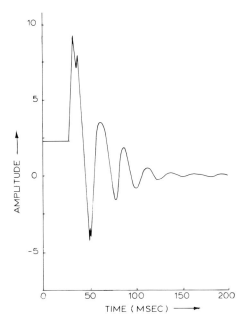

Figure 4-5. Typical Pressure Transducer Pulse Response

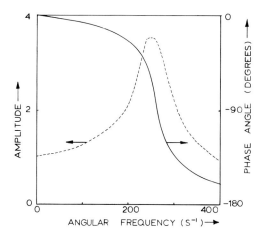

Figure 4-6. Typical Amplitude- and Phase-Frequency Curves

The characteristic values of the system can be determined from the pulse response. With these values, the amplitude-frequency (A-ω) and phase-frequency (α-ω) characteristics can be determined from the following relationships:

$$|F| = [(1-T^2\omega^2)^2 + (2DT\omega)^2]^{-1/2} \qquad (4-2)$$

$$\alpha = -\arctan\frac{2DT\omega}{1-T^2\omega^2} \qquad (4-3)$$

where D = degree of damping
      T = period of oscillation of undamped system

Typical amplitude- and phase-frequency characteristics are shown in Figure 4-6.

It can be seen from the amplitude frequency of the transducer shown in Figure 4-6 that an amplitude increase of 10 percent is reached at a frequency of about 80 rad/sec ($\approx$ 12.5 Hz). This means that pressure fluctuations with a frequency of 12.5 Hz will be measured with an error in amplitude of 10 percent; with a frequency of 30 Hz the error will be about 100 percent!

The limiting frequency shown in Table 4-1 represents the frequency at which the error in amplitude reaches 10 percent. It should be noted that the limiting frequencies listed in the table are considerably lower than the values sometimes claimed by pressure transducer manufacturers. Some transducer suppliers claim a flat frequency response up to 100 Hz; however, the data in Table 4-1 does not substantiate that number. As mentioned before, the dynamic response of piezo pressure transducers is far better (several orders of magnitude) than membrane-type pressure transducers.

4.2.3   COMPARISON OF DIFFERENT TRANSDUCERS

In general, electric transducers are more accurate than pneumatic transducers, while pneumatic transducers, in turn, are more accurate than mechanical transducers. Typical accuracy of electric transducers is ±0.5 percent to 1.0 percent, pneumatic transducers about ±1.5 percent, and mechanical transducers about ±3 percent. Figure 4-7 shows error curves for

TABLE 4-2.   COMPARATIVE DATA OF VARIOUS COMMERCIAL PRESSURE TRANSDUCERS

Company	Model	Measuring Principle	Full Range Sensitivity mV/V	Bridge Resistance Ω	Total Error %	Zero Drift With ΔT @ Diaphragm bar/°C	Zero Drift With ΔT @ Measuring Head %/°C	Sensitivity Change With Temperature %/°C	Hysteresis %	Reproducibility %	Max. Temp. Diaphragm °C	Max. Temp. Meas. Head %°C
Brosa	EBM 0520	USG	>1.85	600	–	±0.01	–	–	±0.5	–	400	150
	EBM 1618								±0.5	±0.1		
	EBM 1810								±0.2			
Dickersbach	A	M	–	–	±3%	–	–	–	–	–	400	–
	AWM	M	–	–	±3%	–	–	–	–	–		–
	D	M	–	–	±2%	–	–	–	–	–		–
Dynisco	PT420A	SGHg	>3.33	350	±0.5	±0.02	±0.02	±0.01	–	±0.1	400	100
	PT460E	SGHg	>3.33	350	±1.0	±0.04	±0.1	±0.04	–	±0.2	400	100
Gentran	Model 100	SGR	>3.0	350		±0.01 (%/°C)	–	±0.02	±0.5	±0.1	400	–
	GT 72/73 & 75	SGR	>3.0	350		±0.02		±0.02	±0.5	±0.1	400	–
	GT 76/77	SGR	>3.0	350		±0.02		±0.02	±1.0	±0.1	400	–
	GT 90	M	–	–	±3					±0.5	400	–
Dr. Staiger, Mohilo & Co.	ENAP 3	SGHg	>2.0	350	±1.0	±0.02	±0.02	±0.007	±0.2	±0.2	400	80
	Pd 1924		–			±0.01 (%/°C)	–					
Rosemount	1401A	PN			±1.5	±0.01	–	–	–	±0.5	400	90
Sensotec	MPT	SGR	>1.5	350	±0.5	±0.01		±0.02	±0.5	±0.1	370	–
Ritzinger	PUW 04	SGR	2.0	350/1000		±0.001		±0.002	±0.2	±0.1	360	–

USG = Unbonded Wire Strain Gauge
  M = Mechanical
SGHg = Strain Gauge with Mercury
  PN = Pneumatic
 SGR = Strain Gauge with Pushrod

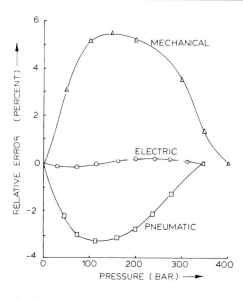

Figure 4–7. Error Curves for Three Different Pressure Transducers

the three systems, determined according to VDI/VDE-Guideline 2184 (4). The testing temperature was 180°C.

The reproducibility for electric systems is about ±0.1 percent to ±0.2 percent, for pneumatic systems about ±0.5 percent, and about ±1 percent to ±2 percent for mechanical pressure transducers. The hysteresis with electric and pneumatic transducers is about 0.1 percent to 0.2 percent, while it is as high as 4 percent to 5 percent with mechanical systems.

Table 4–2 lists several manufacturers of pressure transducers with some of the more important specifications.

4.3   TEMPERATURE MEASUREMENT

Temperature measurement occurs at various locations of the extruder: along the extruder barrel, in the polymer melt, and at the extrudate once it has emerged from the die. The choice of the type of temperature measurement will depend on what is being measured and where. First, the methods of temperature measurement will be reviewed.

4.3.1   METHODS OF TEMPERATURE MEASUREMENT

In extrusion, temperatures can be measured with resistive temperature sensors, thermocouple temperature sensors, and radiation pyrometers. There are two types of resistive temperature sensors: the conductive type and the semiconductor type. Both operate on the principle that the resistance of sensor material changes with temperature.

The conductive-type temperature sensor (RTD) uses a metal element to measure temperature. The resistance of most metals increases with temperature; thus, by measuring resistance, one can determine the temperature. Platinum is used where very precise measure-

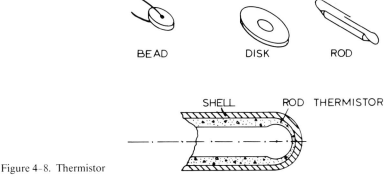

BEAD   DISK   ROD

SHELL   ROD THERMISTOR

Figure 4-8. Thermistor

ments are required and where high temperatures are involved. Platinum is available in highly purified condition; it is mechanically and electrically stable and corrosion-resistant. Most of the RTD sensors have a wound wire configuration; although for some applications, metal-film elements are used.

The semiconductor type sensor utilizes the fact that the resistance of a semiconductor decreases with temperature. The most common type of semiconductor temperature sensor is the thermistor shown in Figure 4-8.

Because of their small size, thermistors can be used where other temperature sensors cannot be used. A typical resistance temperature (RT) curve is generally nonlinear; this is one of the drawbacks of thermistors. However, techniques to deal with the thermistor nonlinearities are now well established (14); thus, the nonlinearity is not a major problem. Another disadvantage is the low operating currents ($<100\,\mu A$) and the tendency to drift over time. An advantage is their quick response time.

Thermocouple (TC) temperature sensors are also known as thermoelectric transducers; a basic TC circuit is shown in Figure 4-9.

A pair of wires of dissimilar metals are joined together at one end (hot junction or sensing junction) and terminated at the other end by terminals (the reference junction) maintained at constant temperature (reference temperature). When there is a temperature difference between sensing and reference junctions, a voltage is produced. This phenomenon is known as the thermoelectric effect.

The amount of voltage produced depends on the temperature difference and the metals used. Figure 4-10 shows the output voltage versus temperature for various metal combinations.

One of the most common TC's is the iron-constantan TC. Thermocouples come in various configurations, with exposed junction, grounded junction, ungrounded junction, surface patch, etc., see Figure 4-11.

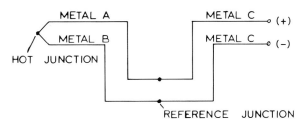

METAL A   METAL C   o (+)

METAL B   METAL C   o (−)

HOT JUNCTION

Figure 4-9. Basic TC Circuit

REFERENCE   JUNCTION

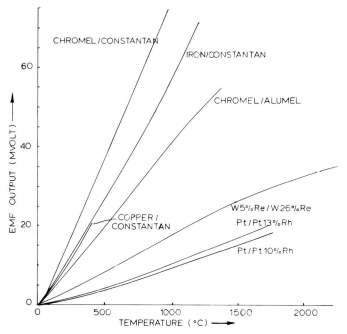

Figure 4-10. V,T Curves
for Various Metal Combi-
nations

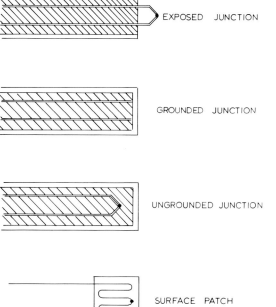

Figure 4-11. Various Thermocouple
Configurations

Detailed information on thermocouples and their use in temperature measurements can be found in the book by Pollock (69), the ASTM publication on thermocouples (70), the NBS monograph on thermocouples (71), and the book by Baker and Ryder (72).

For temperature measurements on the emerging extrudate, contacting-type measurements are not suitable because of damage to the extrudate surface. For noncontacting temperature measurements, infrared (IR) detectors are used. In some of the better IR thermometers, an entire surface can be scanned and isotherms can be determined. With additional instrumentation, quantitative information of the temperature distribution can be obtained (5).

IR probes that can be mounted in an extruder barrel or die are available (6). These probes are used to measure a more or less average stock temperature over a certain depth of the polymer, about 1 mm to 5 mm for most unfilled polymers. The actual depth of the measurement is determined by the optical properties of the polymer melt; in particular, the transmittance. The measurement is affected by variations in the consistency of the polymer melt. Thus, when fillers, additives, or other polymeric components are added, the temperature readings will be affected. A significant advantage of this type of stock temperature measurement is the rapid response, which is in the order of one millisecond. The response of conventional melt thermocouples is several orders of magnitude slower!

A comparison of various temperature sensors is shown in Table 4–3.

TABLE 4–3. COMPARISON VARIOUS TEMPERATURE SENSORS

	Thermocouple	RTD	Thermistor
Reproducibility	1 to 8°C	0.03 to 0.05°C	0.1 to 1°C
Stability	1 to 2°C in 1 year	< 0.1% in 5 years	0.1 to 3°C in 1 year
Sensitivity	0.01 to 0.05 mV/°C	0.2 to 10 ohms/°C	100 to 1000 ohms/°C
Interchangeability	Good	Excellent	Poor
Temperature Range	−250 to 2300°C	−250 to 1000°C	−100 to 280°C
Signal Output	0–60 mV	1–6V	1–3V
Minimum Size	25 µm diam.	3 mm diam.	0.4 mm diam.
Linearity	Excellent	Excellent	Poor
Response Time	Good	Fair	Good
Point Sensing	Excellent	Fair	Excellent
Area Sensing	Poor	Excellent	Poor
Cost	Low	High	Low
Unique Features	Greatest economy, widest range	Greatest accuracy, very stable	Greatest sensitivity

4.3.2   BARREL TEMPERATURE MEASUREMENT

The barrel temperature needs to be measured to provide information on the axial barrel temperature profile and to provide a signal for the controllers of the barrel heaters and cooling devices. The temperature should be measured as close as possible to the inner barrel surface, since the polymer temperature is the primary concern. The worst possible location of the temperature sensor would be in the barrel heater itself. However, there are some commercial extruders where the temperature sensor is placed in the barrel heater to reduce the thermal lag of the system. The major drawback of this approach is that one controls the heater temperature and not the temperature of the polymer in the extruder barrel. Some extruders are equipped with a combination of deep-well and shallow-well temperature sensors to improve the temperature control of the extruder (7, 8). The advantages and disadvantages of deep-well and shallow-well temperature sensors in terms of temperature control are discussed in Section 4.5.2.4, as well as dual sensor temperature control.

In the measurement of barrel temperature, a temperature sensor is pressed into a well in the extruder barrel; the sensor is generally spring-loaded. Most temperature sensors are constructed with a metallic sheath to obtain sufficient mechanical strength. As a result, significant thermal conduction errors can occur.

Figure 4–12 shows how the accuracy of the temperature measurement depends on the depth of the well and the type of temperature sensor (40).

This illustrates quite clearly that the depth of the well should be at least 30 mm to minimize the measurement errors. The characteristics of the temperature sensor itself have a strong effect on the accuracy of the measurement. If special precautions have been taken to minimize heat losses along the stem of the temperature sensor, the measurement error can be greatly reduced as compared to standard temperature sensors.

It is important to realize that barrel temperature measurement with a shallow well can be, and most likely will be, inaccurate. With a well depth of 10 mm, the measured temperature will probably be about 10°C below actual temperature. When air drafts occur around the

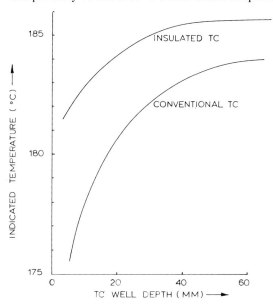

Figure 4–12. Dependence of Temperature Measurement on the Depth of the Well. True barrel temperature 185 °C, measurements made in still air

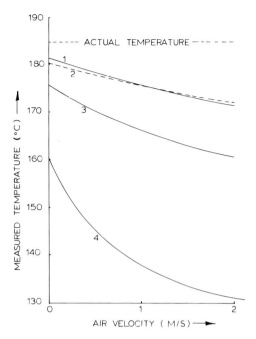

Figure 4-13. The Effect of Air Current on Measured Temperature

1 = nipple bridge, insulated TC
2 = nipple, insulated TC
3 = nipple, regular TC
4 = nipple bridge, regular TC

extruder, the measured temperature can be as much as 50°C below the actual temperature. This is shown in Figure 4-13.

4.3.3   STOCK TEMPERATURE MEASUREMENT

The measurement of the temperature of the polymer melt is of major importance. Unfortunately, several factors complicate the stock temperature measurement substantially. It is very important to be aware of these complications in order to properly appreciate the measured value.

Measurement of stock temperatures along the extruder barrel is difficult because of the rotation of the screw. To measure stock temperatures, the temperature sensor has to protrude into the polymer. The temperature sensors cannot be placed in the barrel because the protruding sensor would be damaged by the screw flight. One alternative is to place the temperature sensors in the screw channel. However, this is a rather complex operation (9-12); it requires special thermocouple mountings and sliding contacts. Another alternative is to place the temperature sensors in the barrel and to machine slots in the screw flight at the location of the protruding sensor (13). The disadvantage of this method is the substantial modification of the flow patterns in the extruder as a result of the slots in the screw flight. This, in turn, will change the actual temperature distribution in the polymer melt.

When considering the stock temperatures along the length of the extruder, it should be realized that the temperature, in most cases, varies much more strongly in radial direction than in axial direction. Radial temperature gradients will be particularly high in the plasticating region of the extruder. In this region, a thin melt film, with a melt film thickness in the order of 1 mm, separates the solid bed from the barrel. The temperature difference across this film is often in the range of 30 to 80°C. Thus, the radial temperature gradient will be of the order of 50,000° C per meter. This is about three orders of magnitude higher than the typical axial temperature gradient in an extruder. Therefore, if one would try to measure stock temperature with a protruding temperature sensor, one would experience an extreme sensitivity to radial location of the sensor. The movement of the solid bed is another consideration. Any temperature sensor should avoid contact with the solid bed, since this would most likely result in damage of the sensor.

The solids conveying and plasticating zone generally extend about two-thirds of the length of the extruder. This means that stock temperature measurements are really only possible in the last one-third of the extruder. The polymer temperature in the solids conveying and melting zone generally is maximum at or close to the barrel wall, as will be shown in Chapter 7. This means that measurement of barrel temperature is a good indication of the maximum stock temperature. Thus, barrel temperature measurement may be more meaningful, and certainly much easier, than stock temperature measurement with a protruding temperature sensor.

For these reasons, stock temperatures along the extruder are generally not measured on production extruders. Only a few, highly instrumented, development extruders are equipped with stock temperature measurement capability along the barrel. The situation is much simpler at the end of the barrel because there the screw flight is no longer present. The temperature sensor can protrude freely into the melt stream without danger of being damaged by the screw. However, even in this situation, there are several complicating factors involved in the temperature measurement.

Numerous detailed studies have been devoted to the measurement of temperature profile in polymer melts flowing through channels. One of the most comprehensive studies on the theoretical and experimental aspects of temperature measurement of polymer melts was carried out by van Leeuwen (15–18). Other studies on melt temperature measurement are listed in the following references (19–23). When a polymer melt flows through a channel, a certain temperature profile will establish itself in the polymer melt. The temperature profile in a steady-state process after some time will become constant with respect to time; this is the so-called fully developed temperature profile. The temperature at any point can be predicted from the equations of mass, momentum and energy. When a temperature sensor, such as a thermocouple, is inserted into the polymer melt stream to measure the temperature of the melt, the steady-state flow is disturbed and a new steady state will develop after a short time. Therefore, the measured temperature will be different from the true, undisturbed melt temperature. Thus, in order to determine the true melt temperature, certain corrections have to be made to the measured (disturbed) melt temperature. The factors that have to be taken into account are: i) heat conduction along the probe, ii) heat convection from the probe, and iii) energy dissipation at the probe due to shear heating. The design of the temperature sensor should be such that the above-mentioned errors are minimized. Various designs of temperature sensors for stock temperature measurement are shown in Figure 4-14.

The flush-mounted temperature sensor does not disturb the flow in the channel. However, the sensor does not protrude into the melt stream. The measured temperature, therefore, will be more representative of the metal wall temperature than the polymer melt temperature. It should be remembered, though, that the temperature of the polymer melt at the wall will equal the metal temperature at the wall. The flush-mounted probe, therefore, will give a reasonably good indication of the temperature of the polymer melt at the interface. The problem with this design is that the maximum temperature of the polymer melt generally does not occur at the wall! Unlike the situation in the solids conveying and melting zone, the maximum temperature of the polymer melt in the melt conveying zone of the extruder (including adaptor and die) generally occurs some distance away from the wall. For this reason, a protruding sensor will yield more meaningful information.

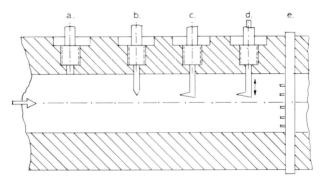

Figure 4-14. Various Temperature Sensor Configurations
a. Flush – mounted
b. Straight protruding
c. Upstream fixed
d. Upstream radially adjustable
e. Bridge with multiple probes

The straight immersion sensor is simple and sturdy. It gives a reasonable indication of the stock temperature in the flow channel. This design, however, results in significant errors in the measured temperature because of shear heating and heat conduction along the probe. This is due to the perpendicular location of the probe with respect to the direction of flow. An improved design was proposed by van Leeuwen (15); this is the upstream probe. The sensor is oriented parallel to the flow, causing only a minimal disturbance to the flow at the point where the temperature is measured. The upstream probe is capable of measuring local temperatures. The parallel portion of the probe should be long and thin to reduce heat conduction errors and to be able to measure rapid changes in temperature. On the other hand, the mechanical strength of the probe should be sufficient to withstand the forces that a melt probe is normally exposed to. Damage can easily occur during start-up or shut-down. To avoid this problem, probes have been made with adjustable depth so that the probe is inserted into the polymer when steady conditions have been reached. At high flow rates and high polymer melt viscosity, the forces that the polymer melt exerts on the immersion probe can be substantial. For these reasons, the parallel-to-flow portion of the probe is often made into a conical shape or a short, small diameter tube.

The depth adjustment capability has another use besides avoiding damage. It allows the measurement of the temperature profile across the depth of the flow channel using only one probe. Adjustable upstream temperature probes are currently commercially available, e. g. by Goettfert. However, their application in commercial extruders is still rather limited.

The suspension bridge with multiple probes has the advantage of being able to monitor various temperatures at different locations at the same time. This allows a very careful monitoring of the thermal conditions in the polymer melt throughout the material. It is often possible to incorporate temperature probes in spider legs, torpedos, etc., to obtain information on the polymer temperature away from the outside wall.

In Germany, the VDMA (Verein Deutsche Maschinenbau Anstalten) has issued a publication (VDMA 24485) to standardize evaluation and testing of temperature sensors. In this publication, a test setup is described that allows determination of the thermal conduction error as well as the transient response of the probe. From the transient response, one can determine the delay time, recovery time and 90 percent time ($T_{0.9}$). Puetz (4) describes evaluation and comparison of five different temperature probes according to the procedure described in VDMA 24485. The upstream temperature probe proved to be much more accurate than the straight protruding temperature probe with a difference in measured temperature of about 10 to 15°C! Of the five probes, only one probe was able to determine temperature fluctuations occurring in less than one minute. This was the upstream probe with a small and thin parallel section. The other temperature sensors could only measure temperature fluctuations occurring in more than one or two minutes. This means that conventional thermocouples are not suited to determine high frequency (t < 1 minute) extrusion instabilities.

4.3.3.1   ULTRASOUND TRANSMISSION TIME

A significant amount of research and development at the IKV in Aachen, West Germany, has shown that the measurement of ultrasound transmission time (UTT) yields a useful quantity to characterize the thermodynamic condition of the polymer melt (68). The UTT is a unique function of temperature and pressure, as shown in Figure 4-15 for HDPE.

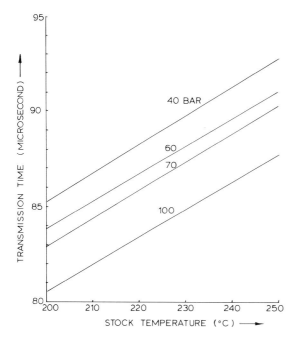

Figure 4-15. UTT as Function of Temperature and Pressure

Advantages of the UTT measurement are:

–   The transducer does not disturb the polymer melt flow.

–   The UTT measurement is a result of linearly integrating measurement across the depth of the flow channel.

–   The measurement is not affected by heat conduction errors or viscous heat dissipation.

Commercial applications of UTT measurement (Omyson) have been in use since about 1976. It has been found that the robust construction of the transducer has held up well in demanding industrial environments. An important application in extrusion is the noncontacting temperature measurement of heat-sensitive materials. Protruding temperature sensors can easily cause degradation with such materials. The UTT measurement, pressure compensated, is often used for stock temperature control in a similar fashion as with the conventional melt temperature sensor (see Section 4.6).

The UTT measurement is also used in the control of entire extrusion lines. The use of UTT measurement reportedly has resulted in considerable technical improvements in product performance in several polymer processing operations (68). The improved temperature uniformity of the polymer melt can result in closer tolerances on the extrudate. It should be noted, however, that the pressure compensation is rather involved and that the accuracy of the resulting temperature is only as good as the pressure measurement. In this respect, the melt temperature measurement by IR radiation is less complicated, see Section 4.3.1.

4.4   OTHER MEASUREMENTS

Pressure and temperature are two process parameters of major importance. There are, however, various other parameters that cannot be ignored.

4.4.1   POWER MEASUREMENT

The basic method to measure electrical power consumed by a load is to connect an ammeter in series with the load and a voltmeter across the load. In a DC circuit, the power is obtained by multiplying current with voltage. The same is true for an AC circuit where only a resistive load is concerned. In this case, the current and voltage are in phase. In an AC circuit, where the load has an inductive or capacitive component, the current and voltage are no longer in phase.

In an AC circuit with an inductive or capacitive component, there are two types of power. One is true or useful power capable of doing useful work, e. g., turn the rotor of a motor. The other power is reactive power that cannot do useful work. Total power, which is known as apparent power, is the vectorial sum of true power and reactive power. The ratio between true power and apparent power is known as the power factor of the circuit.

The power factor, which is always less than 100 percent, is a function of the phase difference between the current and voltage in a circuit. The apparent power in a circuit can be measured by the basic volt and ammeter method. To distinguish apparent power from true power, apparent power is stated in units of volts-amperes. The true power, which is stated in units of watts, is measured by a wattmeter.

The power consumption of regular AC motors can be measured quite easily by using a readily available wattmeter. The power consumption of SCR-controlled DC motors and variable frequency AC motors is more difficult to measure because of the distorted wave form going into the motor. If a conventional wattmeter is used to measure power consumption of a SCR-controlled DC motor, substantial measurement error will be made, depending on the phase angle. On a DC motor, it is easier to measure the AC power going into the drive before the rectifier circuit. This measurement, of course, will include static losses in the drive and rectifier. Extruders with DC drive often have an armature current readout on the instrument panel. The approximate power consumption Z of the drive can be calculated from the armature current I_a by using the following relationship:

$$Z = 0.9 \frac{N_{act}}{N_{max}} I_a V_a \tag{4-4}$$

where N_{act} is the actual rpm, N_{max} the maximum rpm, I_a the armature current, and V_a the armature voltage. This relationship is accurate to approximately 5 percent and does not reflect the static losses in the drive.

A good method to measure the actual mechanical power consumed in the extrusion process is to measure the torque transmitted through the shank of the extruder screw. The actual power is obtained by multiplying the torque with the screw speed. In fact, this is an excellent method to determine the overall energy efficiency of the extruder drive system. This can be done by comparing the screw power to the total power consumption of the drive. Unfortunately, this type of data is not readily available. The torque can be measured with a torque transducer. Torque transducers generally measure the torsion of the shaft by means of a strain gauge on the shaft; the torque is directly proportional to the angular torsion of the shaft. Unfortunately, accurate torque transducers are quite expensive and, thus, can add significantly to the cost of an extruder.

The power consumption of the barrel and die heaters can be determined by measuring voltage and current to the heater. This works well in current proportioned temperature control. It does not work well with on-off control or time-proportioning temperature control. In the latter case, a wattmeter should be used with a power integrating function. In this case, the integrated power over a certain time period can be measured so that the average power consumption of the barrel heater can be established. Commercial extruders generally do not have sufficient instrumentation to accurately determine the power consumption of the heaters; in many cases it cannot be determined at all!

4.4.2   ROTATIONAL SPEED

The magnetic pickup is a very common method for measuring rotational speed. The basic elements are a metallic-toothed wheel connected to the rotating shaft and a magnetic pickup or coil, see Figure 4-16. The teeth of the wheel pass near the pickup coil. The pickup typically consists of a housing containing a small permanent magnet with a coil wound around it. A fixed magnetic field surrounds the pickup. When the teeth of the wheel pass through the field, a voltage pulse is induced in the coil. The frequency of the pulses depends on the number of teeth and the speed of rotation. Since the number of teeth is known, the pulse frequency can be related directly to the rotational speed. The pulses can be measured by a frequency counter or they can be converted to a DC voltage.

Another method for measuring rotational speed is the rotating disk and light sensor. The basic elements are a perforated rotating disk connected to the rotating shaft, a light source, and a light sensor, see Figure 4-17. A fixed light source is placed on one side of the disk in line with the holes. A light sensor is placed on the opposite side of the disk in line with the light source. When the perforated disk rotates, a pulsed output signal is produced. When the number of holes in the perforated disk is known, the pulse frequency can be converted to rotational speed, as with the magnetic pickup.

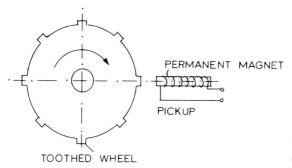

PERMANENT MAGNET

PICKUP

TOOTHED WHEEL

Figure 4-16. Magnetic Pickup to Measure Rotational Speed

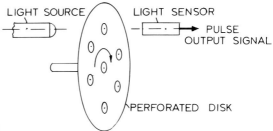

Figure 4-17. Optical Speed Measurement

Another simple method is to use an electrical tachometer. The small DC generator is coupled to the rotating shaft. The output voltage of the generator is fed to a voltmeter. The generator output voltage is directly proportional to the rotational speed of the shaft. Thus, the measured voltage can be directly converted into rpm.

4.4.3   EXTRUDATE THICKNESS

A variety of methods are available to measure extrudate thickness. The methods can be broadly classified into contacting and non-contacting techniques. The contacting thickness measurement techniques are generally simple and inexpensive; however, the contact of the transducer with the extrudate can adversely affect the extrudate surface quality. In cases where the requirements for surface quality are very high, non-contacting thickness measurement is generally preferred.

In the contacting measurement techniques, the micrometer caliper is a common instrument. The micrometer, however, can only be used for spot measurements and this is done manually. A spring-loaded dial gauge can be moved over the extrudate if the thickness variations are small. Thus, the dial gauge can be used to monitor the variation of thickness with time, i.e., in extrusion direction. If an accurate traversing mechanism is constructed, the dial gauge can also measure the thickness variation perpendicular to the extrusion direction. At the point of measurement, the opposite side of the extrudate has to be firmly supported to avoid measurement errors.

The micrometer and dial gauge are simple mechanical devices. In many cases, one would like to have continuous record of the thickness measurement. The LVDT (linear variable differential transformer) provides an electrical signal that can be used to monitor the thickness on a recorder. The LVDT is a device in which the displacement of an iron core changes the inductive coupling between primary and secondary coils, see Figure 4-18.

Movement of the core produces an AC output signal that reflects the amount and direction of movement. If a LVDT is used in thickness measurement, one has to check the effect of temperature on the accuracy because the extrudate is generally at an elevated temperature that may not be constant. The LVDT can be quite accurate; it can measure to an accuracy of about 1 μm.

Another more or less contacting thickness measurement technique is pneumatic gauging. The device consists of a nozzle fixed in position relative to a stop. Air at a constant supply pressure passes through a restriction and discharges through the nozzle, see Figure 4-19.

The nozzle back pressure P depends on the gap between the measured surface and the nozzle opening. If the thickness increases, the gap decreases, restricting the discharge of air, thus increasing pressure P. The pressure gauge indicates deviation of the thickness from

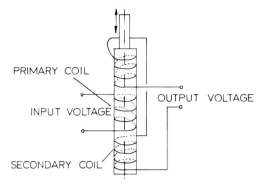

PRIMARY COIL

OUTPUT VOLTAGE

INPUT VOLTAGE

SECONDARY COIL

Figure 4-18. LVDT

some normal value. With the proper design, this pressure is directly proportional to the deviation, limited, however, to a range of about 100 micron. The device is very sensitive, up to 0.0001 mm over a range of 0-2 mm. It is rugged and, with periodic calibration, quite accurate. The gauge is adaptable to automatic line control where the pressure signal is recorded or used to actuate an alarm when the thickness exceeds a certain threshold value.

Another contacting thickness measurement is the capacitance measurement. Metal plates are placed at either side of the polymer film. Thus, the material and plates form a capacitor, with the polymer acting as a dielectric. Since the capacitance depends on the thickness of the dielectric, the material thickness is determined by measuring the capacitance. The problem in applying this technique to extrusion is that it will be difficult to establish good contact between polymer and metal plates, particularly in continuous monitoring of thickness.

The thickness of a polymer extrudate can also be measured ultrasonically. In this type of measurement, the sensor uses mechanical vibrations of high frequencies, beyond the audio range, i.e. more than about 15,000 vibrations per second. The vibrations are produced by a transducer which converts the electrical output of an oscillator to ultrasonic vibrations of corresponding frequencies.

There are two kinds of ultrasonic transducers, one is the magnetostrictive type and the other the piezoelectric type. The former consists of a metal rod placed in a coil driven by oscillator signals. The alternating magnetic field alternately elongates and compresses the rod. With one end of the rod fixed and the other end connected to a diaphragm, ultrasonic sound waves are produced.

The piezoelectric ultrasonic transducer is more common. When a voltage is applied to a piezoelectric material, it will compress or expand. If the voltage is alternating at ultrasonic frequency, the piezoelectric will compress and expand at the same frequency. The mechanical vibrations can be transferred to a diaphragm to produce ultrasonic sound waves.

For thickness measurement, the transducer is placed against the material. Ultrasonic vibrations pass through the material to the surface and are reflected back to the transducer. The

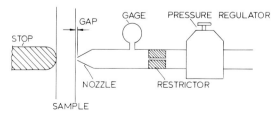

GAP   GAGE   PRESSURE REGULATOR

STOP

NOZZLE   RESTRICTOR

SAMPLE

Figure 4-19. Pneumatic Thickness Gauge

time required for the vibrations to travel through the material depends on the thickness of the material. When resonance occurs, there is a sudden change in the load that the transducer offers the oscillator, producing a corresponding change in oscillator current. By determining the resonant frequency, the thickness of the material can be determined. Obviously, good contact is required between transducer and extrudate. This good contact is difficult to achieve in continuous thickness monitoring of a moving extrudate.

Ultrasonic measurements have also proven useful in the characterization of polymer melts, see Section 4.3.3.1. The ultrasonic transit time ("Laufzeit") is dependent on the elastic properties of the material, pressure, temperature, chemical composition and structure. It has been found (37) that the ultrasonic transit time is a sensitive measure of the condition of the polymer melt; in particular, melt homogeneity. In a process control system, the ultrasonic transit time can provide a more useful feedback control signal than a single melt temperature measurement.

Thus far, the discussion has dealt with contacting thickness measurements. In addition to the fact that there is contact between the sensor and the extrudate, there is another problem in that these methods cannot be applied to continuous thickness monitoring of annular profiles, i.e., tubes and pipes. In automated extrusion lines, non-contacting thickness measurement has become quite popular. Most of the non-contacting thickness measurement techniques are based on a radiation sensor picking up a signal from a radiation source.

Various types of radiation are used: alpha-rays, beta-rays, gamma-rays, x-rays, and infrared radiation. A continuous stream of radiation is emitted from a constant radiation source (x-ray tube or radioisotope), passes through the material whose thickness is being measured, and strikes the radiation sensor. As radiation passes through the extrudate, some of the radiation is absorbed and, as a result, the radiation reaching the sensor is less intense. The amount of absorption depends on the material's density and thickness. If the density is constant, the amount of radiation absorption is a direct measure of thickness.

The absorption of radiation is governed by the following relationship:

$$I(x) = I(o)\exp(-\mu x) \qquad\qquad (4\text{--}5)$$

where $I(x)$ is the intensity after transversing a distance x through a material with absorption coefficient μ; $I(o)$ is the incident intensity. After proper calibration, very high accuracies can be achieved, down to 0.2 μm. Measurements can be made at high speeds. These factors have made the radiation type thickness measurement an almost standard tool on automated extrusion lines.

Nuclear radiation sensors cover a thickness range from 10 μm to about 3 mm. Some sensors come with air gap temperature sensors to compensate for changes in density of the air column as the temperature varies across the sheet. These sensors can be designed to automatically correct for dirt build-up and drift. A disadvantage of nuclear radiation is the potential health hazard. Very high standards have to be applied to the design of the measuring device to ensure that all radiation is contained within the instrument. With thick extrudates, relatively high radiation levels are required because of the exponential decay of radiation intensity with distance. Therefore, for relatively thick flat profiles, the LVDT type sensor may be more appropriate. Infrared sensors can be used for clear thin films in the thickness range of 2 μm to 200 μm. This sensor employs simultaneous valuation of the measurement and reference wave length.

A problem with radiation sensors is that the measurement is affected by density or compositional variations in the material. This is because most of these sensors measure weight per unit area. Thus, when changes in density occur in the extrudate, the accuracy of the thickness measurement will be directly affected. Some sensors have special compensation circuitry to eliminate the sensitivity to compositional changes. These sensors are referred to as "nondiscriminatory", meaning that additives, base resin, and variations in composition are measured equally (73). Of course, the sensitivity to density variations can be used advantageously to detect flaws in the material, such as voids, contaminants, etc.

Gamma-rays and high-frequency x-rays have a relatively high penetrating capability and are almost exclusively used for thick parts, several millimeters up to as high as 40 mm. However, they are also used for thicknesses down to about 100 μm. Beta-rays have less penetrating capability and are used for relatively thin parts, less than about 3 to 5 millimeters. Two isotopes are often used, Krypton-85 and Strontium-90. Krypton-85 is used in a measurement range of about 10 to 1000 g/m^2, with a corresponding thickness range of about 5 μm to 750 μm. Strontium-90 is employed in a measurement range of about 100 to 5000 g/m^2, with a corresponding thickness range of about 100 μm to 5000 μm (79). A disadvantage of beta gauges is that they tend to drift and, thus, require frequent recalibration.

Gamma-ray thickness detectors often employ the "gamma backscattering" technique (80). This type of measurement offers the advantage that the object is measured from one side; thus allowing simple installation and measurement of relatively complicated shapes. The isotope used is generally Americium-241, which has a half-life of about 450 years as opposed to about 10 years for the isotopes used in beta-ray thickness detectors. Another advantage of this technique is that the gauge readings are relatively insensitive to changes in the composition of the material, allowing relatively simple electronics. Low frequency x-rays (soft x-rays) are also used for thickness measurement of relatively thin products.

4.4.4   EXTRUDATE SURFACE CONDITIONS

In the extrusion of sheet and film, it is often very important that the surface conditions of the extrudate are maintained within a narrow range. Small irregularities, such as specks, have to be avoided in the more demanding applications, e.g., recording tapes, high-quality transparent sheet, etc. It is very difficult for the human inspector to accurately detect a small dirt spot of 1-mm diameter moving at a speed of 1 to 5 m/sec., particularly if the web has a substantial width, more than 2 meters. For these demanding applications, automatic inspection systems are available.

An automatic inspection system uses various transducers that produce electrical signals representative of the surface condition of the web. These signals then have to be analyzed and interpreted. If a signal, or a number of signals, exceed a preset detection threshold, an alarm is activated and the area of concern can be further analyzed to determine what corrective action should be taken. Where visible flaws are to be detected, the transducers most commonly chosen are light-sensitive; they produce signals that are parametric measures of such physical phenomena as reflection, transmission, and the like.

Several commercial automatic inspection systems have been developed (24, 25) that utilize a laser scanning system. The web is scanned with a moving light beam from a laser flying spot scanner. The reflected or transmitted beam which has been modified by the characteristics of the web is picked up by a light-sensitive detector, a photomultiplier tube. This design allows a very high scan rate (more than 5,000 scans per second) and permits 100-percent coverage of web as wide as 4 meters, moving at speeds of 1 m/sec and higher.

The measurement of haze and luminous transmittance of transparent plastics is described in ASTM D1003 (American Society for Testing and Materials). The haze of a specimen is defined as the percentage of transmitted light which, in passing through the specimen, deviates more than 2.5 degrees from the incident beam by forward scattering. Luminous transmittance is defined as the ratio of transmitted to incident light. The measurement of luminous reflectance, transmittance and color is described in ASTM E308. The measurement of gloss is described in ASTM D523, also in DIN-Norm 67530 (Deutsche Industrie Norm = German Industrial Standard).

The quantitative measurement of gloss of extruded sheet is described by Michaeli (26). In this study, a goniophotometer was used to measure both the gloss height (maximum intensity of the reflected light beam) and the gloss sharpness expressed as the reciprocal width of the gloss distribution curve; the width is measured in degrees.

Quantitative characterization of the texture of extruded film was studied by Nadav and Tadmor (27, 28). The samples were characterized by measuring light transmission through a film sample. The results were analyzed using the concepts of scale of segregation and intensity of segregation.

In the surface analysis of an extrudate, the irregularities in the size range of μm to mm (10^{-6} to 10^{-3} m) are of primary interest. These irregularities are responsible for the gloss, roughness, and color properties of the surface. Most methods of surface texture determination are based on reflection of light waves (29). In transparent materials, nonuniformities in the composition of the material can cause significant changes in optical properties. These compositional variations cannot be assessed by the measurement of light reflection; however, by measuring light transmission the effect of compositional variations can be determined accurately (30). Compositional variations, such as voids or cracks, can be measured ultrasonically, or by using microwaves or x-rays (31, 32). The measurement with microwaves can be done in a continuous fashion; however, the imperfections have to be rather large (about 1 mm) to be detectable.

Color measurement requires the determination of the spectral distribution of the light reflected from the surface. In continuous color measurement, usually a limited number (e.g., three) of spectral regions are determined by measurement across a filter. Various commercial color measuring instruments, colorimeters and spectrophotometers, are available today (33), including systems for monitoring continuous webs. These systems automatically provide data on color, opacity and yellowing at one or several user-selected web locations.

Color measurement can also be tied in with the compound preparation step. The spectrophotometer thus provides a feedback control signal to adjust the level of colorants in order to obtain the desired color automatically (74, 75). Some of the problems in assessing color and color difference were discussed by Osmer (34). A good basic text on the principles and theory of color is the book by Billmeyer and Saltzman (76).

The orientation of the polymer molecules in the extrudate has a large effect on the physical properties. At the IKV in Aachen, West Germany, a technique was developed to measure the anisotropy of an extrudate in a continuous fashion (35, 36). The measurement is based on the compensation of the orientation birefringence; the phase difference from the transmission of the light wave through the polymer is reduced to zero by the use of a proper crystal. The birefringence value, thus obtained, has to be divided by the sample thickness to determine the optical phase difference. Therefore, the thickness has to be measured simultaneously. For practical purposes, the degree of anisotropy is often used. This is the actual birefringence divided by the maximum birefringence of the polymer. The degree of anisotropy, therefore, is no longer dependent on the sample thickness.

4.5   TEMPERATURE CONTROL

The dynamic behavior of an extruder is, to a significant extent, determined by the temperature control system on the extruder. It is, therefore, important to understand the basic characteristics of the various temperature control systems. Most control systems are closed-loop or feedback systems. The variable to be controlled is measured and this information is sent to a control unit. From the control unit a signal is sent to an actuator that adjusts the process such that the control variable is as close as possible to the desired value, the setpoint. Some systems are non-feedback or open-loop systems. These are used when the effect of the input signal on the process can be accurately predicted. This is generally not the case in extrusion and, as a result, feedback control systems are commonly used in extruders.

There are basically two ways to keep the level of a variable within certain limits: the on-off method and the modulating or continuous adjustment method. The on-off control is probably the simplest type of automatic control.

4.5.1   ON-OFF CONTROL

In on-off control of temperature, the power to the extruder is full-on when the measured temperature is below the setpoint and completely off when the measured temperature is above the setpoint. This situation is shown in Figure 4–20.

The temperature in most homes is controlled by thermostats using the on-off principle. On-off control is widely used in the industry to control temperature and other variables. In fact, some missile guidance systems use on-off control, indicating that accurate control in sophisticated systems can be achieved by on-off control. That is, of course, if the system characteristics lend itself to such type of control.

One of the problems of applying on-off control to the heating and cooling of extruders is the significant thermal lag in these machines. There is an inherent delay between the time the controller calls for heat from the heater band and the time when the heat actually reaches the sensor. The same holds true when the controller turns off. This thermal lag can be several minutes (about 5 minutes for a 90-mm extruder); the larger the extruder, the long-

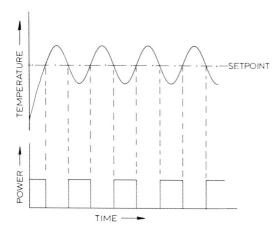

Figure 4–20. On-Off Temperature Control

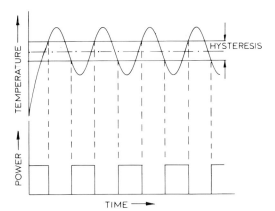

Figure 4–21. Time-Temperature Diagram
for an On-Off Controller With Hysteresis

er the thermal lag will be. Putting the temperature sensor close to the heater (shallow well) reduces this thermal lag, but makes the sensed temperature less representative of the temperature of most importance – the temperature of the polymer inside the barrel.

The temperature resulting from this type of on-off control will be fluctuating around the setpoint. The frequency and amplitude of the temperature fluctuations will be determined by the thermal lag of the particular machine. A problem of on-off control is the possible effect of process disturbances and electrical noise interference which can cause the output to cycle rapidly as the temperature crosses the setpoint. This condition can be detrimental to most final control elements such as contactors. To prevent this, an on-off differential or hysteresis is added to the controller function. This function requires that the temperature exceed the setpoint by a certain amount (half the differential) before the output will turn on again. Hysteresis will prevent the output from chattering if the peak-to-peak noise is less than the hysteresis. The amount of hysteresis determines the minimum temperature variation possible. However, process characteristics will add to this differential. Figure 4–21 shows a time-temperature diagram for an on-off controller with hysteresis.

A different representation of the hysteresis curve is shown in the transfer function of Figure 4–22.

The transfer function describes the power-temperature relationship of a controller. When the temperature is ascending, the power is turned off when the temperature exceeds T_2; when the temperature is descending, the power is turned on when the temperature drops below T_1.

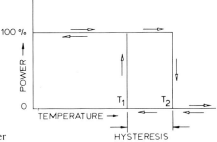

Figure 4–22. Transfer Function of an On-Off Controller

4.5.2   PROPORTIONAL CONTROL

One of the drawbacks of on-off control is that there are only two power input levels possible: fully on and fully off. In essentially all practical cases, the power level required to maintain a certain temperature will be somewhere between 0 percent and 100 percent power. Therefore, application of on-off control will invariably lead to fluctuations of the actual temperature. To avoid this problem, a control system is needed that can adjust the power input level to the exact level required to maintain the temperature at the setpoint. Only then is it possible to maintain a steady temperature.

4.5.2.1   PROPORTIONAL ONLY CONTROL

The proportional controller allows a continuous adjustment of the power input level (from 0 to 100 percent) depending on the actual temperature. The range of temperature over which the power is adjusted from 0 percent to 100 percent is called the proportional band. The proportional band is usually expressed as a percentage of instrument span and is often centered about the setpoint. Thus, in a controller with a 500°C span, a 5 percent proportional band would be 25 degrees about the setpoint. Sometimes, the setpoint is located at the upper temperature limit of the proportional band.

Figure 4–23 shows the transfer function for a reverse-acting controller.

It is called reverse-acting because the output decreases with increasing temperature. If the temperature is below the lower boundary of the proportional band, T_1, the power to the heaters is on 100 percent. Above the upper boundary of the proportional band, T_2, the power to the heaters is completely off. Within the proportional band, the power varies proportionally with temperature, from 0 percent power at T_1 to 100 percent power at T_2.

The proportional band in most controllers is adjustable to obtain stable control under different process conditions. A narrower proportional band would result in a steeper transfer function or power-temperature relationship. In the extreme case when the width of the proportional band is reduced to zero, the proportional controller would act simply as an on-off controller. In that case, all the advantages of the proportional control would be lost.

The proportional band is often expressed as a percent of span in the plastics industry, but it is also expressed as controller gain in other industries. The proportional band width in percent span and controller gain are related inversely:

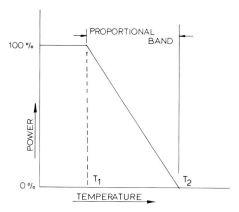

Figure 4–23. Transfer Function for a Reverse-Acting Controller

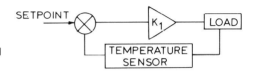

Figure 4–24. Block Diagram of a Proportional
Control System

$$\text{Gain} = \frac{100\%}{\text{Proportional band in percent span}} \qquad (4\text{–}6)$$

Thus, reducing the width of the proportional band will increase the gain. A proportional band of 5 percent corresponds to a gain of 20, a proportional band of 4 percent corresponds to a gain of 25, etc.

A block diagram of a proportional control system is shown in Figure 4–24.

The control system contains a comparator which compares the actual temperature, T_a (measured by the temperature sensor), with the desired, or setpoint temperature, T_s, to provide an error or deviation signal, e. The signal is positive when the process is below setpoint, zero when the process is at setpoint, and negative when the process is above setpoint. The proportional term, K_1, gives an output proportional to the error:

$$\text{Output} = K_1 e \qquad (4\text{–}7)$$

When the setpoint is centered in the proportional band, the power is 50 percent when the error signal is zero, i.e., the process is at setpoint.

In a real process, it is rare that the power input required to maintain setpoint temperature is exactly 50 percent of full power. Therefore, the temperature will increase or decrease, adjusting the power level until an equilibrium condition exists. The temperature difference between the stabilized temperature and the setpoint is called offset or droop. The amount of offset can be reduced by narrowing the proportional band. However, the proportional band can be narrowed only so far before instability occurs. An illustration of a process coming up to temperature with an offset is shown in Figure 4–25.

To understand the mechanism by which offset occurs with a proportional controller, one should look at the controller transfer curve and the process transfer curve at the same time, see Figure 4–26.

The process transfer curve indicates the power-temperature characteristic of the actual system, i.e., the extruder and its surroundings. This curve indicates how much power is re-

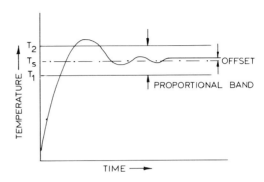

Figure 4–25. Process Coming Up to Temperature With an Offset

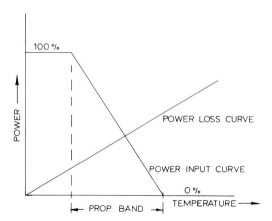

Figure 4-26. Controller Transfer Curve and Process Transfer Curve

quired to maintain a certain temperature level on the machine. The higher the temperature level that needs to be maintained, the more power will be required. For most machines, the relationship between temperature and power requirement will be approximately linear. The power requirement is determined by the heat losses in the system by conduction, convection and radiation. By improving the thermal insulation of the extruder, the heat losses can be reduced. This will directly affect the process transfer curve; adding insulation will reduce the power requirement at a certain temperature, resulting in a reduced slope of the process transfer curve. Obviously, this will also improve the energy efficiency of the entire process.

Figure 4-26 shows the controller transfer curve (power input curve) superimposed on the process transfer curve (power loss curve). The point where the two curves intersect is the temperature where the input power to the heaters is in equilibrium with the power losses. If the point of intersection occurs above the setpoint, there will be a positive droop, if it occurs below the setpoint, the droop will be negative. From Figure 4-26, it is now clear how the offset can be eliminated without changing the width of the proportional band. This is done by shifting the entire proportional band to a higher or lower temperature. Figure 4-27 illustrates the effect of resetting the proportional band.

This resetting can be done manually or automatically. With manual reset, a potentiometer is used to electrically shift the proportional band. The amount of shifting has to be done in small increments until the controller power output matches the process power demand at setpoint temperature.

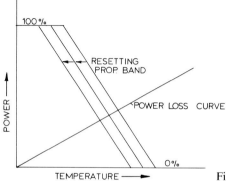

Figure 4-27. Effect of Resetting the Proportional Band

4.5.2.2   PROPORTIONAL+INTEGRAL CONTROL

Automatic reset is done by using an electronic integrator to perform the reset function. The deviation or error is integrated with respect to time and the result is added to the deviation signal to move the proportional band. The block diagram of proportional control with automatic reset is shown in Figure 4-28.

The output now becomes:

$$Output = K_1 e + K_2 \int edt \qquad\qquad (4-8)$$

The K_1 term operates exactly as in a simple proportional controller. The integration term K_2 has to be made long enough so that it will be negligible at the frequencies when the control loop has 180 degrees phase shift. These are the critical frequencies for loop stability. If this is done, the K_2 term will not have an effect on the stability of the loop, but it will accomplish its task, i.e., eliminate the offset.

The integrator keeps adjusting the level of the proportional band until the deviation is zero. When this condition is achieved, the input to the integrator becomes zero and its output stops changing. At this point, the correct amount of reset is held by the integrator. If the process heat requirements should change, a deviation would again occur which the integrator would integrate and corrective action would be applied. This corrective action has to be applied rather slowly to avoid oscillations. The automatic reset term can be thought of as a slow gain, as opposed to the proportional term which can be considered a fast gain.

One problem that can occur in a proportional controller with automatic reset is "reset windup." This occurs when the integrator has acted on the error signal when the temperature is outside the proportional band. The resulting large "woundup" output of the integrator causes the proportional band to move so far that the setpoint is outside the band. The temperature must pass the setpoint before the controller output will change. As the temperature crosses the setpoint, the deviation signal changes sign and the integrator output starts to decrease or unwind. The result can be a large temperature overshoot. This can be prevented by stopping the integrator from acting if the temperature is outside the proportional band. This function is referred to as reset windup inhibit or reset inhibit.

One characteristic of all proportional plus integral (PI) controllers is that the temperature often overshoots the setpoint on startup. This occurs because the automatic reset begins acting when the temperature reaches the lower boundary of the proportional band. As the temperature reaches the setpoint, the reset action has already moved the proportional band higher, causing excess heat output. As the temperature exceeds the setpoint, the sign of the deviation signal reverses and the integrator brings the proportional band back to the position required to eliminate the offset. This situation is illustrated in Figure 4-29.

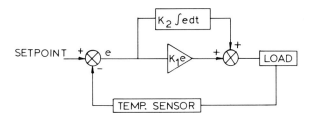

Figure 4-28. Block Diagram PI Control System

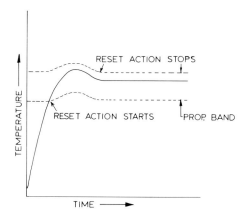

Figure 4-29. Temperature During Startup With PI Control

4.5.2.3   PROPORTIONAL + INTEGRAL + DERIVATIVE CONTROL

One drawback of the automatic reset is its relatively slow response. The response time can be reduced by addition of a second corrective term which acts on the rate of change in temperature. A block diagram of a proportional plus integral plus derivative (PID) control system is shown in Figure 4-30.

The second corrective term yields a signal proportional to the rate of change in the error signal. This rate of change is the derivative of the measured temperature with respect to time; thus, the term derivative control. The output of the PID control is:

$$\text{Output} = K_1 e + K_2 \int e \, dt + K_3 \frac{de}{dt} \qquad (4\text{-}9)$$

The derivative control comes into action when a transient occurs. This action is immediate; it does not wait until the error builds up, but it responds directly to the rate of change of the error. Corrective action will be taken in the shortest possible time and the magnitude of the temperature deviation, caused by an upset, will be greatly reduced.

Derivative control helps to prevent overshooting or undershooting the setpoint. It is an anticipatory function that adjusts the controller output, in advance, to anticipated needs. This reduces the time lag it takes for the controller to respond to a change in the process. Derivative control is more important on machines with a long thermal lag. Therefore, derivative

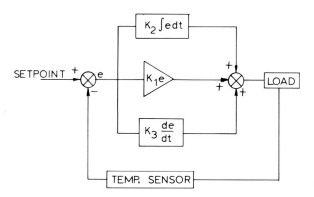

Figure 4-30. Block Diagram PID Control System

control is important for large extruders. Small extruders, with their inherently shorter thermal lag, may not benefit much from a derivative type of temperature control.

A drawback of the rate control is the fact that it has a destabilizing influence on the control loop. Therefore, it has to be sized carefully to maintain adequate stability in the control loop. In spite of this, the derivative control can generally yield an improvement in response time by a factor of 2 to 4.

4.5.2.4   Dual Sensor Temperature Control

A few commercial temperature control systems are based on dual input from two temperature sensors. One temperature sensor is located in a deep well and measures temperature close to the polymer. The other temperature sensor is located in a shallow well and measures temperature close to the heater/cooler. The dual sensor temperature control can combine the advantages of deep-well-only control and shallow-well-only control, but can largely eliminate the drawbacks of these types of control.

The deep-well-only control is well-established and reliable; however, its main drawback is slow response to changes in external conditions, such as changes in ambient temperature, heater line-voltage variations, changes in cooling water temperature, etc. On the other hand, the response to changes in internal conditions is quite rapid. Examples are changes in screw speed, viscosity, changes in the polymer, temperature changes in the polymer, etc. The shallow-well-only control has the advantage of being able to respond quickly to changes in external conditions, but slowly to changes in internal conditions.

One dual sensor control system uses a weighted average of the signals from the two sensors. By doing this, the advantages of both deep-well and shallow-well can be enjoyed to some extent; however, the same is true about the disadvantages. A newer system uses two different temperature control loops. The first one uses only the deep-well sensor and controls the power to the heater. The second loop uses only the shallow-well sensor. This is a cascade loop. It does not control the heaters directly, but acts on the first loop in such a manner as to keep the temperature at the deep well at the setpoint. Both of these dual sensor temperature control systems have been patented (38, 39).

4.5.3   CONTROLLERS

4.5.3.1   TEMPERATURE CONTROLLERS

There are two types of temperature controllers, analog and digital. An analog controller contains a number of discrete components – resistors, capacitors, integrated circuits, operational amplifiers and performs its control algorithm through these components. In a digital controller, a microprocessor replaces the discrete components of the analog unit with integrated circuit chip logic. It takes analog input, converts it into a digital signal, and then performs its control algorithm through a stored computer program, which the microprocessor executes.

Microprocessor-based controllers offer great flexibility in that controller function can be changed readily by simply changing a few steps in the program. Thus, the controller function can be modified by changing the software without having to make any modifications to the hardware.

The temperature controller is not capable of handling the high currents required to power the heater bands. For this reason, a power controller is linked between the temperature controller and the heater band. The temperature controller dictates to the power controller how much power to supply to the heaters.

4.5.3.2   POWER CONTROLLERS

A relatively inexpensive, time-proportioning power controller is the mercury contactor. With this switch, there is no zero-crossing detection, so there will be some noise in the circuit when the contactor is turned on and off.

This problem can be avoided by using a zero-crossover-fired power controller. This is a time-proportioning system controlled by the temperature controller. Being zero-voltage fired, this controller's circuit is free of RF noise. In other words, the switching occurs when no voltage is on the line.

A disadvantage of the contactors is their limited life; they are usually replaced after one million operations. In a typical operation, contactors are operated once every 20 seconds. This gives a reasonable compromise between temperature ripple and load life. Operation every 20 seconds means 4,320 operations per day, or one million operations in about 230 days. One will have to replace the contactor every 230 days. If there are 8 controllers per machine, the machine will be down about once a month on the average.

A newer type of power controller is the true proportional power controller – also known as current proportioning controller or phase-angle-fired proportional controller. The term "true proportional" means that the power to the heater is adjusted on a continuous basis. This provides a smooth, stepless output of power. This is particularly useful for extruder dies where rapid and smooth response is very important. True power proportioning is less wearing on the heater bands because it eliminates the on-off cycling.

True proportional control is obtained with solid state switching of loads through SCR's, triacs, and similar solid state devices. This allows a reduction of cycle time to the millisecond level. If the cycle time is reduced to one-half the power line period (8.3 msec. for 60 Hz), then the proportioning action is referred to as stepless control or phase-fired control. A 50-percent phase-fired output is shown in Figure 4–31.

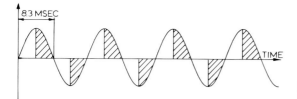

Figure 4–31. A 50-Percent Phase-Fired Output

It is clear that SCR power controllers can also be used for time proportioning power control by increasing the cycle time. However, this type of use does not utilize the inherent advantages of the SCR power controller.

In a true proportional controller, the SCR switches extremely fast – in the order of half a microsecond. This rapid switching causes radio frequency interference (RFI) to be generated on the power lines. This can cause interference with nearby computer and communications equipment. A further problem is that the power drawn from the power lines has a severely distorted wave shape. This can result in extra charges from the power company.

These problems can be avoided by the use of RFI-free or zero-crossing SCR circuitry. A characteristic very much like time proportioning is used, but the power is always turned on and off at the instant when the power line voltage is zero, as shown in Figure 4–32.

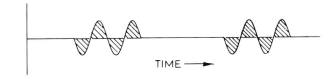

Figure 4–32. RFI-Free Power Proportioning

TIME ⟶

Another type of power controller is the solid-state relay power controller. This controller is reliable and inexpensive. It is zero-crossover fired, thus it does not generate RFI. However, to use the controller at its rated current level, it must be provided with a good heat sink.

4.5.3.3   DUAL OUTPUT CONTROLLERS

Controlling the temperature of an extruder may require adding heat to the machine by the barrel heaters or taking heat out of the machine by cooling. Since a substantial amount of heat is generated internally in the extrusion process, cooling is often required to maintain the desired process temperature. For this reason, it is important to utilize dual output temperature controllers. These are controllers that control both the heating as well as the cooling process.

Thus, if proportional temperature control is used in heating, then proportional temperature control should also be used in cooling. This would seem to be an obvious requirement. However, a large number of commercial extruders come standard with PID heating control and on/off cooling control. The fact that such extruders are built and the fact that customers buy such extruders illustrates that concern about precise temperature control is not very widespread. Such extruders would only be appropriate if the process is designed in such a way that extruder cooling is not used in normal operation – a situation that is unlikely to occur in actual practice. Dual output controllers, nowadays, are only slightly more expensive than single output controllers. Thus, cost is only a minor consideration.

4.5.4   TIME-TEMPERATURE CHARACTERISTICS

Thus far, the discussion has focused on the various types of temperature controllers. However, the real issue is how the actual temperature of the extruder will change as a result of the action of the temperature controller. In order to determine this, one has to consider the thermal characteristics of the actual system, in this case the extruder.

4.5.4.1   THERMAL CHARACTERISTICS OF THE SYSTEM

The thermal characteristics of the system determine how the output signal (temperature) varies with changes in input (heat). One of the simplest and most common methods to determine the system characteristics is to measure the transient response of temperature to a step change in heating power, see Figure 4–33.

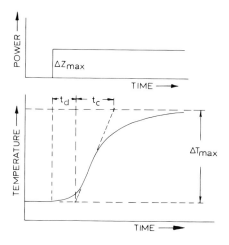

Figure 4-33. Temperature Response of Step Change in Heater Power

This temperature/time curve is often referred to as the process reaction curve or simply the response curve.

From the response curve, several parameters can be established that are important in understanding the behavior of the system. The dead time (t_d) is the time to the intersection of the maximum temperature gradient. The constant K_s indicates the ratio of maximum temperature rise to the step change in power. The time constant (t_c) is the maximum change in temperature divided by the maximum temperature gradient.

The dead time in extruders can range from about 1 to 5 minutes. This constitutes one of the main problems in temperature control, because this means that the effect of a change in power input level is not felt until after at least 1 minute. This thermal lag time is influenced by the depth of the temperature sensor, the thermal conductivity of the barrel, and the design of the barrel heaters. A typical time constant in an extruder can range from 30 to 120 minutes. This value depends on the heating capacity and the specific heat and mass of the extruder barrel.

Although these parameters do not fully describe the system, they can be used to make approximate predictions of the degree of difficulty in temperature control (41).

```
        t_d/t_c ≤ 0.1 easy control
0.1 < t_d/t_c < 0.3 reasonable control
        t_d/t_c ≥ 0.3 difficult control
```

4.5.4.2   MODELING OF RESPONSE IN LINEAR SYSTEMS

In a linear control system, the relationship between input y and output x can be described by a differential equation:

$$A_n \frac{d^n x}{dt^n} + A_{n-1} \frac{d^{n-1} x}{dt^{n-1}} + \ldots + A_1 \frac{dx}{dt} + A_0 x = y \qquad (4-10)$$

The order of the differential equation determines the order of the control system. Thus, a first order control system is described by:

$$A_1 \frac{dx}{dt} + A_0 x = y \tag{4-11}$$

After a long time ($t \to \infty$), the following relationship is valid:

$$A_0 x = y \tag{4-12}$$

Thus, a change in input will produce a proportional change in the output. If $t_c = A_1/A_0$ and $K_s = 1/A_0$, equation 4-11 can be written as:

$$t_c \frac{dx}{dt} + x = K_s y \tag{4-13}$$

The solution to this differential equation, when the input variable is given a step change of 1, is the well-known exponential relationship of a first order control system.

$$x = K_s \, [1 - \exp(-t/t_c)] \tag{4-14}$$

where t_c is the system time constant and K_s the system gain. Various step response functions are shown in Figure 4-34. Figure 4-34a shows a first order system, Figure 4-34b a higher order system, and Figure 4-34c a first order system with lag time.

In reality, most systems do not behave in a truly linear fashion, but the assumption of linearity can usually be justified over a small region near the operating point (linearization).

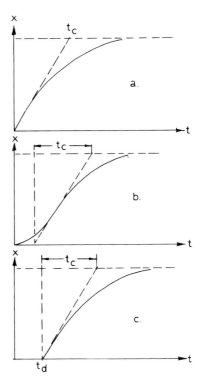

Figure 4-34. Various Step Response Functions

If a mathematical description of the system characteristics is not possible, the system parameters have to be determined experimentally. If the step response function is of a higher order than one (see Figure 4–34b), the response can be approximated by a first order response function with lag time.

If the following control system is now considered, see Figure 4–35:

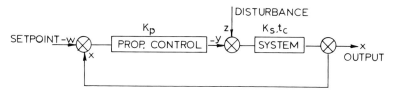

Figure 4–35. First Order Control System

The response function of the system can be described by:

$$t_c \frac{dx}{dt} + x = K_S \ (z - y) \tag{4-15}$$

For the controller:

$$y = K_p \ (x - w) \tag{4-16}$$

If equations 4–15 and 4–16 are combined, then the equation for the closed-loop control system is obtained:

$$\frac{t_c}{1 + K_S K_p} \frac{dx}{dt} + x = \frac{K_S}{1 + K_S K_p} \ (z + K_p w) \tag{4-17}$$

When the effect of a disturbance (step change) at a constant setpoint is considered ($z=1$ and $w=0$):

$$\frac{t_c}{1 + K_S K_p} \frac{dx}{dt} + x = \frac{K_S}{1 + K_S K_p} \tag{4-18}$$

After a very long time ($t \to \infty$), the output will attain the following value:

$$x \ (t \to \infty) = \frac{K_S}{1 + K_S K_p} \tag{4-19}$$

The time constant of the closed-loop system is:

$$t^* = \frac{t_c}{1 + K_S K_p} \tag{4-20}$$

Thus, one can conclude that the remaining deviation $x(t \to \infty)$ reduces with the increased proportionality constant K_p of the controller. The time constant also reduces with increasing K_p.

A step change in setpoint (w = 1 and z = 0) is described by:

$$\frac{t_C}{1 + K_S K_p} \frac{dx}{dt} + x = \frac{K_S K_p}{1 + K_S K_p} \qquad (4-21)$$

After a very long time (t → ∞), the output will attain the following value:

$$x\ (t \rightarrow \infty) = \frac{K_S\ K_p}{1 + K_S K_p} \qquad (4-22)$$

Thus, the change in output increases with the proportionality constant K_p of the controller and the system gain K_S. In this example, it was assumed that there was no dead time in the system. If a dead time is present, then output fluctuations will occur upon changes in input. In this case, the control loop can become unstable, depending on the actual dead time and the proportionality constant of the controller. The effect of changes in setpoint or a disturbance is more difficult to analyze in higher order systems. The controller parameters have to be selected such that the control loop is stable in all cases.

4.5.4.3   TEMPERATURE CHARACTERISTICS WITH ON-OFF CONTROL

The approximate time/temperature characteristics of a simple on-off controller are relatively easy to visualize. If a dual output on-off control is considered, then a typical temperature/time curve can be as shown in Figure 4-36.

If the system heats up from ambient temperature, the heating power will be turned off when the temperature reaches:

$$T_{ho} = T_s - \frac{1}{2} \Delta T_n \qquad (4-23)$$

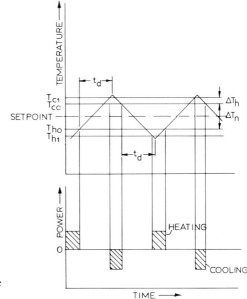

Figure 4-36. Typical Temperature/Time Curve for a Dual Output On-Off Control

where T_s is the symmetrically located setpoint, ΔT_n the neutral temperature zone in which no heating or cooling takes place; it is also referred to as a dead band. If there is a dead time t_d in the system, the temperature will continue to rise for this length of time. When the temperature reaches T_{c1}, cooling is turned on. However, if the dead time t_d is long enough, the actual temperature can continue to rise.

$$T_{c1} = T_s + \frac{1}{2} \Delta T_n + \Delta T_h \qquad (4-24)$$

where ΔT_h is the hysteresis loop width.

When the dead time has elapsed, the temperature will start to drop. The cooling will be turned off when the temperature drops below T_{co}.

$$T_{co} = T_s + \frac{1}{2} \Delta T_n \qquad (4-25)$$

The temperature will continue to drop for a time period of t_d. When the temperature drops below T_{h1}, the heating is turned on again. After another time period of t_d, the temperature will start to increase again. This temperature cycle will repeat itself if the thermal conditions of the process remain the same. The maximum temperature T_{max} is:

$$T_{max} = T_{ho} + t_d\, v_h \qquad (4-26)$$

where v_h is the rate of temperature rise upon heating.

The minimum temperature T_{min} is:

$$T_{min} = T_{co} - t_d\, v_c \qquad (4-27)$$

where v_c is the rate of temperature drop upon cooling.

The maximum temperature swing ΔT_{max} is:

$$\Delta T_{max} = T_{ho} - T_{co} + t_d(v_h + v_c) = \Delta T_n + t_d(v_h + v_c) \qquad (4-28)$$

If a typical value of the neutral zone is $10° C$, a typical value of the dead time 2 minutes, and the rate of temperature change $6°C/minute$, then the maximum temperature swing becomes $34° C$! This assumes that the rate of temperature change in heating is the same as in cooling ($v_h = v_c$). These values are typical of a dual output on-off controller on an extruder. Because of the resulting large fluctuations in temperature, on-off control is not very desirable if accurate temperature control is required.

If the neutral zone is wide enough, the temperature may not reach the point at which the cooling is turned on. Cooling can be avoided if:

$$v_h t_d < \Delta T_n + \Delta T_h \qquad (4-29)$$

This situation is portrayed in Figure 4–37.

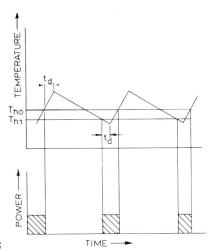

Figure 4–37. Temperature/Time Curve Without Cooling

In this case, the maximum temperature swing is:

$$\Delta T_{max} = \Delta T_h + t_d(v_h + v_c^{-1})$$ \hfill (4–30)

The rate of temperature change upon cooling v_c^{-1} is much slower in this case because it is passive cooling. In other words, the barrel cooling has not been activated; the temperature drops simply as a result of heat losses in the system. If $\Delta T_h = 1°$ C, $t_d = 2$ min., $v_h = 6°$ C/min., and $v_c^{-1} = 2°$ C/min., the maximum temperature swing is $\Delta T_{max} = 17°$ C. Thus, the amplitude of the temperature fluctuation is cut in half as compared to the first case. However, even a temperature fluctuation of $17°$ C would still be considered unacceptable in most extrusion operations.

4.5.5 TUNING OF THE CONTROLLER PARAMETERS

4.5.5.1 PERFORMANCE CRITERIA

The proper selection, design, or tuning of controllers requires certain controller criteria that are most appropriate for the particular application. Some response criteria include over-shoot, decay ratio, rise time, response time, frequency of oscillation, phase and gain margin,

Overshoot $= A/R$
Decay Ratio $= B/A$
Rise Time $= t_r$
Response Time $= t_{95}$
Frequency of
Oscillation $= \omega = 1/t_c$

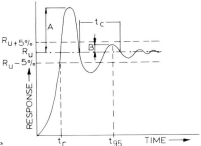

Figure 4–38. Performance Criteria from a Response Curve

error integrals, etc. (42–48). Three common error integrals are integral of square error (ISE), integral of absolute error (IAE), and integral time and absolute error (ITAE). Some of these criteria are illustrated in Figure 4–38 showing the response of a control system to a step change in input.

4.5.5.2   EFFECT OF PID PARAMETERS

Tuning of a PID controller should ideally lead to values of the P, I, and D terms of the controller that result in the most favorable actual control performance. The P-term is described by the proportional gain K_p or the proportional bandwidth X_p. The effect of changing the proportional band is shown in Figure 4–39.

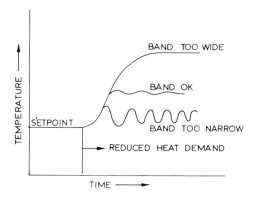

Figure 4–39. Effect of Bandwidth Setting, X_p

A narrow proportional band can result in strong oscillations, while a wide proportional band will result in a large offset, in the absence of reset function. The I-term provides the reset function and is described by the integration time constant t_i or reset time constant. The effect of different values of the reset time constant is shown in Figure 4–40.

When the reset time constant is too long, the process will come back to the setpoint very slowly. On the other hand, when the reset time constant is too short, oscillations will occur. The reset time constant is generally considered optimum when the temperature returns to setpoint as rapidly as possible without overshoot.

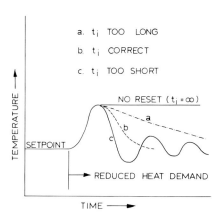

Figure 4–40. Effect of Reset Time Constant, t_i

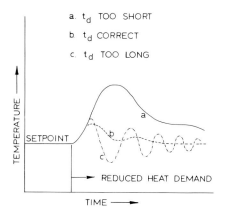

Figure 4–41. Effect of Rate Time Constant, t_d

The derivative term is characterized by the derivative time constant t_d or rate time constant. The effect of different rate time constants is shown in Figure 4-41.

When the rate time constant is too long, the temperature changes too rapidly, resulting in overshoot and oscillations.

When the rate time constant is too short, the temperature will return to setpoint too slowly. The correct rate time constant will return the temperature to setpoint with a minimum of oscillations.

4.5.5.3   TUNING PROCEDURE WHEN PROCESS MODEL IS UNKNOWN

The tuning technique that is applied will depend on whether or not the process model is known. When the process model is not known, the most widely used tuning techniques incorporate the ultimate-period method, the reaction-curve method, and various search methods.

The ultimate-period method, also called Ziegler-Nichols method (42), starts by obtaining dynamic response data. These data are obtained by tuning out the integral and derivative actions of the controller and using only the proportional control mode. The proportional gain is gradually increased until the closed-loop system is forced to cycle continuously at the point of instability. The proportional gain at the point of continuous cycling (ultimate gain) and the period of oscillation (ultimate period) identify the frequency response of the open-loop system at one point. Recommended controller settings are shown in Table 4-4.

TABLE 4-4.   RECOMMENDED CONTROLLER SETTINGS BASED ON ULTIMATE GAIN K_u AND ULTIMATE PERIOD t_u

Control	Criterion	Gain	Reset Time	Rate Time
P	1/4 decay	$0.5\ K_u$	–	–
PI	1/4 decay	$0.45\ K_u$	$0.833\ t_u$	–
PID	1/4 decay	$0.6\ K_u$	$0.5\ t_u$	$0.125\ t_u$
PID	Some overshoot	$0.33\ K_u$	$0.5\ t_u$	$0.33\ t_u$
PID	No overshoot	$0.2\ K_u$	$0.33\ t_u$	$0.5\ t_u$

The reaction-curve method is based on the open-loop response of the process to a step input. This response curve can be used to derive the dynamic characteristics of the process. If

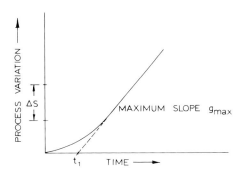

Figure 4-42. Typical Reaction Curve

TABLE 4-5. RECOMMENDED CONTROLLER SETTINGS BASED
ON REACTION CURVE PROCESS TEST

Control	Gain	Reset Time	Rate Time
P	$1/g_1t_1$	–	–
PI	$0.9/g_1t_1$	$3t_1$	–
PID	$1.2/g_1t_1$ to	$2.5t_1$ to	$0.5t_1$ to
	$2/g_1t_1$	$2t_1$	$0.3t_1$

the process can be described by a first-order lag and dead time, the controller settings can be calculated.

The controller is placed on manual control and a step change ΔS is applied. With a recorder, note the size of the step change and time of application. The reaction curve should be approximately as shown in Figure 4-42.

Determine the rate of change per unit step change in the manipulated variable, $g_1 = g_{max}/\Delta S$. The recommended controller parameter settings, based on the quarter-decay performance criterion, are shown in Table 4-5.

The two previous tuning techniques require a reasonably detailed control-loop analysis. In practice, many controllers are tuned by trial-and-error methods based on process experience. Both the Ziegler-Nichols method and the reaction-curve method are based on the assumption that the disturbances enter the process at one particular point. These methods, therefore, do not always give satisfactory results. In these cases, the final adjustments must be made by trial-and-error search method.

4.5.5.4   TUNING PROCEDURE WHEN PROCESS MODEL IS KNOWN

Various techniques are available for the experimental determination of process model. Astron (77) and Eykhoff (78) have given a survey of different identification techniques. The most common technique is to apply a step or impulse perturbation and to evaluate the transfer function from the resulting transient response. A more sophisticated technique is the stochastic identification technique. In this technique, input variations of a known random form are applied. The similar statistical properties in input and output variables are then correlated, eventually yielding the process transfer function. The transfer function is the ratio of the Laplace transform of the responding variable (output) to the Laplace transform of the disturbing variable (input).

The experimental determination of the process model requires good instrumentation with sufficient dynamic response. Once the transfer function is known, the tuning can be based on Bode plots, Nyquist diagrams, or analytical methods. Bode plots and Nyquist diagrams represent the transient performance characteristics based on the open-loop frequency-response function. The logarithmic or Bode plot shows how the phase angle and the magnitude of the direct-transfer function depends on the frequency. In the polar or Nyquist diagram, the magnitude and phase angle of the direct-transfer function are plotted as a vector with the frequency as a parameter. The normal design criterion for control systems using the frequency-response approach is the specification of the gain margin and the phase margin of the open-loop system. Common criteria are 30 degrees for phase margin and 1.7 to 3 for the gain margin. The tuning or design of the controller is accomplished by adding the controller frequency-response characteristics to the system characteristics to achieve the desired phase and gain margins for the combined system.

Analytical techniques generally involve two areas. The first is direct solution of the system differential equations in the Aime domain, usually by state variables. The second area is optimization of a specific performance criterion. The criteria for optimization by analytical techniques usually involve minimum response time or an integral time cost function.

4.5.5.5   PRETUNED TEMPERATURE CONTROLLERS

Several controller manufacturers supply temperature controllers where the parameters are factory-tuned and nonadjustable. The parameter settings are based on long-term experience with temperature controllers in extrusion applications. This is possible because the thermal characteristics of most extruders are rather similar. Advantages of this approach are easier installation and less chance of controller tampering by unqualified personnel.

Controllers with nonadjustable parameters are not well-suited to extrusion operations where changes in temperature are very rapid (water cooling, blown film, etc.) or very slow combined with long dead times, as may occur in very large machines. Typical controller settings for pretuned controllers are shown in Table 4-6.

TABLE 4-6. TYPICAL PARAMETERS FOR PRETUNED CONTROLLERS

Control	Nonadjustable	Adjustable
PD	$X_p = 5\%$	$X_p = 0 - 10\%$
	$t_d = 0.5$ min.	
PID	$X_p = 8\%$	$X_p = 0 - 10\%$
	$t_i = 8$ min.	
	$t_d = 0.5$ min.	

X_p = Proportional Band
t_i = Reset Time Constant
t_d = Rate Time Constant

4.5.5.6   SELF-TUNING TEMPERATURE CONTROLLERS

Self-tuning temperature controllers are a relatively new development in extrusion (49). They have only become commercially available since around 1982. These controllers are microprocessor-based; they are programmed to sense certain process conditions and adjust the controller parameters when necessary. These adjustments are made internally by the controller itself without the help of an operator.

The terminology used to describe these controllers is rather confusing. They are referred to as adaptive-tuning controllers, self-tuning controllers, and automatic-tuning controllers, but these terms mean different things to different suppliers. Clearly, the ability of the controller to retune its parameter settings will be only as good as the software. Some controllers tune their parameter settings during startup, but do not retune when operating conditions have been reached. Other controllers tune their parameter settings during startup, but continue to retune – if deemed necessary by the internal software – once operating conditions have been reached. Again, other controllers require a predetermined process disturbance, other than startup, to tune the controller parameters; they may or may not retune when operating conditions have been reached.

To properly evaluate the goodness of the self-tuning controller, one would have to know the details of the internal controller software. However, suppliers of self-tuning controllers are not likely to give out such information because the software is the main factor that sets one controller apart from another. Thus, the details of the software will likely be treated as proprietary information. Therefore, the best method to evaluate a self-tuning controller is to install the controller on an extrusion line and closely monitor the actual performance.

4.6   TOTAL PROCESS CONTROL

There is a definite trend in extrusion towards total process control. In a total extrusion control system, temperature measurement and control is tied in with pressure control, thickness gauging, motor load and speed, and possibly other process functions. Between the extremes of simple discrete temperature control and total process control, there are many intermediate levels of controls. Melt temperature control systems have been used for quite some time already. In this type of control, the settings of the first two or three zones closest to the point of melt temperature measurement are continuously adjusted to maintain a constant melt temperature. The temperature settings of the zones are changed automatically by a cascade-type control system. Only low frequency changes in melt temperature can be controlled in this fashion because the response of the barrel temperature zones to changes in setpoint is quite slow.

Melt pressure control systems are also quite common in extrusion. The screw speed is varied continuously in order to maintain a constant pressure. Newer commercial systems incorporate in one microprocessor-based unit combined control of melt temperature, melt pressure, extrudate thickness and/or width, and possibly other process variables. Other control systems are geared towards total plant control. In addition to the regular extrusion control, such a control system can handle upstream raw-material handling, metering of regrind and/or additives, auxiliary extruders in a coextrusion system, drive and temperature control on a gear pump, biaxial orientation in a tenter frame, in-line coating of the extruded web, tension control, corona treating stations, slitting, rewinding, etc.

4.6.1   TRUE TOTAL EXTRUSION PROCESS CONTROL

Concentrating on the control of the extrusion process, it is clear from the literature (54–68) that true total extrusion process control is quite complicated and has not been fully achieved in practice. Under true total process control, the process is considered a multivariable system and the interaction between the variables is known and fully taken into account in the control scheme. One can therefore assume that many commercial microprocessor-based controllers that control melt temperature, pressure, and extrudate dimensions are most likely built with more or less independent control loops, each controlling only one variable. These controllers more or less consolidate multiple discrete controllers into one convenient package without major changes to the individual control characteristics. This may reduce cost, but does not necessarily improve overall control performance.

The first requirement in the development of a true extrusion process control is a dynamic process model. The goodness of the process control will depend very strongly on the accuracy of the process model. However, obtaining a good dynamic process model is quite complicated in practice. The dynamic process model can, in principle, be derived from extrusion theory and various attempts in that direction have indeed been made (50–54). As will be discussed in Chapter 7, extrusion theory to date has not been developed to a point that the entire process can be predicted with a sufficient degree of accuracy. Also, the theoretical prediction of extruder performance requires a substantial amount of computation. Thus, following the phenomenological approach, to derive a dynamic process model from extrusion theory is likely to be quite complex, relatively inaccurate, and requires much computation. However, it is possible that a simplified theoretical model in combination with experimental data could yield an accurate dynamic process model.

Most work on the development of dynamic process models has been empirical; this work is usually referred to as process identification. As mentioned earlier, two classes of empirical identification techniques are available; one uses deterministic (step, pulse, etc.) functions, the other stochastic (random) identification functions. With either technique, the process is perturbed and the resulting variations of the response are measured. The relationship between the perturbing variable and the response is expressed as a transfer function. This function is the process model. Empirical identification of process models by the deterministic method has been reported by various workers (55–58). A drawback of this method is the difficulty in obtaining a measurable response while restricting the process to a linear response (small perturbation). If the perturbation is large, the process response will be nonlinear and the representations of the process with a linear process model will be inaccurate.

Stochastic identification techniques, in principle, provide a more reliable method of determining the process transfer function. Most workers have used Box and Jenkins (59) time-series analysis techniques to develop dynamic models. An introduction to these methods is given by Davies (60). In stochastic identification, a low amplitude sequence (usually a pseudorandom binary sequence, PBRS) is used to perturb the setting of the manipulated variable. The sequence generally has an implementation period smaller than the process response time. By evaluating the auto- and cross-correlations of the input series and the corresponding output data, a quantitative model can be constructed. The parameters of the model can be determined by using a least squares analysis on the input and output sequences. Because this identification technique can handle many more parameters than simple first-order plus deadtime models, the process and its related noise can be modeled more accurately.

Identification of the noise and its probable causes usually leads to the most effective method of removing these disturbances.

The time series method also contains the means for determining the goodness of the model fit by examining the cross and autocorrelations between the residual and the input sequence. From Box and Jenkins models, a minimum variance control strategy can be determined, resulting in a minimum deviation of the controlled variable. Tuning procedures can also be accurately determined from the model. The structure and initial parameters for self-tuning regulators can be determined from the minimum variance controller.

A drawback of the stochastic identification technique is its complexity and the substantial computational requirement. Only a limited number of investigators have applied this technique to the extrusion process. Parnaby et al (61–63) did the first work on stochastic identification of extrusion process models. A hierarchal automatic optimal control scheme was developed and evaluated on a laboratory extrusion line (63). The only operator input required is the desired output rate and die inlet melt temperature. A variable die restriction was used to adjust diehead pressure and throughput. Considerable improvements in control were obtained, particularly in the control of diehead pressure. Other applications of stochastic identification to extrusion have been made by Patterson, et al (64, 65), Costin (66, 67) and at the IKV in Aachen (81).

REFERENCES – CHAPTER 4

1. H. Recker et al., Plastverarbeiter, 28, 1, 1–8 (1977).
2. J. D. Lenk, "Handbook of Controls and Instrumentation", Prentice-Hall Inc. (1980).
3. D. B. Hoffman and N. Sarasohn, Plastics Design and Processing, July, 20–23 (1972).
4. H. Puetz in "Messen und Regeln beim Extrudieren", VDI-Verlag GmbH, Duesseldorf (1982).
5. K. H. Scholl in "Der Extruder als Plastifiziereinheit", VDI-Verlag GmbH, Duesseldorf (1977).
6. A. Galskoy and K. K. Wang, Plastics Engineering, Nov., 42–45 (1978).
7. H. E. Harris, Plastic Technology, Febr., 22–29 (1982).
8. S. H. Walton, The Wire Association International (1977).
9. H. Janeschitz-Kriegl, Proc. 4th Int'l Congress on Rheology, Brown Univ., Providence, R. I. (1963).
10. H. Janeschitz-Kriegl and J. Schijf, Plastics & Polymers, 37, Dec. 523–527 (1969).
11. D. I. Marshall, I. Klein, and R. H. Uhl, Soc. Plastics Engrs. J., 20, 329 (1964).
12. L. P. B. M. Janssen, G. H. Noomen, and J. M. Smith, Plastics & Polymers, 43, Aug., 135–140 (1975).
13. N. Wheeler, Techn. Papers ANTEC, Pittsburg, PA, Jan. (1962).
14. H. L. Trietley, EDN, 28, 2, 93–96 (1983).
15. J. van Leeuwen, M. Goren, A. A. Grimbergen, and J. C. Molijn, Plastica, 17, 6, 269–277 (1969).
16. J. van Leeuwen, M. Goren, A. A. Grimbergen, and J. C. Molijn, Plastica, 17, 7, 319–326 (1969).
17. J. van Leeuwen, M. Goren, A. A. Grimbergen, and J. C. Molijn, Plastica, 17, 8, 371–386 (1969).
18. J. van Leeuwen, Polym. Eng. Sci., 7, 98–109 (1967).
19. W. Schlaeffer, J. Schijf, and H. Janeschitz-Kriegl, Plastics & Polymers, 39, June, 193–199 (1971).
20. M. Hulatt and W. L. Wilkinson, Plastics and Rubber Processing, March, 15–18 (1977).
21. H. J. Kim and E. A. Collins, Polym. Eng. Sci., 11, 83 (1971).
22. E. A. Collins and F. E. Filisko, AICHE J., 16, 339 (1970).
23. H. T. Kim and J. P. Darby, SPE J., 26, 31–36 (1970).
24. F. T. Farrace, "Standardization of an Automatic Inspection System", Nondestructive Testing Standards – A Review, ASTM STP 624, Harold Berger, Ed., American Society for Testing and Materials, 231–245 (1977).
25. H. Droscha, Kunststoffe, 72, 5, (257–259) (1982).
26. W. Michaeli, "Zur Analyse der Flachfolien und Tafelextrusions-Prozesse", Ph. D. Thesis IKV, TH Aachen (1975).
27. N. Nadav, M. Sc. Thesis, Technion, Israel Institute of Technology.
28. N. Nadav and Z. Tadmor, Chem. Eng. Sci., 28, 2115–2126 (1973).
29. H. Recker et al, Plastverarbeiter 28, 1, 1–8 (1977) and 28, 3, 137–142 (1977).
30. F. L. Binsberger, Journ. of Polym. Sci. 11, 10, 1915–1929 (1967).
31. H. Hermann, Kunststoffe, 61, 11, 839–842 (1971).
32. G. Wittig, Material Pruefung, 16, 10, 327 (1974).

33. N. N., Plastics Technology, 28, 11, 18 (1982).
34. D. Osmer, Plastics Compounding, Jan/Feb, 14-24 (1983).
35. II. Hensel, "Die Orientierungsdoppelbrechung", Ph. D. thesis IKV, TM Aachen (1975).
36. G. Menges et al., Plastverarbeiter, 24, 2, 73-77 (1973).
37. P. Fischer, Plastverarbeiter, 29, 5, 231-243 (1978).
38. U. S. 3,866,669.
39. U. S. 4,272,466.
40. H. Goermar and H. Puetz, Kunststoffe 69, Heft 9, p. 588-590 (1979).
41. H. Recker, "Messen und Regeln beim Extrudieren", VDI-Verlag GmbH, Duesseldorf, West Germany (1982), p. 46.
42. J. G. Ziegler and N. B. Nichols, Trans. ASME, 64, 11, 759 (1942).
43. K. L. Chien, J. A. Hrones, and S. B. Resweck, Trans. ASME, 74, 175 (1952).
44. E. Pester and E. Kollman, Electrische Ausruestung, 3-8 and 45-52 (1960).
45. J. A. Miller, A. M. Lopez, C. L. Smith and P. W. Murrill, Control Engineering, 14, 12, 72 (1967).
46. P. D. Roberts, Measurements and Control, 9, 227, June (1976).
47. C. L. Smith, Instruments & Control Systems, 43-46, Sept. (1976).
48. R. F. Bayless, Plastics Engineering, 24, 3, 59-62, March (1978).
49. R. Chostner, Plastics Technology, 68-70, Febr. (1983).
50. R. B. Kirby, SPE Journal, 18, 1273 (1962).
51. W. L. Krueger, SPE Journal, 18, 1282 (1962).
52. D. H. Reber, R. E. Lynn Jr., and E. J. Freech, Polym. Eng. Sci., 13, 346 (1973).
53. Z. Tadmor, S. D. Lipshitz, and R. Lavie, Polym. Eng. Sci., 14, 112 (1974).
54. N. Brauner, R. Lavie, and Z. Tadmor, Int. IFAC Conference on Instrumentation and Automation in the Paper, Rubber, and Plastics Ind., 3 rd Proc., Brussels, 6, 353 (1978).
55. N. R. Schott, Ph. D. Dissertation, Univ. of Arizona (1971).
56. W. Fontaine, Ph. D. Dissertation, Ohio State Univ. (1975).
57. S. Dormeier, SPE ANTEC Tech. Papers, 25, 216 (1979).
58. D. Fingerle, J. Elastomers Plast., 10, 293 (1978).
59. G. E. P. Box and G. M. Jenkins, "Time Series Analysis Forecasting and Control", Revised Edition, Holden-Day, San Francisco, CA (1976).
60. W. D. T. Davies, "System Identification for Self-Adaptive Control", Wiley-Interscience, New York (1970).
61. J. Parnaby, A. K. Kochhar, and B. Woods, Polym. Eng. Sci., 15, 594 (1975).
62. A. K. Kochhar and J. Parnaby, Automatica, 13, 177 (1977).
63. A. K. Kochhar and J. Parnaby, Int. Mech. Eng. (Lond.), Proc., 192, 299 (1978).
64. G. A. Hassan and J. Parnaby, Polym. Eng. Sci., 21, 276 (1981).
65. I. Patterson, P. Brandin, and J. Parin, SPE ANTEC Tech. Papers, 25, 166 (1979).
66. M. H. Costin, M. Sc. Thesis, McMaster University, Hamilton, Canada (1981).
67. M. H. Costin, P. A. Taylor, and J. D. Wright, Polym. Eng. Sci., 22, 393 (1982).
68. H. G. Wiegand, "Prozeßautomatisierung beim Extrudieren und Spritzgießen von Kunststoffen", Carl Hanser Verlag (1979).
69. D. D. Pollock, "The Theory and Properties of Thermocouple Elements", ASTM publication STP 492.
70. "Manual on the Use of Thermocouples in Temperature Measurements", ASTM publication STP 470 B.
71. "Thermocouple Reference Tables", NBS Monograph 125.
72. H. Dean Baker, E. A., Ryder, and N. H. Baker, "Temperature Measurement in Engineering.". Vol. I and II, Omega Press.
73. E. L. Sarber, Plastics Technology, 50-55, June (1983).
74. N. N., Plastics Compounding, 21-32, March/April (1980).
75. E. Galli, Plastics Compounding, 18-26, March/April (1983).

76. F. W. Billmeyer and M. Saltzman, "Principles of Color Technology, John Wiley & Sons (1966).
77. K. J. Astrom, "An Introduction to Stochastic Control Theory", Academic Press, NY (1970).
78. P. Eykhoff, "System Identification", Wiley-Interscience (1974).
79. H. Marchand, Plast. Techn., Feb., 67–70 (1984).
80. C. Kirkland, Plast. Techn., Feb., 97–99 (1981).
81. 12. Kunststofftechnische Kolloquium des IKV in Aachen, 176–185 (1984).

PART II

PROCESS ANALYSIS

5 FUNDAMENTAL PRINCIPLES

Before going into a detailed process analysis of extrusion, it may be useful to review the basic principles that will be applied in the analysis. One can think of the basic principles as the tools used in the process analysis. This chapter is meant to be a review, not an exhaustive dissertation on the subject. For more in-depth and detailed information, the reader is referred to general texts, such as the one by Bird et al. (1) and others (2–4).

5.1 BALANCE EQUATIONS

In extrusion, as well as in many other processes, one deals with the transport of mass, momentum, and energy. Balance equations are used to describe the transport of these quantities. They are universal physical laws that apply to all media (solids and fluids). Matter is considered as a continuum. Thus, the volume over which the balance equation is formulated must be large enough to avoid discontinuities.

5.1.1 THE MASS BALANCE EQUATION

The mass balance equation, also referred to as the equation of continuity, is simply a formulation of the principle of the conservation of mass. The principle states that the rate of mass accumulation in a control volume equals the mass flow rate into the control volume minus the mass flow rate out of the control volume. In Cartesian coordinates (x, y, z), the mass balance equation for a pure fluid can be written as:

$$\frac{\partial \rho}{\partial t} + \frac{\partial}{\partial x}(\rho v_x) + \frac{\partial}{\partial y}(\rho v_y) + \frac{\partial}{\partial z}(\rho v_z) = 0 \qquad (5-1)$$

If a steady state process is analyzed, the first term of equation 5-1 (the time derivative of density) will equal zero; the same is true for equation 5-2. The mass balance equation expressed in cylindrical coordinates (r, Θ, z) is:

$$\frac{\partial \rho}{\partial t} + \frac{1}{r}\frac{\partial}{\partial r}(\rho r v_r) + \frac{1}{r}\frac{\partial}{\partial \Theta}(\rho v_\Theta) + \frac{\partial}{\partial z}(\rho v_z) = 0 \qquad (5-2)$$

If the fluid is considered incompressible, the density terms (ρ) disappear in equations 5-1 and 5-2. In one-dimensional steady state flow problems (velocity components in only one direction), the mass balance equation is automatically satisfied and does not enter into the calculations. In two-dimensional flow problems (no velocity component in the third coordinate direction), the mass balance is satisfied by the introduction of a stream function (1).

5.1.2   THE MOMENTUM BALANCE EQUATION

The momentum of a body is the product of its mass and velocity. Since velocity is a vector, momentum is also a vector. The momentum balance equation describes the conservation of momentum; it is also referred to as the equation of motion.

Momentum can be transported by convection and conduction. Convection of momentum is due to the bulk flow of the fluid across the surface; associated with it is a momentum flux. Conduction of momentum is due to intermolecular forces on each side of the surface. The momentum flux associated with conductive momentum transport is the stress tensor. The general momentum balance equation is also referred to as Cauchy's equation. The Navier-Stokes equations are a special case of the general equation of motion for which the density and viscosity are constant. The well-known Euler equation is again a special case of the general equation of motion; it applies to flow systems in which the viscous effects are negligible.

In polymer flow systems, the inertia and body forces are generally negligible. For these systems, the momentum balance equation in Cartesian coordinates can be written as:

$$\frac{\partial P}{\partial x} = \frac{\partial \tau_{xx}}{\partial x} + \frac{\partial \tau_{xy}}{\partial y} + \frac{\partial \tau_{xz}}{\partial z} \tag{5-3a}$$

$$\frac{\partial P}{\partial y} = \frac{\partial \tau_{yx}}{\partial x} + \frac{\partial \tau_{yy}}{\partial y} + \frac{\partial \tau_{yz}}{\partial z} \tag{5-3b}$$

$$\frac{\partial P}{\partial z} = \frac{\partial \tau_{zx}}{\partial x} + \frac{\partial \tau_{zy}}{\partial y} + \frac{\partial \tau_{zz}}{\partial z} \tag{5-3c}$$

or in cylindrical coordinates:

$$\frac{\partial P}{\partial r} = \frac{1}{r}\frac{\partial}{\partial r}(r\,\tau_{rr}) + \frac{1}{r}\frac{\partial \tau_{r\theta}}{\partial \theta} - \frac{\tau_{\theta\theta}}{r} + \frac{\partial \tau_{rz}}{\partial z} \tag{5-4a}$$

$$\frac{1}{r}\frac{\partial P}{\partial \theta} = \frac{1}{r^2}\frac{\partial}{\partial r}(r^2\,\tau_{r\theta}) + \frac{1}{r}\frac{\partial \tau_{\theta\theta}}{\partial \theta} + \frac{\partial \tau_{\theta z}}{\partial z} \tag{5-4b}$$

$$\frac{\partial P}{\partial z} = \frac{1}{r}\frac{\partial}{\partial r}(r\,\tau_{rz}) + \frac{1}{r}\frac{\partial \tau_{\theta z}}{\partial \theta} + \frac{\partial \tau_{zz}}{\partial z} \tag{5-4c}$$

The analysis of many flow problems can be simplified by considering only one component of the equation of motion, the one in the direction of flow. Further simplifying assumptions are often necessary in order to solve the problem. In the analysis of isothermal processes, only two balance equations are needed, the mass and momentum balance. However, there are three unknown variables: velocity, stress, and pressure. In order to solve the problem, additional information is required. This is information on how the fluid deforms under the application of various stresses. This information is described by the constitutive equation of the fluid, see also Section 6.2 on melt flow properties.

An example of the use of momentum balance in pipe flow of a Newtonian fluid is given in Appendix 5-1.

5.1.3   THE ENERGY BALANCE EQUATION

The energy balance equation states that the rate of increase in specific internal (thermal) energy in a control volume equals the rate of energy addition by conduction plus the rate of energy dissipation. The principle of energy conservation is also described by the first law of thermodynamics, see Section 5.2. If a constant density is assumed, the energy equation can be written as:

$$\dot{E}_{acc} + \dot{E}_{conv} = \dot{E}_{cond} + \dot{E}_{diss} \tag{5-5}$$

where

$$\dot{E}_{acc} = \rho C_v \frac{\partial T}{\partial t} \tag{5-5a}$$

$$\dot{E}_{conv} = \rho C_v \left(v_x \frac{\partial T}{\partial x} + v_y \frac{\partial T}{\partial y} + v_z \frac{\partial T}{\partial z} \right) \tag{5-5b}$$

$$\dot{E}_{cond} = \frac{\partial}{\partial x} \left(k \frac{\partial T}{\partial x} \right) + \frac{\partial}{\partial y} \left(k \frac{\partial T}{\partial y} \right) + \frac{\partial}{\partial z} \left(k \frac{\partial T}{\partial z} \right) \tag{5-5c}$$

$$\dot{E}_{diss} = \tau_{xx} \frac{\partial v_x}{\partial x} + \tau_{yy} \frac{\partial v_y}{\partial y} + \tau_{zz} \frac{\partial v_z}{\partial z} + \tau_{xy} \left(\frac{\partial v_x}{\partial y} + \frac{\partial v_y}{\partial x} \right)$$

$$+ \tau_{xz} \left(\frac{\partial v_x}{\partial z} + \frac{\partial v_z}{\partial x} \right) + \tau_{yz} \left(\frac{\partial v_y}{\partial z} + \frac{\partial v_z}{\partial y} \right) \tag{5-5d}$$

$\dot{E}_{acc}$ is the accumulation term, $\dot{E}_{conv}$ the convection term, $\dot{E}_{cond}$ the conduction term, and $\dot{E}_{diss}$ the dissipation term. Equation 5-5c is an expression of Fourier's law of heat transfer, see also Section 5.3.1. In cylindrical coordinates, only the convection, conduction, and dissipation terms change:

$$\dot{E}_{conv} = \rho C_v \left(v_r \frac{\partial T}{\partial r} + \frac{v_\theta}{r} \frac{\partial T}{\partial \theta} + v_z \frac{\partial T}{\partial z} \right) \tag{5-6b}$$

$$\dot{E}_{cond} = \frac{1}{r} \frac{\partial}{\partial r} \left(kr \frac{\partial T}{\partial r} \right) + \frac{1}{r} \frac{\partial}{\partial \theta} \left(k \frac{1}{r} \frac{\partial T}{\partial \theta} \right) + \frac{\partial}{\partial z} \left(k \frac{\partial T}{\partial z} \right) \tag{5-6c}$$

$$\dot{E}_{diss} = \tau_{rr}\frac{\partial v_r}{\partial r} + \tau_{\theta\theta}\frac{1}{r}\left(\frac{\partial v_\theta}{\partial \theta} + v_r\right) + \tau_{zz}\frac{\partial v_z}{\partial z} + \tag{5-6d}$$

$$\tau_{r\theta}\left(r\frac{\partial}{\partial r}\frac{v_\theta}{r} + \frac{1}{r}\frac{\partial v_r}{\partial \theta}\right) + \tau_{rz}\left(\frac{\partial v_z}{\partial r} + \frac{\partial v_r}{\partial z}\right) + \tau_{\theta z}\left(\frac{1}{r}\frac{\partial v_z}{\partial \theta} + \frac{\partial v_\theta}{\partial z}\right)$$

The energy equation has to be used to analyze non-isothermal processes. In these situations, there are generally four unknown variables: velocity, stress, pressure, and temperature. In order to solve such a non-isothermal problem, one more equation is needed in addition to the three balance equations. The missing relationship is the constitutive equation of the fluid; this relationship basically describes the relationship between stress and deformation. In polymer extrusion, the material undergoes large changes in temperature as it is transported along the extruder. Consequently, the energy equation is used extensively in the analysis of the extrusion process.

5.2   BASIC THERMODYNAMICS

Thermodynamics is concerned with energy; the exchange and transformation of energy in a system. The first law of thermodynamics is a statement of the principle of conservation of energy. For a closed (constant mass) system, the first law of thermodynamics is expressed by:

$$\Delta E = Q - W \tag{5-7}$$

where ΔE is the total energy change of the system, Q is heat added to the system, and W is work done by the system. The total energy change ΔE can be split up into several terms, each representing the change in energy of a particular form:

$$\Delta E = \Delta E_k + \Delta E_p + \Delta U \tag{5-8}$$

where ΔE_k is the change in kinetic energy, ΔE_p is the change in gravitational potential energy, and ΔU is the change in internal energy. By definition, the kinetic energy is:

$$E_k = \frac{mv^2}{2} \tag{5-9}$$

where m is mass and v is velocity.

The gravitational potential energy is:

$$E_p = mza_g \tag{5-10}$$

where z is the elevation above a reference level and a_g the local acceleration of gravity. These energy functions are common to both mechanics and thermodynamics.

The internal energy function U, however, is peculiar to thermodynamics. It represents the kinetic and potential energies of the molecules, atoms, and subatomic particles that make up the system on a microscopic scale. There is no known way to determine absolute values of U. Fortunately, only changes of ΔU are needed and these can be found from experiment. Whenever the state of the system is fixed, the internal energy U is fixed. If equation 5-8 is used in equation 5-7, the first law of thermodynamics can be written as:

$$\Delta E_k + \Delta E_p + \Delta U = Q - W \tag{5-11}$$

If the sum of the kinetic and potential energies of the system does not change, this equation becomes

$$\Delta U = Q - W \tag{5-12}$$

or in differential form:

$$dU = \delta Q - \delta W \tag{5-13}$$

If, in addition, the process is adiabatic (Q=0), equation 5-12 becomes

$$\Delta U = - W \tag{5-14}$$

Note that in equation 5-13 the differential signs on Q and W are written δ, whereas the differential sign of U is d. There is a fundamental difference between a property like U and the quantities Q and W. A property like U always has a value dependent only on the state. A process which changes the state of a system changes U. Thus, dU represents an infinitesimal change in U, and integration gives a difference between two values of the property:

$$\int_{U_1}^{U_2} dU = U_2 - U_1 = \Delta U$$

On the other hand, Q and W are not properties of the system and depend on the path of the process. Thus δ is used to denote an infinitesimal quantity. Integration gives a finite quantity and not a difference between the two values:

$$\int \delta Q = Q \quad \text{and} \quad \int \delta W = W$$

Thus, integration of equation 5-13 yields equation 5-12.

Special thermodynamic functions are defined as a matter of convenience. The simplest such function is the enthalpy H, explicitly defined for any system by the mathematical expression:

$$H = U + PV \tag{5-15}$$

where P is pressure and V is volume.

The enthalpy has units of energy and is a system property like U, P and V. Whenever a differential change occurs in a system, its properties change by:

$$dH = dU + d(PV) \tag{5-16}$$

The amount of heat in a closed PVT system that must be added to accomplish a given change of state depends on how the process is carried out. Only for a reversible process where the path is fully specified is it possible to relate the heat to a property of the system. On this basis, the specific heat at a constant volume can be defined:

$$C_V = \left(\frac{\delta Q}{dT}\right)_V \tag{5-17}$$

It represents the amount of heat required to increase the temperature by dT when the system is held at a constant volume. C_V is a property of the system; this follows from equation 5-12. For a constant volume, reversible process $dU = \delta Q$, because no work can be done without volume changes. Thus, the specific heat at a constant volume can be related to the internal energy by:

$$C_V = \left(\frac{\partial U}{\partial T}\right)_V \tag{5-18}$$

Thus, for a constant volume process

$$dU = \delta Q = C_V dT \tag{5-19}$$

Equation 5-19 is a useful relationship. If specific heat and temperature change are known, then the amount of heat required to accomplish this change in temperature can be determined. If the amount of heat and the specific heat are known, then the resulting change in temperature can be calculated. This relationship is indispensable in the analysis of the extrusion process. For instance, if the amount of viscous heat generation in a certain amount of polymer is known, then the resulting adiabatic temperature rise can be determined if the specific heat of the polymer is known.

The specific heat at constant pressure is defined as:

$$C_p = \left(\frac{\delta Q}{dT}\right)_p \tag{5-20}$$

It represents the amount of heat required to increase the temperature by dT when the system is heated in a reversible process at a constant pressure. By using equations 5-13 and 5-16 it can be shown that:

$$C_p = \left(\frac{\partial H}{\partial T}\right)_p \tag{5-21}$$

Thus, C_p is also a property of the system. Further

$$dH = \delta Q = C_p dT \tag{5-22}$$

In polymer melts, the material is generally considered to be incompressible; in this case $C_V = C_p$. The actual relationship between C_V and C_p is given by equation 6-94.

Entropy is an intrinsic property of a system. For a reversible process, changes in entropy are given by:

$$dS = \frac{\delta Q}{T} \qquad (5-23)$$

The second law of thermodynamics states that the entropy change of any system and its surroundings, considered together, is positive and approaches zero for any process which approaches reversibility. The mathematical expression of the second law is simply:

$$\int_1^2 \frac{\delta Q}{T} \leqslant S_2 - S_1 \qquad (5-24)$$

The left term is the entropy transfer, $\delta Q/T$; it forms a direct link with the heat transfer (δQ) if the temperature at the system boundary is T. Entropy transfer, as a concept, makes the distinction between heat transfer and work transfer as parallel forms of energy transfer. Only the transfer of energy as heat is accompanied by entropy transfer. The work transfer interaction is not accompanied by entropy transfer. The term on the right of the inequality sign is the entropy change; it is a thermodynamic property. The inequality sign in equation 5-24 expresses the essence of the second law of thermodynamics. The change from state 1 to state 2 can occur over various paths. The difference between possible paths is described by the strength of the inequality sign. The entropy generation expresses this difference quantitatively:

$$S_{gen} = S_2 - S_1 - \int_1^2 \frac{\delta Q}{T} \geqslant 0 \qquad (5-25)$$

Thus, any thermodynamic process is accompanied by entropy generation. When $S_{gen} > 0$, the process is considered irreversible; when $S_{gen} = 0$, the process is considered reversible. It should be noted that the entropy generation S_{gen} is, of course, path dependent and, therefore, not a thermodynamic property as opposed to the entropy change S_2-S_1. In fluid flow the volumetric rate of entropy generation is:

$$\dot{S}_{gen} = \frac{1}{T}\left[\frac{\partial}{\partial x}\left(k\frac{\partial T}{\partial x}\right) + \frac{\partial}{\partial y}\left(k\frac{\partial T}{\partial y}\right) + \frac{\partial}{\partial z}\left(k\frac{\partial T}{\partial z}\right)\right] +$$

$$+ \frac{1}{T}\left[\tau_{xx}\frac{\partial v_x}{\partial x} + \tau_{yy}\frac{\partial v_y}{\partial y} + \tau_{zz}\frac{\partial v_z}{\partial z}\right] + \qquad (5-26)$$

$$+ \frac{1}{T}\left[\tau_{xy}\left(\frac{\partial v_x}{\partial y} + \frac{\partial v_y}{\partial x}\right) + \tau_{xz}\left(\frac{\partial v_x}{\partial z} + \frac{\partial v_z}{\partial x}\right) + \tau_{yz}\left(\frac{\partial v_y}{\partial z} + \frac{\partial v_z}{\partial y}\right)\right]$$

The first bracketed (square) term is the conductive energy flux, which is the same used in the energy equation 5-5c. The other bracketed terms represent the dissipative energy flux, as used also in the energy equation 5-5d. Thus, the volumetric rate of entropy generation can be written as:

$$\dot{S}_{gen} = \frac{1}{T}\left(\dot{E}_{cond} + \dot{E}_{diss}\right) \tag{5-27}$$

5.2.1   RUBBER   ELASTICITY

The theory of rubber elasticity is, to a large extent, based on thermodynamic considerations. It will be briefly discussed as an example of how thermodynamics can be applied in polymer science. For more detailed information, the reader is referred to the various textbooks (10-13). It is assumed that there is a three-dimensional network of chains, that the chain units are flexible and that individual chain segments rotate freely, that no volume change occurs upon deformation and that the process is reversible (i. e., true elastic behavior). Another usual assumption is that the internal energy U of the system does not change with deformation. For this system the first law of thermodynamics can be written as:

$$dU = TdS - \delta W \tag{5-28}$$

If the equilibrium tensile force is F and the displacement dl then the work done by the system is:

$$\delta W = PdV - Fd\,l \tag{5-29}$$

The change in U with respect to l at a constant temperature and volume is:

$$\left(\frac{\partial U}{\partial l}\right)_{T,V} = T\left(\frac{\partial S}{\partial l}\right)_{T,V} + F \tag{5-30}$$

Thus, the equilibrium tensile force F is determined by the change in internal energy with deformation and the change in entropy with deformation. An ideal rubber is defined as a material for which the change in internal energy with deformation equals zero. Thus, the only contribution to the force F is from the entropy term:

$$F = -T\left(\frac{\partial S}{\partial l}\right)_{T,V} \tag{5-31}$$

When a rubber is deformed, its entropy S changes. The long chains tend to adopt a most probable configuration; this is a highly coiled configuration. When the material is stretched the chains uncoil, resulting in a less probable chain configuration. When the force is removed, the system wants to return to the more probable coiled-up state. Thus, the entropy of the system increases. The basic problem in the theory of rubber elasticity is a statistical mechanical problem of determining the change in entropy in going from an undeformed state to a deformed state. The extension ratio α is defined as:

$$\alpha = \frac{l}{l_0} = \frac{l_0 + \Delta l}{l_0} = 1 + \varepsilon \tag{5-32}$$

where l_o is the original length, Δl the increase in length, and ε the elongational strain.

If a cube of unit length dimension is considered, the force will equal the stress σ. Equation 5-31 can now be written as:

$$\sigma = -T\left(\frac{\partial S}{\partial \alpha}\right)_{T,V} \tag{5-33}$$

The most probable state of a system is the state that has the greatest number of ways, Y, of being realized. The entropy S is related to this number of complexions Y by the Boltzman relation:

$$S = C_B \ln Y \tag{5-34}$$

where C_B is the Boltzman constant (1.38E-23 J/° K).

The change in entropy upon extension can be expressed as:

$$\Delta S = -\frac{1}{2} NC_B \left(\alpha^2 + \frac{2}{\alpha} - 3 \right) \tag{5-35}$$

where N is the number of freely orienting chain segments. Substituting this expression into equation 5-33 yields the stress-extension ratio relationship:

$$\sigma = NC_B T \left(\alpha - \frac{1}{\alpha^2} \right) \tag{5-36}$$

The number of chains per unit volume can be related to the density ρ and the average molecular weight between crosslinks M_c:

$$N = \frac{\rho N_A}{M_c} \tag{5-37}$$

where N_A is Avogadro's number (6.025E23 mol^{-1}).

Considering that $N_A C_B$ is the gas constant R (8.314 J/mol° K), equation 5-36 can be written as:

$$\sigma = \frac{\rho RT}{M_c} \left(\alpha - \frac{1}{\alpha^2} \right) \tag{5-38}$$

The elastic modulus can be determined by:

$$E = \frac{d\sigma}{d\varepsilon} = \frac{d\sigma}{d\alpha} = \frac{\rho RT}{M_c} (1 + 2\alpha^{-3}) = \frac{\rho RT}{M_c} [1+2(1+\varepsilon)^{-3}] \tag{5-39}$$

For small extensional strains the elastic modulus becomes:

$$E = \frac{3\rho RT}{M_C} \qquad (5-40)$$

For a linear and isotropic material the shear modulus G is related to the extensional modulus by:

$$G = \frac{E}{2(1+\nu)} \qquad (5-41)$$

where ν is Poisson's ratio, $\nu = \frac{1}{2}$ for ideal rubbers.

Thus, the shear modulus can be obtained from:

$$G = \frac{\rho RT}{M_C} \qquad (5-42)$$

5.2.2   STRAIN INDUCED CRYSTALLIZATION

It has been observed that the crystallization behavior of polymers is modified when the material is strained. This behavior has been found in rubbers and in thermoplastics (14). In thermoplastics, the effect of strain on crystallization behavior has been studied quite extensively in solutions, in melt, and in solid state. For instance, it has been found that flowing polymer melts can crystallize at temperatures that are substantially above the crystallization temperature of the same material in a quiescent state. This strain induced crystallization is generally explained in terms of thermodynamic processes. During a phase change, the Gibbs free energy is:

$$\Delta G = \Delta H - T\Delta S \qquad (5-43)$$

where S is the conformational entropy.

For an equilibrium process $\Delta G = 0$. This would be the case for crystallization or melting at the melting point. The melting point T_m, therefore, is:

$$T_m = \frac{\Delta H}{\Delta S} \qquad (5-44)$$

When the polymer is being deformed, the molecules are oriented to some extent and this reduces the conformational entropy. If the material is considered to be entropy-elastic, the energy expended in the deformation of the polymer will reduce the entropy but not affect the internal energy. This point was discussed in some detail by Astarita (15). Thus, if the enthalpy is unaffected by orientation and the entropy reduced, the melting point will increase with increasing orientation. This melting point elevation will increase the degree of supercooling, the driving force of crystallization. The anisotropy of the oriented polymer favors crystallization in the direction of orientation and discourages it orthogonally. This explains the change in crystal growth mechanism from the three-dimensional (spherulitic) to the unidimensional (fibrillar) growth.

5.3 HEAT TRANSFER

Heat transfer takes place by three mechanisms: conduction, convection and radiation. In conductive heat transfer, the heat flows from regions of high temperature to regions of low temperature. The transfer takes place due to motion at the molecular level. Matter must be present in order for conduction to occur. The material itself does not need to be in motion for conduction to take place; in fact, many times the conducting medium will be stationary. In a solid material, the only mode of heat transfer is conduction (16). In convection, heat transfer is due to the bulk motion of the fluid. Convective heat transfer only occurs in fluids. In radiation, heat or radiant energy is transferred in the form of electromagnetic waves.

5.3.1 CONDUCTIVE HEAT TRANSFER

The most important relationship in conductive heat transfer is Fourier's law; for conduction in x-direction:

$$\dot{Q}_x = - k_x A_x \frac{\partial T}{\partial x} \tag{5-45}$$

where $\dot{Q}_x$ is the heat flow (rate of conduction), k_x the thermal conductivity, A_x the area normal to heat flow, and T temperature.

For conduction in y and z direction, Fourier's law is simply:

$$\dot{Q}_y = - k_y A_y \frac{\partial T}{\partial y} \tag{5-46}$$

and

$$\dot{Q}_z = - k_z A_z \frac{\partial T}{\partial z} \tag{5-47}$$

Fourier's law states that the heat will flow from high to low temperatures. The heat flow is proportional to the thermal conductivity, the temperature gradient, and the cross-sectional area normal to heat flow. Thus, in order to calculate the heat flow, one has to know the thermal conductivity of the material and the temperature distribution within the material. The temperature distribution has to be determined by solving the energy equation (5-5) as discussed in Section 5.1.3.

5.3.2 CONVECTIVE HEAT TRANSFER

Analysis of convective heat transfer is considerably more difficult than analysis of conductive heat transfer in a stationary material. This is simply due to the fact that more terms have to be carried in the energy equation (5-5) that has to be solved in order to find the temperature distribution. Many practical problems encountered in polymer processing are described by equations that do not allow simple analytical solutions. In many cases, therefore, one has to use numerical techniques to obtain solutions to the problem.

5.3.3   DIMENSIONLESS NUMBERS

It is quite common in process engineering to describe certain phenomena by relating dimensionless combinations of physical variables. These combinations are referred to as dimensionless groups or dimensionless numbers. This approach offers a number of advantages. One is assured that the equations are dimensionally homogeneous. By using dimensionless numbers, the number of variables describing the problem can be reduced. One can predict the effect of a change in a certain variable even if the problem cannot be completely solved. If the dimensionless numbers describing the problem remain the same, then the solution to the problem will remain unchanged, even if individual variables are varied. This latter characteristic is very useful in scale-up problems, see also Section 8.8.

Dimensionless numbers can be derived by making the appropriate balance equations dimensionless when the problem can be fully described. (See the example on heat transfer in Newtonian fluid between two plates later in this section.) Dimensionless numbers can also be derived from dimensional analysis; this approach is used if the problem cannot be completely described mathematically. An example of dimensional analysis is the determination of the force F that a sphere of diameter D encounters in a fluid with viscosity η when the relative velocity between sphere and fluid is v. From physical arguments, it can be deduced that F has to be a function of diameter, velocity, fluid density and viscosity:

$$F = F(D, v, \rho, \eta) \tag{5-48}$$

Even though the actual form of the equation is not known, the equation has to be dimensionally homogeneous. This is also true if the force is written in the following form:

$$F = C D^a v^b \rho^c \eta^d \tag{5-49}$$

The dimensions of these variables can be expressed in length L, time t, and mass M; thus, $F(Lt^{-2}M)$, $D(L)$, $v(Lt^{-1})$, $\rho (L^{-3}M)$, and $\eta (L^{-1}t^{-1}M)$. The requirement for dimensional homogeneity yields:

$$L = L^a L^b L^{-3c} L^{-d} \tag{5-50a}$$

$$t^{-2} = t^0 t^{-b} t^0 t^{-d} \tag{5-50b}$$

$$M = M^0 M^0 M^c M^d \tag{5-50c}$$

From the three equations 5-50a, 5-50b, and 5-50c, three of the unknowns (a, b, c, and d) can be expressed as a function of the fourth. If a, b, and c are expressed as a function of d, this yields:

$$a = 2-d \tag{5-51a}$$

$$b = 2-d \tag{5-51b}$$

$$c = 1-d \tag{5-51c}$$

With this relationship, equation 5-49 can be rewritten as:

$$\frac{F}{\rho v^2 D^2} = C\left(\frac{\eta}{\rho v D}\right)^d \tag{5-52}$$

The original number of variables has now been reduced from 5 to 2 dimensionless numbers. The dimensionless number on the right-hand side of equation 5-52 is the well-known Reynolds number:

$$N_{Re} = \frac{\rho v D}{\eta} \tag{5-53}$$

The Reynolds number represents the ratio of inertia forces ($\rho v D$) to viscous forces (η). In the flow of a fluid through a flow channel, turbulent flow will occur when the Reynolds number is above the critical Reynolds number, which is about 2100. Below the critical Reynolds number, laminar flow takes place; this is also referred to as streamline flow or viscous flow. In polymer processing, the polymer melt viscosity is generally very high. As a result, the Reynolds number in polymer processing is very small, typically 10^{-3}. Therefore, in polymer processing, the polymer melt flow is always laminar. The low Reynolds number generally allows one to neglect the effect of inertia and body force, as was done in the formulation of the momentum balance equation, equation 5-3.

Several important dimensionless numbers in combined heat and momentum transfer in fluids can be derived by considering the simple flow of a Newtonian fluid between two flat plates, one stationary and one moving at velocity v, see Figure 5-1.

If it is assumed that only conduction and viscous dissipation play a role of importance, the energy balance can be written as:

$$k\frac{\partial^2 T}{\partial y^2} + \tau_{xy}\frac{\partial v_x}{\partial y} = 0 \tag{5-54}$$

When the fluid is Newtonian, the shear stress can be written as (see Section 6.2.1):

$$\tau_{xy} = \mu\frac{\partial v_x}{\partial y} \tag{5-55}$$

Thus, equation 5-54 becomes:

$$k\frac{\partial^2 T}{\partial y^2} + \mu\left(\frac{\partial v_x}{\partial y}\right)^2 = 0 \tag{5-56}$$

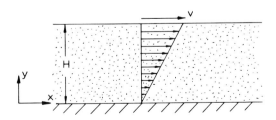

Figure 5-1. Flat Plate Flow Problem

If the pressure gradient in flow direction is assumed to be zero, i.e. only drag flow, the velocity gradient becomes:

$$\frac{\partial v_x}{\partial y} = \frac{v}{H} \tag{5-57}$$

Thus, equation 5-56 becomes

$$\frac{\partial^2 T}{\partial y^2} = -\frac{\mu v^2}{H^2} \tag{5-58}$$

By integrating twice, the temperature profile is obtained:

$$T(y) = -\frac{\mu v^2}{2kH^2} y^2 + C_1 y + C_2 \tag{5-59}$$

If the temperature at the stationary wall is T_0, $T(y=0)=T_0$, and T_1 at the moving wall, $T(y=H)=T_1$, the integration constants can be calculated:

$$C_2 = T_0 \tag{5-60a}$$

$$C_1 = \frac{T_1-T_0}{H} + \frac{\mu v^2}{2kH} \tag{5-60b}$$

The temperature profile can now be written as:

$$T = -\frac{\mu v^2}{2kH^2} y^2 + \left(\frac{T_1-T_0}{H} + \frac{\mu v^2}{2kH}\right) y + T_0 \tag{5-61}$$

This is essentially the same equation as equation 7-92 describing the temperature profile in the melt film in the melting region of an extruder. The equation can now be written in the following dimensionless form:

$$\frac{T-T_0}{T_1-T_0} = \frac{y}{H} + \frac{\mu v^2}{2k(T_1-T_0)}\left(1 - \frac{y}{H}\right)\frac{y}{H} \tag{5-62}$$

If the dimensionless temperature is T^0 and dimensionless distance y^0, then equation 5-62 can be written as:

$$T^0 = y^0 + \frac{1}{2} N_{Br} (1-y^0) y^0 \tag{5-63}$$

The Brinkman number N_{Br} is a measure of the importance of viscous heat generation relative to the heat conduction resulting from the imposed temperature difference $\Delta T (=T_1-T_0)$:

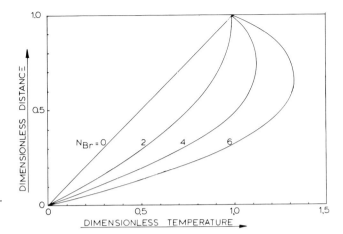

Figure 5-2. Temperature Profiles at Various Brinkman Numbers

$$N_{Br} = \frac{\mu v^2}{k \Delta T} \tag{5-64}$$

If the Brinkman number is larger than 2, there is a maximum temperature at a position intermediate between the two walls, see Figure 5-2.

In the previous problem, only conduction and dissipation were considered to play a role. If the analysis is now extended to include the effect of convection, the energy equation becomes:

$$\rho C v_x \frac{\partial T}{\partial x} = k \frac{\partial^2 T}{\partial y^2} + \tau_{xy} \frac{\partial v_x}{\partial y} \tag{5-65}$$

This equation can be made dimensionless by introducing the following dimensionless variables:

$$T^0 = \frac{T}{\Delta T} \tag{5-66a}$$

$$y^0 = \frac{y}{H} \tag{5-66b}$$

$$x^0 = \frac{x}{L} \tag{5-66c}$$

$$v_x^{\ 0} = \frac{v_x}{v} \tag{5-66d}$$

This results in the following expression:

$$\frac{\rho C v H^2}{kL} v_x^{\ 0} \frac{\partial T^0}{\partial x^0} = \frac{\partial^2 T^0}{\partial (y^0)^2} + \frac{\mu v^2}{k \Delta T} \left(\frac{\partial v_x^{\ 0}}{\partial y^0} \right)^2 \tag{5-67}$$

The first dimensionless group is the Graetz number:

$$N_{Gz} = \frac{\rho C v H^2}{kL} = \frac{v H^2}{\alpha L} \qquad (5-68)$$

where α is the thermal diffusivity, see Section 6.3.5. Equation 5-67 can now be written as:

$$N_{Gz} \ v_x^0 \ \frac{\partial T^0}{\partial x^0} = \frac{\partial^2 T^0}{\partial (y^0)^2} + N_{Br} \left(\frac{\partial v_x^0}{\partial y^0} \right)^2 \qquad (5-69)$$

This equation does not have a simple analytical solution and is generally solved by some numerical technique. However, irregardless of the actual solution, as long as the Greatz and Brinkman numbers are constant, the solution to the problem will remain unchanged.

The Graetz number can be considered to be a ratio of two time values, one being the time required to reach thermal equilibrium through conduction in crossflow direction (dimension H), the other time being the average residence time in the flow channel of length L. Thus, the Graetz number is a measure of the importance of conduction normal to the flow relative to the thermal convection in the direction of flow. If the Graetz number is large, the conduction normal to the flow is large relative to the convection in flow direction. This situation often occurs in extruders in the flow through the screw channel and die flow channels.

A dimensionless number closely related to the Graetz number is the Peclet number:

$$N_{Pe} = \frac{vH}{\alpha} = N_{Gz} \frac{L}{H} \qquad (5-70)$$

The Peclet number provides a measure of the importance of thermal convection relative to thermal conduction. The Peclet number in polymer processing is often quite large, typically of the order of 10^3 to 10^5. This indicates that convective heat transport is often quite important in polymer melt flow.

Another important dimensionless number is the Nusselt number:

$$N_{Nu} = \frac{hL}{k} \qquad (5-71)$$

where h is the interfacial heat transfer coefficient.

The Nusselt number is basically a dimensionless temperature gradient averaged over the heat transfer surface. The Nusselt number represents the ratio of the heat transfer resistance estimated from the characteristic dimension of the object (L/k) to the real heat transfer resistance (1/h). In many convective heat transfer problems, the Nusselt number is expressed as a function of other dimensionless numbers; e.g., the Reynolds number and the Prandtl number.

The Prandtl number is:

$$N_{Pr} = \frac{C_p \eta}{k} = \frac{\eta}{\alpha \rho} \qquad (5-72)$$

The Prandtl number is simply the ratio of kinematic viscosity (η/ρ) to thermal diffusivity. Physically, the Prandtl number represents the ratio of the hydrodynamic boundary layer to the thermal boundary layer in the heat transfer between fluids and a stationary wall. In simple fluid flow, it represents the ratio of the rate of impulse transport to the rate of heat transport. It is determined by the material properties; for high viscosity polymer melts, the number is of the order of 10^6 to 10^{10}.

The Nahme number or Griffith number is:

$$N_{Na} = \frac{\alpha_T v^2 \eta}{k} \tag{5-73}$$

where α_T is the temperature coefficient of viscosity η as defined in equation 6-40.

The Nahme number can be considered to be the ratio of viscous dissipation to thermal conduction in the direction perpendicular to flow. Large values of the Nahme number (> 1) indicate that the temperature nonuniformities created by the viscous dissipation have a substantial effect on the resulting velocity profile. Thus, if the Nahme number is large, a nonisothermal flow analysis has to be made in order to maintain sufficient accuracy. If the Nahme number is small, an isothermal analysis can yield relatively accurate results. The Nahme number is also referred to as the Griffith number.

The Biot number is:

$$N_{Bi} = \frac{H}{T_m - T_w} \left(\frac{\partial T}{\partial x} \right)_w \tag{5-74}$$

The Biot number is a dimensionless heat transfer number. When the Biot number is zero, there is no exchange of heat; i.e., adiabatic conditions. When the Biot number is infinitely large, the wall temperature T_w equals the melt temperature T_m; this corresponds to isothermal conditions. Normal values for the Biot number in extrusion dies range from 1 to 100, depending strongly on the presence and amount of insulation.

The Fourier number is:

$$N_{Fo} = \frac{\alpha t}{H^2} \tag{5-75}$$

The Fourier number is a convenient number in the analysis of transient heat transfer problems; i.e., where the temperature changes with time. The Fourier number can be considered to be a ratio of two time values; one is the actual time value, the other is the time necessary to reach thermal equilibrium in the sample by conduction. If the Fourier number is large (> 1), the sample will reach thermal equilibrium within the considered time frame. If the Fourier number is small (< 0.1), only the skin of the sample will have changed in temperature, while the bulk of the material will be largely unaffected. In many cases, the temperature distribution is about 90 percent uniform when the Fourier number equals unity.

The thermal diffusivity of most polymers is about 10^{-7} m^2/s. If a polymer slab 1 mm thick is heated from two sides (H = 0.0005 m), after approximately 2.5 seconds the temperature in the slab will be quite uniform. If the slab is 10 mm thick, the uniform temperature conditions will not be approached until after 250 seconds or 4.17 minutes. This explains, at least partly, why it is difficult to obtain uniform melt temperatures in an extruder where the screw has very deep channels. In fact, if the channels are too deep, it is unlikely that the melting process can be completed in the extruder. A typical average residence time in a single screw extruder is about one to three minutes.

5.3.4   VISCOUS HEAT GENERATION

Viscous heat generation is the dissipation of mechanical energy in a viscous fluid. The last term in the energy equation 5-5d shows the viscous dissipation in the most general case. In the simpler case of unidirectional shear, the viscous heat generation per unit volume is:

$$\dot{E}_d = \tau \dot{\gamma} \tag{5-76}$$

If the flow properties of the fluid can be described by a power law equation (see Section 6.2.2), the viscous heat generation per unit volume is:

$$\dot{E}_d = m \dot{\gamma}^{n+1} \tag{5-77}$$

where m is the consistency index and n the power law index. The power law index is unity for a Newtonian fluid and between one and zero for a pseudoplastic fluid, such as polymer melts. From equation 5-77, it can be seen that the viscous heat generation increases more than proportionally with the shear rate. This has important implications in the extrusion process; this will be discussed in Chapter 7.

Viscous heat generation occurs throughout a fluid. The local rate of heat generation depends on the local shear rate. If the shear rate is constant throughout the entire volume of a fluid, the viscous heat generation will be uniform throughout the fluid. This is the case in pure drag flow (Couette flow); i.e., flow without the presence of pressure differences in the flow direction, see Section 6.2.1. If the shear rate is not uniform throughout the volume, the viscous heat generation will not be uniform either. This is the case in pure pressure flow (Poiseuille flow) through a pipe. In this flow situation, the shear rate in the center is zero and maximum at the wall. Consequently, the viscous heat generation in the center is zero and maximum at the wall just as with the shear rate. Since viscous heat generation occurs throughout a fluid, it is an effective way of heating a polymer melt because it will result in a relatively uniform temperature increase if the shear rate is approximately constant throughout the fluid.

5.3.5   RADIATIVE HEAT TRANSPORT

Heat radiation consists of electromagnetic waves with a wavelength (λ) range of 0.5 to 10 microns. All bodies emit electromagnetic waves as a result of the thermal agitation of the molecules. The rate at which a body emits radiant energy depends mostly on its temperature. Between two surfaces, an exchange of radiation can take place if the intermediate space is transparent to the radiation spectrum. If the temperatures of the two surfaces are different, the sum of the two opposite heat flows generally will not be equal to zero.

The rate of emission of radiant energy is given by the Stefan-Boltzmann law:

$$\dot{Q} = e C_{SB} A T^4 \tag{5-78}$$

where e is the emissivity, C_{SB} the Stefan-Boltzmann constant, A the area of the object, and T the absolute temperature ($^\circ$ K). For a perfectly "black" body, the emissivity equals unity. The precise value of the Stefan-Boltzmann constant C_{SB} can be derived from other physical constants by the relationship:

$$C_{SB} = \frac{2\pi^5 C_B^4}{15 v_1^2 C_{pl}^3} = 5.6697E-8 \left[\frac{W}{m^{2\circ} K^4} \right] \tag{5-79}$$

where C_B is the Boltzmann constant (1.38054E-23 J/$^\circ$ K), v_1 speed of light (2.997925E8 m/sec), and C_{pl} the Planck constant of action (6.6256E-34 Jsec).

For a "black" body, the spectral distribution of energy flux is given by Planck's law of radiation. The wavelength at which this intensity is maximum is inversely proportional to the absolute temperature. This is Wien's law; this can be formulated as:

$$\lambda_{max} T = 2.898E-3 \; [m^\circ K] \tag{5-80}$$

At room temperature $\lambda_{max} = 10\mu m$ (infrared) and at 6000 $^\circ$ K $\lambda_{max} = 0.5$ µm (green). The fact that the color of a body depends on its temperature is used in optical temperature measurements. This is often referred to as infrared temperature measurement even though some measurements may occur outside the infrared region of the spectrum. The infrared region ranges from a wavelength of 0.7 µm to about 400 µm.

A perfectly black body emits the maximum amount of radiation based on its temperature; its emissivity is unity. According to Kirchhoff's law, its absorptivity will also be unity. In reality, surfaces have emissivities and absorptivities for infrared radiation that are less than unity. The actual value will depend on the material, the surface roughness, the temperature and the wavelength of the radiation.

Absorptivity values are usually given as average values. For most non-metallic surfaces, this value is larger than 0.8. Clean and polished metal surfaces have absorptivities ranging between 0.05 and 0.20.

The radiative heat transfer between two surfaces is primarily determined by their emissivities, absorptivities, and temperatures. For each surface, the amount of radiation leaving the surface is the sum of the emitted radiation ($eC_{SB}T^4 = eq_b$) plus the reflected radiation $(1-a)q_i$:

$$q = eq_b + (1-a)q_i \tag{5-81}$$

where a is the absorptivity and q the heat flux; q_b is the black body radiation and q_i is the incident radiation.

Also, the net energy flux leaving the surface equals its own emission minus the fraction of the incident radiation that is absorbed by the wall:

$$q_n = eq_b - aq_i \tag{5-82}$$

If the wall is at the same temperature as its surroundings, the net energy flux will be zero and $q_b = q_i = C_B T^4$. In this case, the emissivity has to equal the absorptivity (e = a); this is known as Kirchoff's law. From expression 5-81, it can be deduced that the energy flux through any plane in a closed space equals $q_b = C_{SB}T^4$.

In order to calculate the incident radiation q_{ik} reaching wall k, one has to know the fractions f_{jk}, indicating what part of the total radiation of wall j reaches wall k. The geometrical factors, called view factors, are calculated by integration. In doing this, one has to take into account the fact that the radiation intensity depends on the angle with which the radiation hits the surface. In the simple case of two large parallel surfaces (1 and 2), both fractions are equal to unity, $f_{12} = f_{21} = 1$. Consider the case where one body (body 1 with area A_1 and temperature T_1) is totally enclosed by another body (body 2 with area A_2 and temperature T_2). The incident radiation on wall 1 ($A_1 q_{i1}$) equals f_{21} multiplied with the total radiation of wall 2 ($f_{21}A_2 q_2$). Thus, the radiation heat flux reaching body 1 (q_{i1}) equals the total radiation heat flux leaving body 2 (q_2). Therefore:

$$q_{n1} = q_1 - q_{i1} = q_1 - q_2 \qquad (5\text{-}83\,a)$$

and

$$q_{n1} = q_{b1} - \frac{1-a_1}{a_1} q_{n1} - q_{b2} + \frac{1-a_2}{a_2} q_{n2} \qquad (5\text{-}83\,b)$$

The net radiation transport from body 1 ($q_{n1}A_1$) and from body 2 ($q_{n2}A_2$) are related by:

$$q_{n1}A_1 + q_{n2}A_2 = 0 \qquad (5\text{-}84)$$

This is valid because the energy lost by body 1 is gained by body 2. The net radiation flux leaving body 1 can be expressed as:

$$q_{n1} = C_{SB}(T_1^4 - T_2^4) \left(\frac{a_1 a_2 A_2}{a_1 A_1 + a_2 A_2 - a_1 a_2 A_1} \right) \qquad (5\text{-}85)$$

The heat transfer coefficient for radiation h_s can now be expressed as:

$$h_s = C_{SB} \left(\frac{T_1^4 - T_2^4}{T_1 - T_2} \right) \left(\frac{a_1 a_2 A_2}{a_1 A_1 + a_2 A_2 - a_1 a_2 A_1} \right) \qquad (5\text{-}86)$$

At room temperature and relatively small temperature differences, the value of h_s will be about 5 W/m$^{2\circ}$ C. This value is large enough that it cannot be neglected relative to free convection. Obviously, at higher temperatures the contribution of the radiative heat transport increases substantially.

Since radiative heating at elevated temperatures (above 300° C) generally occurs in the infrared region, it is often referred to as infrared (IR) heating. The applications of IR heating in polymer processing are numerous: thermoforming, film extrusion, orientation, embossing, coating, laminating, ink drying, and fusing. It is also used in curing filament wound structures and in the manufacture of slit polypropylene yarns out of polypropylene film. Paint drying and baking is the largest single use for infrared heating. In polymer processing, thermoforming is probably the largest outlet for infrared heating. A good series of papers on infrared heating of plastics was written by Kraybill (19–21).

5.3.5.1 DIELECTRIC HEATING

Dielectric heating occurs when a dielectric material is placed in an electric field that alternates at high frequency. A dielectric material is an electric insulator and it has a low conductivity (high resistivity). Materials with a resistivity higher than 10^9 ohm-cm are generally considered to be dielectric; most polymers fall into this category. Dielectric energy, also referred to as radio-frequency energy, occupies the frequency spectrum of about 10 to 100 megahertz.

In dielectric heating, heat is generated throughout the mass of material. In plastics, this can be a very beneficial feature, considering the low thermal conductivity of plastics. Since heat will be transferred away most quickly at the walls, dielectric heating often results in a temperature profile where the highest temperature occurs in the center and the lowest temperature at the wall. This is opposite to conductive heating where the highest temperature will occur at the wall. Also, because heat is generated throughout the material, temperature gradients are likely to be small in dielectric heating as compared to conductive heating. Dielectric heating is uniform throughout a mass of material because all of the polar molecules are oriented by an electric field. The oscillations of the molecules, resulting from the alternating field, produce heat through molecular friction. The rate at which electrical energy can be dissipated in a dielectric material per unit volume $\dot{Q}$ is proportional to the frequency of the electric field f and to the square of the electric field strength E.

$$\dot{Q} = 2\pi E^2 f \varepsilon_0 \varepsilon \tan\delta \qquad\qquad (5\text{–}87)$$

where ε_0 is absolute permittivity of free space (8.854E-12 farad/m), ε relative permittivity or dielectric constant of the material, and $\tan\delta$ the loss tangent or dissipation factor. From equation 5–87, it is clear that fast heating can be accomplished most easily by increasing the field strength; the heating rate increases with the field strength squared! The maximum field strength that can be applied is determined by the dielectric strength of the material to be heated. If the field strength is too high, dielectric breakdown will occur. This will result in sparking and can cause severe damage to the material. If the heating rate at the maximum allowable field strength is too slow, further increases can be obtained by increasing the frequency. Most polymers have a dielectric strength that ranges between 100 and 200 kV/cm, a dielectric constant that ranges between 2 and 4, and a dissipation factor that ranges between 0.01 and 0.0001.

Dissipation factor, power factor, loss angle, etc. are important terms in dielectric heating. They are defined as follows:

```
loss angle = δ = 90 – φ

phase angle = φ

power factor = cosφ = sinδ

dissipation factor = cotanφ = tanδ (also loss tangent)

loss factor = εtanδ = εcotanφ
```

For most polymers, the loss angle is quite small, thus $\sin\delta \simeq \tan\delta$; in other words the power factor and dissipation factor are almost equal.

When components of a material have different loss factors, selective heating will occur. The loss factor of most materials increases with moisture content. Regions with high moisture content will heat faster than others; thus, more water will be removed from high moisture regions. This will result in a uniform moisture distribution in the material. Non-polar polymers such as polyethylene will not heat well at all in a high frequency field. The relative response of various polymers to dielectric heating is shown in Table 5-1. Polymers with poor response to dielectric heating can be made to respond better by adding additives to the polymer. An example of this is the radio frequency heating of ultrahigh molecular weight polyethylene (UHMWPE), containing small amounts of Frequon (18).

TABLE 5-1. RELATIVE RESPONSE OF VARIOUS POLYMERS TO DIELECTRIC HEATING

Polymer	Loss Factor	Good	Fair	Poor	None
ABS	0.025		X		
Acetal copolymer	0.025		X		
Cellulose acetate	0.15		X		
Epoxy resins	0.12		X		
Polyamide	0.16		X		
Polycarbonate	0.03		X		
Polyester	0.05		X		
Polyethylene	0.0008				X
Polyimide	0.013			X	
Polymethyl methacrylate	0.09		X		
Polypropylene	0.001				X
Polystyrene	0.001				X
Polytetrafluoroethylene	0.0004				X
Polyvinylchloride	0.4	X			
Rubber, compounded	0.13		X		
Silicones	0.009				X
Urea-formaldehyde	0.2	X			
Water	0.4	X			

Dielectric or radio frequency heating is used in various parts of the polymer processing industry. Some examples are preheating for molding, curing of thermosetting resins, heat sealing of film, drying of coatings on web substrates, flow molding, etc.

5.3.5.2   MICROWAVE HEATING

Microwave heating is a close cousin of dielectric heating; the main difference being the higher frequencies of microwaves, ranging from about 1000 to 100,000 megahertz. This is about two to three orders of magnitude higher than the frequency spectrum of dielectric energy. By definition, the wavelength of microwave energy must lie in the range of the spectrum between 1 meter and 1 millimeter. This corresponds to a frequency range of 300 megahertz (3E8Hz) to 300 gigahertz (3E11Hz). The frequencies that can be used in the U.S. are controlled by the Federal Communications Commission. For industrial applications, the two most important microwave frequencies are 915 MHz and 2450 MHz. The lower frequency is generally used for high-powered systems (over 200 kW) where the power factor of the material is reasonably high. The higher frequency is used for low-power systems (less than 100 kW) where the material has a relatively low power factor. Consumer microwave ovens operate at 2450 MHz.

When a dielectric material is placed in a microwave field, the dipolar molecules will tend to align their dipole moment along the field intensity vector. When the field intensity vector varies sinusoidally with time, the direction of the vector will reverse every half cycle. This will cause a realignment of the polar molecules. The internal friction that has to be overcome involves a loss of energy from the electromagnetic wave. This results in the conversion of a portion of the electromagnetic energy into thermal energy. In this case, the heat generation is proportional to the number of reversals of the electric field vector; i.e., the frequency. The amount of displacement that occurs during each reversal is determined by the electric field strength. Thus, the heat generation is also a function of the electric field strength; just as with RF heating.

The rate of heating by microwave energy is described by the same equation used for radio frequency heating, equation 5-87. Thus, the amount of heating depends on the field strength, frequency, and loss factor. The latter factor is a material property; the first two factors are dependent on the details of the hardware.

The heat is generated throughout the material; however, the power level reduces with depth of penetration. The depth at which the power is reduced to one-half is given by:

$$H_{0.5} = \frac{0.11 \lambda_0}{\varepsilon \tan \delta} \tag{5-88}$$

where λ_0 is the wavelength in free space, ε the permittivity or dielectric constant of the material, and δ the loss angle. Considering that wavelength is the speed of light divided by the frequency, equation 5-88 can also be written as:

$$H_{0.5} = \frac{3.3E7}{f \, \varepsilon \tan \delta} \tag{5-89}$$

From these expressions, one can see that the depth of penetration reduces with increasing frequency and with increasing loss factor.

If heat losses due to conduction, convection, radiation, or change of state are neglected, the rate of increase in temperature $\dot{T}$ from the absorption of microwave energy can be determined from the following equation:

$$\dot{T} = \frac{\dot{Q}}{C_p \rho} \tag{5-90}$$

where $\dot{Q}$ is the rate of energy dissipation per unit volume from equation 5-87, C_p is the specific heat and ρ the material density.

The penetrating action of microwave energy enables rapid and uniform heating of large cross-sections. With conventional heating methods (hot air, steam, infrared, fluidized bed, etc.), the rate of heating is limited by the poor thermal conductivity of polymers. This is not the case in microwave heating; very short heating chambers can be used.

Applications of microwave heating are drying, continuous curing of polymers (rubbers, filled polyethylenes, etc.), preheating for compression or transfer molding, bonding, etc.

5.4   BASICS OF DEVOLATILIZATION

In devolatilization, one or more volatile components are extracted from the polymer. The polymer can be either in the solid state or in the molten state. Two processes occur in the devolatilization process. First, the volatile components diffuse to the polymer-vapor interface; secondly, the volatile components evaporate at the interface and are carried away. Thus, the first part of the process is diffusional mass transport and the second part a convective mass transport. If the diffusional mass flow rate is less than the convective mass flow rate, the process is diffusion-controlled. In polymer-volatile systems, the diffusion constants are generally very low and, therefore, in many polymer devolatilization processes, the process is diffusion-controlled.

The important relationship in diffusional mass transport is Fick's law. It states that in a one-dimensional diffusion, the positive mass flux of component A is related to a negative concentration gradient. It can be written as:

$$J_A = - D'_{AB} \frac{dC_A}{dy} \qquad\qquad (5\text{--}91)$$

where J_A is the diffusional mass flow rate, C_A the local concentration of component A and D'_{AB} the binary diffusivity.

Fick's law is valid for constant densities and for relatively low concentrations of component A in component B. The term binary mixture is used to describe a two-component mixture. A binary diffusivity is the diffusion constant of one component of a binary mixture. The diffusional mass transport is driven by a concentration gradient, as described by Fick's law. This is very similar to Fourier's law which relates heat transport to a temperature gradient, see equation 5-45. It is also very similar to Newton's law which relates momentum transport to a velocity gradient, see equation 6-16. Because of the similarities in diffusional mass transport, heat transport, and momentum transport, many problems in diffusion are described with equations of the same form as used in heat transfer problems or momentum transfer problems. Also, several of the dimensionless numbers that are used in heat transfer problems, see Section 5.3.3, are also used in diffusional mass transfer problems.

For a binary system of constant density, where a low concentration component A is diffusing through the other component, the equation of continuity for component A can be written as:

$$\frac{\partial C_A}{\partial t} + v_x \frac{\partial C_A}{\partial x} + v_y \frac{\partial C_A}{\partial y} + v_z \frac{\partial C_A}{\partial z} = D'_{AB} \left[\frac{\partial^2 C_A}{\partial x^2} + \frac{\partial^2 C_A}{\partial y^2} + \frac{\partial^2 C_A}{\partial z^2} \right]$$

$$(5-92)$$

This equation of continuity which incorporates Fick's law is used to describe diffusional transport problems. In most analyses of diffusion processes, it is assumed that the concentration at the liquid-vapor or solid-vapor interface is the equilibrium concentration between the vapor phase and the liquid or solid phase. When a liquid phase of a mixture is in equilibrium with a vapor phase of that mixture, the partial pressure of one component depends on the temperature, pressure, and entire composition of the mixture. Partial pressure $\overline{P}_A$ of component A is defined as:

$$\overline{P}_A = P x_A \qquad (5-93)$$

where x_A is the mole fraction of component A in the gas mixture and P is the total pressure on the mixture.

The partial pressure for ideal gasses is described by Dalton's law:

$$\overline{P}_A = \frac{x_A R_0 T n}{V} \qquad (5-94)$$

where R_0 is the gas constant, T is the absolute temperature, n is the number of weight moles of gas, and V is the volume.

For a binary mixture, the composition is completely specified by x'_A, which is the mole fraction of component A in the liquid in equilibrium. In this case, the partial pressure of component A will be a function of pressure, temperature and x'_A If the properties of the liquid are pressure-independent, and if the gasses behave as ideal gasses, then the partial pressure of component A at constant temperature can be written as a function of only x'_A. If the liquid phase consists of only one component, the partial pressure of A equals the vapor pressure of pure A. The partial pressure of component A, $\overline{P}_A$, is described as a function of x' by Henry's law. It states that $\overline{P}_A$ is directly proportional to x'_A at low concentrations of component A:

$$\overline{P}_A = H_A x'_A \qquad (5-95)$$

where H_A is the Henry's law constant.

The Henry's law constant depends on the temperature, the volatile and the pressure. It is not valid for substances such as electrolytes, which dissociate in solution. For ideal solutions, Henry's law is valid over the entire range of concentrations (0% – 100%) and the Henry's law constant equals the vapor pressure of that component.

Polymer-solvent mixtures are highly non-ideal. Because of the very long polymer molecules, the polymer exerts an influence far in excess of its molar concentration. This behavior is often described by the Flory-Huggins relationship (22, 23):

$$\ln \frac{\overline{P}}{P_0} = \ln(1 - V_p) + \left(1 - \frac{1}{DP} \right) V_p + \chi V_p^2 \qquad (5-96)$$

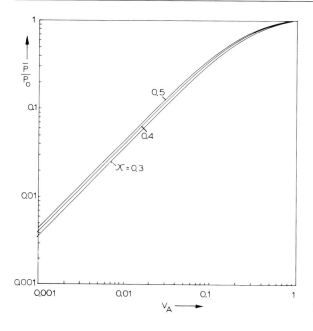

Figure 5-3. $\overline{P}/P_O$ Ratio Versus V_A

where $\overline{P}$ is the effective partial pressure of the volatile component, P_O is the vapor pressure of the pure volatile component, DP is the degree of polymerization, V_p is the volume fraction of the polymer and χ is an interaction parameter.

For polymer-solvent systems where the solvent is chemically similar, the interaction parameter generally falls within the range of 0.3 to 0.5. If the volatile component is fully miscible, a first approximation of the interaction parameter is $\chi = 0.4$. If the volatile component is not fully miscible, the interaction parameter $\chi > 0.5$.

For polymers with a high degree of polymerization and relatively small concentrations of the volatile component A, the Flory-Huggins relationship can be simplified to:

$$\ln\frac{\overline{P}}{P_O} = \ln V_A + 1 + \chi \qquad (5\text{-}97)$$

where V_A is the volume fraction of the volatile component A. According to equation 5-97, the ratio of partial pressure to vapor pressure is directly proportional to V_A and depends exponentially on the interaction parameter χ. The $\overline{P}/P_O$ ratio as a function of V_A at various values of the interaction parameter is shown in Figure 5-3.

It can be seen that where the concentration of the volatile component is below 5 percent, the relationship between $\overline{P}$ and V_A is essentially linear. In this range, Henry's law can be applied with reasonable accuracy. At concentrations above 5 percent, considerable deviations from Henry's law (linear behavior) occur and the Flory-Huggins (F-H) relationship, or a similar relationship, should be used. In many cases, it has been observed that the interaction parameter χ of the F-H relationship is concentration dependent to the extent that this concentration dependence cannot be neglected. This indicates that the basic assumptions underlying the F-H relationship are not fulfilled.

An improved theory for vapor-liquid equilibrium of mixtures based on free-volume considerations was proposed by Prigogine (24, 25). This theory has been further developed by var-

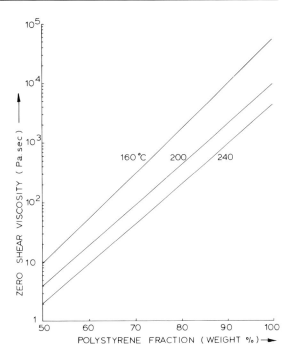

Figure 5-4. Viscosity of PS as a Function of Solvent Concentration

ious workers, e. g., Flory (26). Bonner and Prausnitz (27) discuss the new theory in detail and describe its application with a number of examples.

The viscosity of a polymer melt generally reduces with increased amounts of volatile component. Figure 5-4 shows the viscosity of polystyrene as a function of the solvent concentration. In this example the solvent is ethylbenzene (28).

It can be seen that in this example the viscosity reduces exponentially with the solvent concentration. A 10 percent change in solvent concentration causes approximately a 5 x change in viscosity. Thus, if the initial solvent concentration is 20 percent and the final concentration almost zero, the viscosity increase as a result of devolatilization will be about 25 x! This indicates that when substantial amounts of volatiles are removed from a polymer melt, very large increases in viscosity can occur as a result of the devolatilization.

The flow of concentrated polymer solutions and polymer melts is essentially always laminar as a result of the high viscosity. Heat transfer in such flow systems is quite poor as the heat transfer occurs primarily by conduction and the thermal conductivity in most cases is very low, see also Section 6.3.1.

The diffusion in concentrated polymer solutions is very much slower than in low viscosity (low molecular weight) liquids. The diffusion coefficients for concentrated polymer solutions range from about 10^{-8} m^2/s to 10^{-12} m^2/s. For low viscosity liquids, the diffusion coefficients generally range from about 10^{-6} m^2/s to 10^{-7} m^2/s. The difference is several orders of magnitude! The diffusion rate is very much temperature dependent. At higher temperatures, the vibration of segments of the polymer molecules becomes more pronounced and the density of the polymer reduces. As a result, diffusion of a volatile component will occur at a higher rate. The rate of diffusion will generally also depend on the actual concentration of the volatile component. The presence of a low molecular weight component increases the mobility of the polymer molecules. Thus, the rate of diffusion will

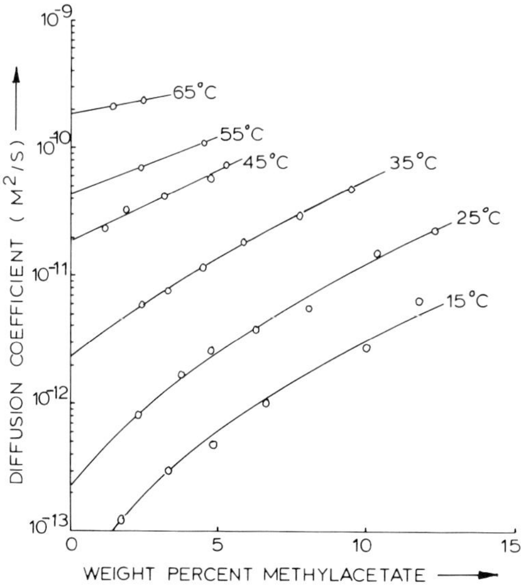

Figure 5-5. Diffusion Coefficient as Function of Solvent Concentration

tend to be higher at large concentrations of the volatile component. Figure 5-5 shows the diffusion coefficient as a function of the solvent concentration at various temperatures for a system of PMA and methylacetate (29).

This figure clearly shows the temperature and concentration dependence of the diffusion coefficient.

5.4.1   DEVOLATILIZATION OF PARTICULATE POLYMER

Theoretical description of devolatilization of particulate polymer can generally be achieved with a relatively high degree of accuracy. In most cases, the process will be diffusion controlled. The diffusion coefficients in solid polymers are very low, ranging from about 10^{-12} m^2/s to about 10^{-14} m^2/s. The temperature in the polymeric particle can usually be taken as constant since the thermal diffusivity ($\alpha \simeq 10^{-7}$ m^2/s) is many orders of magnitude higher than the diffusion coefficient.

In the case of spherical particles with low concentrations of volatile components for which the concentration dependence of the diffusion coefficient can be neglected, the diffusion equation in sperical coordinates can be written as:

$$\frac{\partial C}{\partial t} = \frac{2D'}{r} \frac{\partial C}{\partial r} + D' \frac{\partial^2 C}{\partial r^2} \qquad (5-98)$$

where C is the concentration of the volatile component and D' the diffusion coefficient.

If C_e is the equilibrium concentration at the interface and C_0 the initial concentration, then the solution to equation 5-98 can be written in terms of the average concentration $\overline{C}$ as a function of time (30):

$$\frac{\bar{C} - C_e}{C_0 - C_e} = \frac{6}{\pi^2} \sum_{i=1}^{\infty} \frac{1}{i^2} \exp\left(-\frac{\pi^2 i^2 D' t}{R^2}\right) \tag{5-99}$$

where R is the radius of the spherical particle.

The equilibrium concentration C_e is usually very small relative to the initial concentration C_0, therefore, the C_e term is often neglected. In this case:

$$\frac{\bar{C} - C_e}{C_0 - C_e} \approx \frac{\bar{C}}{C_0} \tag{5-99a}$$

If the temperature cannot be assumed constant, then the equations have to be solved numerically. The same is true if the diffusion coefficient is dependent on the concentration.

In many cases, however, one can reasonably assume a linear dependence of the diffusion coefficient on concentration:

$$D' = D'_0 (1 + \alpha_c C) \tag{5-100}$$

where α_c is the coefficient describing the concentration dependence of the diffusion coefficient.

Figure 5-6 shows the dimensionless concentration $\bar{C}/C_0$ as a function of dimensionless time $D'_0 t / R^2$ at various values of the parameter $\alpha_c C_0 / D'$ as determined by Meier (31).

The top curve, for which this parameter is zero, represents the case for which the diffusion coefficient is independent of concentration and is described by equation 5-99.

The dimensionless $D'_0 t / R^2$ can be considered a Fourier number for diffusion.

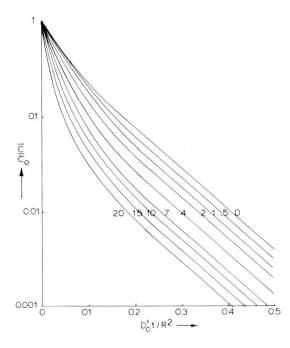

Figure 5-6. Dimensionless Concentration Versus Time

5.4.2   DEVOLATILIZATION OF POLYMER MELTS

In the devolatilization of polymer solutions and polymer melts, the diffusion of the volatile component is in many cases the rate-controlling part of the process. It is generally assumed that the concentration at the interface is at the equilibrium concentration corresponding to the partial pressure of the volatile component in the vapor. A concentration gradient will form in the melt film and the diffusion rate will be determined by the slope of the concentration gradient. If the volatile concentration is large, the viscosity of the liquid will be relatively low and the mass transport of the volatile component will often occur by bubble transport. This is frequently referred to as foam devolatilization. This causes a rather rapid reduction in volatile concentration and results in a rapid increase in viscosity of the liquid. The increasing viscosity inhibits the formation of bubbles and as the volatile concentration becomes low, the mass transport will be governed solely by molecular diffusion. The devolatilization process of polymer melts is usually analyzed as a diffusion-controlled process. Relatively little work has been done to study and analyze foam devolatilization (32, 35, 36, 37). Devolatilization in single screw extruders generally occurs at relatively low levels of volatiles; therefore, foam devolatilization is usually not considered to play a role of importance in devolatilizing extrusion. However, some recent research at Farrel Corp. indicates that foam devolatilization occurs quite readily and may determine the devolatilization process to a large extent (38, 39).

A technique that is often employed in devolatilization of polymer solutions is flash devolatilization (34). In this technique, the polymer solution is delivered to a flash point under high pressure and at temperatures above the boiling point of the volatile component. The solution is then expanded through a nozzle; large amounts of volatiles can thus be extracted rather quickly. The foamy liquid that results from this operation is often exposed to another devolatilization step to remove residual amounts of volatiles. This second step is generally a conventional melt film devolatilization, where the material in the film is continuously renewed to obtain an effective extraction of the volatiles.

Stripping agents, such as water, are often added to the polymer to enhance the devolatilization process. The improvement is obtained by bubble formation, which substantially improves the devolatilization process.

Consider a liquid polymer film with surface area A and depth H moving in direction x in plug flow with a volumetric flow rate $\dot{V}_f$. The film is losing a volatile solute by evaporation in the y direction at a rate of $\dot{E}$. If the mass transport in y direction occurs by molecular diffusion only, and if dispersion in x direction and changes in $\dot{V}_f$ due to loss of volatile are both neglected, an expression for $\dot{E}$ can be developed. The exposure time of the film λ_f is defined as:

$$\lambda_f = \frac{HA}{\dot{V}_f} \qquad\qquad (5\text{--}101)$$

The characteristic time for diffusion is defined as:

$$\lambda_D = \frac{H^2}{D'} \qquad\qquad (5\text{--}102)$$

where D' is the molecular diffusivity of the volatile solute in the liquid polymer.

It can be shown (30) that if $\lambda_f/\lambda_D \leqslant 0.1$, the film can be considered of infinite depth. In this case, the layer in which the concentration is varying is much thinner than the total thickness of the melt film H. The stage efficiency X for this situation can be expressed as:

$$X = \left(\frac{4\,\lambda_f}{\pi\,\lambda_D}\right)^{1/2} \tag{5-103}$$

The stage efficiency is the actual rate of evaporation divided by the maximum possible rate of evaporation:

$$X = \frac{\dot{E}}{\dot{V}_f(C_0 - C_e)} \tag{5-104}$$

where C_0 is the initial volatile concentration and C_e the concentration of the volatile component in the liquid phase which is in equilibrium with the vapor phase.

If $\lambda_f/\lambda_D > 0.1$, the melt film cannot be considered infinite. The stage efficiency in this case can be described by:

$$X = 1 - \frac{8}{\pi^2}\,\exp\left(-\frac{\pi^2\,\lambda_f}{4\,\lambda_D}\right) \tag{5-105}$$

If the film can be considered infinite, the concentration profile can be described by:

$$\frac{\partial C}{\partial t} = D'\,\frac{\partial^2 C}{\partial y^2} \tag{5-106}$$

With boundary conditions $C(0) = C_e$ and $C(\infty) = C_0$, the actual concentration profile as a function of time becomes:

$$\frac{C - C_e}{C_0 - C_e} = \mathrm{erf}\left(\frac{y}{\sqrt{4D't}}\right) \tag{5-107}$$

The same problem in conductive heat transport will be discussed in Section 6.3.5. The error function erf(x) is defined by equation $6-99$. Figure $5-7$ shows the concentration profile at various values of the parameter $\sqrt{D't}$.

The penetration depth is about 0.1 mm when $\sqrt{D't} = 25 \cdot 10^{-6}$. If the diffusion coefficient is assumed to be $D' \simeq 10^{-8}$ m²/s, then the corresponding time is 0.0625 seconds. If the diffusion coefficient is assumed to be $D' = 10^{-6}$ m²/s, then the corresponding time is $6.25 \cdot 10^{-4}$ seconds. Since these are typical values of the diffusion coefficient, the assumption of infinite melt film thickness may not be valid in the analysis of devolatilization in extrusion equipment. A typical exposure time of the melt film in an extruder is of the order of one second; a typical melt film thickness is 0.1 mm ($\simeq 0.004$ inch).

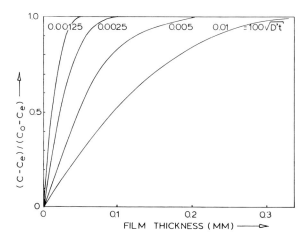

Figure 5-7. Concentration Profile at Various Values of $\sqrt{D't}$

The rate of diffusion J at the interface per unit area and per unit time is determined by:

$$J = \frac{-1}{t} \int_0^t D' \left(\frac{\partial C}{\partial y}\right)_{y=0} dt = 2\left(C_0 - C_e\right)\left(\frac{D'}{\pi t}\right)^{1/2} \qquad (5\text{–}108)$$

Thus, the diffusional transport reduces with $1/\sqrt{t}$. The initial diffusional mass transport will be the highest; thereafter, it will reduce with time according to equation 5-108. Thus, in order to maintain a high devolatilization efficiency, it is very important that the surface through which the volatile is escaping is frequently renewed. This can be achieved by feeding the film into a mixer. The material leaving the mixer will have a homogeneous composition with the same average volatile level as the film entering the mixer. The material leaving the mixer can be spread into a new film with more volatiles diffusing out and evaporating. The process can be repeated many times. It can be demonstrated (33), assuming ideal mixing, that for the n-th stage:

$$\frac{C_n - C_e}{C_0 - C_e} = (1 - X)^n \qquad (5\text{–}109)$$

The improved devolatilization efficiency with surface renewal can be assessed by comparing the residual concentration C_n after n ideal surface renewals with exposure time between film renewal λ_f to the residual concentration $C(n\lambda_f)$ with the same total exposure time but without surface renewal by:

$$\frac{C_n - C_e}{C(n\lambda_f) - C_e} = \frac{[1 - X(\lambda_f)]^n}{1 - X(n\lambda_f)} \qquad (5\text{–}110)$$

Figure 5-8 shows the ratio $C_n/C(n\lambda_f)$ for various values of λ_f/λ_D, assuming that the value of C_e is very small and negligible.

It is clear from Figure 5-8 that surface renewal can yield a considerably lower residual concentration compared to the case without surface renewal. The improvement in devolatilization efficiency becomes more pronounced when the ratio λ_f/λ_D be-comes larger. When the

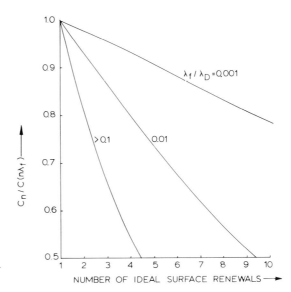

Figure 5-8. Ratio $C_n/C(n\lambda_f)$ for Various Values of λ_f/λ_D

ratio $\lambda_f/\lambda_D > 0.1$, a single curve represents the $C_n/C(n\lambda_f)$ versus n relationship. In this case, equation 5-110 can be written as:

$$\frac{C_n - C_e}{C(n\lambda_f) - C_e} = \left(\frac{8}{\pi^2}\right)^{n-1} \tag{5-111}$$

When the ratio λ_f/λ_D is very small (< 0.001), the benefits of surface renewal are relatively minor. In actual polymer processing equipment, the surface renewal process will generally not be ideal, because only a fraction of the bulk material will be spread out into a thin film. Therefore, the actual devolatilization efficiency would be expected to be less than predicted by equations 5-109 through 5 111.

APPENDIX 5 – 1

EXAMPLE: PIPE FLOW OF NEWTONIAN FLUID

In pipe flow, the fluid moves as a result of a pressure gradient along the pipe, see Figure 5-9.

The velocity at the wall is zero and maximum at the center. The velocity gradient (shear rate) is zero in the center and maximum at the wall. The momentum balance for this case can be determined by taking a force balance on a small fluid element as shown in Figure 5-9; this gives:

$$dP[\pi(r+dr)^2 - \pi r^2] = d\tau 2\pi(r+dr)dz \tag{1}$$

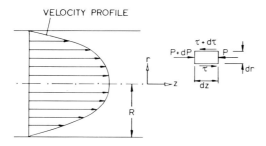

Figure 5-9. Flow Through a Circular Pipe

This results in:

$$\frac{d\tau}{dr} = \frac{1}{2}\frac{dP}{dz} \qquad (2)$$

This same result can be obtained directly from momentum balance equation 5-4c by eliminating the stress derivatives in the tangential (θ) and axial direction (z); this yields:

$$\frac{dP}{dz} = \frac{1}{r}\frac{d(r\tau)}{dr} \qquad (3)$$

Verify that equations 2 and 3 are the same when $\tau/r = d\tau/dr$, which is true in this case. For a Newtonian fluid, the shear stress can be related to the velocity gradient by:

$$\tau = \mu\,\frac{dv}{dr} \qquad (4)$$

where μ is the Newtonian viscosity, see also Section 6.2. Inserting equation 4 into equation 2 or 3 gives:

$$\frac{d^2v}{dr^2} = \frac{g_z}{2\mu} \qquad (5)$$

where g_z is the axial pressure gradient $\left(g_z = \dfrac{dP}{dz}\right)$.

By integrating once, the velocity gradient is obtained:

$$\frac{dv}{dr} = \frac{g_z r}{2\mu} + C_1 \qquad (6)$$

Since the velocity gradient is zero at the center ($r = 0$), the integration constant C_1 has to be zero. The velocity profile is obtained by integrating expression 6:

$$v = \frac{g_z r^2}{4\mu} + C_2 \qquad (7)$$

The integration constant C_2 can be determined from the condition that the velocity at the wall is zero, i.e. $v(R)=0$. This yields:

$$C_2 = - \frac{g_z R^2}{4\mu} \tag{8}$$

Thus, the velocity profile becomes:

$$v = \frac{g_z}{4\mu} (r^2 - R^2) \tag{9}$$

The velocity at the center $(r=0)$ is maximum and is given by:

$$v_{max} = - \frac{g_z R^2}{4\mu} \tag{10}$$

The velocity is positive when the pressure gradient is negative. The velocity profile can now be expressed as:

$$v = v_{max} \left(1 - \frac{r^2}{R^2} \right) \tag{11}$$

The volumetric flow rate can now be determined by integrating the velocity over the cross-sectional area of the pipe:

$$\dot{V} = \int_0^R v 2\pi r dr \tag{12}$$

With equation 9, the flow rate can be determined to be:

$$\dot{V} = - \frac{\pi g_z R^4}{8\mu} \tag{13}$$

This is the well-known Poiseuille equation first published in 1840 (40). The flow rate is directly proportional to the pressure gradient and inversely proportional to the fluid viscosity. The flow rate depends strongly on the radius; it increases with the radius to the fourth power.

REFERENCES – CHAPTER 5

1. R. B. Bird, W. E. Stewart and E. N. Lightfoot, "Transport Phenomena", Wiley, New York (1960).
2. J. R. Welty, C. E. Wicks and R. E. Wilson, "Fundamentals of Momentum, Heat and Mass Transport", Wiley, New York (1969).
3. C. Truesdell and R. A. Toupin, "The Classical Field Theories", in Handbuch der Physik, Vol. III, Springer, Berlin (1960).
4. W. J. Beek and K. M. Muttzall, "Transport Phenomena", Wiley, New York (1975).
5. L. E. Sisson and D. R. Pitts, "Elements of Transport Phenomena", McGraw-Hill, New York (1972).
6. W. C. Reynold and H. C. Perkins, "Engineering Thermodynamics", 2nd Edition, McGraw-Hill, New York (1977).
7. G. J. Van Wylen and R. E. Sonntag, "Fundamentals of Classical Thermodynamics", 2nd Edition, Wiley, New York (1973).
8. R. W. Haywood, "Equilibrium Thermodynamics", Wiley, New York (1980).
9. A. Bejan, "Entropy Generation through Heat and Fluid Flow", Wiley, New York (1982).
10. P. J. Flory, "Principles of Polymer Chemistry", Cornell University Press, Ithaca, New York, (1953).
11. L. R. G. Treloar, "The Physics of Rubber Elasticity", 2nd Edition, Oxford Univ. Press, Oxford (1958).
12. F. Bueche, "Physical Properties of High Polymers", Wiley - Interscience, New York (1962).
13. A. V. Tobolsky, "Properties and Structure of Polymers", Wiley, New York (1960).
14. R. L. Miller, editor, "Flow-Induced Crystallization in Polymer Systems", Gordon and Breach Science Publishers, New York (1979).
15. G. Astarita, Polym. Eng. Sc., 14, 730–733 (1974).
16. H. S. Carslaw and J. C. Jaeger, "Conduction of Heat in Solids", 2nd Edition, Oxford Univ. Press (1959).
17. R. Siegel and J. R. Howell, "Thermal Radiation Heat Transfer", 2nd Edition, McGraw-Hill (1981).
18. B. Miller, Plastics World, March, 99–104 (1981).
19. R. R. Kraybill, SPE Antec, Vol. 27, 590–592 (1981).
20. R. R. Kraybill and W. J. Hennessee, SPE Antec, Vol. 28, 826–829 (1982).
21. R. R. Kraybill, SPE Antec, Vol. 29, 466–468 (1983).
22. P. J. Flory, J. Chem. Phys., 10, 51 (1942).
23. M. L. Huggins, Ann. NY Acad. Sci., 43, 9 (1942).
24. I. Progogine, N. Trappeniers and V. Mathot, Disc. Farad. Soc., 15, 93 (1953); J. Chem. Phys., 21, 559 (1953).
25. I. Progogine, "The Molecular Theory of Solutions", North Holland, Amsterdam (1957).
26. P. J. Flory, J. Am. Chem. Soc., 87, 1833 (1965).
27. D. C. Bonner and J. M. Prausnitz, AIChE J., 19, 943 (1973).
28. E. Schumacher, M. Sc. Thesis, Univ. of Stuttgart, W. Germany (1966).
29. H. Fujita, A. Kishimoto and K. Matsumoto, Trans. Faraday Soc., 56, 424 (1960).
30. J. Crank, "The Mathematics of Diffusion", 2nd Edition, Clarendon Press, Oxford (1975).
31. E. Meier, Chemie Ing. Techn., 42, 20 (1970).
32. R. E. Newman and R. H. M. Simon, 73rd Annual AIChE Meeting, Chicago (1980).
33. J. A. Biesenberger, Polym. Eng. Sci., 20, 1015–1022 (1980).
34. M. H. Pahl in "Entgasen von Kunststoffen", VDI-Verlag GmbH, Duesseldorf (1980).

35. K. G. Powell and C. D. Denson, Paper No. 41 a presented at the Annual Meeting of the AIChE in Washington, D. C. (1983).
36. H. J. Yoo and C. D. Han, Paper No. 41 b presented at the Annual Meeting of the AIChE in Washington, D. C. (1983).
37. M. Amon and C. D. Denson, Polym. Eng. Sci., 24, 1026–1034 (1984).
38. M. A. Rizzi, P. Hold, M. R. Kearney, and A. D. Siegel, Paper No. 41 e presented at the Annual Meeting of the AIChE in Washington, D. C. (1983).
39. P. S. Mehta, L. N. Valsamis, and Z. Tadmor, Polym. Process Eng., 2, 103–128 (1984).
40. J. L. Poiseuille, Compte Rendus, 11, 961 and 1041 (1840); 12, 112 (1841).

6   IMPORTANT POLYMER PROPERTIES

To understand the extrusion process, it is not enough just to know the hardware aspects of the machine. To fully understand the entire process, one also has to know and appreciate the properties of the material being extruded. The characteristics of the polymer determine, to a large extent, the proper design of the machine and the behavior of the process. There are two main classes of properties that are important in the extrusion process. These are the rheological properties and the thermal properties. The rheological properties describe how the material deforms when a certain stress is applied. The rheological properties of the bulk material are of importance in the feed hopper region of the extruder. The rheological properties of the polymer melt are important in the plasticating zone, the melt conveying zone, and the die forming region.

6.1   PROPERTIES OF BULK MATERIALS

Some of the most important properties of the bulk material are the bulk density, coefficient of friction, and particle size and shape. From these properties, the transport behavior of the bulk material can be described with reasonable accuracy. These properties will be discussed in more detail in the following sections.

6.1.1   BULK DENSITY

The bulk density is the density of the polymeric particles, including the voids between the particles. It is determined by filling a container of certain volume (1 liter or more) with the bulk material without applying pressure or tapping. The content is then weighed and the bulk density is obtained by dividing the material weight by the volume. In order to get reproducible results, the dimensions of the container should be several orders of magnitude larger than the particle dimensions.* Low bulk density materials (ρ_b < 0.2 gr/cc) tend to cause solids conveying problems, either in the feed hopper or in the feed section of the extruder. Materials with irregularly shaped particles tend to have a low bulk density; examples are fiber scrap or film scrap (flakes). When the bulk density is low, the mass flow rate will be low as well. Thus, the solids conveying rate may be insufficient to supply the downstream zones (plasticating and melt conveying) with enough material. Special devices and special extruders have been designed to deal with these low bulk density materials. A crammer feeder, see Figure 6-1, is a device used to improve the solids transport from the feed hopper into the extruder barrel.

Special extruders have been designed with the diameter of the feed section larger than the transition and metering section. Two possible configurations, both found commercially, are shown in Figure 6-2.

---

* ASTM D1895 describes standard test methods for apparent density (bulk density), bulk factor (ratio bulk density to actual density) and pourability of plastic materials.

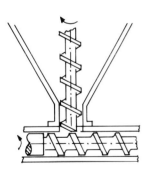

Figure 6-1. Crammer Feeder

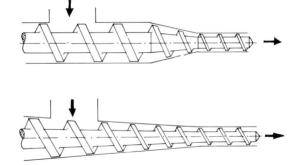

Figure 6-2. Extruders Designed to Handle Low Bulk Density Feed

Since scrap or regrind is more difficult to handle, it is often blended with the virgin material to reduce the handling problems.

The bulk density at atmospheric pressure is useful but limited information. It is very important to know how the bulk density changes with pressure, because the compressibility of the bulk material determines, to a large extent, the solids conveying behavior. Compaction occurs by a rearrangement of the particles and an actual deformation of the particles. The difference between the loosely packed (untapped) bulk density and the packed or tapped bulk density is sometimes referred to as compressibility. This is actually the compression due to rearrangement of the particles. Thus, it would be better to refer to this property as rearrangement compressibility. When the rearrangement compressibility is large, this indicates that the material is prone to packing in storage. This can result in discharge problems. The difference between free-flowing and non-free-flowing is said to occur at a rearrangement compressibility of approximately 20 percent (1). Other workers (2) put the boundary between free-flowing and non-free-flowing at an angle of repose of about 45 degrees, with the non-free-flowing material having an angle repose of more than 45 degrees and the free-flowing materials an angle of repose less than 45 degrees. The angle of repose is the included angle formed between the side of a cone-shaped pile of material and the horizontal plane, see Figure 6-3.

The above-mentioned boundaries between free-flowing and non-free-flowing are only approximate indicators. They are used because of their simplicity and ease of measurement. However, neither rearrangement compressibility nor the angle of repose are true measures of the flowability of particulate materials. Free-flowing materials are also referred to as non-cohesive materials and non-free-flowing materials as cohesive materials. The shear stress at incipient internal shear deformation in non-cohesive materials can be uniquely related to the normal stress. Consequently, the coefficient of cohesion, see equation 6-6, is zero for non-cohesive particulate materials.

When the rearrangement compressibility is above 40 percent, the material will have a very strong tendency to pack in the feed hopper and the chance of discharge problems will be

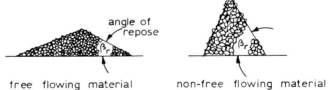

Figure 6-3. Angle of Repose   free flowing material   non-free flowing material

very high. The tendency towards packing can be assessed in a qualitative fashion by the hand squeeze test. Material is squeezed in the hand and the condition of the material is observed after squeezing. If the material has formed a hard clump that cannot be easily broken up, this indicates a highly compressible material. If the material remains clumped but is easily broken up, this indicates a moderately compressible material. If the material does not clump at all but flows after squeezing, this indicates a low compressibility and a relatively free-flowing material.

Many investigators have studied the compression characteristics of bulk materials (3–14); a number of these studies have dealt with polymeric bulk materials. The compaction process is quite complicated for a number of reasons. The distribution of stresses in the material during compaction is rather complex and depends very much on the geometry and surface conditions of the compression apparatus and the detailed characteristics of the bulk material. The effect of pressure on bulk density is often described by an empirical relationship:

$$\zeta = \zeta_0 \exp(-\chi P) \qquad (6-1)$$

where ζ is the porosity, ζ_0 the porosity at zero pressure, P the pressure, and χ the compressibility coefficient.

It should be realized that equation 6-1 is an approximate relationship. The actual compaction behavior will depend strongly on the details of the compaction apparatus and the compacting procedures and conditions.

6.1.2   COEFFICIENT OF FRICTION

The coefficient of friction of the bulk material is another very important property. One can distinguish both internal and external coefficient of friction. The internal coefficient of friction is a measure of the resistance present when one layer of particles slides over another layer of particles of the same material. The external coefficient of friction is a measure of the resistance present at an interface between the polymeric particles and a wall of a different material of construction. The coefficient of friction is simply the ratio of the shear stress at the interface to the normal stress at the interface. Friction itself is the tangential resistance offered to the sliding of one solid over another.

In discussing the coefficient of friction, one has to specify whether it is a static or dynamic coefficient of friction. The static coefficient of friction, f^*, is determined by

$$f^* = \frac{\tau_{ij}^*}{\tau_{ii}} \qquad (6-2)$$

where τ_{ij}^* is the maximum shear stress just before sliding occurs and τ_{ii} the corresponding normal stress.

The dynamic coefficient of friction f is determined by

$$f = \frac{\tau_{ij}}{\tau_{ii}} \qquad (6-3)$$

where τ_{ij} is the actual shear stress during sliding motion and τ_{ii} the corresponding normal stress.

A common way to determine the static coefficient (external) is the measurement of the angle of slide. The object is put on a surface and the angle of the surface with the horizontal plane is increased until the object just begins to slide. The angle that corresponds to the on-set of sliding is slide angle and the coefficient of static friction equals the tangent of the angle of slide. Thus,

$$f^* = \tan\beta_S \qquad\qquad\qquad (6-4)$$

where f^* is the external coefficient of static friction and β_S the angle of slide.

Bowden and Tabor (23) attributed friction to two factors; one factor being the adhesion that occurs at the regions of real contact. The actual area of contact is several orders (about four) of magnitude smaller than the apparent contact area. If sliding is to take place, the local regions of adhesion have to be sheared. The second factor is the plowing, grooving or cracking of one surface by the asperities of the other. In static friction, the only factor of importance is the adhesion at the contact sites. In dynamic friction, the plowing factor starts to play a role; whereas, the adhesion factor reduces in significance.

Measurement of the external coefficient of friction of particulate polymers is very difficult because of the very large number of variables that influence the coefficient of friction. Many investigators have made elaborate measurements on the external coefficient of friction (24–32). The result of this work is that many variables have been identified that affect the frictional behavior; however, most measurement techniques do not yield accurate and reproducible results that can be used in the analysis of the extrusion process. The most elaborate measurements and the most meaningful results have probably been obtained at the DKI in Darmstadt, West Germany (95). It is possible to obtain reproducible results by very careful experimental techniques and special surface preparation of the metal wall. However, the frictional coefficients determined in this fashion are hardly representative of the frictional process conditions occurring in an extruder. Some of the variables that affect the coefficient of friction are temperature, sliding speed, contact pressure, metal surface conditions, particle size of polymer, degree of compaction, time, relative humidity, polymer hardness, etc. The coefficient of friction is very sensitive to the condition of the metal surface. The coefficient of friction of a polymer against an entirely clean metal surface is very low initially, as low as 0.05 or less. However, after the polymer has been sliding on the surface for some time, the coefficient of friction will increase substantially and may stabilize at a value about an order of magnitude higher than the initial value. This behavior was described in detail by Schneider (24, 25) for a variety of polymers. This effect is attributed to the transfer of polymer to the metal surface. Instead of pure polymer-metal friction, the actual situation is a polymer-metal/polymer friction. It has also been found that the measured coefficient of friction is changed if the metal surface is accidentally touched by hand. The finger grease actually changes the metal surface conditions and the resulting coefficient of friction. The unavailability of accurate and appropriate data on coefficient of friction is one of the main stumbling blocks in being able to accurately predict extruder performance. Predictions of the solids conveying rate and pressure development from theory are very sensitive to the actual values of the coefficient of friction, e. g. see Figure 7–17. Thus, for accurate prediction, the coefficient of friction should be known to at least a one-percent accuracy; however, this is usually not feasible.

A comprehensive survey of the work on polymer friction was put together by Bartenev and Lavrentev (32). An exhaustive study on frictional properties was undertaken at the DKI

(Deutsches Kunststoff-Institut) in Darmstadt, West Germany, sponsored by the VDMA (Verband Deutscher Maschinen- und Anlagenbau). The data was compiled in a book (95) that contains frictional properties of 27 different polymers, with the coefficient of friction given as a function of temperature, sliding velocity, and normal pressure. The measurements were made on the univeral disk-Tribometer (see Section 11.2.1.2, Figure 11–9), using conditions that closely resemble the friction process in a screw extruder. This publication is probably the most complete compilation of frictional properties determined under controlled and meaningful conditions with good reproducibility (better than 10 percent).

A typical plot of external coefficient of friction versus temperature at various pressure is shown in Figure 6–4a.

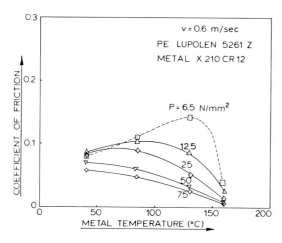

Figure 6–4a. Coefficient of Friction versus Temperature

The material is polyethylene (Lupolen 5261 Z) and the sliding velocity is 0.60 m/sec. At low pressures, the coefficient of friction increases with temperature, reaches a peak at the melting point, and then starts to drop rapidly. At high pressures, the coefficient of friction drops monotonically with temperature.

If the transport of particulate polymer occurs by plug flow, then the only frictional coefficient of importance is the external coefficient of friction. This condition is usually assumed in the solids conveying zone of an extruder with a smooth barrel surface. However, if there is any internal deformation occurring within the particulate material, the internal coefficient of friction will also start to play a role of importance. This is the case when one analyzes the flow of material through a feed hopper or in the solids conveying zone of an extruder when the barrel surface is grooved at the feed section or when the channel is not fully filled so that no pressure increase and no compacting can take place.

The flowability of a particulate material is determined by its shear properties. When internal shear deformation is just about to occur, the local shear stress is called the shear strength. The shear strength is a function of the normal stress; this functional relationship is referred to as the yield locus (YL). For a free-flowing material, the yield locus under fully mobilized friction conditions is

$$\tau = f_i^* \sigma = \tan\beta_i \, \sigma \qquad\qquad (6-5)$$

where f_i^* is the internal static coefficient of friction, β_i the angle of internal friction ($\beta_i = \text{arc} \tan f_i^*$), τ the shear strength and σ the normal stress.

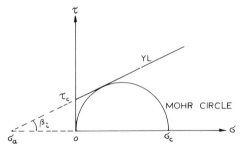

Figure 6-4b. Unconfined Yield Strength of a Cohesive Material

The shear strength of non-free-flowing (cohesive) materials is not a unique function of the normal stress. The shear strength of these materials increases with pressure. The YL is a function of consolidation pressure and consolidation time. Thus, the shear strength has to be described by a series of yield loci, each curve representing a certain consolidation pressure and time. These curves can often be described by:

$$\tau = (\sigma + \sigma_a)\tan\beta_i = \tau_c + \sigma\tan\beta_i \qquad (6-6)$$

where σ_a is an apparent tensile strength, which is obtained by extrapolating the yield locus to zero shear stress, see Figure 6-4b.

The actual tensile strength is usually less than the apparent tensile strength. The value of the shear stress at zero normal stress is often referred to as the coefficient of cohesion $\tau_c = \sigma_a\tan\beta_i$. This coefficient is a measure of the magnitude of the cohesive forces in the particulate material that must be overcome for internal shearing to occur.

In a state of incipient failure, the yield locus is tangent to the Mohr circle. The Mohr circle graphically represents the equilibrium stress condition at a particular point at any orientation for a system in a condition of static equilibrium in a two-dimensional stress field. The equilibrium static conditions can also be applied to sufficiently slow steady flows. The maximum principal stress σ_c in Figure 6-4b is called the unconfined yield strength. This is the maximum normal stress, under incipient failure conditions, at a point where the other principal stress becomes zero. Such a situation occurs on the exposed surface of an arch or dome in a feed hopper at the moment of failure, see Figure 7-4b. In the analysis of bridging in feed hoppers, the unconfined yield strength becomes a very important parameter. The magnitude of the unconfined yield strength is determined by the YL and depends, therefore, on the consolidation pressure and time.

The principal stresses in cohesive materials can be related by:

$$\sigma_{max} = \frac{1+\sin\beta_i}{1-\sin\beta_i}\sigma_{min} + \sigma_c \qquad (6-7)$$

where σ_{max} is the maximum normal stress and σ_{min} the minimum normal stress.

For a cohesive particulate material, each YL curve ends at a point where the normal stress equals the consolidation pressure. Mohr circles can now be drawn that are tangent to the end point of the various yield loci. The envelope of these circles is called the effective yield locus (EYL). This is generally a straight line passing through the origin, see Figure 6-5.

The angle between the EYL and the normal stress axis is called the effective angle of friction, β_e. The EYL describes the shear stress-normal stress characteristics of a particulate

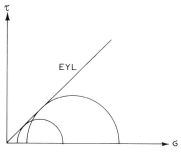

Figure 6-5. Effective Yield Locus

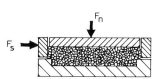

Figure 6-6. Schematic of a Jenike Shearing Cell

material that is consolidated and being sheared under the same stress conditions. This applies directly to a steady flow situation because, in this case, shearing takes place throughout the particulate material. The Mohr circle describing the stress condition at any point must be tangent to the EYL. When the stress field is such that the Mohr circle describing the stress field is below the EYL, no shearing flow will occur. The effective yield locus for a non-cohesive material will coincide with the yield locus. Thus, the effective angle of friction will equal the internal angle of friction ($\beta_e = \beta_i$ for a non-cohesive material).

A shearing cell was developed by Jenike (42) to measure shear properties of particulate solids, see Figure 6-6.

In addition to the determination of the various YL curves and the EYL, the shearing cell can also be used to measure the YL curve between the particulate solids and the confining wall. This is referred to as the wall yield locus (WYL) and it generally lies considerably below the YL. If the WYL is a straight line, it can be described by:

$$\tau_w = \tau_{wa} + \sigma_w \tan \beta_w = \tau_{wa} + f_w^* \sigma_w \qquad (6\text{-}8)$$

where β_w is the wall angle of friction, f_w^* the static coefficient of friction at the wall ($f_w^* = \tan\beta_w$), τ_w the shear stress at the wall, and τ_{wa} the adhesive shear stress at the wall.

Rautenbach and Goldacker (43, 44) described an apparatus developed to measure the internal frictional properties of particulate solids under steady shear. They found that after exceeding the shear strength, the deformation occurred in one or more discrete shear planes. The thickness of these planes was in the order of only a few particle diameters. Thus, they did not observe the development of a continuous velocity gradient throughout the material. These observations were made with non-cohesive polymeric powders. Based on these observations, a dynamic internal coefficient of friction f_i was defined, representing the ratio of steady shear stress to steady normal stress. It was found that the dynamic internal coefficient of friction was independent of velocity and pressure, slightly dependent on temperature and particle size, and strongly dependent on particle shape. In all cases, the dynamic internal coefficient of friction was much higher (about five times) than the dynamic external coefficient of friction. The dynamic angle of internal friction ($\varphi_i = \arctan f_i$) was found to relate reasonably well with the angle of repose β_r according to the expression:

$$\varphi_i \simeq 0.75 \beta_r \qquad (6\text{-}9)$$

This expression is empirical and should only be used to estimate the approximate value of φ_i.

6.1.3   PARTICLE SIZE AND SHAPE

The range of polymeric particles used in extrusion is quite wide, from about 1 micron to about 10 millimeter. Figure 6-7 shows the nomenclature generally used to describe particulate solids of a certain particle size range.

Particle size can be determined by a variety of techniques, microscopic measurement being one of the most common techniques. If the particles have a considerable particle size distribution, one would like to measure this distribution. In this case, microscopic measurement becomes very time consuming, unless it is tied in to an automatic or semi-automatic image analyzer that can generate the distribution curve. Sieving is a simple and popular technique; however, particles should be larger than about 50 micron. Values obtained by sieving of non-spherical particles must be modified to conform with those obtained by methods that yield data on equivalent spherical diameter (esd). The esd is the diameter of a sphere having the same volume as the non-spherical particle. Sedimentation methods are used for particles less than 50 micron. Light transmission or scattering is another method used for particle size measurement.

Two other important parameters in particle analysis are surface area, pore size and volume. The basic method for measuring surface area involves determining the quantity of an inert gas, usually nitrogen, required to form a layer one molecule thick on the surface of a sample at cryogenic temperature. Many techniques are used for pore size measurement: impregnation with molten metal, particle beam transmission, water absorption, freezing point depression, microscopy, mercury intrusion, and gas condensation and evaporation. The last three techniques are most often utilized.

The particle shape can generally be established by simple visual observation or by using a microscope. The transport characteristics of particulate solids are quite sensitive to the particle shape. Both the internal and external coefficient of friction can change substantially with variations in particle shape even if the major particle dimensions remain unchanged. Small differences in the pelletization process can cause major problems in a downstream extrusion process. Variations in the ratio of regrind to virgin polymer can cause variations in the extrusion process.

The ease of solids transport is often determined by the particle size. Pellets are generally free-flowing and do not have a strong tendency to entrap air. From a solids conveying point

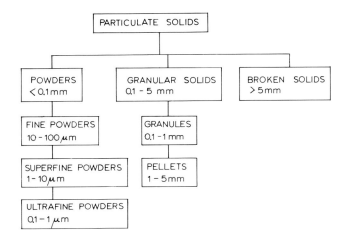

Figure 6-7. Nomenclature
in Particulate Materials

of view, pelletized materials are the easiest to work with. Granules are often free-flowing, sometimes semi-free-flowing; they are more likely to entrap air. Semi-free-flowing granules may require special feeding devices (such as a vibrating pad on the hopper) to ensure steady flow. Powders tend to be cohesive and also tend to entrap air. Therefore, in most cases, special precautions have to be taken to successfully extrude powder material. The degree of difficulty in extrusion of powders generally increases with reducing particle size. Broken solids usually consist of fiber or film scrap. The particles are generally of irregular shape and the bulk density is often very low. This type of particulate solid is also problematic from a solids conveying point of view, because the particles tend to interlock and resist vibration. Static build-up can also be a problem in these materials; this can be solved by the use of a static eliminator.

6.1.4 OTHER PROPERTIES

A variety of other properties can affect the conveying characteristics of the bulk material. Hygroscopic materials tend to absorb moisture; this may cause agglomeration and reduce the flowability of the material. Additives that act as external lubricants can change the frictional characteristics and adversely affect the solids transport in the extruder.

6.2 MELT FLOW PROPERTIES

Knowledge of the flow properties of the polymer melt is very important in the analysis of the extrusion process. The first traces of melt generally appear only a few diameters from the feed opening of the extruder. The metering end of the extruder is, in many cases, completely filled with polymer melt. The polymer melt flow properties determine to a large extent the characteristics of the extrusion process. Knowledge of the melt flow properties allows accurate optimization of the screw design and the process operating conditions. If the melt flow properties are not known, the selection of the extruder screw and the determination of the process operating conditions becomes a trial and error process at best.

6.2.1 BASIC DEFINITIONS

Before going into detail on the flow behavior of polymer melts, it may be useful to describe and define some of the basic terminology used in fluid flow.

DRAG FLOW: Flow caused by the relative motion of one or more boundaries with respect to the other boundaries that contain the fluid.

This is also referred to as Couette flow, although Couette flow is only a specific type of drag flow. Drag flow is important in extrusion. The two major boundaries that contain the polymer in the extruder are the barrel surface and the screw surface. Since the screw is rotating in a stationary barrel, one boundary is moving relative to the other; this causes drag flow to occur.

PRESSURE FLOW: Flow caused by the presence of pressure gradients in the fluid; in other words, local differences in the pressure.

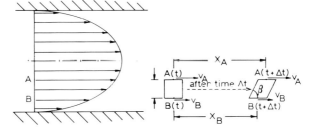

Figure 6-8. Shearing of a Fluid in Pipe Flow

One of the most common examples of pressure flow (pressure driven flow) is the flow of water that occurs when one opens a water faucet. This flow occurs because the pressure upstream is higher than the pressure at the faucet. There is no relative motion of the fluid boundaries (wall of the water pipe); thus, this is pure pressure flow. In most extruder dies, the flow through the die is a pure pressure driven flow. The polymer melt flows through the die as a result of the fact that the pressure at the die inlet is higher than the pressure at the outlet. The flow rate is determined by the pressure at the die inlet, often referred to as diehead pressure. In some extruder dies, the polymer coats a part that moves through the die; e. g., a wire coating die. In such a die, the flow is not a pure pressure flow, but a combination of drag flow (as a result of the moving wire) and pressure flow (as a result of the diehead pressure).

SHEAR: Occurrence of velocity differences in a direction normal to flow.

A fluid is sheared when velocity differences in normal direction occur in the fluid, as shown in Figure 6-8.

ELONGATION: Occurrence of velocity differences in the direction of flow.

Elongational deformation of a fluid occurs when the velocity changes in the direction of flow, as shown in Figure 6-9.

PLUG FLOW: A flow situation where all fluid elements move at the same velocity; i.e., flow without shear.

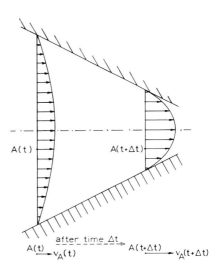

Figure 6-9. Elongational Flow in a Converging Flow Channel

Plug flow generally does not happen in polymer melts, except in the case of wall slip (PVC). However, it does occur with granular polymeric solids. The solids conveying theory of single screw extruders is based on the assumption of plug flow of the solid polymer.

SHEAR RATE ($\dot{\gamma}$): The difference in velocity per unit normal distance (normal to the direction of flow).

The rate of shearing or shear rate is one of the most important parameters in polymer melt processing. If the process is to be described quantitively, the shear rate in the fluid at any location needs to be known. The shear rate is generally written with the Greek letter gamma, $\dot{\gamma}$, with the dot above the gamma indicating a time derivative $\left(\dot{\gamma} = \dfrac{d\gamma}{dt}\right)$. In terms of Figure 6–8, the shear rate can be written as:

$$\dot{\gamma}_{AB} \simeq \frac{v_A - v_B}{AB} \tag{6-10}$$

Equation 6-10 is only valid for very small values of the normal distance AB. More accurately, the shear rate is:

$$\dot{\gamma}_{AB} = \lim_{AB \to 0} \frac{v_A - v_B}{AB} = \frac{dv(x)}{dy} \tag{6-11}$$

From equation 6-11, it can be seen that the local shear rate equals the local gradient of the velocity profile. Thus, if the velocity profile is known, the shear rate at any location can be determined.

SHEAR STRAIN (γ): Displacement (in the direction of flow) per unit normal distance over a certain time period.

The shear strain is generally written with the Greek letter gamma (γ), this time without the dot! The relationship between shear rate ($\dot{\gamma}$) and shear strain (γ) is:

$$\dot{\gamma} = \frac{d\gamma}{dt} \quad \text{and} \quad \gamma = \int \dot{\gamma} dt \tag{6-12}$$

In terms of Figure 6–8, the shear strain can be written as:

$$\gamma_{AB} \simeq \frac{x_A - x_B}{AB} = \frac{\Delta x}{\Delta y}\Big|_{AB} = \tan\beta \tag{6-13}$$

The units of shear rate are $\sec^{-1}$ and the shear strain is a dimensionless number.

SHEAR STRESS (τ): The stress required to achieve a shearing type of deformation.

When a fluid is sheared, a certain force will be required to bring about that deformation. This force divided by the area over which it works is the shear stress. The shear stress is generally written with the Greek letter tau (τ). In a simple example, shown in Figure 6–10, the shear stress is:

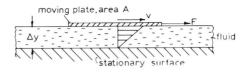

Figure 6-10. Simple Shear Deformation

$$\tau = \frac{F}{A} \qquad (6-14)$$

and the shear rate is:

$$\dot{\gamma} = \frac{V}{\Delta y} \qquad (6-15)$$

SHEAR VISCOSITY (η): The resistance to shear flow. Quantitatively, the shear viscosity is determined from the ratio of shear stress and shear rate.

$$\eta = \frac{\tau}{\dot{\gamma}} \qquad (6-16)$$

The shear viscosity is generally written with the Greek letter eta (η); the units of viscosity are stress x time. The viscosity is usually expressed in Poise ($=$ dyne-sec/cm^2) or Pa-sec ($=10$ Poise). In order to determine the shear viscosity of a fluid, one has to determine the shear rate in a certain shear deformation and the corresponding shear stress. Special instruments are available to determine the viscosity of polymer melts; these are referred to as rheometers.

NEWTONIAN FLUID: A fluid whose viscosity is independent of the shear rate.

Most low viscosity liquids and gases behave as a Newtonian fluid. In a plot of shear stress versus shear rate, a Newtonian fluid will exhibit a linear relationship, see Figure 6-11, curve b. Therefore, Newtonian fluids are also referred to as linear fluids.

A plot of shear stress versus shear rate is generally referred to as a "flow curve."

NON-NEWTONIAN FLUID: A fluid whose viscosity is dependent on the shear rate.

High viscosity polymer melts behave as non-Newtonian fluids, with the viscosity reducing with increasing shear rate.

Another type of non-Newtonian fluid is a dilatant fluid. The viscosity of a dilatant fluid increases with increasing shear rate, see Figure 6-11, curve a.

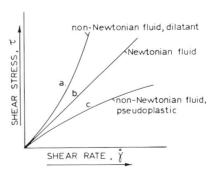

Figure 6-11. Flow Curves of a Dilatant Fluid, a Newtonian Fluid, and a Pseudoplastic Fluid

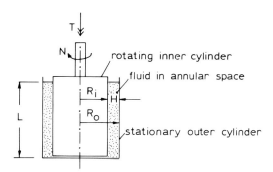

Figure 6-12. Coaxial Cylinders

The reduction of viscosity with increasing shear rate is called pseudoplastic behavior, see Figure 6-11, curve c. The shear stress-shear rate relationship of non-Newtonian fluids is non-linear. Therefore, non-Newtonian fluids are also referred to as non-linear fluids.

The concepts of shear rate, shear stress, and viscosity are extremely important in developing a thorough understanding of the extrusion process (and other polymer processing operations). Therefore, a few examples will be given to illustrate how shear rate and shear stress can be determined in simple geometries.

Example: Coaxial Cylinders, see Figure 6-12.

The fluid is contained in the annular space, with one boundary formed by the inner cylinder and the other boundary formed by the outer cylinder. Since the inner boundary is moving with respect to the outer boundary, a drag flow will be set up in the fluid and the fluid will be sheared. The shear rate in the fluid will be the difference in velocity divided over the normal distance. Thus,

$$\dot{\gamma} \approx \frac{v_o - v_i}{R_o - R_i} = \frac{2\pi R_i N}{H} \qquad (6\text{--}17)$$

This expression is reasonably accurate as long as the radial clearance (H) is small relative to the radius. The shear rate equals the circumferential velocity of the inner cylinder (velocity of the outer cylinder is zero, $v_o = 0$) divided by the radial clearance. Thus, if the geometry and rotational speed are known, the shear rate can be directly determined from that information. The shear rate will be high when the diameter of the inner cylinder is large, when the rotational speed is high, or when the radial clearance is small.

The shear stress acting on the fluid is obtained from the torque T that is necessary to rotate the inner cylinder. The total shear force F acting on the inner cylinder is the shear stress τ multiplied with the area of the inner cylinder ($2\pi R_i L$):

$$F = \frac{T}{R_i} = 2\pi R_i L \tau \qquad (6\text{--}18)$$

Thus, the shear stress is:

$$\tau = \frac{T}{2\pi R_i{}^2 L} \qquad (6\text{--}19)$$

Measurement of the torque, therefore, allows the determination of the shear stress. The shear viscosity can now be determined as well:

$$\eta = \frac{\tau}{\dot{\gamma}} = \frac{TH}{4\pi^2 R_i{}^3 LN} \qquad (6\text{-}20)$$

The coaxial cylinder geometry can thus be used to determine the viscosity of a fluid. In practice, this geometry is only used to determine flow properties of low viscosity liquids.

Example: Screw Extruder, see Figure 6–13.

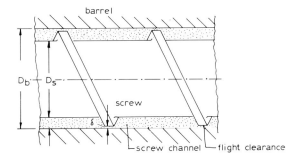

Figure 6-13. Screw Extruder

The geometry of a screw extruder is quite similar to the coaxial cylinder set-up. The difference is the presence of the helical flight wrapped around the core of the screw. The shear rate in the screw channel is approximately:

$$\dot{\gamma}_{ch} \approx \frac{\pi(D-2H)N}{H} = \pi N \left(\frac{D}{H} -2 \right) \qquad (6\text{-}21)$$

where D is the OD of the screw, H the channel depth, and N the rotational speed of the screw in rev/sec.

This is essentially the same expression as found for the coaxial cylinder problem, equation 6-17. The polymer melt between the screw flight and the barrel is exposed to a different shear rate:

$$\dot{\gamma}_{cl} \approx \frac{\pi(D-2\delta)N}{\delta} = \pi N \left(\frac{D}{\delta} -2 \right) \qquad (6\text{-}22)$$

The channel depth H is, in most cases, much larger than the radial flight clearance δ. Therefore, the shear rate in the clearance will be much higher than the shear rate in the screw channel. A typical value of D/H is 20 and a typical value of D/δ is 1000. Thus, the shear rate in the flight clearance will be approximately 50 times higher than the shear rate in the screw channel. This has important implications for the operation of the extruder, as will be discussed in more detail in the next two chapters.

6.2.2   POWER LAW FLUID

In the previous section, it was discussed that polymer melts are pseudoplastic fluids. The fact that the polymer melt viscosity reduces with shear rate is of great importance in the extrusion process. It is, therefore, important to know the extent of change that will occur in a particular polymer. The general shape of the viscosity-shear rate curve for a pseudoplastic polymer melt will look as shown in Figure 6-14.

The viscosity at very low shear rates is essentially independent of shear rate. Thus, the fluid behaves as a Newtonian fluid at low shear rates. The low shear rate plateau value η_0 is often referred to as the low shear limiting Newtonian viscosity. The high shear rate plateau value η_∞ is often referred to as the high shear limiting Newtonian viscosity. This value is difficult to determine experimentally because the effects of pressure and temperature become very pronounced at these high shear rates (over $10^6 \sec^{-1}$).

The range of shear rates encountered in most polymer processing operations is approximately 1 to 10,000 $\sec^{-1}$. It can be seen in Figure 6-14 that within this range the viscosity-shear rate curve can be reasonably approximated with a straight line relationship. This is true for most polymers. It should also be noted that Figure 6-14 uses a double logarithmic scale. The log-log scale is convenient because the viscosity changes about 4 to 5 orders of magnitude over more than 10 orders of magnitude change in shear rate. A straight line relationship on a log-log plot indicates that the variables can be related by a power law equation. This is generally written as:

$$\eta = m\dot\gamma^{n-1} \quad \text{or} \quad \tau = m\dot\gamma^{n} \tag{6-23}$$

where m is the consistency index and n the power law index. This law is often referred to as the power law of Ostwald and de Waele (15, 16). The power law index indicates how rapidly the viscosity reduces with shear rate. For pseudoplastic fluids, the power law index ranges from 1 to 0. When the power law index is unity, the fluid is Newtonian and the consistency index becomes the Newtonian viscosity. The power law index indicates the degree of non-Newtonian behavior. If the power law index ranges from 0.8 to 1.0, the fluid is almost Newtonian. If the power law index is less than 0.5, the fluid is strongly non-Newtonian. It turns out that most large volume commodity polymers fall into this latter category; e.g., polyethylene, polyvinylchloride, polypropylene, polystyrene, styrene acrylonitrile, acrylonitrile butadiene styrene, etc. Examples of polymers with a relatively high power law index are

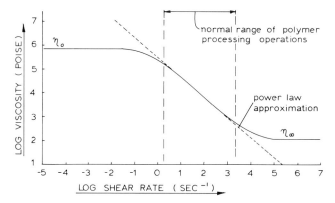

Figure 6-14. General Pseudoplastic Behavior

polycarbonate, polyamide, polyethylene terephthalate, polysulfone, and polyphenylene sulphide. The approximate power law index for a number of polymers is shown in Table 6-1 at the end of this chapter.

Expression 6-23 can be used if the shear rate is positive throughout the flow channel being considered. If the shear rate changes sign at some point in the flow channel, a more general power law equation should be used:

$$\tau = m|\dot{\gamma}|^{n-1}\dot{\gamma} \qquad (6-24a)$$

Another form of the power law equation that is used quite often is:

$$\dot{\gamma} = \varphi|\tau|^{s-1}\tau \qquad (6-24b)$$

where φ is the specific fluidity and s the pseudoplasticity index. The pseudoplasticity index s is the reciprocal of the power law index $n = 1/s$. The specific fluidity φ is related to the consistency index by:

$$\varphi = m^{-1/n} = m^{-s} \qquad (6-24c)$$

Power law expressions 6-23 through 6-24 can be used to describe simple viscometric flow; i.e., flow with velocity components in only one direction. For more complicated flow situations, a more general power law expression should be used. In order to do this, the rate of deformation tensor Δ_{ij} has to be introduced. The components of Δ_{ij} in Cartesian coordinates are:

$$\Delta_{xx} = 2\frac{\partial v_x}{\partial x} \qquad \Delta_{xy} = \Delta_{yx} = \frac{\partial v_x}{\partial y} + \frac{\partial v_y}{\partial x} \qquad (6-25a)$$

$$\Delta_{yy} = 2\frac{\partial v_y}{\partial y} \qquad \Delta_{xz} = \Delta_{zx} = \frac{\partial v_x}{\partial z} + \frac{\partial v_z}{\partial x} \qquad (6-25b)$$

$$\Delta_{zz} = 2\frac{\partial v_z}{\partial z} \qquad \Delta_{yz} = \Delta_{zy} = \frac{\partial v_y}{\partial z} + \frac{\partial v_z}{\partial y} \qquad (6-25c)$$

The components of Δ_{ij} in cylindrical coordinates are:

$$\Delta_{rr} = 2\frac{\partial v_z}{\partial r} \qquad \Delta_{r\theta} = \Delta_{\theta r} = r\frac{\partial}{\partial r}\left(\frac{v_\theta}{r}\right) + \frac{1}{r}\frac{\partial v_r}{\partial \theta}$$

$$(6-25d)$$

$$\Delta_{\theta\theta} = 2\left(\frac{1}{r}\frac{\partial v_\theta}{\partial \theta} + \frac{v_r}{r}\right) \qquad \Delta_{\theta z} = \Delta_{z\theta} = \frac{\partial v_\theta}{\partial z} + \frac{1}{r}\frac{\partial v_z}{\partial \theta} \qquad (6-25e)$$

$$\Delta_{zz} = 2 \frac{\partial v_z}{\partial z} \qquad\qquad \Delta_{zr} = \Delta_{rz} = \frac{\partial v_z}{\partial r} + \frac{\partial v_r}{\partial z} \qquad (6\text{-}25\,f)$$

Because the viscosity is a scalar, it can be a function only of the scalar invariants of the rate of deformation tensor. There are three combinations of the components of the rate of deformation tensor (Δ_{ij}) which are scalar invariants. They define any state of deformation rate independently of the coordinate system. They are referred to as the principal invariants of the rate of deformation tensor:

$$I_1 = \Delta_{ii} \qquad\qquad (6\text{-}26\,a)$$

$$I_2 = \Delta_{ij}\,\Delta_{ij} \qquad\qquad (6\text{-}26\,b)$$

$$I_3 = \det \Delta \qquad\qquad (6\text{-}26\,c)$$

where 'det' means the determinant of the enclosed matrix.

Equation 6–26 uses the summation convention on repeated subscripts.

In Cartesian coordinates:

$$I_1 = \Delta_{xx} + \Delta_{yy} + \Delta_{zz} \qquad\qquad (6\text{-}27\,a)$$

$$I_2 = \Delta_{xx}{}^2 + \Delta_{yy}{}^2 + \Delta_{zz}{}^2 + 2\Delta_{xy}{}^2 + 2\Delta_{xz}{}^2 + 2\Delta_{yz}{}^2 \qquad (6\text{-}27\,b)$$

$$I_3 = \begin{vmatrix} \Delta_{xx} & \Delta_{xy} & \Delta_{xz} \\ \Delta_{yx} & \Delta_{yy} & \Delta_{yz} \\ \Delta_{zx} & \Delta_{zy} & \Delta_{zz} \end{vmatrix} \qquad\qquad (6\text{-}27\,c)$$

If a fluid can be considered incompressible, the first principle invariant of the rate of deformation tensor will be zero, $I_1 = 0$. The third principle invariant vanishes in many simple flow situations, like axial flow in pipe, tangential flow between concentric cylinders, etc. In more general terms, the third invariant is zero in rectilinear flow and in two-dimensional flow.

The power law expression can now be written in general terms:

$$\tau_{ij} = \left(m \; |\sqrt{0.5I_2}| \;^{n-1} \right) \Delta_{ij} \qquad\qquad (6\text{-}28)$$

In Cartesian coordinates, $0.5I_2$ is obtained by using equations 6–27b and 6–25.

$$0.5I_2 = 2 \left(\frac{\partial v_x}{\partial x}\right)^2 + 2 \left(\frac{\partial v_y}{\partial y}\right)^2 + 2 \left(\frac{\partial v_z}{\partial z}\right)^2 + \left(\frac{\partial v_y}{\partial x} + \frac{\partial v_x}{\partial y}\right)^2 +$$

$$\left(\frac{\partial v_z}{\partial y} + \frac{\partial v_y}{\partial z}\right)^2 + \left(\frac{\partial v_x}{\partial z} + \frac{\partial v_z}{\partial x}\right)^2 \qquad\qquad (6\text{-}29)$$

Similarly, in cylindrical coordinates:

$$0.5 I_2 = 2 \left(\frac{\partial v_r}{\partial r} \right)^2 + 2 \left(\frac{1}{r} \frac{\partial v_\theta}{\partial \theta} + \frac{v_r}{r} \right)^2 + 2 \left(\frac{\partial v_z}{\partial z} \right)^2 +$$

$$\left[r \frac{\partial}{\partial r} \left(\frac{v_\theta}{r} \right) + \frac{1}{r} \frac{\partial v_r}{\partial \theta} \right]^2 + \left(\frac{1}{r} \frac{\partial v_z}{\partial \theta} + \frac{\partial v_\theta}{\partial z} \right)^2 + \left(\frac{\partial v_r}{\partial z} + \frac{\partial v_z}{\partial r} \right)^2$$

$$(6-30)$$

With these more general expressions (6-28 through 6-30), more complicated flow situations can be described; i.e, flow with velocity components in two or three directions. It should be remembered that the power law description is an approximation; it is not accurate over the entire range of shear rate. However, in most practical polymer processing problems, the use of the power law equation yields sufficiently accurate results. The major advantage of the power law equation is its simplicity, despite the appearance of equations 6-28 through 6-30. The relationship between stress and rate of deformation can be described with only two fluid properties, the consistency index m and power law index n. A drawback of the power law is that it does not allow construction of a time constant from the constants m and n. This is a problem in the analysis of transient flow phenomena where a characteristic time constant is necessary to describe the flow situation.

The truncated power law of Spriggs (17) allows a more accurate description. It is written as:

$$\eta = \eta_0 \qquad\qquad \text{for } \dot{\gamma} \leq \dot{\gamma}_0 \qquad\qquad (6-31\,a)$$

$$\eta = \eta_0 \left(\frac{\dot{\gamma}}{\dot{\gamma}_0} \right)^{n-1} \quad \text{for } \dot{\gamma} > \dot{\gamma}_0 \qquad\qquad (6-31\,b)$$

In this model, there are three constants: a zero shear rate viscosity η_0, a characteristic time $1/\dot{\gamma}_0$, and a dimensionless power law index n. This model contains the horizontal asymptote for small $\dot{\gamma}$ and the power law for large $\dot{\gamma}$.

6.2.3 OTHER FLUID MODELS

The sinh law was proposed by Ehring (18) and can be written as:

$$\eta = t_0 \tau_0 \left(\frac{\text{arcsinh} t_0 \dot{\gamma}}{t_0 \dot{\gamma}} \right) \qquad\qquad (6-32)$$

where τ_0 is a characteristic stress and t_0 a characteristic time. Other workers have modified the Ehring model to improve its flexibility and accuracy in describing stress-deformation rate relationships:

$$\eta = t_0 t_0 \left(\frac{\text{arcsinh} t_0 \dot{\gamma}}{t_0 \dot{\gamma}} \right)^{A_1} + A_2 \tag{6-33}$$

A polynomial relationship was proposed by Rabinowitsch and Weiszenberg; it can be written as (20a):

$$\dot{\gamma} = \alpha_1 \tau + \alpha_3 \tau^3 \tag{6-34}$$

where α_1 and α_3 are rheological constants depending on the nature of the fluid.

The Carreau model (19) has the useful properties of the truncated power law model but avoids the discontinuity in the first derivative; it can be written as:

$$\frac{\eta - \eta_\infty}{\eta_0 - \eta_\infty} = [1 + (\lambda \dot{\gamma})^2]^{\frac{n-1}{2}} \tag{6-35}$$

where η_0 is the zero shear rate viscosity, η_∞ is the infinite shear rate viscosity, λ is a time constant, and n is the dimensionless power law index. The Ellis model (20b) describes the viscosity as a function of shear stress:

$$\frac{\eta_0}{\eta} = 1 + \left(\frac{\tau}{\tau_{1/2}} \right)^{\alpha - 1} \tag{6-36}$$

where η_0 is the zero shear rate viscosity, $\tau_{1/2}$ the value of the shear stress at which $\eta = \eta_0/2$ and $\alpha-1$ is the slope of $(\eta_0/\eta) -1$ versus $\tau/\tau_{1/2}$ on log-log paper. The Ellis model is relatively easy to use and many analytical results have been obtained with this model. Actually, the Ellis model is a more general form of the Rabinowitsch equation. The latter is a special case of the Ellis model when the constant $\alpha = 3$.

Another model in which the viscosity is described as a function of shear stress is the Bingham model (20). This model is used for fluids with a yield stress τ_0. Below this yield stress, the viscosity is infinite (no motion), above the yield stress, the viscosity is finite (motion occurs). The Bingham Fluid model is written as:

$$\eta = \infty \qquad \text{for } \tau \leqslant \tau_0 \tag{6-35a}$$

$$\eta = \mu_0 + \frac{\tau_0}{\dot{\gamma}} \quad \text{for } \tau > \tau_0 \tag{6-35b}$$

This model is primarily used for slurries and pastes. The parameters τ_0 and μ_0 can be related empirically to the volume fraction of solids φ, the particle diameter D_p, and the viscosity of the suspending fluid:

$$\tau_0 = 312.5 \frac{\varphi^3}{D_p^2} \qquad\qquad (6-36)$$

$$\mu_0 = \mu_s \exp\left[\varphi\left(\frac{b}{2} + \frac{14}{\sqrt{D_p}}\right)\right] \qquad\qquad (6-37)$$

where D_p is in μm and τ_0 in Pascal.

Many other fluid models have been proposed. For a more detailed discussion, the reader is referred to the literature (e.g., refs. 17, 20–22, 54, 94).

6.2.4   EFFECT OF TEMPERATURE AND PRESSURE

The effect of shear rate on viscosity has been discussed in some detail in the previous sections. However, there are some other variables that also affect the viscosity. Two important variables that influence the viscosity are temperature and pressure. The effect of these variables is generally not as strong as the effect of shear rate; however, in many cases, the effect of temperature and/or pressure on viscosity cannot be neglected.

When the viscosity is plotted against shear rate at several temperatures, the curve generally lowers with increasing temperature, see Figure 6-15.

This is a result of the increased mobility of the polymer molecules. For the time being, it is assumed that no irreversible changes occur as a result of degradation. However, whenever experiments or processes are conducted at elevated temperature, the possible effects of degradation have to be taken into account. There will be more on degradation in Section 11.3.

It is convenient to plot viscosity as a function of shear stress to evaluate the effect of temperature. This is shown in Figure 6-16 where the same data shown in Figure 6-15 is plotted in terms of viscosity and shear stress.

It can be seen that the shape of the viscosity-shear stress curve does not change appreciably with temperature. With many polymers, the curves can be shifted along lines of constant shear stress to produce a master curve. By comparing Figure 6-15 to Figure 6-16, it is clear that a shift along lines of constant shear rate would not produce a good fit. Figure 6-15 shows a line of constant shear stress; it makes an angle of 45° with both axes. The curves should be shifted in the direction of this constant shear stress line to produce a good master curve. Figure 6-16 also shows lines of constant shear rate. It can be seen in Figure 6-16 that the effect of temperature is greater at lower temperature; this is true for many polymers. It should be mentioned that most polymers do not have as strong a temperature dependence as the polymer shown in Figures 6-15 and 6-16.

The shift factor a_T is a function of the temperature. For polyolefins, the relationship can be written as:

$$a_T = \exp\left[\frac{E}{R}\left(\frac{1}{T} - \frac{1}{T_r}\right)\right] \qquad\qquad (6-38)$$

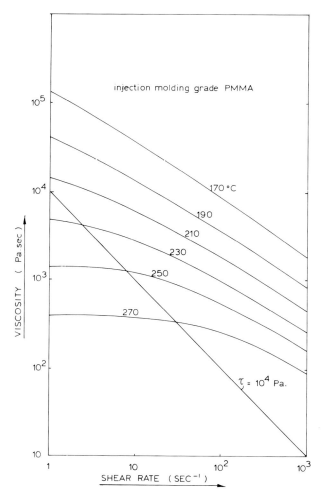

Figure 6–15. Viscosity versus Shear Rate at Various Temperatures

where E is the activation energy, R the universal gas constant, and T_r the reference temperature in degrees Kelvin. Equation 6–38 is known as Andrade's Law (98). It is applicable to semi-crystalline and amorphous polymers above $T_g + 100\,°C$.

For amorphous polymers, the Williams-Landel-Ferry (WLF) equation is often used:

$$a_T = \exp\left[\frac{-C_1(T-T_r)}{C_2+(T-T_r)}\right] \qquad (6-39\,a)$$

where C_1 and C_2 are material constants.

If the reference temperature T_r is taken about $43\,°K$ above the glass transition point T_g, the constants C_1 and C_2 are essentially the same for a large number of amorphous polymers ($C_1 = 8.86$ and $C_2 = 101.6$). This results in the following equation:

$$a_T = \exp\left[\frac{-8.86(T-T_g-43)}{101.6+(T-T_g-43)}\right] \qquad (6-39\,b)$$

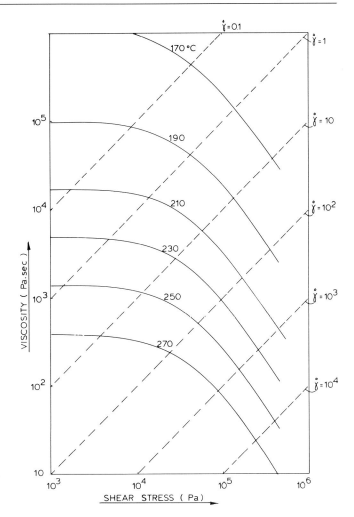

Figure 6-16. Viscosity versus Shear Stress at Various Temperatures

The glass transition temperature of a number of polymers is shown in Table 6-1. Equation 6-39 gives a reasonable description of the temperature dependence of the viscosity in the range of T_g to $T_g + 100°$ C.

A popular empirical form of the temperature dependence of viscosity is:

$$a_T = \exp[\alpha_T(T_r - T)] \qquad (6\text{--}40)$$

where α_T is a temperature coefficient that can be considered constant as long as the temperature range considered is relatively small.

The power law equation including the temperature effect can then be written as:

$$\tau = m_r a_T \dot{\gamma}^n \qquad (6\text{--}41\,a)$$

or

$$\eta = m_r a_T \dot{\gamma}^{n-1} \qquad (6\text{--}41\,b)$$

The temperature sensitivity of the viscosity varies widely for different polymers. As a general rule, amorphous polymers have a high temperature sensitivity, while semicrystalline polymers have a relatively low temperature sensitivity. Polyvinylchloride (PVC) and polymethyl methacrylate (PMMA) are two polymers with a very high temperature sensitivity of the viscosity. Polyethylene and polypropylene both have quite low temperature sensitivity.

The relative change in viscosity per degree of temperature can be determined from:

$$\frac{\partial \eta(T)}{\eta(T_r)\partial T} = \frac{\partial a_T}{\partial T} \tag{6-42}$$

If the expression for a_T is used for amorphous polymers (equation 6-39), one obtains:

$$\frac{\partial \eta(T)}{\eta(T_r)\partial T} = \frac{-C_1 C_2}{[C_2 + (T-T_r)]^2} \exp\left[\frac{-C_1(T-T_r)}{C_2 + (T-T_r)}\right] \tag{6-43}$$

This relationship is shown in Figure 6-17.

It can be seen that the temperature sensitivity drops dramatically (several decades) when $T-T_r$ increases. The closer a polymer is to its glass transition temperature, the larger the temperature sensitivity of the viscosity. This explains why polymers whose normal process temperatures are close to their glass transition temperature exhibit a high temperature sensitivity in processing. Examples are polystyrene, polyvinylchloride and polymethyl methacrylate. In general, polymers that are processed considerably above their glass transition temperature (more than 150°C above T_g) show a relatively small temperature sensitivity. Examples are polyethylene, polypropylene and polyamide.

The effect of pressure on viscosity is relatively insignificant in most polymer processing operations, where pressures generally do not exceed 35 MPa (5000 psi). It has been found, however, that the effect of pressure on viscosity becomes quite significant at pressures substantially above 5000 psi. In fact, in careful rheological measurements, the effect of pressure on both viscosity and density has to be considered even at pressures around 35 MPa (5000 psi).

Special rheometers have been constructed to measure the effect of pressure on viscosity. Various workers have presented data on the pressure dependence on viscosity (33–40). The viscosity as a function of pressure is generally written as:

$$\eta(P) = \eta(P_r)\exp[\alpha_p(P-P_r)] \tag{6-44}$$

The values of the pressure sensitivity term α_p vary considerably from one polymer to another. For polystyrene, increases in viscosity at fixed shear stress and about 150°C have been reported (34, 36, 40) of 200 to 1000 times over a pressure rise of 100 MPa (15,000 psi). For polyethylene at the same temperature and pressure conditions, the viscosity increased only 4 to 5 fold. At a temperature of 200°C and a pressure rise of 100 MPa (15,000 psi), the increase in viscosity of polystyrene was reported to be about 30 to 50 fold; about 10 to 20 times lower than at 150° C!

Data on the pressure sensitivity of the viscosity is quite scarce. It has been found empirically (41) that the relative change in viscosity with pressure divided by the relative change in viscosity with temperature is approximately constant for many polymers:

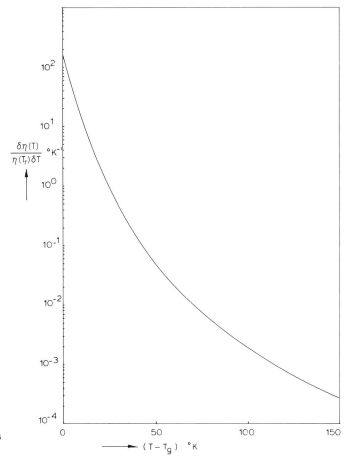

Figure 6-17. Relative
Change in Viscosity versus
T-T$_g$

$$\frac{\dfrac{1}{\eta}\left(\dfrac{\partial \eta}{\partial P}\right)}{\dfrac{1}{\eta}\left(\dfrac{\partial \eta}{\partial T}\right)} \approx -5\text{E--}7 \ [°\text{C/Pa}] \qquad\qquad (6\text{--}45)$$

The numerator can be determined from equations 6-38 through 6-43. Thus, equation 6-45 provides a convenient, though approximate, method to determine the pressure sensitivity of the viscosity of a polymer. From equation 6-45, it is clear that a polymer with a high temperature sensitivity of the viscosity will also have a high pressure sensitivity of the viscosity. This explains the large differences in pressure sensitivity of the viscosity between polystyrene and polyethylene, as mentioned earlier.

6.2.5   VISCOELASTIC BEHAVIOR

Thus far, the polymer melt has been considered as a purely viscous fluid. In a purely viscous fluid, the energy expended in deformation of the fluid is immediately dissipated and is non-recoverable. The other extreme is the purely elastic fluid where the energy expended in deformation of the fluid is not dissipated at all; the deformation is completely reversible and the energy completely recoverable.

Polymers are partly viscous and partly elastic. In the molten state, polymers are primarily viscous but will be elastic to some extent. This behavior is generally referred to as viscoelastic behavior. This characteristic is responsible for the swelling of the extrudate as it emerges from an extruder die. The swelling is caused by elastic recovery of strain imparted to the polymer in and before the die. The swelling is not instantaneous, but takes a finite time to fully develop. This indicates that the rearrangement of the polymer structure takes a certain amount of time; this can range from a fraction of a second to several minutes or even hours, depending on the polymer and the temperature. The polymer properties, therefore, are a function of time and depend on the deformation history of the polymer. The deformation history is often referred to as the shear history; however, it is not only shearing deformation that affects the polymer properties but elongational deformation as well.

In fluids with time dependent behavior, the effects of time can be either reversible or irreversible. If the time effects are reversible, the fluids are either thixotropic or rheopectic. Thixotropy is the continuous decrease of apparent viscosity with time under shear and the subsequent recovery of viscosity when the flow is discontinued. Rheopexy is the continuous increase of apparent viscosity with time under shear; it is also described by the term anti-thixotropy. A good review on thixotropy was given by Mewis (45). Polymer melts do exhibit some thixotropic effects; however, thixotropy can also occur in inelastic fluids. The time scale of thixotropy is not necessarily associated with the time scale for viscoelastic relaxation.

For a proper description of the flow of a polymer melt, the viscoelastic properties have to be taken into account, including the dependence on deformation history. Some experimental and theoretical work on time dependent effects is covered in the following references (46–53); many other publications have been devoted to this subject. Unfortunately, the viscoelastic models that include memory effects (i. e., the dependence on deformation history) are quite complex and difficult to apply. Also, there does not seem to be any model that is widely accepted as being able to describe polymer melt flow accurately over a wide range of flow geometries and conditions. Practicing process engineers probably will find these models difficult to apply to actual extrusion problems. As a result, the workers in this field generally are specialists in rheology.

In the quantitative analysis of most extrusion problems, the polymer melt generally is considered to be a viscous, time-independent fluid. This assumption is, of course, a simplification, but it usually allows one to find a relatively straightforward solution to the problem. This assumption will be used throughout the rest of this book, unless indicated otherwise. In the analysis of any flow problem, however, it should be remembered that elastic effects may play a role. Also, some flow phenomena, such as extrudate swell, clearly cannot be analyzed unless the elastic behavior of the polymer melt is taken into account. For more information on the rheology of viscoelastic fluids, the reader is referred to the literature (e. g., 17, 20–22, 54–63). The last six references (58–63) do not go into great mathematical complexities and are relatively easy to understand for people not specialized in rheology.

6.2.6 MEASUREMENT OF FLOW PROPERTIES

Whenever a process engineer uses flow property data, he should know on what instrument and how these data were determined in order to properly assess the validity of the data. Instruments to determine flow properties are generally referred to as rheometers. The rheometers that will be briefly described in the next few sections are the capillary rheometer, the melt indexer, the cone and plate rheometer, the slit die rheometer, and dynamic mechanical rheometers. For a more detailed description of these and other rheometers, the reader is referred to the literature (e. g., 64, 65, 66). A brief but good survey of commercial rheometers was presented by Dealy (92).

6.2.6.1 CAPILLARY RHEOMETER

A capillary rheometer is basically a ram extruder with a capillary die at the end, see Figure 6-18.

As the piston moves down, it forces the molten polymer through the capillary. The shear stress in the capillary at the wall (τ_{cw}) can be related to the pressure drop along the capillary (ΔP_c) by the following equation:

$$\tau_{CW} = \frac{\Delta P_c D_c}{4 L_c} \qquad (6\text{--}46)$$

If the piston diameter is much larger than the capillary diameter ($D_p \gg D_c$) and if entrance effects are neglected, then:

$$\Delta P_C \approx \frac{4 F_p}{\pi D_p{}^2} \qquad (6\text{--}47)$$

By inserting equation 6-47 into equation 6-46, the wall shear stress in the capillary can be related to the force on the piston. Thus, by measuring the force on the piston, the wall shear

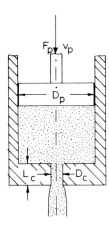

Figure 6-18. Capillary Rheometer

stress in the capillary can be determined. To avoid problems with entrance effects, it is good practice to do measurements with a long capillary (high L/D, 20 to 40). It is even better to do measurements with two capillaries of the same diameter but different length, one having a length of almost zero. The actual pressure drop along the capillary of length L_c is now:

$$\Delta P_c = \frac{4}{\pi D_p^2} [F_p(L_c) - F_p(L_c=0)] \qquad (6-48)$$

The apparent shear rate at the capillary wall can be determined from the flow rate $\dot{V}$ through the capillary. This can be determined from equations 6 and 13 in Appendix 5 – 1.

$$\dot{\gamma}_{aw} = \frac{32\dot{V}}{\pi D_c^3} \qquad (6-49)$$

The flow rate is determined by the area and the velocity of the piston:

$$\dot{V} = v_p \frac{\pi}{4} D_p^2 \qquad (6-50)$$

The apparent shear rate at the capillary wall can now be expressed as a function of the piston velocity:

$$\dot{\gamma}_{aw} = \frac{8 D_p^2 v_p}{D_c^3} \qquad (6-51)$$

Thus, by measuring the piston velocity one can determine the apparent shear rate at the capillary wall. At this point, the apparent viscosity can be determined by dividing the shear stress by the apparent shear rate:

$$\eta_a = \frac{\tau_{cw}}{\dot{\gamma}_{aw}} \approx \frac{F_p D_c^3}{2 \pi D_p^4 v_p} \qquad (6-52)$$

The terms apparent shear rate and apparent viscosity are used because expression 6-49 is valid only for Newtonian fluids. Therefore, if the fluid is non-Newtonian, the actual value of the shear rate at the capillary wall will be different. If the fluid behaves as a power law fluid with power law index n, the actual shear rate at the capillary wall is:

$$\dot{\gamma}_w = \frac{3n+1}{4n} \dot{\gamma}_{aw} = \frac{(3n+1)8\dot{V}}{n \pi D_c^3} \qquad (6-53)$$

Thus, for a power law fluid, the actual viscosity is related to the apparent viscosity by:

$$\eta = \frac{4n}{3n+1} \eta_a \qquad (6-54)$$

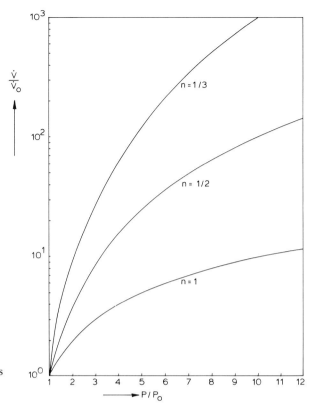

Figure 6-19. Relative Flow Rate as a Function of Relative Pressure Drop

If the capillary rheometer is used to compare different polymers, it is not necessary to go through the various correction procedures. However, if one wants to know the absolute values of the viscosity, it is important to apply the various correction factors. The most important corrections are the correction of the shear rate for non-Newtonian fluid behavior (often referred to as Rabinowitsch correction) and the correction of the shear stress for entrance effects (often referred to as Bagley correction). These are the most common corrections applied to capillary rheometers. Other corrections that are sometimes considered are corrections for viscous heating, corrections for the effect of pressure on viscosity, corrections for compressibility, correction for time effects, etc. If many corrections are applied to the data, the whole measurement and data analysis procedure can become very complex and time consuming.

Figure 6-19 shows how the relative flow rate $(\dot{V}/\dot{V}_0)$ through a capillary depends on the relative pressure drop along the capillary die (P/P_0) for three different values of the power law index: $n = 1$, $n = 1/2$ and $n = 1/3$.

In all cases, the flow rate increases as the die pressure increases. However, there is a very large difference in behavior between fluids of different power law index. For a Newtonian fluid $(n = 1)$, a $10 \times$ increase in pressure results in a $10 \times$ increase in flow rate. For a power law fluid with $n = 1/2$, a $10 \times$ increase in pressure results in a $100 \times$ increase in flow rate. For a power law fluid with $n = 1/3$, a $10 \times$ increase in pressure results in a $1000 \times$ increase in flow rate! Essentially the same results are valid for any extrusion die. It is clear, therefore, that the power law index of a polymer melt, to a large extent, will determine its extrusion behavior. It is very important to know the power law index of a material. This is why the viscosity has to be determined over a wide range of shear rates. The shear rate range should be representative of the shear rates encountered in polymer processing equipment, which is usually 0 to 10,000 sec^{-1}.

Advantages of the capillary rheometer are:

1. Ability to measure very high shear rates ($\simeq 10^6$ sec^{-1}).
2. Ability to measure extrudate swell characteristics.
3. Ability to measure melt fracture characteristics.
4. Relatively easy to use.

Disadvantages of the capillary rheometer are:

1. The polymer is not exposed to a uniform shear rate.
2. Various corrections have to be applied to the data.
3. It does not yield an accurate description of viscoelastic behavior.
4. It is unreliable at high shear rates (temperature effects).

6.2.6.2   MELT INDEX TESTER

The melt index tester is essentially a simple capillary rheometer. The piston is pushed down by placing a weight on top of it, see Figure 6–20a (99).

The melt index is the number of grams of polymer extruded in a time period of 10 minutes. Details of the geometry and test procedures are described in ASTM D1238.

The melt index tester is used in many companies to quickly test the polymer melt. Unfortunately, many times MI data is the only information available on the polymer melt flow properties.

From the dimensions of the MI apparatus, the weight on the plunger and the MI value, one can determine the approximate shear stress, shear rate and viscosity. By using equations 6–46 and 6–47, the shear stress at the capillary wall can be determined from:

$$\tau_w \simeq \frac{F_p D_c}{\pi L_c D_p^{\,2}} = 8.97 F_p \quad [\text{Pa}] \qquad (6\text{--}55)$$

where F_p is the weight on the plunger in grams. The flow rate through the capillary can be expressed as:

$$\dot{V} = \frac{MI}{600 \rho} \quad [\text{gr/sec}] \qquad (6\text{--}56)$$

where MI is expressed in grams per 10 minutes and the density ρ in grams/cc. The apparent shear rate can now be determined by using equation 6–49:

$$\dot{\gamma}_{aw} = 1.845 \frac{MI}{\rho} \quad [\text{sec}^{-1}] \qquad (6\text{--}57)$$

The apparent shear rate is thus directly proportional to the MI value and inversely proportional to the polymer melt density. The apparent viscosity can now be determined from:

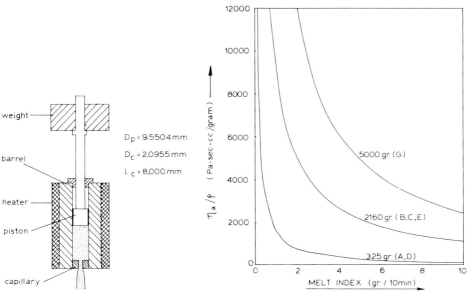

Figure 6-20a. Melt Index Tester

Figure 6-20b. Viscosity/Density
Ratio as a Function of MI at Various
Weights of the Plunger

$$\eta_a \simeq \frac{4.86\,\rho F_p}{MI} \quad [\text{Pa-sec}] \qquad (6\text{-}58)$$

From equation 6-58 it can be seen that the polymer melt viscosity is inversely proportional
to the MI value, see Figure 6-20b.

The melt index can be measured at a number of different conditions as far as load and temperature are concerned. A number of standardized conditions are listed in the following table:

TABLE 6-1. STANDARDIZED CONDITIONS FOR MELT INDEX TESTING

Condition	Temp. [°C]	Load [gr]	Condition	Temp. [°C]	Load [gr]
A	125	325	K	275	325
B	126	2160	L	230	2160
C	150	2160	M	190	1050
D	190	325	N	190	10000
E	190	2160	O	300	1200
F	190	21600	P	190	5000
G	200	5000	Q	235	1000
H	230	1200	R	236	2160
I	230	3800	S	236	5000
J	265	12500	T	250	2160

A high MI value corresponds to a low polymer melt viscosity and a low MI value corresponds to a high polymer melt viscosity. The term "fractional melt index material" refers to
a polymer with a melt index less than one. These are materials with high melt viscosities and
they generally have higher power consumption and diehead pressure in extrusion as compared to polymers with higher melt index values.

As an example, consider a polymer with MI = 0.2 gr/10 min and density $\rho = 1.0$ gr/cc; the MI is determined under condition E ($F_p = 2160$ gram). The approximate viscosity of this material is 52 488 Pa-sec at a shear rate of about 0.37 sec^{-1}. If the MI = 20 gr/10 min with everything else the same, the viscosity would be about 525 Pa-sec at a shear rate of about 37 sec^{-1}. It should be noted that these figures are very approximate and should be regarded as estimates rather than firm numbers. Considering that the standard MI capillary is quite short (about 4 L/D), the entrance effect will be considerable. Therefore, the error in the expression for the wall shear stress (6-55) can be considerable. Consequently, the error in the expression for the apparent viscosity (6-58) can also be considerable.

A large drawback of the MI test is that it yields single point data. Thus, it does not give any idea about the degree of non-Newtonian behavior of the material. This drawback can be negated by running the melt indexer with several different weights on the plunger; however, this is not often done in practice. Another drawback is the relatively poor reproducibility of the melt indexer; under the best circumstances it is about 15 percent (plus or minus).

An advantage of the melt indexer is its low cost and ease of operation. It should be noted that the melt indexer generally operates at low shear rates, see equation 6-57, that are not representative of shear rates usually encountered in the extrusion process. Thus, the MI value is not a good indicator of the processing behavior of a polymer.

Advantages of the melt indexer are:

1. The instrument is simple and inexpensive.
2. Its ease of operation.
3. It is widely available.

Disadvantages of the melt indexer are:

1. It yields only single point data.
2. It has limited accuracy and reproducibility.
3. It is not a good indicator of processability.
4. It is not an accurate description of viscoelastic behavior.

6.2.6.3 CONE AND PLATE RHEOMETER

In a cone and plate rheometer, the polymer melt is situated between a flat and a conical plate. In most rheometers, the cone is rotating and the plate is stationary; however, this is not absolutely necessary. The basic geometry is shown in Figure 6-21.

If the cone rotates with an angular velocity ω and the cone angle β is very small, the shear rate in the fluid is given by (57):

$$\dot{\gamma} = \frac{\omega}{\beta} \tag{6-59}$$

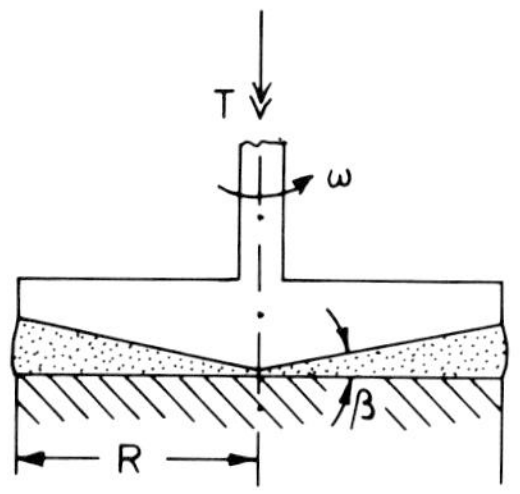

Figure 6-21. Cone and Plate Rheometer

Because of the conical geometry, the shear rate is uniform throughout the fluid. The torque necessary to rotate the cone is related to the shear stress by:

$$T = \frac{2}{3}\pi R^3 \tau \qquad (6\text{--}60)$$

The viscosity can now be determined from:

$$\eta = \frac{\tau}{\dot{\gamma}} = \frac{3T\beta}{2\pi R^3 \omega} \qquad (6\text{--}61)$$

Thus, the viscosity can be directly determined from measurement of torque and rotational speed. If the fluid between the cone and plate is viscoelastic, the plates will be pushed apart when the fluid is sheared. The force with which the plates are pushed apart F is related to the first normal stress difference N_1 in the fluid:

$$F = \frac{\pi R^2}{2} N_1 \qquad (6\text{--}62)$$

The first normal stress difference is an accurate indicator of the viscoelastic behavior of a fluid. Thus, with the cone and plate rheometer, one can accurately determine some viscoelastic characteristics of a fluid.

The cone and plate rheometer is susceptible to irregularities at the liquid-air interface and to secondary flows. As a result, the shear rate in steady shear measurements has to be quite low to avoid the above-mentioned problems. In general, the shear rate should not exceed $1 \sec^{-1}$; data above this rate should be regarded with caution.

Advantages of the cone and plate rheometer:

1. Its ease of operation.
2. It has uniform shear rate distribution.
3. It also measures first normal stress difference.
4. Its uniform temperature distribution.

Disadvantages of the cone and plate rheometer:

1. It is limited to low shear rates ($< 1 \sec^{-1}$).
2. These instruments tend to be expensive.

The disadvantage of the low shear rate limitation can be negated by applying an oscillatory rotary motion to the moving plate. Thus, one can measure complex viscosity as a function of frequency, see also Section 6.2.6.5 on dynamic analysis. In this fashion, frequency levels up to about 500 radians/second are possible; this corresponds to shear rates of about $500 \sec^{-1}$.

6.2.6.4   SLIT DIE RHEOMETER

A slit die rheometer is an extruder die with a rectangular flow channel with provisions to measure pressures at various axial locations. The slit die is either directly connected to an extruder or to a gear pump, which, in turn, is connected to an extruder. A typical slit die geometry is shown in Figure 6–22.

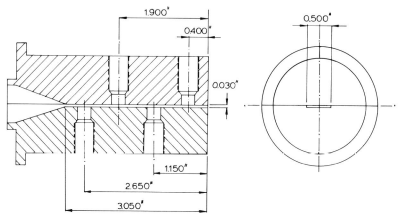

Figure 6-22. Typical Slit Die Geometry

By measuring the flow rate $\dot{V}$ through the flow channel, the apparent shear rate at the wall can be determined:

$$\dot{\gamma}_{aw} = \frac{6\dot{V}}{H^2W} \tag{6-63}$$

where H is the slit height and W the slit width ($W \gg H$).

For a power law fluid with power law index n, the actual shear rate at the wall is:

$$\dot{\gamma}_w = \frac{2n+1}{3n}\dot{\gamma}_{aw} \tag{6-64}$$

The shear stress at the wall can be determined from the gradient of the measured pressure profile (dP/dz):

$$\tau_w = -\frac{dP}{dz}\frac{H}{2} \tag{6-65}$$

The apparent viscosity can be determined from:

$$\eta_a = \frac{\tau_w}{\dot{\gamma}_{aw}} = \frac{H^3W}{12\dot{V}}\frac{dP}{dz} \tag{6-66}$$

For a power law fluid, the actual viscosity is:

$$\eta = \frac{3n}{2n+1}\eta_a = \frac{nH^3W}{4(2n+1)\dot{V}}\frac{dP}{dz} \tag{6-67}$$

These equations are valid provided the ratio of width to height is large; in most cases W/H > 10. If the pressure gradient is constant, the pressure can be extrapolated to the exit of the die. Several workers have found positive exit pressure by using this procedure (54). This exit pressure can be related to the first normal stress difference provided the velocity profile remains fully developed right up to the die exit:

$$N_1 \ (\dot{\gamma}_w) \ = \ P_{exit} \ + \ \tau_w \left(\frac{dP_{exit}}{d\ \tau_w} \right) \tag{6-68}$$

However, the theoretical justification of this relationship seems to be questionable, as discussed by Boger and Denn (67). From a practical point of view, it is very difficult to extrapolate from pressure readings that range from several MPa up to about 30 or 40 MPa down to exit pressures that range from less than 0.1 MPa to about 0.2 MPa. This is particularly true if only two pressure transducers are situated in the fully developed region, as appears to have been the case in several experimental results reported in the literature (68–70). Another problem is the assumption of a constant pressure gradient. It has been shown (71) that the effects of temperature and pressure will generally cause a significant non-linearity. This raises serious doubts about the validity of a linear extrapolation of the pressure profile.

An interesting aspect of the slit die rheometer is the fact that the polymer has a significant temperature and shear history by the time it reaches the slit die. This can affect the rheological properties, as reported by the author (71). On the other hand, if viscosity data is to be used for die design purposes, the slit die viscometer is most likely to produce pertinent viscosity data.

Advantages of the slit die rheometer:

1. It operates in useful shear rate range.
2. It yields accurate shear viscosity versus shear rate data.
3. It yields data representative of behavior during actual extrusion.

Disadvantages of the slit die rheometer:

1. Commercial rheometers tend to be expensive.
2. Proper and frequent calibration of pressure transducers is important to ensure accurate results.
3. It is difficult to determine data that can be related to the viscoelastic behavior of the polymer.

6.2.6.5   DYNAMIC ANALYSIS

Several rheometers do not subject the polymer to a steady rate of deformation but to an oscillatory deformation, usually sinusoidal simple shear. If the angular frequency is ω, and the shear strain amplitude γ_0, the shear strain γ can be written as a function of time:

$$\gamma(t) \ = \ \gamma_0 \sin(\omega t) \tag{6-69}$$

The shear rate is determined by differentiating the shear strain with respect to time. The shear rate is:

$$\dot{\gamma}(t) \ = \ \gamma_0 \omega \cos(\omega t) \ = \ \dot{\gamma}_0 \cos(\omega t) \tag{6-70}$$

Dynamic analysis is generally used to study the linear viscoelastic properties of polymers. The region of linear viscoelastic behavior is where a material function, such as shear modulus or shear viscosity, is independent of the amplitude of the strain or strain rate. Polymers follow linear viscoelastic behavior when the strain or strain rate is sufficiently small. Thus, if the strain amplitude is sufficiently small, the shear stress can be written as:

$$\tau(t) = \tau_0 \sin(\omega t + \delta) \qquad (6-71)$$

where τ_0 is the amplitude of the shear stress and δ the phase angle between stress and strain. This phase angle is often referred to as loss angle. For a purely elastic material, there is no phase shift between stress and strain, thus δ is zero. For a purely viscous material, there will be a maximum phase shift between stress and strain, or δ is 90°.

Equation 6-71 can be rewritten by introducing an in-phase modulus G′ (real) and a 90° out-of-phase modulus G″ (imaginary):

$$\tau(t) = \gamma_0 [G' \sin(\omega t) + G'' \cos(\omega t)] \qquad (6-72)$$

The storage modulus G′ represents the elastic contribution associated with energy storage; it is a function of the stress and strain amplitude and the phase angle:

$$G' = \frac{\tau_0}{\gamma_0} \cos \delta \qquad (6-73)$$

Similarly, the loss modulus G″ represents the viscous contribution associated with energy dissipation; it is:

$$G'' = \frac{\tau_0}{\gamma_0} \sin \delta \qquad (6-74)$$

Both G′ and G″ are components of the complex modulus G*. The magnitude of the complex modulus, as shown in Figure 6-23, is:

$$|G^*| = [(G')^2 + (G'')^2]^{1/2} \qquad (6-75)$$

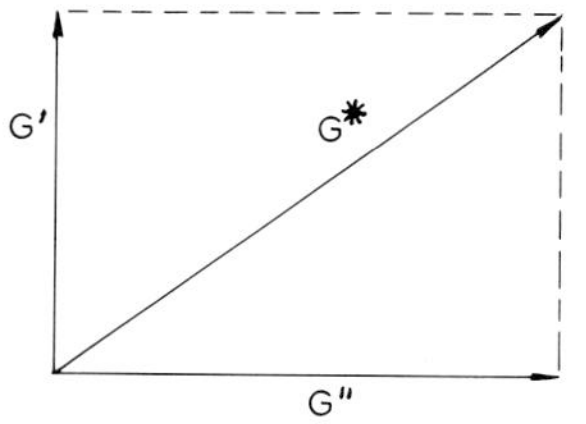

Figure 6-23. Components of the Complex Modulus

A complex notation is frequently used to describe the relationship between stress and strain. If stress and strain are written as:

$$\gamma = \gamma_0 \exp(i\omega t) \qquad (6-76a)$$

$$\tau = \tau_0 \exp[i(\omega t + \delta)] \qquad (6-76b)$$

then the complex modulus can be written as:

$$G^* = \frac{\tau}{\gamma} = \frac{\tau_0}{\gamma_0}(\cos\delta + i\sin\delta) = G' + iG'' \tag{6-77}$$

Equation 6-71 can be rewritten in yet a different form by introducing an in-phase viscosity η' and an out-of-phase viscosity η''. The real part η', the dynamic viscosity, represents the viscous contribution associated with energy dissipation. The imaginary part η'' represents the elastic contribution associated with energy storage. The shear stress can be written as a function of η' and η'' as follows:

$$\tau(t) = \dot{\gamma}_0[\eta'\cos(\omega t) + \eta''\sin(\omega t)] \tag{6-78}$$

The dynamic viscosity is related to the loss modulus by:

$$\eta' = \frac{G''}{\omega} = \frac{\tau_0}{\dot{\gamma}_0}\sin\delta \tag{6-79}$$

The imaginary part of the viscosity is related to the storage modulus by:

$$\eta'' = \frac{G'}{\omega} = \frac{\tau_0}{\dot{\gamma}_0}\cos\delta \tag{6-80}$$

Both η' and η'' are components of the complex viscosity η^*. The magnitude of the complex viscosity is:

$$|\eta^*| = [(\eta')^2 + (\eta'')^2]^{1/2} \tag{6-81}$$

In complex notation, the complex viscosity can be expressed as a function of η' and η'':

$$\eta^* = \frac{\tau}{\dot{\gamma}} = \frac{\tau_0}{\dot{\gamma}_0}(\sin\delta - i\cos\delta) = \eta' - i\eta'' \tag{6-82}$$

The tangent of the phase angle is often used in characterization of viscoelastic material. The "tan delta" can be determined from:

$$\tan\delta = \frac{G''}{G'} = \frac{\eta'}{\eta''} \tag{6-83}$$

The attractiveness of dynamic analysis is that an accurate determination of the viscoelastic behavior can be made. A common geometry for dynamic measurements is the cone and plate rheometer. In dynamic analysis, the viscosity components can be measured up to an angular frequency of about 500 radians/sec. Cox and Merz (72) found empirically that the steady shear viscosity corresponds to the complex viscosity if the shear rate in sec^{-1} is plotted on the same scale as the angular frequency in radians/sec. This can be stated as:

$$| \eta^*(\omega) | = \eta(\dot\gamma) |_{\dot\gamma=\omega}$$ (6-84)

This empirical rule seems to hold up quite well for most polymers. Using this rule, it is possible to determine viscosity data up to 500 sec^{-1} with a cone and plate rheometer by applying an oscillatory motion to the cone. This would be impossible if a steady rotational motion was applied to the cone. In steady shear measurements on a cone and plate rheometer, the maximum shear rate that can be measured is around 1 sec^{-1}, which is much too low for applications to extrusion problems. The same is true for measurements in the parallel plate test geometry. Thus, the dynamic measurement extends the shear rate measurement range considerably, while still being able to take advantage of the cone and plate geometry.

Dynamic mechanical analysis is not limited to just shear deformation, it is also used with elongational deformation. Further, dynamic mechanical analysis is employed in the characterization of solids as well as liquids. In 1982, a new standard was established, ASTM D4065, to standardize procedures for testing all types of materials.

6.3   THERMAL PROPERTIES

By the nature of the plasticating extrusion process, thermal properties are very important. In the early portion of the extruder, solid polymer particles are heated to the melting point. In the mid-portion of the extruder, the molten polymer is raised in temperature to a level considerably above the melting point while the remaining solid particles continue to heat up and melt. In the last portion of the extruder, the molten polymer has to reach a thermally homogeneous state. When the extrudate leaves the extruder die, it has to be cooled down, usually to room temperature. Through this whole process, the polymer experiences a complicated thermal history. The thermal properties of the polymer are crucial to being able to describe and analyze the entire extrusion process.

6.3.1   THERMAL CONDUCTIVITY

The thermal conductivity of a material is essentially a proportionality constant between the conductive heat flux and the temperature gradient driving the heat flux. The thermal conductivity of polymers is quite low, about two to three orders of magnitude lower than most metals. From a processing point of view, the low thermal conductivity creates some real problems. It very much limits the rate at which polymers can be heated up and plasticated. In cooling, the low thermal conductivity can cause non-uniform cooling and shrinking. This can result in frozen-in stresses, deformation of the extrudate, delamination, shrink voids, etc.

The thermal conductivity of amorphous polymers is relatively insensitive to temperature. Below the T_g, the thermal conductivity increases slightly with temperature; above the T_g, it reduces slowly with temperature. The thermal conductivity above the T_g as a function of temperature can be approximated by (41):

$$k(T) = \frac{6T_g-T}{5T_g} \, k(T_g) \qquad\qquad (6\text{--}85)$$

where the temperature is expressed in $^\circ$ K.

In most extrusion problems, however, the thermal conductivity of an amorphous polymer can be assumed to be independent of temperature.

The thermal conductivity of semi-crystalline polymers is generally higher than amorphous polymers. Below the crystalline melting point, the thermal conductivity reduces with temperature; above the melting point, it remains relatively constant. The thermal conductivity increases with density and, thus, with the level of crystallinity. The thermal conductivity at constant temperature as a function of density can generally be written as:

$$k(\rho) = k_r + C_\rho(\rho - \rho_r) \qquad\qquad (6\text{--}86)$$

The change in thermal conductivity with temperature is relatively linear at temperatures above 0° C. The thermal conductivity as a function of temperature can be described with:

$$k(T) = k_0 - C_T T \qquad\qquad (6\text{--}87)$$

where T is in $^\circ$C and $k_0=k(T=0)$.

The temperature coefficient of the thermal conductivity C_T for a particular polymer seems to be relatively independent of the actual density. Thus, the combined density and temperature dependence can be described by:

$$k\ (\rho,T) = [k_r+C_\rho(\rho-\rho_r)]_{T_0} - C_T T \qquad\qquad (6\text{--}88)$$

For polyethylene, the thermal conductivity as a function of density and temperature can be described by:

$$k\ (\rho,T) \approx 0.17+5(\rho-0.90)-0.001T \quad [J/ms^\circ C] \qquad\qquad (6\text{--}89)$$

where T is in $^\circ$C and ρ in gr/cc.

Expressions 6-87 through 6-89 become less accurate as the temperature approaches the melting point because non-linearities become significant. Therefore, these expressions should be used as approximations. Above the melting point, the thermal conductivity of polyethylene is about 0.25 J/ms$^\circ$ C.

The thermal conductivity is dependent on the orientation of the polymer. If the polymer is highly oriented, substantial differences in thermal conductivity can occur in the direction of orientation and perpendicular to it. The difference can be as high as almost 100 percent in PMMA as shown by Eiermann and Hellwege (73). Hansen and Bernier (74) found differences in thermal conductivity as much as 20 $\times$ in HDPE.

Compacted polymer particles have a lower thermal conductivity because of the presence of voids between the particles. Based on experimental data, Yagi and Kunii (75) proposed a model for the thermal conductivity of the bed. For fine particles and low temperatures, the thermal conductivity of the bed can be written as:

$$k_b = \frac{k_g{}^{\rho_b/\rho_p}}{F+k_g/k_p} \qquad\qquad (6\text{--}90)$$

where k_g is the thermal conductivity of the gas occupying the voids, k_p the thermal conductivity of the polymer particles, ρ_b the bulk density, ρ_p the density of the polymer particles, and F a function of the density ratio ρ_b/ρ_p (a power law relationship). Langecker (96, 97) performed extensive measurements of the thermal conductivity of isotropically compressed polymeric powders. He found an essentially linear relationship between the thermal conductivity and the bulk density. In the range of ρ_b/ρ_p from 0.5 to 1.0, the data on polyethylene can be approximated by:

$$k_b = \frac{1}{3} k_p \ (5\rho_b/\rho_p - 2) \qquad\qquad (6-91)$$

The dependence of thermal conductivity and diffusivity of polyethylene on temperature, density and molecular parameters was investigated by Kamal, Tan and Kashani (93). Their publication contains a good review of prior experimental work on thermal conductivity.

6.3.2   SPECIFIC VOLUME AND MORPHOLOGY

The polymer density ρ is a function of pressure, temperature and cooling rate. Specific volume $\hat{V}$ is the reciprocal of density, $\hat{V}=1/\rho$. The general P, $\hat{V}$, T diagram of an amorphous polymer is shown in Figure 6-24.

If the material is cooled very slowly, the specific volume will reach a lower value than at a relatively high cooling rate. In simple terms, at a low cooling rate the polymer molecules, because of their thermal motion, have more opportunity to position themselves closer together. This reduces the free volume of the polymer; i.e., the volume fraction not occupied by polymer molecules. Below the glass transition temperature, the thermal motion of the

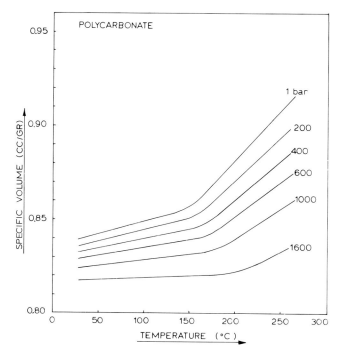

Figure 6-24. P, $\hat{V}$, T Diagram of an Amorphous Polymer

polymer molecules is drastically reduced and the free volume remains approximately constant. Therefore, the change in specific volume with temperature is much larger above T_g than below T_g. The reduction in specific volume below T_g is primarily due to the reduced thermal motion of the polymer molecules.

Increasing the molecular weight increases the glass transition temperature as shown in the Fox and Flory equation:

$$Tg = Tg\infty - \frac{K}{\bar{M}_n} \qquad\qquad (6-91a)$$

where $Tg\infty$ is the glass transition temperature for infinite molecular weight and constant K is a parameter of the polymer.

When the $\hat{V}$,T curve is determined at a higher pressure, the specific volume will reduce. The reduction above T_g will be more significant than below T_g because of the larger free volume above T_g. Another interesting phenomenon is the shift of the T_g to a higher temperature when the pressure is increased. In the normal range of processing pressures (P < 100 MPa), the pressure dependence of T_g can generally be neglected.

When the cooling rate is fast, the material goes through the transition region at a higher temperature and has a larger specific volume at temperatures below this region. This is because the polymer molecules have not had sufficient time to position themselves in a preferred configuration. As the polymer relaxes, it will reduce in specific volume until it eventually reaches the specific volume corresponding to a low cooling rate.

A general P, $\hat{V}$, T diagram of a semi-crystalline polymer is shown in Figure 6–25.

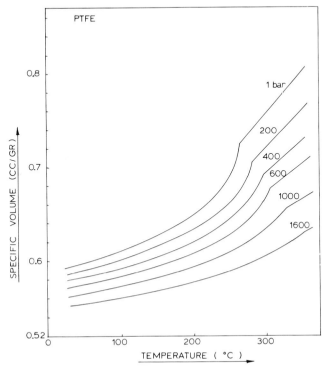

Figure 6-25. P, $\hat{V}$, T Diagram of a Semi-Crystalline Polymer

The behavior in the liquid state is essentially the same as amorphous polymers. In the transition region, an abrupt change in slope occurs as crystallization begins to take place. This is the crystallization temperature T_c. If the material is cooled very rapidly, the crystallization rate can be depressed, depending on the crystallization kinetics. In fact, in some materials with sufficiently slow crystallization kinetics, the crystallization can be almost completely suppressed by rapid cooling. A well-known example is polyethylene terephthalate (PET). If PET is rapidly quenched, it is almost completely amorphous with a density of about 1.33 gr/cc. If it is cooled slowly, it will crystallize with a resulting higher density of about 1.40 gr/cc.

The percent crystallinity of polymers with rapid crystallization kinetics is not much affected by the cooling rate; however, the crystallite morphology may be strongly affected. These differences in morphology can cause significant changes in physical properties. As the extrudate is cooled, the skin will cool most rapidly and the interior region of the extrudate will cool more slowly. This will cause corresponding changes in polymer morphology. Annealing can also modify the polymer morphology, particularly the crystalline regions. Annealing is the process of exposing the polymer to an elevated temperature for a certain period of time. This is sometimes done as a post-extrusion operation to control the polymer morphology and physical properties. In polymers with a high level of crystallinity, annealing causes thickening of the lamellae and an increase of the melting point. The relationship between crystallite size and melting point is given by Hoffman and Lauritzen's equation (81):

$$T_m = T_m^0 \left(1 - \frac{2\sigma_e}{\Delta H_f L} \right) \tag{6-92}$$

where σ_e is the surface free energy, T_m^0 the equilibrium melting temperature, ΔH_f the heat of fusion, and L the lamella thickness.

The application of stress can further cause significant changes in the polymer morphology, see also Section 5.2.2. In flow-induced crystallization of dilute polymer solutions, a shish kebab crystal morphology develops. In melt crystallized polymers, a spherulitic crystal morphology develops, made up of folded chain lamellae. Deformation of the polymer below the melting point can cause very effective orientation of the polymer molecules. This can result in a high degree of anisotropy in the material with very good mechanical properties in the orientation direction. Examples are solid state extrusion and fiber drawing; many other examples are available.

The change in specific volume below T_c is primarily due to the increasing degree of crystallinity. Therefore, the volume changes in semi-crystalline polymers are considerably larger than those experienced with amorphous polymers. When the semi-crystalline polymer is cooled from the melt at elevated pressure, the T_c shifts to a higher temperature. This phenomenon can be described by the Clausius-Clapeyron equation which relates the equilibrium melting point at any pressure to the melting point at atmospheric pressure:

$$T_m(P) = T_m(P_{atm}) \exp\left[\frac{(\hat{V}_a - \hat{V}_c)(P-1)}{\Delta H_f} \right] \tag{6-93}$$

where $\hat{V}_a$ and $\hat{V}_c$ are the amorphous and crystalline specific volumes, P is the hydrostatic pressure in atmospheres, and ΔH_f is the heat of fusion of the polymer at atmospheric pressure. The melting point elevation can be significant in polymer melts; i.e., for polyethylene about $7°C$ per $20\,MPa$ (≈ 3000 psi). A polymer melt under high pressure is thus under a higher degree of supercooling. This will affect the crystallization behavior of the material. This effect will be most noticeable in processes where high pressure commonly occurs, such

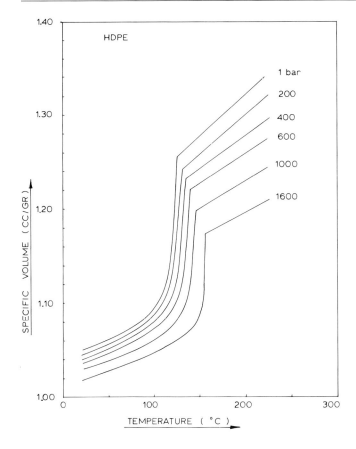

Figure 6-26. P, V̂, T Diagram of HDPE

as injection molding. As mentioned earlier and also in Section 5.2.2, this effect can be further magnified if the polymer melt is under stress. The melting point elevation with pressure explains the large compressibility values for polymer melts found by some investigators (76). Actually, the sudden increase of compressibility at elevated pressures indicates the onset of crystallization.

The P,V̂,T diagram of HDPE is shown in Figure 6-26.

Various workers have developed empirical relationships between V̂, P, and T. Among those are the equations of Spencer and Gilmore (77), Breuer and Rehage (78), Kamal and Levan (79), and Simha and Olabisi (80).

6.3.3   SPECIFIC HEAT AND HEAT OF FUSION

The specific heat is the amount of energy required to raise a unit mass of a material one degree in temperature. In S.I. units, specific heat is expressed in J/kg° K. It can be measured either at constant pressure C_p or at constant volume C_v. The two values are related by:

$$C_p = C_v + T\hat{V}\,\frac{\alpha_v^2}{\kappa} \qquad (6-94a)$$

where the expansion coefficient α_V is:

$$\alpha_V = \frac{1}{\hat{V}} \left(\frac{\partial \hat{V}}{\partial T} \right)_p$$

(6–94b)

and the compressibility:

$$\kappa = - \frac{1}{\hat{V}} \left(\frac{\partial \hat{V}}{\partial P} \right)_T$$

(6–94c)

The specific heat at constant pressure is larger than the specific heat at constant volume because additional energy is required to bring about the volume change against external pressure P.

The specific heat of amorphous polymers increases with temperature in approximately a linear fashion below and above T_g. A steplike change occurs around the glass transition temperature as shown in Figure 6–27a. With semi-crystalline polymers, the step change at T_g is much less pronounced; however, a very distinct maximum occurs at the crystalline melting point. At the melting point, the specific heat is theoretically infinite for a material with a perfectly uniform crystalline structure, as shown in Figure 6–27b. Since this is not the case in semi-crystalline polymers, these materials exhibit a melting peak of certain width as shown in Figure 6–27c. The narrower the peak, the more uniform the crystallite morphology.

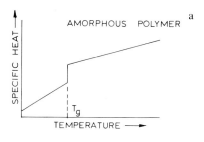

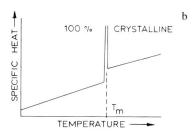

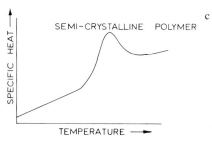

Figure 6-27. Specific Heat as a Function of Temperature

The specific heat above the melting point increases slowly with temperature. The area under the melting peak of the C_p,T curve equals the heat of fusion ΔH_f multiplied with the crystalline weight fraction. Both the heat of fusion and the percent crystallinity are dependent on the thermo-mechanical history of the polymer, as discussed in the previous section.

6.3.4   SPECIFIC ENTHALPY

A very useful thermal property in polymer processing is the specific enthalpy. It is defined
by:

$$\hat{H} = \int_{T_1}^{T_2} C_p(T)\,dT \qquad\qquad (6\text{--}95)$$

If T_1 is taken as ambient temperature and T_2 as the process temperature, the specific enthalpy indicates how much energy is required to accomplish this temperature rise. This can
be considered to be the theoretical minimum specific energy requirement in the extrusion
process.

Figure 6-28 shows $\hat{H}$, T curves for several amorphous and semi-crystalline polymers.

The first observation is that amorphous polymers have a continuous rise in $\hat{H}$, while semi-
crystalline polymers exhibit an abrupt change in slope at the melting point. The second observation is that amorphous polymers generally have much lower $\hat{H}$ values than semi-crys-

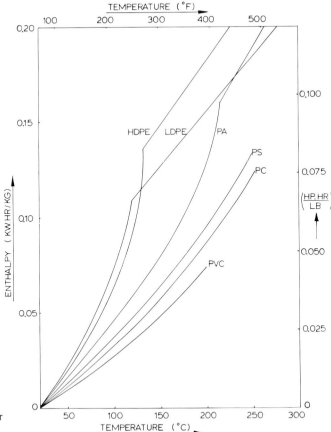

Figure 6-28. $\hat{H}$, T Curves for
Several Polymers

talline polymers over the same ΔT. If one compares PVC to LDPE from $T_1 = 20°C$ to $T_2 = 150°C$, the $\hat{H}(PVC) \simeq 0.05$ kWhr/kg (0.03 hphr/lb), while $\hat{H}$ (LDPE) $\simeq 0.13$ kWhr/kg (0.08 hphr/lb). Thus, if the throughput of the extrusion process is 100 kg/hr and the process temperature 150°C, the theoretical power requirement is 5 kW for PVC and 13 kW for LDPE.

Thus, based on the thermal properties of the polymers, there is a very significant difference in power requirement between PVC and LDPE; about a 3 : 1 ratio! A standard 24 L/D extruder that runs fine with LDPE is likely to have too high a power consumption for PVC. This will result in high stock temperatures and increased chance of degradation, particularly in the case of PVC, and especially with rigid PVC. The thermal properties of PVC dictate an extrusion process with low specific energy consumption and short residence times to minimize high melt temperatures and high temperature exposure time. This is the main reason that closely intermeshing twin screw extruders have become so popular for PVC. These machines have low specific energy consumption, short residence times, narrow residence times distribution, and good control over stock temperationes. Long single screw extruders are generally not well suited for PVC extrusion. As a result, a higher level of stabilizers has to be added to the polymer to enable it to survive the extrusion process without too much degradation. However, this increases the compound cost and may be more expensive in the long run than processing the material on a more suitable machine.

6.3.5   THERMAL DIFFUSIVITY

The thermal diffusivity is a property derived from thermal conductivity, specific heat and density. The relationship is:

$$\alpha = \frac{k}{\rho C_p} \tag{6-96}$$

The thermal diffusivity is a very useful quantity in transient heat transfer problems, as discussed in Section 5.3.1.

The thermal diffusivity can be calculated from the values of k, ρ, and C_p; however, in most cases it is measured directly. In fact, the thermal diffusivity can be measured more easily and accurately than the thermal conductivity. If a thick slab of material, initially at T_0, is suddenly exposed to an elevated temperature T_1 at one wall and maintained at this temperature, the temperature distribution can be described by:

$$\frac{\partial T}{\partial t} = \alpha \frac{\partial^2 T}{\partial y^2} \tag{6-97}$$

if it is assumed that α is independent of temperature.

Equation 6-97 is a shortened version of the general energy balance equation (equation 5-5) valid for simple unidirectional conduction. This is a standard handbook problem; the solution is (see reference 1 of Chapter 5):

$$\frac{T(y,t)-T_0}{T_1-T_0} = 1 - erf\left(\frac{y}{\sqrt{4\alpha t}}\right) \tag{6-98}$$

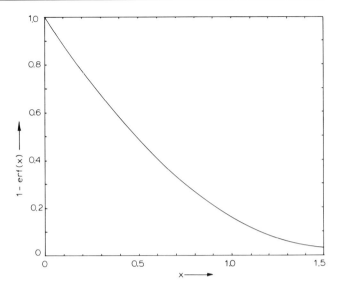

Figure 6-29. Complimentary Error Function

In equation 6 98 erf(x) stands for error function; this is defined as:

$$\mathrm{erf}(x) = \frac{2}{\sqrt{\pi}} \int_0^X \exp(-t^2)\,dt \qquad (6\text{-}99)$$

This integral cannot be solved easily; however, tabular or graphical representations of the error function are available in various handbooks (e.g. 82). Figure 6-29 shows the function 1-erf(x), the complimentary error function.

Equation 6-98 describes conductive heating of a semi-infinite slab. It is valid as long as the thermal penetration thickness δ_t is less than the slab thickness. Essentially all temperature change (99 percent) takes place within the thermal penetration thickness, which is given by:

$$\delta_t = 4\sqrt{\alpha t} \qquad (6\text{-}99\,a)$$

The time, temperature profiles in a slab with double-sided heating are given by equation 7-96.

With equations 6-98 and 6-99, the temperature as a function of time and distance is completely described. Thus, by doing simple temperature measurements on a slab of material with one wall suddenly exposed to a higher temperature, the thermal diffusivity can be determined in a relatively straightforward fashion.

This example illustrates how the thermal diffusivity can be measured and how it is used to describe heat conduction problems. For amorphous polymers and polymer melts, diffusivity is approximately linear with the velocity of sound v_s; the proportionality constant is about 6 E-13 (41):

$$\alpha = C v_s \qquad (6\text{-}100)$$

The sound velocity is related to the molar sound velocity function F_R or Rao function, the molar volume per structural unit V_m, and the Poisson ratio v. It can be written as:

$$v_S = \left(\frac{F_R}{V_m}\right)^3 \left[\frac{3(1-v)}{1+v}\right]^{\frac{1}{2}} \qquad (6\text{-}101)$$

The Poisson ratio for liquids is 1/2 and for isotropic solids 1/3. The Poisson ratio for polymers below T_g is approximately 1/3 and above T_g about 1/2. The ratio of the Rao function and the molar volume is approximately constant for many amorphous polymers (41); the ratio is about 55. Thus, the thermal diffusivity can be approximated with:

$$\alpha = \left[\frac{3(1-v)}{1+v}\right]^{\frac{1}{2}} 10^{-7} \qquad (6\text{-}102)$$

with α expressed in m^2/s.

The ratio F_R/V_m is related to the compressibility κ and density ρ by:

$$\rho\kappa = \left[\frac{F_R}{V_m}\right]^{-6} \qquad (6\text{-}103)$$

With equation 6-103, one can write the expression for the sound velocity as follows:

$$v_S = \left[\frac{3(1-v)}{\rho\kappa(1+v)}\right]^{\frac{1}{2}} \qquad (6\text{-}104)$$

The thermal diffusivity of most polymers is around 10^{-7} m^2/s. In the analysis of most extrusion problems, the thermal diffusivity is considered to be constant. In reality, however, the thermal diffusivity depends on pressure, temperature and orientation. The anisotropy of the thermal diffusivity of uniaxially stretched polyethylene was studied by Kilian and Pietralla (83). They found large differences between the thermal diffusivity in the orientation direction and perpendicular to the orientation direction, as high as 20 : 1. The pressure dependence of the thermal conductivity, thermal diffusivity and specific heat of some polymers was studied by Andersson and Sundqvist (84). They found that the thermal conductivity and thermal diffusivity increase with pressure. At very high pressures ($\simeq$ 4000 MPa), the thermal conductivity and thermal diffusivity about double. However, at pressures within the range of normal polymer processing, pressures less than 100 MPa ($\simeq$ 15,000 psi), the changes are less than 5 percent. At 30 MPa ($\simeq$ 5000 psi), the expected change in thermal conductivity and thermal diffusivity is about 1 to 2 percent; this will be negligible in most cases. The specific heat reduces with increasing pressure; however, in the polymer processing range, the changes are quite small – less than 0.5 percent at 100 MPa ($\simeq$ 15,000 psi).

6.3.6   MELTING POINT

The melting point is the temperature at which the crystallites melt. Since the crystallites are not perfectly uniform, there is really not one single melting point but a melt temperature range. The melting point is often taken as the temperature at the peak of a DSC curve, see Section 6.3.8.

The melting point is dependent on the pressure and crystallite morphology as discussed in Section 6.3.3. It can be measured quite easily and accurately. The melting point dictates, to some extent, the process temperatures necessary in extrusion. As a general rule, the process temperatures are about $50°C$ above the melting point. If the process temperature is too close to the melting point, the polymer melt viscosity will be too high, resulting in excessive power consumption. If the process temperature is too far above the melting point, the polymer may degrade.

For homopolymers, the melting temperature depends on the molecular weight of the polymer:

$$\frac{1}{T_m} - \frac{1}{T_{m_\infty}} = \frac{2\,RM_0}{\Delta H M_n} \qquad\qquad (6\text{--}105)$$

where T_{m_∞} is the melting temperature for an infinite length polymer molecule and M_0 is the molecular weight of the monomer.

6.3.7   INDUCTION TIME

The induction time of a polymer is a very useful quantity in process design, process optimization and troubleshooting. It represents the amount of time elapsed at a certain temperature and in a certain atmosphere before the effects of degradation become measurable. Essentially, the induction time indicates how long a polymer can be exposed to a certain temperature before it starts to degrade.

In extrusion, one would like to make sure that the longest residence time in the machine at a certain process temperature is less than the induction time at the same temperature. Thus, if one knows the residence times to be expected in the extrusion process and if one knows how the induction time varies with temperature, the process temperature at which degradation will be avoided can be accurately determined. This is a very useful tool in process engineering, particularly if one deals with a polymer of limited thermal stability.

If degradation occurs during the extrusion process, there are two approaches that one can take to the problem. One is to modify the process so as to reduce the chance of degradation. The other approach is to modify the polymer to improve its thermal stability. The changes to the process should result in lower stock temperatures, and/or reduced exposure time to elevated temperatures, and possibly exclusion of degradation promoting substances (e.g., oxygen, certain additives, certain metal components of the tooling or substrate, etc.). If the changes to the process cannot alleviate the degradation problem, one has to consider the polymer itself. The thermal stability (induction time) of most polymers can be improved by adding stabilizers to the polymer, such as antioxidants. If process changes cannot solve the problem of degradation, the thermal stability of the polymer should be improved by adding stabilizers to it. Of course, one may opt to select a different polymer altogether.

In some cases, the thermal stability of a polymer or a compound is so poor that it cannot be extruded without degradation under any conditions. This can be conclusively determined if induction time data are available. The process engineer can then go back to the polymer chemist and explain exactly what changes should be made to the polymer. This procedure eliminates the question of whether the problem is caused by the polymer or by the process. Thus, induction time data can act as a bridge between the process engineer and polymer chemist and allow them to communicate and cooperate in a useful fashion.

It is clear that the induction time is a strong function of temperature. For many polymers, a plot of induction time against temperature will form approximately a straight line on semi-log paper, as shown in Figure 6-30.

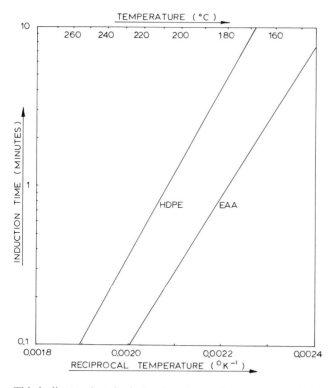

Figure 6-30. Induction
Time Versus Temperature

This indicates that the induction time reduces exponentially with temperature. Curve A in figure 6–30 is a HDPE as received from the manufacturer and curve B is a ethylene acrylic acid (EAA) as received from the manufacturer. The HDPE can be processed at 200° C without noticeable degradation; however, EAA shows clear signs of degradation at that temperature. It should be processed at about 160° C to avoid degradation.

The relationship between induction time t_{ind} and temperature T can generally be expressed as an Arrhenius equation:

$$t_{ind} = A \exp\left(\frac{E}{RT}\right) \qquad (6-106)$$

where A is a time constant, E the activation energy, and R the universal gas constant.

The induction time can be conveniently determined on a TGA, but other instruments can be used as well, for instance a DSC. Thermal characterization will be discussed in the next section.

6.3.8   THERMAL CHARACTERIZATION

In thermal characterization, a controlled amount of heat is applied to a sample and its effect measured and recorded. In isothermal operations, the effect is recorded as a function of time at constant temperature. In a programmed temperature operation, the temperature is changed in a predetermined fashion, e.g., at a certain rate, and the effect is recorded as a function of temperature. General texts on thermal characterization have been written by Wendlandt (85), Daniels (86), and Turi (87).

6.3.8.1   DTA AND DSC

Differential thermal analysis (DTA) and differential scanning calorimetry (DSC) are similar techniques. They measure change in the heat capacity of a sample. These techniques can be used to determine various transition temperatures (T_m, T_g, T_α, T_β, etc.), specific heat, heat of fusion, percent crystallinity, onset of degradation temperature, induction time, reaction rate, crystallization rate, etc. A DSC instrument operates by compensating electrically for a change in sample heat. The power for heating is controlled in such a way that the temperature of the sample and the reference is the same. The vertical axis shows the heat flow in cal/sec.

A DTA instrument operates by measuring the change in sample temperature with respect to an inert reference. Newer DTA instruments with externally mounted thermocouple and reproducible heat path have a precision comparable to the DSC. Older DTA instruments with the thermocouple placed in the sample were less accurate and reproducible.

6.3.8.2   TGA

A thermogravimetric analyzer measures the change in weight of a sample due to volatilization, reaction, or absorption from the gas phase. With polymers, the TGA is used to measure the amount and loss of moisture or diluent, and rates and temperatures of reactions. It is a convenient instrument to determine the polymer induction time, as discussed in Section 6.3.7. Sample size is usually less than one gram, thus the amount of polymer required for characterization is minimal.

6.3.8.3   TMA

In thermomechanical analysis (TMA), the change in mechanical properties is measured as a function of temperature and/or time. A probe in contact with the sample moves as the sample undergoes dimensional changes. The movement of the probe is measured with an LVDT. The sample deformations that can be measured are compression, penetration, extension, and flexure or bending.

6.3.8.4   OTHER THERMAL CHARACTERIZATION TECHNIQUES

While TMA refers to a measurement of a static mechanical property, there are also techniques that employ dynamic measurement. In the torsional braid analysis (TBA), a sample is subjected to free torsional oscillation. The natural frequency and the decay of oscillations are measured. This provides information about the viscoelastic behavior of materials. However, these measurements are elaborate and time consuming. In dynamic mechanical analysis (DMA), a sample is exposed to forced oscillations. A large number of useful properties can be measured by this technique, see also Section 6.2.6.5.

In thermal optical analysis (TOA), the conversion of plane polarized light to elliptically polarized light is measured in semi-crystalline polymers. The intensity of the depolarized light transmitted through a sample is a function of the level of crystallinity. Melting and recrystallization phenomena can be analyzed; the technique does not appear to be sensitive to glass transitions (88). The TOA technique is also referred to as thermal depolarization analysis (TDA) and depolarized light intensity method (DLI).

TABLE 6-2. USEFUL PROPERTIES OF A NUMBER OF GENERIC POLYMERS

	Thermal Cond. k [J/ms°C]	Specific Heat Cp [kJ/kg°C]	Density ρ [gr/cc]	Glass Trans. Tg [°C]	Melting Point Tm [°C]	Power Law Index n [-]	η–T Dependence $\frac{1}{\eta}\frac{\partial\eta}{\partial T}$ [°C^{-1}]
PS	0.12	1.20	1.06	101	–	0.30	0.08
PVC	0.21	1.10	1.40	80	–	0.30	0.20
PMMA	0.20	1.45	1.18	105	–	0.25	0.20
SAN	0.12	1.40	1.08	115	–	0.30	0.20
ABS	0.25	1.40	1.02	115	–	0.25	0.20
PC	0.19	1.40	1.20	150	–	0.70	0.05
LDPE	0.24 (0.34 s)*	2.30	0.92	-120/-90	120	0.35	0.03
LLDPE	0.24 (0.35 s)	2.30	0.92	-120/-90	125	0.60	0.02
HDPE	0.25 (0.49 s)	2.25	0.95	-120/-90	130	0.50	0.02
PP	0.15	2.10	0.91	-10	175	0.35	0.02
PA–6	0.25	2.15	1.13	50	225	0.70	0.02
PA–6.6	0.24	2.15	1.14	55	265	0.75	0.03
PET	0.29	1.55	1.35	70	275	0.60	0.03
PBT	0.21	1.25	1.35	45	250	0.60	0.03
PVF2	0.16	1.38	1.76	-40	170	0.38	0.03
FEP	0.20	1.18	2.15	70	275	0.60	0.04

* Data without parenthesis refers to thermal conductivity of polymer melt, data within parenthesis refers to thermal conductivity of solid polymer.

6.4   POLYMER PROPERTY SUMMARY

In Table 6-1, a number of rheological and thermal properties have been tabulated for several important generic polymers. These data have been gathered from numerous sources, including the author's own measurements. The data should be used as estimates only, because measurement techniques may differ and because considerable differences in properties can occur in one particular polymer as a result of variations in molecular weight distribution, additives, thermo-mechanical history, etc. Actual measurement of polymer properties should always be preferred above published data. However, actual measurement is not always possible, in this case the table may provide useful information.

Finally, some useful references should be mentioned containing data on polymer properties. Nielson's book (89) on polymer properties is an exhaustive survey of a large number physical properties. The book by van Krevelen (41) is an excellent book on polymer properties and their relationship to chemical structure. The VDMA series on properties for polymer processing (90,91,95) contains a large amount of data on thermal properties (90), melt flow properties (91), and frictional properties (95).

Other useful data can be found in the yearly issues of the International Plastics Selector Books, the yearly Modern Plastics Encyclopedia, Plastics Technology Manufacturing Handbook and Buyers' Guide, etc.

REFERENCES - CHAPTER 6

1. R.C. Wahl, Plastics World, 64–67, Nov. (1978).
2. J. R. Mitchell, Chem. Eng. (U. K.), 177–183, Feb. (1977).
3. D. Train and C.J. Lewis, 3 rd Congress European Federation of Chemical Engineers, London, June 20–29 (1962).
4. A. W. Jenike, P.J. Elsey, and R. H. Woolley, Proc. Am. Soc. Test. Mater., 60, 1168 (1960).
5. R. S. Spencer, G. D. Gilmore, and R. M. Wiley, J. Appl. Phys., 21, 527–531 (1950).
6. D. Train, Trans. Inst. Chem. Eng., 35, 262–265 (1957).
7. W. M. Long, Powder Metall., No. 6, 73–86 (1980).
8. K. Schneider, Chem. Eng. Techn., 41, 142 (1969).
9. E. Goldacker, Ph.D. thesis, IKV, Aachen, West Germany (1971).
10. K. Umeya and R. Hara, Polym. Eng. Sci., 18, 366–371 (1978).
11. K. Umeya and R. Hara, Polym. Eng. Sci., 20, 778–782 (1980).
12. N. M. Smith and J. Parnaby, Polym. Eng. Sci., 20, 830–833 (1980).
13. K. Kawakita and K. H. Luedde, Powder Techn., 4, 61 (1970).
14. R. J. Crawford, Polym. Eng. Sci., 22, 300–306 (1982).
15. W. Ostwald, Kolloid-2, 36, 99–117 (1925).
16. A. de Waele, Oil and Color Chem. Assoc. J., 6, 33–88 (1923).
17. R. B. Bird, R.C. Armstrong, and O. Hassager, "Dynamics of Polymeric Liquids", Vol. I, p. 209, Wiley, New York (1977).
18. H. Eyring, Ind. Eng. Chem., 50, 1036–1040 (1958).
19. P. J. Carreau, Ph.D. thesis, Univ. of Wisconsin, Madison (1968).
20. M. Reiner, "Deformation Strain and Flow", a) p. 258, b) p. 246, Interscience Publishers, New York (1960).
21. J. M. McKelvey, "Polymer Processing", Wiley, New York (1962).
22. A. H. P. Skelland, "Non-Newtonian Flow and Heat Transfer", Wiley, New York (1967).

23. F. P. Bowden and D. Tabor, "Friction and Lubrication of Solids", Oxford Univ. Press, London (1950).
24. K. Schneider, Ph. D. thesis, IKV, Aachen, West Germany (1968).
25. K. Schneider, Kunststoffe, 59, 97–102 (1969).
26. C. I. Chung, W. J. Hennessee, and M. H. Tusim, Polym. Eng. Sci., 17, 9–20 (1977).
27. J. Huxtable, F. N. Cogswell and J. D. Wriggles, Plast. Rubber Proc. Appl., 1, 87-93 (1981).
28. H. Chang and R. A. Daane, SPE 32nd ANTEC, San Francisco, p. 335, May (1974).
29. R. B. Gregory, SPE Journal, 25, 55-59 (1969).
30. G. M. Gale, SPE 39th ANTEC, Boston, p. 669, May (1981).
31. J. A. D. Emmanuel and L. R. Schmidt, SPE 39th ANTEC, Boston, p. 672, May (1981).
32. G. M. Bartenev and V. V. Lavrentev, "Friction and Wear of Polymers", Elsevier, New York (1981).
33. B. Maxwell and A. Jung, Modern Plastics, 35, 3, 174-180 (1957).
34. R. F. Westover, SPE Trans. 1, 14-20 (1962).
35. V. Semjonov, Rheologica Acta, 2, 138-142 (1962); 4, 133-137 (1965); 6, 165-170 (1967).
36. R. C. Penwell, R. S. Porter, and S. Middleman, Journal of Polymer Science A2, 9, 4, 731-745 (1971).
37. P. H. Goldblatt and R. S. Porter, Journal of Applied Polymer Science, 20, 1199-1208 (1976).
38. I. J. Duvdevani and I. Klein, SPE Journal, Dec., 41-45 (1967).
39. S. T. Choi, Journal of Polymer Science A2, 6, 2043-2049 (1968).
40. F. N. Cogswell, Plastics and Polymers, 41, 30-43 (1973).
41. D. W. van Krevelen, "Properties of Polymers, Correlations with Chemical Structure", Elsevier, New York (1972).
42. A. W. Jenike, "Gravity Flow of Bulk Solids", Bulletin No. 108 of the Utah Engineering Experimental Station, Univ. of Utah, Salt Lake City (1961).
43. R. Rautenbach and E. Goldacker, Kunststoffe, 61, 104-107 (1971).
44. E. Goldacker and R. Rautenbach, Chemie Ing. Techn. 44, 405-410 (1972).
45. J. Mewis, J. Non-Newtonian Fluid Mech., 6, 1-20 (1979).
46. J. M. Dealy and W. K. W. Tsang, J. Appl. Polym. Sci., 26, 1149-1158 (1981).
47. T. Y. Liu, D. S. Soong, and M. C. Williams, Polym. Eng. Sci., 21, 675-687 (1981).
48. J. L. White and W. Minoshima, Polym. Eng. Sci., 21, 1113-1121 (1981).
49. C. M. Vrentas and W. W. Graessley, J. Non-Newtonian Fluid Mech., 9, 339-355 (1981).
50. G. De Cleyn and J. Mewis, J. Non-Newtonian Fluid Mech., 9, 91-105 (1981).
51. J. Mewis and G. De Cleyn, AIChE J., 28, 900-907 (1982).
52. J. Mewis and M. M. Denn, J. Non-Newtonian Fluid Mech., 12, 69-83 (1983).
53. D. Acierno, F. P. LaMantia, G. Marrucci, and G. Titomanlio, J. Non-Newtonian Fluid Mech., 1, 125-146 (1976).
54. C. D. Han, "Rheology in Polymer Processing", Academic Press, New York (1976).
55. J. D. Ferry, "Viscoelastic Properties of Polymers", 3rd edition, Wiley, New York (1980).
56. J. J. Aklonis, W. J. MacKnight, and M. Shen, "Introduction to Polymer Viscoelasticity", Wiley-Interscience, New York (1972).
57. K. Walters, "Rheometry", Wiley, New York (1975).
58. R. S. Lenk, "Polymer Rheology", Applied Science Publ. LTD, London (1978).
59. J. A. Brydson, "Flow Properties of Polymer Melts", 2nd edition, George Godwin Limited, London (1981).
60. F. N. Cogswell, "Polymer Melt Rheology", George Godwin Limited, London (1981).
61. "Praktische Rheologie der Kunststoffe", VDI-Verlag GmbH, Duesseldorf (1978).
62. O. Plajer, "Praktische Rheologie fuer Kunststoffschmelzen", Zechner & Huehig Verlag GmbH, Speyer, W. Germany (1970).
63. L. E. Nielsen, "Polymer Rheology", Marcel Dekker, New York (1977).

64. "Rheometry: Industrial Applications", K. Walters editor, Research Studies Press, Chichester, England (1980).
65. J. R. Van Wazer, J. W. Lyons, K. Y. Kim, and R. E. Colwell, "Viscosity and Flow Measurement, A Laboratory Handbook of Rheology", Interscience Publishers, New York (1963).
66. J. M. Dealy, "Rheometers for Molten Plastics", Van Nostrand Reinhold Co., New York (1982).
67. D. V. Boger and M. M. Denn, J. Non-Newtonian Fluid Mech., 6, 163-185 (1980).
68. R. D. Pike, D. E. Baird, and M. D. Read, SPE 41st ANTEC, Chicago, 312-315 (1983).
69. C. D. Han, J. Appl. Polym. Sci., 15, 2567-2577 (1971).
70. J. L. S. Wales, J. L. den Otter, and H. Janeschitz-Kriegl, Rheol. Acta., 4, 146-152 (1965).
71. C. Rauwendaal and F. Fernandez, SPE 42nd ANTEC, 282-287, New Orleans (1984).
72. W. P. Cox and E. H. Merz, J. Polym. Sci., 28, 619-622 (1958).
73. K. Eiermann and K. H. Hellwege, J. Polym. Sci., 57, 99 (1962).
74. D. Hansen and G. A. Bernier, Polym. Eng. Sci., 12, 204 (1972).
75. S. Yagi and D. Kunii, AIChE J., 3, 373 (1957).
76. B. Maxwell et al, SPE Trans., 4, 165 (1964).
77. R. S. Spencer and G. D. Gilmore, J. Appl. Phys., 20, 502 (1949), 21, 523 (1950).
78. H. Breuer and G. Rehage, Kolloid Z. Z. Polym., 216, 166 (1967).
79. M. R. Kamal and N. T. Levan, Polym. Eng. Sci., 13, 131 (1973).
80. O. Olabisi and R. Simha, Macromolecules, 8, 206 (1975); 211 (1975).
81. W. Thompson, Phil. Mag., 42, 448 (1971).
82. H. B. Dwight, "Tables of Integrals and Other Mathematical Data", MacMillan, New York (1957), 3rd edition, p. 275.
83. H. G. Kilian and M. Pietralla, Polymer, 19, 664-672 (1978).
84. P. Andersson and B. Sundqvist, J. Polym. Sci., 13, 243-251 (1975).
85. W. W. Wendlandt, "Thermal Methods of Analysis", Wiley, New York (1979).
86. T. Daniels, "Thermal Analysis", Anchor Press, London (1973).
87. E. A. Turi, editor, "Thermal Characterization of Polymeric Materials", Academic Press, New York (1981).
88. G. W. Miller and R. S. Porter, "Analytical Calorimetry", Vol. 2, Plenum Press, New York (1970), p. 407.
89. L. E. Nielson, "Mechanical Properties of Polymers and Composites", Vol. 1 and 2, Marcel Dekker, New York (1974).
90. "Kenndaten fuer die Verarbeitung Thermoplastischer Kunststoffe, Teil I, Thermodynamik", Carl Hanser, Munich (1979).
91. "Kenndaten fuer die Verarbeitung Thermoplastischer Kunststoffe, Teil II, Rheologie", Carl Hanser, Munich (1982).
92. J. M. Dealy, Plast. Eng., March, 57-61 (1983).
93. M. R. Kamal, V. Tan, and F. Kashani, Adv. Polym. Tech., 3, 89-98 (1983).
94. C. J. S. Petrie, "Elongational Flows", Pitman (1979).
95. "Kenndaten fuer die Verarbeitung Thermoplastischer Kunststoffe, Teil III, Tribologie." Carl Hanser, Munich (1983).
96. G. R. Langecker and R. Rautenbach, Powder Techn., 15, 39-42 (1976).
97. G. R. Langecker, Dissertation, IKV, Aachen, West Germany (1977).
98. H. Schott, J. Appl. Polym. Sci., 6, 529 (1962).
99. A. V. Shenoy and D. R. Saini, Adv. Polym. Techn., 6, 1-58 (1986).

7   FUNCTIONAL PROCESS ANALYSIS

Chapters 5 and 6 on fundamental principles and important polymer properties have mostly been a preparation for this chapter. In this chapter, the six main functions of an extruder will be described and analyzed. The main functions are solids conveying, plasticating or melting, melt conveying or pumping, devolatilization, mixing, and die forming.

Each function will be discussed separately. The mechanism behind each function will be described in detail. The emphasis will be on developing a thorough understanding and quantitative description of the entire extrusion process from a functional point of view. In later chapters, this knowledge will be applied to practical aspects, such as screw design, die design, troubleshooting, etc. It should be realized that in dividing the extruder into functional zones, these zones are not discrete, but to some extent, overlapping. Their boundaries can shift when polymer properties or operating conditions change. For instance, in the melt conveying zone, there will be a certain amount of mixing taking place as well; thus, the mixing zone will overlap with the melt conveying zone. On the other hand, the geometrical sections of the screw are fixed, they are discrete and non-overlapping. The geometrical sections of a standard screw are the feed, compression, and metering sections.

7.1   BASIC SCREW GEOMETRY

The analysis of the functional zones of the extruder requires knowledge of the basic relationships describing the geometry of an extruder screw. The geometry of the flight along the

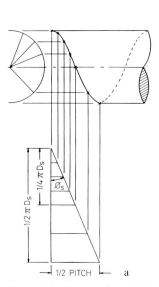
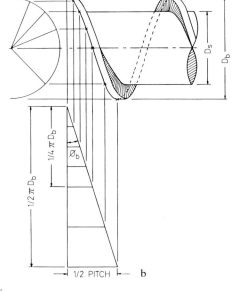

Figure 7–1. Construction of a Basic Screw Geometry

screw surface can be constructed by unrolling the flight onto a flat plane. On a flat surface, the flight along the screw surface will form a right triangle, see Figure 7-1a. The base of the triangle is one-half of the flight pitch if the height of the triangle is half the circumference at the screw surface. The flight lead or pitch S is the axial distance between two points on the flight separated by a full turn of the flight. The flight pitch is the same as the flight lead for a single flighted screw. However, for a multi-flighted screw the pitch is the lead divided by the number of flights. The top angle of the triangle is the helix angle of the screw surface φ_S. If points along the screw surface are connected to corresponding points along the altitude of the triangle, one constructs the helical geometry of the flight along the surface of the root of the screw.

The same procedure can be used to construct the helical geometry of the flight along the OD of the screw as shown in Figure 7-1b.

It should be noted that the helix angle at the OD of the screw φ_b is different from the helix angle at the root of the screw φ_S. The screw geometry is usually represented by straight flights, as shown in Figure 7-2.

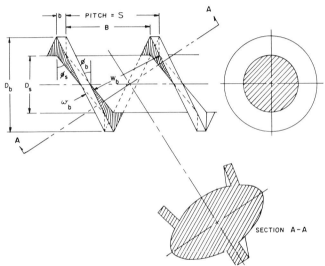

Figure 7-2. Screw   Drawn
With Straight Flights

However, in reality the flights are S-shaped, as seen in Figure 7-1. If a cross-section is made perpendicular to the flights, it can be seen that the screw channel is not a true rectangle. The bottom and top surface of the screw channel are slightly curved and the flight flanks diverge. Thus, the channel width is larger at the screw OD than at the root of the screw. The important geometrical relationships are:

$$S = (B + b)p \qquad\qquad (7-1)$$

$$\varphi_b = \arctan\left(\frac{S}{\pi D_b}\right) \qquad\qquad (7-2)$$

$$\varphi_m = \arctan\left(\frac{S}{\pi D_m}\right) \qquad\qquad (7-3)$$

$$\varphi_S = \arctan\left(\frac{S}{\pi D_S}\right) \qquad\qquad (7-4)$$

$$W_b = B\cos\varphi_b \tag{7-5}$$

$$W_m = B\cos\varphi_m \tag{7-6}$$

$$W_s = B\cos\varphi_s \tag{7-7}$$

$$W_b + w = S\cos\varphi_b \tag{7-8}$$

$$W_m = \pi D_m \sin\varphi_m - w \tag{7-9}$$

Subscript m refers to the dimensions at the midpoint of the channel.

Equation 7-9 is valid only for single flighted screws ($p = 1$).

If the screw has p parallel flights, then:

$$W_m = \frac{\pi D_m \sin\varphi_m}{p} - w \tag{7-10}$$

The channel dimensions used without subscripts are related to the OD of the screw.

7.2   SOLIDS CONVEYING

The solids conveying zone extends from the feed hopper down several diameters into the extruder barrel. The first traces of molten polymer generally do not appear until about three or four diameters into the barrel, measured from the feed port. Since the conveying process in the feed hopper is considerably different from the conveying process in the screw channel, the solids conveying zone will be divided into the gravity induced solids conveying zone (feed hopper) and the drag induced solids conveying zone (extruder screw).

Some feed hoppers are equipped with crammer feeders. This is a feed hopper with a rotating screw in the discharge region. The screw is incorporated to augment the solids conveying rate by the action of the rotating screw. In such a feed hopper, gravity induced solids conveying and drag induced solids conveying occur simultaneously at the same location. In this case, the two types of solids conveying cannot be separated, but must be analyzed together.

7.2.1   GRAVITY SOLIDS CONVEYING

Most feed hoppers have a cylindrical top section and a truncated conical section at the bottom, see Figure 7-3.

The driving force for solids conveying is gravity. Based on the bulk properties and the hopper geometry, one would like to be able to calculate the stress distribution in the hopper, the velocity profiles in the hopper, and finally the discharge rate. Unfortunately, the analysis of flow of granular materials in hoppers is rather complicated as a result of the complex flow

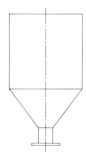

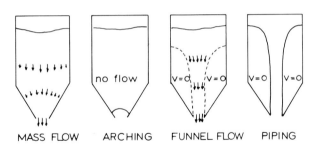

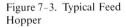

MASS FLOW   ARCHING   FUNNEL FLOW   PIPING

Figure 7-3. Typical Feed
Hopper

Figure 7-4. Various Types of Flow

behavior of granular material, as discussed in Section 6.1. In most cases, the results of the analysis are approximate and apply only to a limited number of cases. Even the relatively simple problem of a non-cohesive bulk material flowing through a hopper has not been completely solved. Not surprisingly, the situation is worse for the more realistic problem of a cohesive particulate material flowing through a hopper. However, the theoretical and experimental work on the transport of bulk material did not really get started until around 1960. Therefore, most of the work is quite recent and it is to be expected that with the increasing interest in the rheological behavior of bulk materials, improved analyses with wider applicability will be developed. A collection of relatively recent experimental and theoretical work was presented at a joint ASME-CSME Conference on Mechanics Applied to the Transport of Bulk Materials (1) and the U. S. - Japan seminars on Continuum Mechanical and Statistical Approaches in the Mechanics of Granular Materials (2). Books on particulate solids have been written by Orr (3) and Brown and Richards (4). Review articles have been published by Richards (5), Wieghardt (6), and Savage (7). Some of the pioneering work on flow of bulk solids was done by Jenike (8, 9).

In the flow of bulk materials through a hopper, one generally distinguishes between two types of flow. In mass flow or hopper flow, the entire volume of particulate solids moves down toward the exit; there are no stagnating regions. The other type of flow is funnel flow. In funnel flow, the bulk material flows out through a flow channel; the wall of the flow channel being formed by stationary particles of the bulk material. In funnel flow, therefore, there is at least one stagnating region, see Figure 7-4.

A common flow problem in hopper flow is arching or bridging. The particles form a natural bridge able to support the material above it. As mentioned in Section 6-1, highly compressible materials have a strong tendency towards bridging. With such materials, the hopper is often equipped with a vibrating pad to dislodge any bridges that might form by a continuous mechanical vibration of the hopper. It is possible to derive criteria to avoid arching; these can be used in the design of feed hoppers and will be discussed later. Piping is a disturbance that occurs in funnel flow. An annular ring of stationary bulk material is formed. The material in this stagnating region is able to support the material above it and the exposed surface of the internal, empty channel.

Both in arching and in piping, the material is consolidated to the extent that it can support the material above it and form an exposed surface. Thus, both these flow problems are typical of cohesive (non-free-flowing) particulate solids. This is particularly true for materials with a high unconfined yield strength σ_c, see Section 6.1.2 and Figure 6-4b.

In the analysis of flow of granular material, two types of flow can be distinguished. The first is slow frictional flow where the particles remain in continuous contact with each other; the internal forces result from Coulomb friction between contacting particles. The second type of flow is much more rapid; the particles are not in constant contact with their neighbors. The energy associated with the velocity fluctuations is comparable to that of the mean motion. In this type of flow, the internal forces arise because of the transfer of momentum during collisions between particles. The constitutive relations for this rapid flow are rate dependent. This type of flow, therefore, is referred to as viscous flow (sometimes just rapid flow). Steady, viscous flows are generally described by elliptic partial differential equations (10–13).

In gravity flow through feed hoppers, it can be assumed that the flow is sufficiently slow that the particles are in constant contact and that momentum transfer by collisions between particles is negligible. The flow through a feed hopper can thus be considered to be frictional flow. Various workers (14–21) have made experimental studies of the flow patterns in feed hoppers. In some studies, dyed particles were used (e.g. 17), in others, x-ray techniques were employed to determine the flow patterns in the hopper (15, 16). In other studies (18, 19, 21), a stereoscopic technique was used. This involves taking photographs of the flow field at short time intervals. Sequential photographs are analyzed using a stereocomparator; this results in a three-dimensional model of the displacement field from which the velocity field can be determined. This technique is limited to two-dimensional flows, but does not require tracers and enables determination of the entire flow field.

The flow of bulk material in a feed hopper is generally quite different in the various sections of the hopper, see Figure 7–5.

The flow in the cylindrical hopper section tends to be plug flow. The plug flow zone is bounded by a rupture zone at the bottom of the plug flow zone. The rupture zone is situated approximately at the cylinder-cone transition. This zone is characterized by intense relative deformation of the granular material. Below the rupture zone, further down in the conical section, there may be local regions of plug flow. These local plug flow regions are generally bounded by the hopper wall and the rupture zone. Finally, in the bottom part of the conical section the particles flow out freely; this is referred to as the free-flow zone.

Unfortunately, at this point in time, the understanding of the mechanics of flow of bulk solids is not well enough developed that this flow behavior can be predicted theoretically. The main area of uncertainty is the proper constitutive equation for bulk materials relating stresses and strain rates. Because the theory is not well developed, only a few aspects of gravity flow in hoppers will be discussed further. These are pressure distribution in feed hoppers, criteria to avoid flow disturbances, and flow rate predictions. It should be remembered that the following relationships have limited applicability and accuracy in terms of their predictive ability.

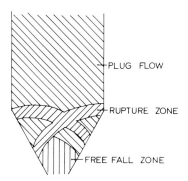

Figure 7–5. Various Flow Regimes in a Feed Hopper

7.2.1.1   PRESSURE DISTRIBUTION

In the cylindrical portion of a hopper, the pressure distribution can be derived if the following assumptions are made: i)   the vertical compressive stress is constant over any horizontal plane, ii) the ratio of horizontal and vertical stresses is constant and independent of depth, iii) the bulk density is constant, and iv) the wall friction is fully mobilized, meaning that the particulate material is in incipient slip conditions at the wall.

A force balance over a differential element, see Figure 7–6, gives:

$$\rho_b g \pi R^2 dh + \pi R^2 dP = (\tau_{wa} + f_w^* kP) 2\pi R dh \qquad (7-11)$$

where ρ_b is bulk density, g the gravitational acceleration, f_w^* the coefficient of friction at the wall, and k the ratio of compressive stress in horizontal direction to compressive stress in vertical direction. The shear stress at the wall is determined by equation 6–8. An expression for the ratio k can be found if it is assumed that the maximum principal stress is in the vertical direction. The ratio k can be determined from equation 6–7.

$$k = \frac{1 - \sin\beta e}{1 + \sin\beta e} \qquad (7-12)$$

where β_e is the effective angle of internal friction.

Equation 7–12 applies if the particulate solids are in a condition of steady flow; it also applies to cohesionless materials in a condition of incipient flow. Integration of equation 7–11 yields the pressure distribution:

$$P = \frac{\rho_b g R - 2\tau_{wa}}{2 f_w^* k} + \frac{1}{2 f_w^* k} \exp\left(\frac{2 f_w^* kh}{R} + constant\right) \qquad (7-13)$$

If the pressure at h = H is taken as zero and if the adhesive wall shear stress is zero, equation 7–13 reduces to the wellknown Janssen equation derived in 1895 (22):

$$P = \frac{\rho_b g R}{2 f_w^* k} \left[1 - \exp\frac{2 f_w^* k (h-H)}{R} \right] \qquad (7-14)$$

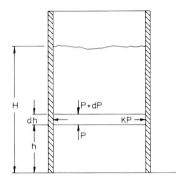

Figure 7–6. Force Balance

If the value of H is sufficiently large, the pressure becomes independent of vertical distance; this limiting pressure value is:

$$P_{max} = \frac{\rho_b g R}{2 f_w^* k} \tag{7-15}$$

The maximum pressure is directly proportional to the bulk density and cylinder radius and inversely proportional to the coefficient of friction at the wall and the ratio k.

Walker (23) made a more rigorous analysis of the pressure distribution in vertical bins. He assumed a plastic equilibrium in the particulate solids with the Mohr circles representing the stress condition at a certain level touching the effective yield locus. Walker derived the following expression for the pressure profile in a vertical cylinder:

$$P = \frac{\rho_b g R}{2 BD^*} \left[1 - \exp \frac{2 BD^* (h-H)}{R} \right] \tag{7-16}$$

where D* is defined as a distribution factor relating the average vertical stress with the vertical stress near the wall. In principle, this distribution function D* can be evaluated by solving the entire stress field, as discussed by Walters and Nedderman (24). However, as a first approximation, the distribution factor can be assumed to be unity. The ratio of shear stress to normal stress at the wall, B, is given by:

$$B = \frac{\sin \beta_e \sin 2\beta^*}{1 - \sin \beta_e \cos 2\beta^*} \tag{7-17}$$

where β^* is the angle between the major principal plane and the cylinder wall.

The angle β^* is related to the effective angle of friction β_e and the wall angle of friction $\beta_w = \arctan f_w^*$ by:

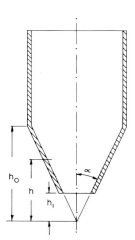

Figure 7-7. Conical Hopper Configuration

$$2\beta^* = \beta_w + \arcsin\frac{\sin\beta_w}{\sin\beta_e} \qquad\qquad (7\text{-}18)$$

where the arcsin value is to be larger than 90°.

Walker also derived equations for the stress distribution in a conical hopper section. These equations have clear practical importance because most feed hoppers are designed with conical sections. The pressure distribution for mass flow conditions is given by:

$$P = \left(\frac{h}{h_0}\right)^c P_0 + \frac{\rho_b gh}{c-1}\left[1 - \left(\frac{h}{h_0}\right)^{c-1}\right] \qquad \text{for } c \neq 1 \qquad (7\text{-}19)$$

and

$$P = \frac{h}{h_0}P_0 + \rho_b gh\ln\left(\frac{h_0}{h}\right) \qquad\qquad \text{for } c = 1 \qquad (7\text{-}20)$$

where h_0 is the height where the vertical pressure is P_0. This height can be taken as the height of the conical hopper section, as shown in Figure 7-7.

The coefficient c for conical hoppers is:

$$c = \frac{2B'D^*}{\tan\alpha} \qquad\qquad (7\text{-}21)$$

where α is the hopper half-apex angle, see Figure 7-7.

The coefficient c for wedge shaped hoppers is:

$$c - \frac{B'D^*}{\tan\alpha} \qquad\qquad (7\text{-}22)$$

The stress ratio B′ is given by:

$$B' = \frac{\sin\beta_e\sin(2\alpha+2\beta^*)}{1 - \sin\beta_e\cos(2\alpha+2\beta^*)} \qquad\qquad (7\text{-}23)$$

In the convergent hopper section angle β^* is given by:

$$2\beta^* = \beta_w + \arcsin\frac{\sin\beta_w}{\sin\beta_e} \qquad\qquad (7\text{-}24)$$

where the arcsin value is to be less than 90° !

If the initial pressure in the conical section is zero, the maximum pressures will occur somewhere along the conical section. In most cases, however, a cylindrical hopper section is placed on top of the conical hopper section. In these situations, the initial pressure distribution in the conical section will be determined by the final pressure distribution in the cylindrical section. If the stress distributions do not match, rupture zones may form in the transition region as observed by Lee, et al. (15). Instabilities of stress conditions at the transition region have been discussed by Bransby and Blair-Fish (25).

7.2.1.2   FLOW RATE

The flow rate of granular materials is independent of the head if the head is sufficiently large. This experimental result was known as early as 1852, when Hagen presented an equation for the flow rate through a circular opening (26):

$$\dot{M} = J_a \frac{\pi}{4} \rho_b \left[g(D_a - f_c d_p)^5 \right]^{0.5} \tag{7-25}$$

where J_a is a parameter called the nondimensional axisymmetric flow rate, D_a is the diameter of the aperture, d_p is the particle diameter, and f_c is a correction factor of the order one. This result can be obtained from a dimensional analysis, see Section 5.3.3. For a two-dimensional slot of length L and width W, the flow rate is:

$$\dot{M} = J L \rho_b \left[g(W - f_c d_p)^3 \right]^{0.5} \tag{7-26}$$

where J is the nondimensional flow rate for slot outlets.

The flow rate seems to be determined, to a large extent, by what happens in the vicinity of the discharge opening. This concept forms the basis of most of the early analyses of flow rate. Brown (27) derived the following expression for the nondimensional flow rate for a slot outlet in case of mass flow:

$$J = \frac{\int_0^\alpha (\cos\alpha')^{1/2} d\alpha'}{(2 \sin^3\alpha)^{1/2}} \tag{7-27}$$

Johanson (28) developed the following expression for steady flow of noncohesive bulk materials:

$$J = (2 \tan\alpha)^{-1/2} \tag{7-28}$$

This result is obtained by assuming the flow in the hopper to be one-dimensional and further assuming that at the orifice, the convective acceleration in the upper converging flow is equal to the gravitational acceleration g, appropriate for the freely falling particles below the orifice level.

Savage (29) derived flow rate equations for a frictional Coulomb material by assuming radial body forces and neglecting wall friction:

$$J = \left[\frac{1 + k}{2(1-2k)\sin\alpha} \right]^{1/2} \tag{7-29}$$

where k is given by equation 7-12.

Expression 7–29 generally overestimates the flow rate by about 40 percent to 100 percent, as determined by comparison to experimental results. Savage (30) extended his analysis to include the effect of wall friction by solving the equations of motion by the method of integral relations. The flow rates were lower than predicted with equation 7–29, but still higher than experimental values. Savage and Sayed (31) improved the earlier analysis (30) and derived an expression for the flow rate in a two-dimensional wedge shaped hopper:

$$J^2 = \frac{(2B - 4A - C)(Ga^2 + 3E)}{2g\sin\alpha(A + B + C)(Ha^2 + 3F)} \qquad (7-30)$$

where the constants are given in Appendix 7–1.

The closed-form solution of the flow rate was derived by using a mean normal stress averaged for the width of the hopper. A more accurate analysis considered the detailed variations of the stresses throughout the hopper. Predictions based on equation 7–30 agreed well with the numerical results from the more accurate analysis; the differences were generally less than one percent. An interesting theoretical prediction is the increase in flow rate when the wall friction is increased at large hopper half-angles. This result is not intuitively obvious, but has been experimentally observed (32–34).

7.2.1.3   DESIGN CRITERIA

Jenike (8, 9) and his coworkers did extensive work on gravity flow of bulk solids, both experimental and theoretical. He developed design methods and criteria for hoppers and bins with steady mass flow without disturbances. In determining various flow criteria, Jenike used a function termed "flow factor." This flow factor is the ratio of the consolidating pressure σ_1 to the stress acting on an exposed surface $\bar{\sigma}_1$:

$$ff = \frac{\sigma_1}{\bar{\sigma}_1} \qquad (7-31)$$

The stress acting on an exposed surface is also the only nonzero principal stress, because the exposed surface is assumed self-supporting and traction-free (i.e., no shear stresses acting on the surface). The flow factor is determined by the geometry of the hopper and the properties of the bulk material. Another function used by Jenike is the "flow function." This flow function is the ratio of the consolidating pressure σ_1 to the unconfined yield strength σ_c, as defined in Section 6.1.2.

$$FF = \frac{\sigma_1}{\sigma_c} \qquad (7-32)$$

The flow function is a material property; it gives an indication of the flowability of a bulk material:

FF > 10	free flowing material
10 > FF > 4	easy flowing material
4 > FF > 1.6	cohesive material
FF < 1.6	very cohesive, non-flowing material

As a general rule, solids which do not contain particles smaller than about 0.2 mm are free flowing; thus, most granular solids are free flowing and most powders are to some extent cohesive.

The exposed surface, whether it is an arch or pipe, is stable when the unconfined yield strength σ_c is higher than the stress acting on the exposed surface $\bar{\sigma}_1$ and unstable when σ_c is less than $\bar{\sigma}_1$. The condition for no arching or piping, therefore, is:

$$\sigma_c < \bar{\sigma}_1 \quad \text{or} \quad FF > ff \tag{7-33}$$

In order to obtain quantitative results, the flow factor has to be determined; this requires knowledge of the stress field in the hopper. Closed form expressions for ff are not available, except for the simplest case of flow through a straight cylinder. Results from numerical analysis are given by Jenike (8, 9) in graphical form for plane symmetry and axial symmetry.

If steady flow is desired, one should design the hopper in such a geometry that mass flow will take place; no stagnating regions should occur. In this case, the solids flow along the walls of the hopper. The wall, therefore, must be sufficiently steep and the flow channel must not have any sharp corners, abrupt transitions, or discontinuities in frictional properties at the wall. As a rule, the hopper half-angle α should not exceed α_{max}, with α_{max} determined from:

$$\alpha_{max} = 55° - \frac{1}{2} \beta_e \tag{7-34}$$

where β_e is the effective angle of internal friction.

The minimum outlet dimension to avoid doming was formulated by Jenike (8) as:

$$D \geqslant c \, \frac{\sigma_c}{\rho_b} \tag{7-35}$$

with $c=2$ for circular outlets and $c=1.8$ for square outlets. The critical yield stress σ_c is determined from the point of intersection of the appropriate flow function and flow factor. Walker (35) and Eckhoff (36) published experiments showing that the Jenike method considerably overdesigns the critical outlet dimensions, as much as a hundred percent or more. Engstad (37) has formulated a relationship for the critical outlet dimension that is claimed to be more accurate and results in less overdesign. The critical outlet dimension, according to Engstad's analysis, is given by:

$$D = \frac{(2 - 2\sin\beta_i)}{\sin\beta_i} \, \frac{\sigma_{co}}{\rho Y_b} \tag{7-36}$$

where the various terms of equation 7-36 are given in Appendix 7-2.

From an analysis of the flow field in a hopper, criteria can be formulated for preventing arching and funneling. These criteria place restrictions either on the dimensions of the outlet or the slope of the walls. Another approach is to modify the hopper geometry into a non-linear, curved shape in order to obtain optimum flow conditions. Lee (38) designed a hyperbolic hopper by making certain assumptions about the rate of change of the horizontal cross-section with respect to axial distance. Richmond (39) suggested that in an optimum hopper, the material would be on the verge of arching at any point. By using a one-dimensional analysis, this condition could be achieved, resulting in an exponential profile. Gardner (40) proposed a solution based on having a single surface at any level in the hopper on the verge of arching. Richmond and Morrison (41) used a modified procedure based on arching being imminent only along the axis of the hopper. For this case, positive pressures exist at all other points in the hopper. This approach leads to smoothly curved, funnel-shaped hoppers. However, the optimum shape will have to be modified if the converging hopper section is connected to a cylindrical hopper section. A practical drawback of this approach is that such a carefully tailored curved hopper geometry is difficult to manufacture and could become quite expensive.

7.2.2   DRAG INDUCED SOLIDS CONVEYING

Once the particulate solids have reached the feed port of the extruder, the material will flow down until it is situated in the screw channel. At this point, the gravity induced flow mechanism will essentially cease. In most extruders, the screw and barrel are placed in a horizontal direction and the role of gravity becomes a very minor one. In fact, in most analyses of solids conveying in single screw extruders, the effect of gravity is assumed to be negligible. When the material is in the screw channel, forward motion of the solids occurs as a result of the relative motion between boundaries of the solids; i.e., the screw channel surface and the barrel surface. The flow rate of the solids is determined by the forces acting on the solids. These forces are, to a large extent, determined by the frictional forces acting on the solids at the boundaries.

It has been found experimentally that, in most cases, polymeric particulate solids compact readily in the early portion of the screw channel. As a result, the solids form into a solid bed and the solids move down the screw channel in plug flow; thus, at any cross-section of the solid bed all elements move at the same velocity. In other words, there is no internal deformation taking place inside the solid bed. This compaction of the particulate solids into a solid bed can occur only if there is a sufficient amount of pressure generation in the screw channel.

If there is not enough pressure generation in the screw channel, the particulate solids will not form a solid bed. In this case, plug flow will not occur; there will be internal deformation in the solid material. As a result, the solids conveying process will be less steady compared to plug flow. The type of non-plug flow that occurs when the channel is only partially filled with polymer particles has been referred to as "Archimedean transport" (42). Archimedean transport will occur if the intake rate of the extruder is less than the plug flow solids conveying rate. In metered starve fed extrusion, the Archimedean transport is created intentionally. In some cases, Archimedean transport occurs accidentally if the flow rate through the feed hopper is too low or if the feed port geometry of the extruder is too small. Archimedean solids transport is also likely to occur if the pressure at the end of the solids conveying zone is low. An example would be a screw type solids conveying device (a screw feeder) where the discharge pressure at the end of the solids conveying zone is zero. Archimedean transport is essentially always associated with a partially filled screw channel, see Figure 7–8.

Pressure generation cannot take place if the screw channel is not fully filled and consequently compaction and plug flow cannot occur.

It has been observed (44) that even under normal extrusion conditions, Archimedean transport can occur for a short length. Observations of the filling action in the feed port have

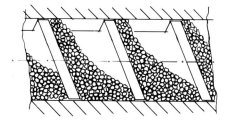

Figure 7-8. Archimedean Transport Zone

shown that the flow of material from the hopper to the screw channel is such that little pressure is likely to be transmitted. During some parts of the filling cycle, empty spaces in the screw channel have been observed. This indicates that virtually no pressure exists for some time periods. Thus, Archimedean transport can occur over a short distance in the feed port region.

The solids transport resulting from the relative motion of the boundaries of the solid material is referred to as drag induced solids conveying. For optimum solids conveying, the particulate solids should compact easily and move in plug flow. It is important to note that properties of the bulk material that are advantageous in the gravity flow in the feed hopper can be detrimental in the drag induced solids conveying in the screw channel. An example is the flow function, defined by equation 7-22. For optimal gravity flow in the feed hopper, one would like to have a bulk material with a large value of the flow function. However, such a material will not form a strong solid bed because the yield strength will be low relative to the compacting pressure. Thus, such a material may well be problematic in the drag induced solids conveying because of internal deformation of the solid bed.

The first comprehensive analysis of solids transport in single screw extruders was made by Darnell and Mol in 1956 (43). Later, numerous workers extended the work of Darnell and Mol; however, the basic analysis has remained relatively unchanged. In order to come to a quantitative description of the drag induced solids conveying process, the following assumptions were made by Darnell and Mol:

1. The particulate solids behave like a continuum.

2. The solid bed is in contact with the entire channel wall; i.e., the barrel surface, the root of the screw, the flank of the active flight, and the flank of the passive flight.

3. The channel depth is constant.

4. The flight clearance can be neglected.

5. The solid bed moves in plug flow.

6. The pressure is a function of down channel distance only.

7. The coefficient of friction is independent of pressure.

8. Gravitational forces are neglected.

9. Centrifugal forces are neglected.

10. Density changes in the plug are neglected.

The first five assumptions are made in most analyses of solids conveying. The last five assumptions have been relaxed by various workers. The general approach to solids conveying analysis is to consider an element of the solid bed in the screw channel and determine all forces that are acting on this solid bed element. The most important forces are the frictional forces at the boundaries and the forces resulting from pressure gradients in the solid bed.

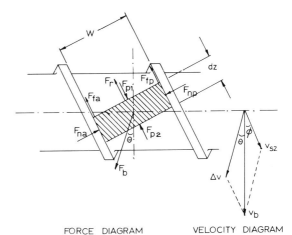

Figure 7-9. Forces Acting on a Solid
Plug Element and Velocity Diagram   FORCE DIAGRAM   VELOCITY DIAGRAM

Figure 7-9 shows the various forces acting on the solid bed element; the screw is considered stationary and the barrel rotating.

F_r is the frictional force between the solid bed and the root of the screw:

$$F_r = PWdzf_s \qquad (7-37)$$

where f_s is the dynamic coefficient of friction on the screw surface.

F_{na} is the normal force on the solid bed on the active flight flank:

$$F_{na} = PHdz + F^* \qquad (7-38)$$

where F^* is an extra normal force, which is unknown.

F_{np} is the normal force on the solid bed on the passive flight flank:

$$F_{np} = PHdz \qquad (7-39)$$

F_{fa} is the frictional force between the solid bed and the active flight flank:

$$F_{fa} = F_{na}f_s \qquad (7-40)$$

F_{fp} is the frictional force between the solid bed and the passive flight flank:

$$F_{fp} = PHdzf_s \qquad (7-41)$$

F_{p_1} is the force against the face of the solid plug element resulting from the local pressure P:

$$F_{p_1} = WHP \qquad (7-42)$$

F_{p_2} is the force against the face of the solid plug element resulting from the local pressure P+dP:

$$F_{p2} = WH(P+dP) \tag{7–43}$$

Obviously, if the pressure gradient in the down channel direction is zero, then $F_{p_1} = F_{p_2}$. The final force F_b is the frictional force between the solid bed and the barrel surface:

$$F_b = PWdzf_b \tag{7–44}$$

where f_b is the dynamic coefficient of friction on the barrel surface.

The force F_b makes an angle θ with the plane perpendicular to the screw axis, see Figure 7–9. The direction of force F_b may be a little surprising. However, it should be realized that the direction of F_b is determined by the direction of the vectorial velocity difference between the barrel and the solid bed:

$$\Delta \vec{v} = \vec{v}_b - \vec{v}_{sz} \tag{7–45}$$

where $\vec{v}_b$ is the barrel velocity and $\vec{v}_{sz}$ the solid bed velocity vector. From the velocity diagram in Figure 7–9, the direction of $\Delta \vec{v}$ and F_b becomes clear. The angle θ is the solids conveying angle. If the solids conveying angle can be determined, then the solid bed velocity can be calculated directly from it:

$$v_{sz} = v_b \left[\cos\varphi - \frac{\sin\varphi}{\tan(\theta+\varphi)} \right] \tag{7–46a}$$

Equation 7–46a can be rewritten as follows:

$$v_{sz} = v_b \frac{\sin\theta}{\sin(\theta+\varphi)} \tag{7–46b}$$

The vectorial velocity difference Δv between the barrel velocity v_b and the solid bed velocity v_{sz} is:

$$\Delta v = (v_b^2 + v_{sz}^2 - 2|v_b||v_{sz}|\cos\varphi)^{1/2} \tag{7–47}$$

Δv as a function of v_b, θ, and φ is given in equation 7–61.

Once the solid bed velocity is known, the solids conveying rate is simply determined from:

$$\dot{M}_S = \rho HpWv_{sz} = \rho HWpv_b \frac{\sin\theta}{\sin(\theta+\varphi)} \tag{7–48}$$

where ρ is the solid bed density and p the number of parallel flights.

At this point, there are two unknowns; the extra force F^* and the solids conveying angle θ. The solution procedure followed by Darnell and Mol was to break up all forces into their axial and tangential components. The sum of all forces in the axial direction is taken to be zero, assuming that acceleration is negligible. The tangential force components are used in a torque balance with the net torque also assumed to be zero. The extra force F^* is then eliminated from the two balance equations and an expression for the solids conveying angle results. Darnell and Mol also included the dependence of the helix angle and flight width on radial distance. The same procedure has been followed by many other workers; e.g., Tadmor and Broyer (45, 46). This approach leads to a rather lengthy expression for the solids conveying angle as given in Appendix 7–3.

Considering that in most extruder screws the screw diameter is much larger than the channel depth (D/H >> 1, usually about 10), the change in channel width and helix angle over the depth of the channel will be rather small. If it is assumed that the channel curvature can be neglected, the screw channel can be unrolled onto a flat plane. The slight error that is made in this process is relatively insignificant considering the limited accuracy and reproducibility of most data on the coefficient of friction, as discussed in Section 6.1.2. Two simplifications result from this assumption. The first one is that now the channel width and helix angle are constant over the depth of the channel. The second simplification is that the extra force F* can be determined directly from a force balance in the cross-channel direction:

$$F^* = F_b \sin(\theta+\varphi) \tag{7-49}$$

An expression for the solids conveying angle θ is obtained from a force balance in the down channel direction:

$$PWdzf_b[\cos(\theta+\varphi) - f_s\sin(\theta+\varphi)] = WHdP + Pdzf_s(W+2H) \tag{7-50}$$

Equation 7-50 can be integrated to give the pressure as a function of down channel distance. If the pressure at $z=0$ is taken as $P(z=0)=P_0$, the solution is:

$$\ln \frac{P}{P_0} = \left\{ \frac{f_b}{H} \left[\cos(\theta+\varphi) - f_s\sin(\theta+\varphi) \right] - f_s\left(\frac{W+2H}{WH} \right) \right\} z \tag{7-51}$$

Equation 7-51 indicates that at a certain solids conveying angle the pressure will increase exponentially with down channel distance. This means that very high pressures can be generated in the solids conveying zone, at least theoretically. Equation 7-51 can be worked out further to yield a closed form expression for the solids conveying angle:

$$\theta = \arcsin \left[\frac{(1 + f_s^2 - k^2)^{1/2} - f_s k}{1 + f_s^2} \right] - \varphi \tag{7-52}$$

where

$$k = \frac{H}{f_b z} \ln \frac{P}{P_0} + \frac{f_s}{f_b} \left(1 + \frac{2H}{W} \right) \tag{7-52a}$$

Equation 7-52 is considerably more compact than the more elaborate solution given in Appendix 7-3. Figure 7-10 compares the two solutions for a 50-mm (2-inch) extruder with a square pitch and a channel depth of 5 mm (0.2 inch).

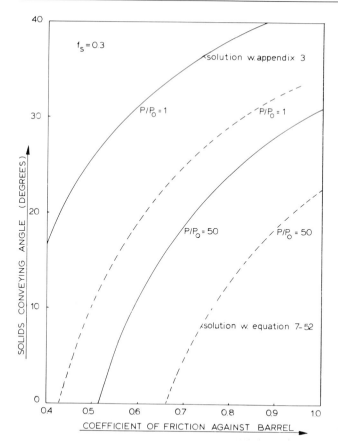

Figure 7-10. Comparison Results of Equation 7-52 and Equations in Appendix 7-3

If the dependence of φ and W on radial distance is neglected, the two solutions yield similar results. If the dependence is included, the solids conveying angle is considerably larger, about 10° in this example. However, the effect of this larger solids conveying angle is not as large as it might seem. If one solids conveying angle is 20° and the other 30°, the resulting difference in the solid bed velocity is only about 15 percent. Thus, a 50 percent difference in the solids conveying angle results in a difference in the solid bed velocity and solids conveying rate of only about 15 percent. Equation 7-52 should be sufficiently accurate for most calculations, particularly since the actual values of the coefficient of friction are generally not precisely known.

Figure 7-10 shows that the solids conveying angle increases when the coefficient of friction on the barrel increases relative to the coefficient of friction on the screw. This is to be expected, considering that the frictional force on the barrel constitutes the driving force on the solid bed. If the pressure gradient increases, the solids conveying angle will decrease because a positive pressure gradient causes a retarding force on the solid bed. It can also be seen in Figure 7-10 that the solids conveying angle reaches very small values or even negative values when the value of the coefficient of friction on the barrel approaches the coefficient of friction on the screw. This makes sense when one considers that the frictional stress on the screw acts on a larger surface than the frictional stress on the barrel when one assumes the flat plate model. In the absence of a pressure gradient and if $f_b = f_s$, the retarding force on the solid bed will be larger than the driving force; thus, no forward transport will occur. The situation is even worse if there is a positive pressure gradient. Thus, in order for forward transport to occur, the coefficient of friction on the barrel should be larger than the coefficient of friction on the screw.

From equation 7-52, the solids conveying angle and thus the solids conveying rate can be calculated if the pressure gradient is known. This process can also be reversed. If the actual solids conveying rate is known, then the pressure gradient can be calculated using equation 7-51. However, the solids conveying angle in equation 7-51 must be expressed as a function of the solid bed velocity. The following relationship can be derived from the velocity diagram, see Figure 7-9.

$$\tan(\theta+\varphi) = \frac{v_b \sin\varphi}{v_b \cos\varphi - v_{sz}} = a \qquad (7\text{--}53)$$

The pressure profile can now be written as:

$$P(z) = P_0 \exp\left[\left(f_b \frac{1 - af_s}{(a^2 + 1)^{1/2}} - f_s \frac{W + 2H}{W}\right)\frac{z}{H}\right] \qquad (7\text{--}54)$$

From equation 7-54 it can be seen that the exponential term will increase with f_b and decrease with f_s. Thus, the pressure rise will be most rapid when f_b is large and f_s is small. The exponential term is inversely proportional to the channel depth H; thus, the pressure will rise more slowly when the channel depth is increased.

The transport of the solids down the screw channel can be compared to a nut located on a long threaded rotating shaft. If the nut can freely rotate with the shaft, it will not move in the axial direction. However, if the nut is kept from rotating with the shaft, it will move in the axial direction. In the extruder, the frictional force on the barrel wall will keep the solid bed from freely rotating with the screw. The frictional force on the barrel, therefore, constitutes the driving force of the solid bed. The frictional force on the screw surface constitutes a retarding force on the solid bed. If the frictional force on the barrel is zero, no forward transport will occur. If the frictional force on the screw is zero, maximum forward transport will occur. From the velocity diagram of this extreme case, Figure 7-9, it can be seen that the maximum solids conveying angle in this case is:

$$\theta_{max} = \frac{\pi}{2} \qquad (7\text{--}55)$$

The maximum solid bed velocity in this case is:

$$v_{sz-max} = \frac{v_b}{\cos\varphi} \qquad (7\text{--}55\,a)$$

And thus the maximum solids conveying rate is:

$$\dot{M}_{s-max} = \frac{\rho W H v_b}{\cos\varphi} \approx \rho \pi^2 D^2 H N \tan\varphi \qquad (7\text{--}55\,b)$$

The approximately equal sign is used because the flight width is neglected in the right-hand expression. For optimum solids conveying, the frictional force on the barrel should be max-

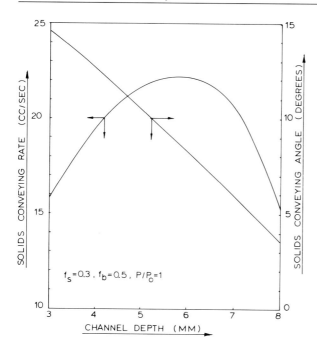

Figure 7-11. The Effect of Channel Depth on Solids Conveying Rate

imum and the frictional force on the screw should be minimum. This is clear from simple qualitative arguments without any elaborate analysis or equations. One would like, therefore, to have a low coefficient of friction on the screw and a high coefficient of friction on the barrel. In many instances, the screw is chrome-plated or nickel-plated and highly polished to minimize the friction on the surface. Special platings are available where the surface is impregnated with a fluoropolymer to give a low coefficient of friction, see also Section 11.2.1.4. On the other hand, the surface of the barrel should be rough to increase the frictional force on the barrel. Many extruders have grooves machined into the internal barrel surface in the solids conveying zone to improve the solids conveying performance; more on this in Section 7.2.2.

With the equations developed so far, the solids conveying performance can be analyzed as a function of screw geometry and polymer properties. The effect of channel depth on solids conveying rate is shown in Figure 7-11.

The curve shown is for a 50-mm extruder running at 100 rpm, assuming $f_s = 0.3$, $f_b = 0.5$, and the pressure gradient as zero. At low values of the channel depth, the solids conveying rate increases with channel depth. However, if the channel depth is further increased, the solids conveying rate reaches a maximum and then reduces with further increases in the channel depth. This result may be somewhat surprising, particularly since there is no pressure increase in this case. The result can be explained by considering the forces acting on the solid bed. If the channel depth is increased at a condition of zero pressure development, the frictional force on the screw will increase, while the frictional force on the barrel will remain unchanged. Thus, the retarding force increases while the driving force stays the same. This has to result in a reduction of the solids conveying angle. The reducing of the solids conveying angle causes a reduction in the solids conveying rate, but the increase in channel depth increases the cross-sectional area of the screw channel and this causes an increase in the solids conveying rate. This explains why the solids conveying rate increases at first and then decreases. The solids conveying angle as a function of channel depth is also shown in

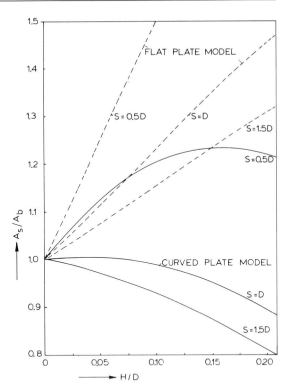

Figure 7-12. Ratio A_s/A_b for Flat Plate and Curved Plate Model

Figure 7-11. It is clear that the solids conveying angle drops monotonically with channel depth.

It should be noted, however, that the effect of the channel depth on the solids conveying rate as determined from a flat plate analysis will be different when the curvature of the channel is taken into account. In reality, when the channel depth increases, the area of the flight flanks will increase but the area of the root of the screw will decrease; this decrease is not taken into account in the flat plate model. The contact area A_b between the a differential element of the solid bed and the barrel is:

$$A_b = (\pi D \sin\varphi_b - w) \Delta z_b \qquad (7-56)$$

In the flat plate model, the contact area between the solid bed and the screw is:

$$A_s = [\pi(D-2H)\sin\varphi_b - w + 2H]\Delta z_b \qquad (7-57\,a)$$

while in the curved plate model the contact area is:

$$A_s = [\pi(D-2H)\sin\varphi_s - w + H(\Delta z_b/\Delta z_s + 1)]\Delta z_b \qquad (7-57\,b)$$

Both the helix angle φ_s and the down channel incremental distance Δz_s are different in the curved plate model. The ratio of A_s/A_b for both the flat plate model and the curved plate model is shown in Figure 7-12 for three values of the flight pitch.

It is clear that the flat plate model considerably overestimates the contact area between the solid bed and the screw. Therefore, when the curvature is taken into account, the solids conveying rate will increase in a monotonic fashion with the channel depth as long as the pressure rise is sufficiently small. This demonstrates that the assumptions underlying a model have to be critically evaluated each time the model is used to analyze the influence of a certain parameter.

Equation 7-57a and b assumes a zero radius of curvature between the flight flank and the screw root. If the radius R_c is taken into account, the screw contact area is reduced by $R_c(4-\pi)$. Therefore, one would like to use the largest possible radius; i.e., $R_c = H$. The resulting flight geometry is shown in Figure 8-4a of Section 8.2.3. In this case, the screw contact area in the flat plate model becomes:

$$A_S = [\pi(D-2H)\sin\varphi_b - w + H(\pi-2)]\Delta z_b \qquad (7-57c)$$

Considering that the ratio of channel depth to channel width is usually about 1/10, using a radius of curvature equal to the channel depth can reduce the contact area between solid bed and screw by about 6 to 7 percent, while the contact area between solid bed and barrel remains unchanged. Another advantage of the large flight radius is that the flight width at the screw OD can be reduced so that the channel cross-sectional area can be at least as large as with a small flight radius. If the flight width at the screw OD is maintained at the same value, a flight curvature of $R_c = H$ will reduce the channel cross-sectional area by $2H^2(\pi/4-1)$. With a usual ratio of channel width to channel depth, this will result in a reduction in cross-sectional area of about 4 percent. Thus, the beneficial effect of a large flight radius (reduced screw contact area) is larger than the adverse effect (reduced cross-sectional area of the screw channel). An additional benefit of the large flight radius is the reduced chance of solid bed deformation, see Section 8.2.3.

Another method to reduce the screw contact area is to use flat slanted flight flanks; i.e., a trapezoidal flight geometry, see also Figure 8-4b. If the flight flank angle is 45 degrees, the screw contact area in the flat plate model becomes:

$$A_S = [\pi(D-2H)\sin\varphi_b - w + H(2\sqrt{2}-2)]\Delta z_b \qquad (7-57d)$$

When the channel width is about ten times the channel depth, the screw contact area will be reduced by about 10 percent. However, the cross-sectional area of the screw channel will be reduced by the same percentage. Thus, the net effect will be less beneficial than the curved flight geometry.

A multiple-flighted screw geometry will adversely effect the solids conveying performance for two reasons. An additional flight will reduce the open cross-sectional area of the screw channel, causing a reduction in solids conveying rate at a constant solid bed velocity. Also, each channel will have a much larger H/W ratio; therefore, the solids conveying angle (see equation 7-52) and the solid bed velocity (see equation 7-46) will be reduced. The combined effect can cause a substantial reduction in solids conveying rate. This is shown in Figure 7-13 for a 50-mm extruder screw under the same conditions as described in Figure 7-11.

From Figure 7-13 it can be seen that the difference in solids conveying rate can be substantial. At larger values of the channel depth, the difference between single or double-flighted screw geometry can mean the difference between output or no output at all. This has been experimentally observed by the author in unpublished studies on solids conveying in single

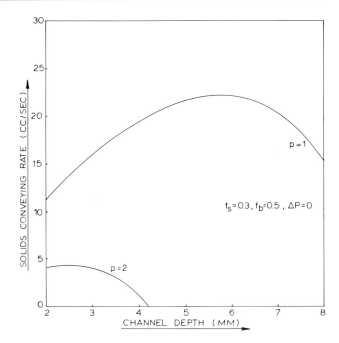

Figure 7-13. The Effect of Multiple Flights on Solids Conveying Performance

screw extruders. A short 4 L/D 19-mm extruder, used just for solids conveying, was outfitted with a single-flighted screw and yielded reasonable output with open discharge and with moderate discharge pressures. The same extruder with a double-flighted screw with all other dimensions the same did not yield any output at all, even with open discharge.

Clearly, this adverse effect of multiple flights on solids conveying performance will be less severe as the screw diameter increases. Nevertheless, the adverse effect is there and it should be taken into account. It can also be seen in Figure 7-13 that the optimum channel depth becomes smaller as the geometry is changed from a single-flighted to a double-flighted screw geometry. The effect of the number of flights will reduce as the coefficient of friction on the barrel becomes larger relative to the friction on the screw. In extruders with grooved barrel sections, therefore, one would expect a much smaller adverse effect due to multiple flights as compared to an extruder with a standard smooth barrel.

The predicted solids conveying rate is very sensitive to the values of the coefficient of friction; changes in coefficient of friction of 20 percent can cause changes in rate of 100 to 1000 percent. This has several important implications. If accurate predictions are required for application to an actual extrusion problem, very accurate data on coefficient of friction will be required. However, measurement of accurate and meaningful data on coefficient of friction is very difficult, as discussed in Section 6.1.2. The reproducibility of measured data on the coefficient of friction generally ranges from 10 to 50 percent. This means that the solids conveying rate predicted from these data cannot be very accurate in a quantitative sense. The predicted results should be analyzed in a qualitative sense. In that respect, the solids conveying theory is very useful in analyzing extrusion problems and in screw design. Thus, the theory should be used to uncover the important trends.

If the actual solids conveying performance is as sensitive to the coefficient of friction as the theory indicates, small changes in the actual coefficient of friction would have a substantial effect on the entire extrusion process. An external means of altering the coefficient of friction is the temperature setting. By changing the barrel temperature, the coefficient of friction on the barrel will change and by changing the screw temperature the coefficient of friction on the screw will change. Thus, small changes in the temperature in the feed section should be able to have a large effect on the overall extruder performance. This behavior has actually been observed by several workers in the field. Kessler, Bonner, Squires, and Wolf (47) reported a rather convincing case concerning the extrusion of nylon on a 82.55-mm (3.25-inch) diameter extruder. A 28°C (50° F) increase in the rear barrel temperature reduced the diehead pressure fluctuation from about 2.8 MPa (400 psi) down to about 0.4 MPa (60 psi).

Ideally, the barrel temperature should be set to the temperature at which f_b is maximum and the screw temperature to the temperature at which f_s is minimum. If the coefficient of friction is known as a function of temperature, then the optimum barrel and screw temperature can be determined directly from the friction data. Unfortunately, in most practical situations, data on coefficient of friction as a function of temperature are not known. Thus, in most cases, the optimum barrel and screw temperatures have to be determined by trial and error. The effect of the rear barrel temperature on extruder performance is generally larger than any other temperature zone. In practice, therefore, considerable attention should be paid to the proper fine-tuning of the rear barrel temperature zones.

7.2.2.1   FRICTIONAL HEAT GENERATION

If one would want to determine optimum temperature settings from data on the coefficient of friction at various temperatures, there is an additional complication that should be taken into account. This complication is the frictional heat generation that occurs in the solids conveying zone. As a result of this frictional heat generation, the temperature at the interface between solid bed and barrel may be substantially higher than the barrel temperature setting indicates.

In any sliding motion where a frictional force is operational, there will be frictional heat generation between the two bodies. The skeptic can experimentally verify this fact by climbing up into a rope for a reasonable distance and then sliding down while keeping the hands tightly clamped around the rope. In a very short distance, the effects of frictional heat generation between the rope and the hands will become noticeable. Most likely, the experimenter will develop a healthy respect for what frictional heat generation is capable of doing. The rate of frictional heat generation equals the product of the frictional force F_f and the relative velocity Δv:

$$\dot{Q} = F_f \Delta v = f F_n \Delta v \tag{7-58}$$

where F_n is the normal force and f the coefficient of friction.

On the screw surface, the relative velocity between the solid bed and screw is simply the solid bed velocity. Thus, the frictional heat generation on the screw is:

$$\dot{Q}_s = P f_s (W+2H) v_{sz} dz + F^* v_{sz} \tag{7-59}$$

where $F^* = P f_b W \sin(\theta + \varphi) dz$.

Written in different form:

$$\dot{Q}_s = P v_{sz} dz [f_s (W+2H) + f_b W \sin(\theta+\varphi)] \tag{7-60}$$

On the barrel surface, the relative velocity between solid bed and barrel can be written as:

$$\Delta v = v_b \frac{\sin\varphi}{\sin(\theta+\varphi)} \qquad\qquad (7\text{--}61)$$

The frictional heat generation on the barrel surface is:

$$\dot{Q}_b = Pf_b W v_b \frac{\sin\varphi}{\sin(\theta+\varphi)} \, dz \qquad\qquad (7\text{--}62)$$

In most cases, the solid bed velocity v_{sz} will be small compared to the relative velocity between solid bed and barrel Δv. Therefore, the frictional heat generation will generally be higher on the barrel surface than on the screw surface. The frictional heat generation of the barrel surface is dissipated into two fluxes, one conducting the heat into the solid bed and the other conducting the heat into the extruder barrel. The actual temperature profile of the solid bed will depend strongly on the heat flux in the barrel. If the barrel is intensely cooled, the majority of the frictional heat generation will conduct away through the barrel.

This tends to slow down the temperature development at the interface and extends the length of the solids conveying zone. When the temperature at the interface reaches the melting point, the solids conveying zone will terminate because polymer melt will form at the interface and the solid-to-solid frictional mechanism will cease to operate.

It should be noted in equations 7-59 through 7-62 that the frictional heat generation is directly proportional to the local pressure. Earlier it was determined that the local pressure increases exponentially with down channel distance, see equations 7-51 and 7-54. Therefore, the frictional heat generation will increase exponentially with down channel distance. As a result, the temperature at the interface will closely follow the local pressure. This is shown qualitatively in Figure 7-14.

Thus, when the local pressure becomes sufficiently high, the temperature at the interface will reach the melting point. At this location, the exponential rise in pressure will terminate because the solid-to-solid frictional mechanism breaks down. In fact, this pressure/temperature relationship constitutes a built-in safety mechanism against the development of very high pressures. The temperature rise resulting from the pressure rise limits the maximum pressure that can develop in the solids conveying zone. Thus, the extruder self-regulates the maximum pressure that develops in the solids conveying zone. By intensely cooling the barrel, the maximum pressure can be increased considerably. Obviously, the maximum pressure that can develop will also depend on the actual coefficients of friction and the screw geometry.

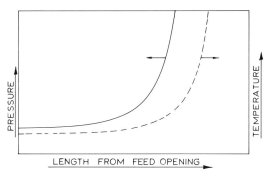

Figure 7-14. Typical Pressure and Temperature Profiles Along the Extruder

LENGTH FROM FEED OPENING

Because of the inherent non-isothermal nature of the solids conveying process, accurate prediction of the actual solids conveying process becomes quite difficult. Not because the mathematics are so complicated, they are relatively straightforward, but because the coefficient of friction should be known as a function of temperature and pressure. This information is generally not available. Unless this information is available, a complete non-isothermal solids conveying analysis would not be very useful. Detailed calculations of non-isothermal solids conveying were performed by Tadmor and Broyer (46). Their numerical calculations of pressure and temperature profiles showed that the temperature rise very closely follows the pressure rise.

7.2.2.2   GROOVED BARREL SECTIONS

In Section 7.2.2, it was discussed how solids conveying can be improved by increasing the roughness of the internal barrel surface. This conclusion can be reached without detailed theoretical analysis; it is obvious from simple qualitative arguments. The desirability of a large coefficient of friction at the barrel was recognized as early as 1941 by Decker (48) as a result of a very simplified analysis. One of the simplest methods of ensuring a high coefficient of friction on the barrel is to machine grooves into the barrel surface.

In the late sixties, several theoretical and experimental studies were made on the effect of grooved barrel sections on solids conveying performance and overall extruder performance (e.g., 49–51). This work was mostly done in West Germany. It was soon realized that substantial improvements could be made to the extruder performance by using grooved barrel sections. The main benefits of the grooved barrel section were found to be: i) substantially improved output, ii) substantially improved extrusion stability, and iii) lower pressure sensitivity of the output. Since these benefits appeal to most extrusion processors, grooved barrel sections have become quite popular. Also, the use of grooved barrel sections has allowed a number of processors to extrude materials that could not be processed on conventional extruders; e.g., very high molecular weight polyethylenes, powders, etc.

A typical grooved barrel section is shown in Figure 7–15.

The effective groove length from the feed port is about 3 D to 5 D. The groove depth generally reduces in a linear fashion, reaching zero depth at the end of the grooved section. Cooling channels are located relatively close to the internal barrel surface. This is important because the cooling capacity of the grooved barrel section has to be high. The high cooling capacity is necessary to avoid too high a temperature rise at the internal barrel surface. Melting should be avoided if the effectiveness of the grooved barrel section is to be main-

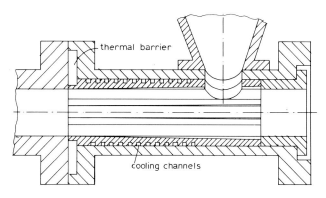

Figure 7-15. Typical Grooved Barrel Section

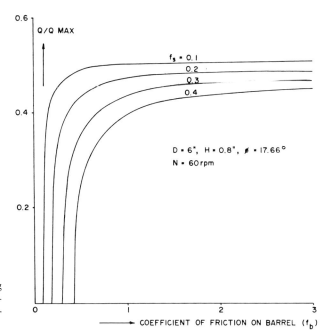

Figure 7-16. Solids Conveying Rate as a Function of the Coefficient of Friction Against the Barrel

tained. From the discussions in the previous section it is clear that with a grooved barrel section there will be a very large frictional heat generation at the barrel surface. Thus, good cooling is crucial to the proper operation of a grooved barrel section. For the same reason, a thermal barrier is generally designed between the grooved barrel section and the smooth barrel section.

In applying the grooved barrel concept to existing extruders, a word of caution is in order. The high effective coefficient of friction at the barrel surface results in a rapid rise in pressure; this is obvious from equation 7-54. However, when the solids conveying section is intensely cooled, the built-in safety mechanism against high pressures can break down. In fact, these grooved barrel sections are designed to ensure that the safety mechanism breaks down because melting has to be avoided in the grooved barrel section. As a result, extremely high pressures will develop in the grooved barrel section. Pressures of 100 MPa to 300 MPa (15,000 to 45,000 psi) are not uncommon. Therefore, these extruders have to be specially designed to withstand these high pressures. Otherwise, mechanical failure of the barrel would occur. Thus, if a grooved barrel section is used in an existing extruder, it is prudent to keep the length of the grooved section reasonably short (1D to 2D) in order to avoid excessive pressures. Obviously, this will limit the benefits that one can derive from a grooved barrel section.

Another practical consideration is the wear of the grooves. Since the active edge of the groove is exposed to very high stresses, considerable wear can occur, particularly if the polymer contains abrasive components. Therefore, the grooved barrel section is generally made out of a strong wear-resistant material in order to maintain optimum performance over a long period of time.

The benefits of a grooved barrel section can be analyzed from the theory of drag induced solids conveying developed in Section 7.2.2. Figure 7-16 shows how the solids conveying rate varies with the coefficient of friction on the barrel f_b, for a situation where the pressure gradient is zero.

The figure shows curves of constant coefficient of friction on the screw f_s. Four typical values are shown $f_s = 0.1$, $f_s = 0.2$, $f_s = 0.3$, and $f_s = 0.4$. The predictions shown in Figures 7-16 and 7-17 are based on the equations in Appendix 7-3. It can be seen that the curve rises steeply when f_b is small, but later reaches a plateau at high f_b values. Two major problems are evident when $f_b \simeq f_s$. One is that the solids conveying rate is considerably below the theoretical maximum value. The second, more important problem is that small changes in f_b will result in very large changes in the solids conveying rate when $f_b \simeq f_s$. This will lead directly to surging of the extruder. Since small variations in the friction on the barrel are bound to occur by the nature of the process, there is a high possibility of extrusion instabilities when f_b is approximately equal to f_s. Thus, this constitutes an unstable operating point. When f_b is increased to a value about two or three times f_s, the solids conveying rate increases substantially, but, at the same time, the slope of the curve is much reduced. In this situation, small variations in the barrel friction will result only in small changes in the solids conveying rate. Thus, the process will be inherently much more stable when f_b is much larger than f_s. This explains how grooved barrel sections can substantially improve extrusion stability.

Figure 7-17 shows how the solids conveying rate varies with the pressure gradient along the solids conveying zone.

When $f_b = f_s$, the solids conveying rate at a zero pressure gradient is quite small and the rate drops off quickly as the pressure gradient increases. At a relatively small pressure gradient, the rate becomes zero. As f_b is increased at constant f_s, the rate at zero pressure gradient increases and the fall-off with pressure gradient becomes less severe. At relatively high values of f_b, the fall-off with pressure gradient becomes very small; in fact, the output becomes essentially independent of back pressure. This indicates a high degree of positive displacement behavior and this is quite unusual in conventional extruders. However, many workers (e.g., 52, 53) have experimentally verified the fact that the output is essentially independent of back pressure when an extruder is equipped with a grooved barrel section. Thus, the

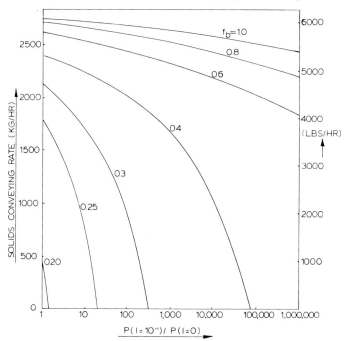

Figure 7-17. Solids Conveying Rate as a Function of the Pressure Gradient

three main benefits of grooved barrel sections, high output, good stability, and pressure independent output, can be predicted directly from theory.

Besides the high pressures and the wear problems, there are a few other disadvantages of grooved barrel sections. The main disadvantage is probably the fact that a substantial amount of energy is lost through the intensive cooling of the grooved barrel section. Nowadays, with the increasing cost and concern about energy, this energy loss is more of a factor than it was 20 years ago. Detailed measurements of energy consumption in various sections of the grooved barrel extruder were made by Menges and Hegele (54). They found that as much as 30 percent to 40 percent of the mechanical energy is lost through the cooling water; about 60 percent of the mechanical energy is dissipated in the solids conveying zone. In the worst case, the specific energy loss through the cooling water is about 150 KJ/kg = 0.042 kWhr/kg (0.07 Hphr/lb). By improving the thermal barrier between the grooved barrel section and smooth barrel section, the specific energy loss can be reduced to about 100 kJ/kg = 0.028 kWhr/kg (0.046 Hphr/lb). However, considering that the specific enthalpy, see Section 6.3.4, of the polymer is generally around 0.06 kWhr/kg (0.10 Hphr/lb), the losses in the grooved barrel section are quite substantial. In a later publication, Menges (55) reports a total mechanical energy loss through the cooling water of about 14 percent. The mechanical energy is the energy supplied by the screw which is transformed into heat by frictional and viscous heat generation. Helmy (229) reported lower specific energy consumption with grooved barrel extrusion with HMWPE and MMWPE than with smooth barrel extrusion. He also found lower melt temperatures with grooved barrel extrusion. However, with normal polyethylene, Helmy found the specific energy consumption of the grooved barrel extruder to be about 10 to 25 percent higher than the smooth barrel extruder.

Energy losses in the grooved barrel section can be reduced by reducing the amount of cooling. This can be achieved by a closed-loop temperature control of the grooved barrel section, as discussed by Menges, Feistkorn, and Fischbach (237). They found that the energy efficiency in the grooved feed section could be increased from 45 percent to 80 percent by increasing the cooling water temperature from 5° C to 70° C.

Another related drawback of the grooved barrel extruder is its higher torque requirement. On a modification of an existing extruder, this generally requires a gear change; in some cases, a new high torque drive may be necessary. However, this should not be a problem on a new extruder designed for operation with a grooved barrel section.

Another disadvantage of the grooved barrel section is that material can accumulate in the grooves; this can cause problems with a product changeover. Also, the screw geometry has to be adapted to the presence of a grooved barrel section. Screw design rules that work for conventional extruders do not work for extruders with grooved barrel sections. In extruders with grooved barrel sections, the extruder screws generally have a much lower compression ratio (if at all), a shallower feed section and a deeper metering section. The melting and mixing capability must also be greater because complete melting and thermal homogeneity are more difficult to achieve; higher output at the same screw speed means shorter residence time, thus, less time for completion of melting and for mixing (lower shear strain).

Despite the disadvantages, extruders with grooved barrel sections have found widespread acceptance in Europe. In many cases, the grooved barrel section has become a standard instead of an option. Somewhat surprising, the acceptance of grooved barrel sections in the U.S. has been quite slow. One reason may be that U.S. extruder manufacturers have not promoted the grooved barrel concept vigorously. Only since the early eighties have extruders with grooved barrel sections become more common in the U.S.

Most grooved barrel sections used in the past had axial grooves running parallel to the axis of the screw. A relatively recent development is the barrel section with helical grooves. One of the first publications on the subject was an article by Langecker et al. (56). They claimed a higher conveying efficiency as compared to axial grooves and a 20 percent reduction in energy consumption; this corresponded to a 45 percent reduction in energy loss through cooling of the grooved barrel section. Langecker et al. also found that screws with a very small compression ratio were most suitable for use with helically grooved barrel sections; in some cases, compression ratios of unity or slightly less than unity (actually a decompression screw) gave optimal performance. Langecker filed for a patent (57) on the helical groove idea as far back as 1972. Another patent on barrel sections with helical grooves was issued to Maillefer (58) in 1979.

Gruenschlosz (59) presented an analysis of a barrel section with helical grooves, attempting to explain the advantages of the helical grooves over axial grooves. Gruenschlosz examines a situation where the grooves in the barrel are quite wide and deep, similar to the channel in the screw. Thus, transport occurs in the barrel channel as well as in the screw channel. He also assumes no shearing between the bulk in the screw channel and in the barrel channel.

The effect of the helical grooves can be explained by considering the velocity diagram, shown earlier in Figure 7-9. Figure 7-18a shows a typical velocity diagram with a smooth barrel. The frictional heat generation is directly determined by the relative velocity between the barrel and the solid bed Δv and the coefficient of friction between the barrel and the solid bed f_b. Figure 7-18b shows a velocity diagram with axial grooves in the barrel.

With the grooves in the axial direction, there will be no movement of the material in the barrel grooves. The frictional heat generation in this situation will be determined again by Δv and the effective coefficient of friction on the barrel surface f_{eb}. With axial grooves, the relative velocity difference between the barrel and solid bed will be somewhat smaller than with a smooth barrel. However, the effective coefficient of friction on the barrel surface f_{eb} will be much higher as a result of the grooves. The net effect is that the frictional heat generation will be much higher than with a smooth barrel.

The effective coefficient of friction will be partially determined by the contact of the solid bed in the screw channel and the barrel flight tip surface and partially by the contact of the solid bed in the screw channel and the solid bed in the barrel grooves. Friction at the barrel flight tip surface is similar to friction at a smooth barrel wall. However, friction at the solid bed in the barrel grooves will be of an entirely different nature. To some extent, the friction will be determined by the internal friction of the polymer, but the friction will be augmented by the action of the active edge of the barrel groove. If the groove is relatively wide compared to the barrel flight width, one can assume that an effective coefficient of friction is roughly determined by the internal coefficient of friction of the polymer. Since the internal

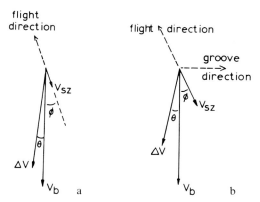

Figure 7-18. Velocity Diagram with Smooth Barrel (a) and Axially Grooved Barrel (b)

coefficient of friction is generally two to three times higher than the external coefficient of friction, it can be assumed that the effective coefficient of friction with axial grooves is about two to three times as high as compared to a smooth barrel. This explains the high frictional heat generation in axially grooved barrels.

The frictional heat generation can be reduced by reducing the effective coefficient of friction or by reducing the velocity difference between the solid bed in the screw channel and the barrel. Reducing the effective coefficient of friction will adversely affect the solids conveying rate and pressure generating capability. However, the velocity difference can be reduced by using helical grooves. This is explained by the velocity diagrams in Figure 7-19.

The velocity diagram in the case of no material movement in the barrel grooves is shown in Figure 7-19 a. If the material in the barrel grooves is stationary with respect to the barrel, the effective barrel velocity equals the actual barrel velocity. The effective velocity difference Δv_1 in the case of no material movement in the barrel grooves is:

$$\Delta \vec{v}_1 = \vec{v}_b - \vec{v}_{sz} \qquad (7\text{-}63)$$

However, the situation will change significantly if there is movement of the material in the barrel grooves. Figure 7-19b shows the situation where the velocity of the material in the barrel grooves is v_{sg}. In this case, the effective barrel velocity becomes v_{be}; this is determined from vectorial addition of the barrel velocity v_b and the solid bed velocity in the barrel grooves v_{sg}:

$$\vec{v}_{be} = \vec{v}_b + \vec{v}_{sg} \qquad (7\text{-}64)$$

The magnitude of v_{be} is determined by the following relationship:

$$v_{be} = (v_b^2 + v_{sg}^2 - 2|v_b||v_{sg}|\cos\varphi_b)^{1/2} \qquad (7\text{-}65)$$

where φ_b is the barrel groove helix angle.

The relative velocity between the solid bed in the screw channel and the material in the barrel grooves Δv_2 is determined from the vectorial difference between v_{be} and v_{sz}:

$$\Delta \vec{v}_2 = \vec{v}_{be} - \vec{v}_{sz} = \vec{v}_b + \vec{v}_{sg} - \vec{v}_{sz} \qquad (7\text{-}66)$$

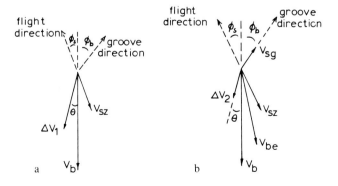

Figure 7-19. Velocity Diagram with a Helically Grooved Barrel Without (a) and With (b) Transport in Barrel Groove

From figure 7-19b it is clear that the effective relative velocity can be reduced substantially if there is movement of the material in the barrel grooves. The magnitude of the effective velocity difference Δv_2 is:

$$\Delta v_2{}^2 = v_b{}^2 + v_{sz}{}^2 + v_{sg}{}^2 + 2v_b(v_{sg}\cos\varphi_b - v_{sz}\cos\varphi_s) +$$
$$2v_{sz}v_{sg}(\sin\varphi_s\sin\varphi_b - \cos\varphi_s\cos\varphi_b) \qquad (7\text{-}67)$$

where φ_s is the helix angle of the screw and φ_b the helix angle of the barrel.

It can be easily seen from Figure 7-19b that the effective velocity difference Δv_2 is minimized when:

$$v_{sg}\sin\varphi_b = v_{sz}\sin\varphi_s \qquad (7\text{-}68)$$

This is the case when the helix angle on the barrel φ_b equals the solids conveying angle of the solid bed in the screw channel θ_s. Thus, the effective velocity difference and the frictional heat generation is minimized when:

$$\varphi_b = \theta_s \qquad (7\text{-}69)$$

If this condition can be achieved, the conveying efficiency of the solid bed in the screw channel will be high because of a large frictional force acting on it at the barrel surface. At the same time, however, the velocity difference between the solid bed in the screw channel and the material in the barrel groove is minimized, resulting in a substantially reduced frictional heat development in the grooved section. This approach to the conveying mechanism in grooved barrel sections demonstrates that the claimed advantages of helical grooves can be confirmed by an engineering analysis. Therefore, it does seem to make sense to use helical grooves instead of axial grooves. Helical grooves have the potential to eliminate one of the main drawbacks of axial grooves; that is, the substantial loss of energy as a result of the very high frictional heat generation and the need for intensive cooling, causing loss of energy through cooling of the grooved barrel section.

In the actual extrusion process, the solids conveying angle θ_s will not be absolutely constant along the length of the screw. In the initial portion of the channel, a substantial amount of compacting will often take place, resulting in corresponding reductions in the solid bed velocity, and, thus, in the solids conveying angle θ_s, see equation 7-48. The theoretically optimum barrel helix angle φ_b, therefore, will vary along the axial length of the grooved barrel section. This would be quite difficult to machine, however, and could make the grooved section relatively expensive. From a practical point of view, it would seem reasonable to make the barrel helix angle equal to the solids conveying angle based on a fully compacted solid bed. With a fully compacted solid bed, the pressure and frictional heat generation are very high and, thus, more of a concern.

In order to obtain an expression for the optimum helix angle of the barrel groove(s), the forces acting on the solid bed in the screw channel and on the solid bed in the barrel groove can be analyzed in similar fashion as done for the smooth barrel. Figure 7-20a shows the forces acting on the solid bed in the screw channel and the corresponding velocity diagram.

The frictional force acting on the barrel surface F_{bs} makes an angle β with the tangential direction. This angle is determined by the direction of the velocity vector Δv_2. If it is assumed

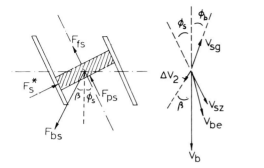

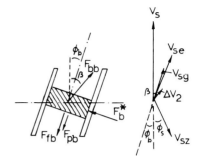

Figure 7-20a. Force Diagram of a Solid Bed Element in the Screw Channel

Figure 7-20b. Force Diagram of the Solid Bed Element in a Barrel Groove

that the frictional force F_{bs} is determined by the internal friction of the bulk material, then the force can be expressed as:

$$F_{bs} = Pf_iW_sdz_s \qquad (7-70)$$

The extra force F_s^* can be determined from a force balance in cross channel direction:

$$F_s^* = F_{bs}\sin(\varphi_s+\beta) = Pf_iW_s\sin(\varphi_s+\beta)dz_s \qquad (7-71)$$

The frictional force on the screw surface F_{fs} can be determined from:

$$F_{fs} = Pf_s(W_s+2H_s)dz_s \qquad (7-72)$$

The net pressure force acting on the solid bed element is:

$$F_{ps} = \frac{dP}{dz_s} W_sH_sdz_s \qquad (7-73)$$

The relationship between angle β and the pressure gradient in the screw down channel direction is obtained by a force balance in the screw down channel direction:

$$F_{bs}\cos(\varphi_s+\beta) - F_s^*f_s = F_{fs} + F_{ps} \qquad (7-74)$$

This yields the following equation:

$$Pf_i[\cos(\varphi_s+\beta) - f_s\sin(\varphi_s+\beta)] = Pf_s\left(1+ \frac{2H_s}{W_s}\right) + \frac{dP}{dz_s} H_s \qquad (7-75)$$

Figure 7-20b shows the forces acting on the solid bed in the barrel groove and the corresponding velocity diagram.

In using the velocity diagram, it should be kept in mind that the barrel is now taken as being stationary and the screw moving at a tangential velocity $v_s = -v_b$. The solid bed is moving at velocity v_{sg}. The velocity of the solid bed in the screw channel with respect to the stationary barrel is v_{se}; this is determined by the vectorial addition of v_s and v_{sz}:

$$\vec{v}_{se} = \vec{v}_s + \vec{v}_{sz} \tag{7-75a}$$

The frictional force acting on the screw surface F_{bb} makes the same angle β with the tangential direction. The angle, in this case, is determined by the vectorial difference of v_{se} and v_{sg}:

$$\Delta\vec{v}_2 = \vec{v}_{se} - \vec{v}_{sg} = \vec{v}_s + \vec{v}_{sz} - \vec{v}_{sg} \tag{7-76}$$

The frictional force F_{bb} is determined from:

$$F_{bb} = Pf_i W_b dz_b \tag{7-77}$$

The extra force F_b^* is determined from a force balance in cross-groove direction:

$$F_b^* = F_{bb} \sin(\beta - \varphi_b) \tag{7-78}$$

The frictional force on the barrel surface is:

$$F_{fb} = Pf_b (W_b + 2H_b) dz_b \tag{7-79}$$

The force resulting from the pressure gradient is:

$$F_{pb} = \frac{dP}{dz_b} W_b H_b dz_b \tag{7-80}$$

As before, the relationship between angle β and the pressure gradient in the barrel down groove direction is obtained by a force balance on the solid bed element in the down groove direction:

$$Pf_i[\cos(\beta - \varphi_b) - f_b \sin(\beta - \varphi_b)] = Pf_b\left(1 + \frac{2H_b}{W_b}\right) + \frac{dP}{dz_b} H_b \tag{7-81}$$

The optimum barrel helix angle φ_b^* is the one for which the velocity difference Δv_2 is minimized; this occurs when $\varphi_b = \theta_s$ and $\beta = 0$. At this point, the pressure gradient in the down channel direction can be expressed in known quantities:

$$\frac{dP}{dz_s} = \frac{Pf_i}{H} (\cos\varphi_s - f_s \sin\varphi_s) - \frac{Pf_s}{H_s}\left(1 + \frac{2H_s}{W_s}\right) \tag{7-82}$$

This pressure gradient can now be inserted into equation 7–81 to yield an expression for the optimum barrel helix angle. It should be remembered that the relationship between the down channel coordinate z_s and the down groove coordinate z_b is:

$$z_b \sin \varphi_b = z_s \sin \varphi_s \qquad (7\text{--}83)$$

The expression for the optimum barrel helix angle now takes the following form:

$$\cos \varphi_b^* = A_1 \sin \varphi_b^* + A_2 \qquad (7\text{--}84)$$

where:

$$A_1 = \frac{H_b}{H_s \sin \varphi_s} \left(\cos \varphi_s - f_s - \frac{f_s}{f_i} - \frac{2 H_s f_s}{f_i W_s} \right) - f_b \qquad (7\text{--}84\,a)$$

and

$$A_2 = \frac{f_b}{f_i} \left(1 + \frac{2 H_b}{W_b} \right) \qquad (7\text{--}84\,b)$$

The solution to equation 7–84 can be expressed in the now familiar form:

$$\varphi_b^* = \arcsin \left[\frac{(1 + A_1^2 - A_2^2)^{1/2} - A_1 A_2}{1 + A_1^2} \right] \qquad (7\text{--}85)$$

Equation 7–85 allows the calculation of the optimum barrel helix angle if the screw geometry is known and if the various coefficients of friction, internal and external, are known as well. The solution for the optimum helix angle is not completely analytical because A_2 contains a term W_b that is dependent on the barrel helix angle. This can be solved by initially guessing a value of W_b, then calculate φ_b^* according to equation 7–85. Then W_b can be calculated with:

$$W_b = \frac{\pi D \sin \varphi_b^*}{p_b} - w_{bg} \qquad (7\text{--}85\,a)$$

where w_{bg} is the perpendicular barrel flight width and p_b the number of parallel grooves in the barrel. The calculated value of W_b can then be used to calculate φ_b^* again. This process can be repeated until the initial value of W_b and the calculated value of W_b are within a certain tolerance. Convergence is very rapid and accurate values of φ_b^* are generally obtained in two or three iterations. Figure 7–21a shows φ_b^* as a function of the screw helix angle when $f_b = f_s = 0.2$, $f_i = 0.6$, and $H_s = 15.24$ mm (0.6 inch).

The optimum barrel helix angle increases with the screw helix angle and with reducing barrel groove depth. The optimum barrel helix angle is relatively insensitive to the internal coefficient of friction as shown in Figure 7–21b.

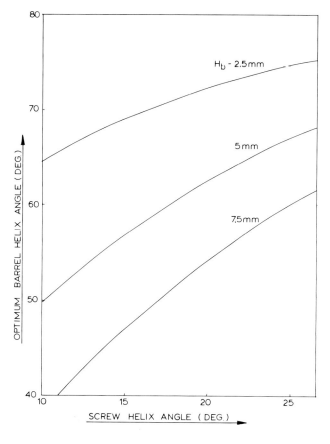

Figure 7-21a. Optimum Barrel Helix Angle as a Function of the Screw Helix Angle

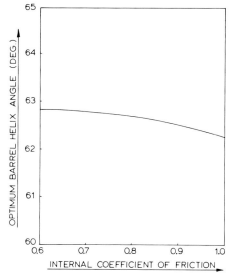

Figure 7-21b. Optimum Barrel Helix Angle versus the Internal Coefficient of Friction

7.3   PLASTICATING

The second functional zone in the extruder is the plasticating zone or melting zone. The melting zone starts as soon as melt appears; usually after 3 to 5 diameters from the feed opening. Since most of the frictional heat generation in the solids conveying zone generally occurs at the barrel-solid bed interface, the first traces of melt usually appear at the barrel surface. It should be noted that the start of melting does not necessarily occur where the measured barrel temperature profile exceeds the melting point. The temperature at the barrel-solid bed interface can be quite different from the measured barrel temperature. In fact, melting can often be initiated without any external heating simply by the action of the frictional heat generation.

As melting proceeds, the initial melt film at the barrel surface will grow in thickness. This is particularly true in the very early stages of melting because the thin melt film is subjected to a high rate of shearing causing a rapid temperature rise in the material and a high melting rate. Once the thickness of the melt film exceeds the radial flight clearance, the melt will flow into the screw channel displacing the solid bed. In most cases, the solid bed will be pushed against the passive flight flank and the melt will start to accumulate in a melt pool between the solid bed and the active flight flank. Maddock (60) was the first worker to accurately describe the melting behavior in single screw extruders. His observations were based on screw extraction experiments, where the screw is stopped with material still in the extruder and the machine is rapidly cooled. The screw is then pushed out of the extruder barrel, generally with a hydraulic piston. The material in the screw channel can then be analyzed at various axial locations along the screw. Maddock's description was accurate but only qualitative; he did not attempt to model the melting process to allow a quantitative description of the melting process.

Several years later, Tadmor and coworkers did extensive experimental work on melting in single screw extruders (61). In addition to the experimental work, Tadmor performed a theoretical analysis of the plasticating process and developed the now classic Tadmor melting model (62). This was a major contribution to the extrusion theory, particularly since melting was the one major extruder function for which a theoretical model had not been developed. Theoretical models for solids conveying and melt conveying or pumping had already been developed in the fifties or earlier. However, the melting theory was not developed until the mid-sixties.

7.3.1   THE TADMOR MELTING MODEL

An idealized cross-section of the screw channel in the melting zone is shown in Figure 7-22.

It is assumed that the screw is stationary; thus, in a crosssection perpendicular to the screw flights, the barrel moves towards the active flight at velocity v_{bx}, which is the cross channel component of the barrel velocity v_b. As a result of this motion, the thin melt film between the solid bed and the barrel is sheared at a high rate. This will cause substantial viscous heat generation in the melt film, see also Section 5.3.4. Since the melt film is quite thin, the effect of pressure gradients on the velocity profile in the melt film will be quite small. Therefore, the flow in the melt film will be essentially drag flow and the shear rate and viscous heat generation will be relatively uniform along the depth of the melt film.

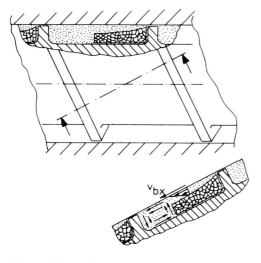

Figure 7-22. Tadmor Melting Model

The melt flows from the melt-film towards the active flight flank. Only a small fraction of the material can flow through the clearance. As a result, the majority of the melt will flow into the melt pool. A circulating flow will be set up in the melt pool as a result of the barrel velocity. Since most of the viscous heat generation occurs in the upper melt film, it is generally assumed that all melting takes place at the upper solid bed-melt film interface. As melting proceeds, the cross-sectional area of the solid bed will reduce and the cross-sectional area of the melt pool will tend to increase. The melt pool, therefore, will exert considerable pressure on the solid bed. This reduces the width of the solid bed, while the melt film between the solid bed and the barrel remains relatively constant. In order for this mechanism to work, the solid bed has to be continuously deformable. The width of the solid bed reduces as more material melts away from the barrel side of the solid bed. Thus, the material in the solid bed close to the solid-melt interface moves towards the interface as melting proceeds. This occurs at a velocity v_{sy}, the solid bed melting velocity. This velocity determines the melting rate of the solid bed.

There are basically two sources of energy utilized for melting in the extruder. The first and generally the most important one is the mechanical energy supplied by the screw which is transformed into heat by a process of viscous heat generation. The second source of energy is the heat supplied by the external barrel heaters and possibly screw heaters. In most extruders, the majority of the energy will be supplied by the screw, about 80 to 90 percent or more. There is a very good reason for this. Energy supplied by the screw will be dissipated primarily in the melt film. The resulting viscous heat generation will be relatively uniform throughout the material. Thus, the temperature rise in the polymer melt will be relatively uniform and the heat transfer distances will be small. On the other hand, heat supplied by the barrel heaters has to be conducted through the entire thickness of the barrel and through the entire thickness of the melt film before it can reach the solid bed. Problems with this energy transport are considerable heat losses by conduction, convection and radiation. Another, probably more severe, problem is the low thermal conductivity of the polymer. The heat has to be transferred across the entire melt film thickness and, therefore, the conductive heat flux will be small. This heating process can be sped up by increasing the barrel temperature; however, this temperature is limited by the possibility of degradation taking place in the polymer.

If melting only occurs by heating from the barrel heaters without viscous heat generation, the rate of melting would be unacceptably slow. This is precisely why ram extruders have such a poor plasticating capability; there is little or no viscous heat generation in these type of extruders, see Section 2.4. This also explains why reciprocating single screw extruders have become so popular on injection molding machines, even though they are much more complex than ram extruders. Thus, it should be clear that the key to the plasticating ability of screw extruders is the viscous heat generation in the polymer melt.

In order to be able to predict the melting rate, the amount of heat flowing towards the solid-melt interface has to be known. This can be determined if the temperature profile in the melt film and in the solid bed are known. The temperature profile in the melt film can be determined from the velocity profile in the melt film. In order to derive an expression for the temperature profile in the melt film, the following assumptions are made:

1) The process is a steady-state process.

2) The polymer melt density and thermal conductivity are constant.

3) Convective heat transfer is neglected.

4) Conductive heat transfer only in a direction normal to interface.

With these assumptions, the general energy equation (5–5) can be simplified to:

$$k_m \frac{d^2T}{dy^2} + \tau_{x'y} \frac{\partial v_{x'}}{\partial y} = 0 \qquad (7-86)$$

The shear stress can be determined from the equation of motion. In order to derive an expression for the shear stress, additional assumptions are made:

5) The polymer melt flow is laminar.

6) Inertia and body forces are negligible.

7) There is no slip at the walls.

8) There are no pressure gradients in the melt film.

9) The temperature dependence of the viscosity can be neglected.

With these assumptions, the general equation of motion (5–3) can be simplified to:

$$\frac{d\,\tau_{x'y}}{dy} = 0 \qquad (7-87)$$

The direction of coordinate x' is determined by the vectorial velocity difference between the barrel and the solid bed, see Figure 7–23a.

The magnitude of this velocity difference Δv, the relative velocity, is given by equation 7–47 and the angle θ with the tangential direction is:

$$\theta = \arctan\left(\frac{v_b \sin\varphi}{v_b \cos\varphi - v_{sz}}\right) - \varphi \qquad (7-88)$$

This angle θ is essentially the same angle as the solids conveying angle discussed in Section 7.2.2. If the polymer melt behaves as a Newtonian fluid, equation 7–87 becomes:

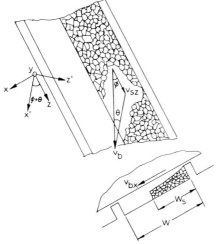

Figure 7-23a. Velocity Diagram and Coordinate System in the Melting Model

$$\frac{d^2 v_{x'}}{dy^2} = 0 \qquad\qquad (7\text{-}89)$$

By integrating twice and taking the boundary conditions $v_x'(0)=0$ and $v_x'(H_m)=\Delta v$, the following linear velocity profile is obtained:

$$v_x' = \frac{y}{H_m}\, \Delta v \qquad\qquad (7\text{-}90)$$

Equation 7-86 now becomes:

$$k_m \frac{d^2 T}{dy^2} + \mu \left(\frac{\Delta v}{H_m}\right)^2 = 0 \qquad\qquad (7\text{-}91)$$

The boundary conditions are $T(0)=T_m$ and $T(H_m)=T_b$. By integrating twice, a quadratic temperature profile is obtained:

$$T(y) = \left(\frac{-\mu \Delta v^2}{2 k_m H_m^{\,2}}\right) y^2 + \left(\frac{2 k_m \Delta T_b + \mu \Delta v^2}{2 k_m H_m}\right) y + T_m \qquad (7\text{-}92)$$

where $\Delta T_b = T_b - T_m$.

From equation 7-92, the heat flow into the interface Q_{in} can now be determined. By using Fourier's law, equation 5-45, the heat flow per unit area (heat flux) becomes:

$$-\dot{q}_m = \frac{-\dot{Q}_{in}}{A} = k_m \frac{dT(o)}{dy} = \frac{2 k_m \Delta T_b + \mu \Delta v^2}{2 H_m} \qquad\qquad (7\text{-}93)$$

The temperature profile in the solid bed can be determined from the energy equation (5-5) applied to a moving solid slab:

$$\rho_s C_s v_{sy} \frac{dT}{dy} = k_s \frac{d^2T}{dy^2} \qquad (7-94)$$

If the temperature at the interface is taken as the melting point, $T(o) = T_m$, and the temperature far away from the interface is taken as a reference temperature, $T(-\infty) = T_r$, then the temperature profile in the solid bed becomes:

$$T(y) = \Delta T_r \exp\left(\frac{y v_{sy}}{\alpha_s}\right) + T_r \qquad (7-95)$$

where $\Delta T_r = T_m - T_r$.

The reference temperature T_r is typically the temperature at which the solid polymer particles are introduced into the extruder. In equation 7-95, it is assumed that the bulk of the solid bed is at T_r and that only a relative thin skin is heating up. The validity of this assumption depends primarily on the total thickness of the solid bed and the residence time of the solid bed in the extruder. The validity of the assumption can be tested by analyzing the Fourier number, as discussed in Section 5.3.3. If the thickness of the solid bed is about 10 mm and the residence time less than one or two minutes, then the assumption is probably valid. However, if the solid bed thickness is in the range of 3 mm to 4 mm at the same residence times, then the assumption is clearly not valid. Thus, the assumption should not be used when the thickness of the solid bed is less than about 5 mm (0.2 inch); because, in reality, the solid bed is heated from all sides. This means that equation 7-95 should not be used for small diameter extruders with shallow channels and should not be used towards the end of melting if the solid bed thickness has reduced to less than 5 mm.

The actual temperature profile in the solid bed will change with axial location. The developing temperature profile can be described with equation 6-98 for one-sided heating, as discussed in Section 6.3.5. The temperature profile in the center for two-sided heating is shown in Figure 7-23 b. The vertical axis shows the dimensionless temperature and the horizontal axis shows the Fourier number, as discussed in Section 5.3.3.

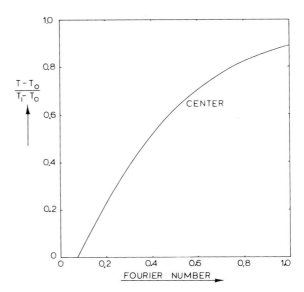

Figure 7-23b. Temperature Profile in a Solid Slab With Two-Sided Heating

Figure 7-23b is a graphical representation of the equation describing the developing temperature profile in a finite slab heated from both sides when $y=0$ (reference 16 of Chapter 5):

$$\frac{T_1-T}{T_1-T_0} = 2\sum_{n=0}^{\infty} \frac{(-1)^n}{(n+0.5)\pi} \exp\left[-\left(n+\frac{1}{2}\right)^2 \pi^2 \alpha t/b^2\right] \cos\frac{\pi y(2n+1)}{2b} \qquad (7\text{-}96)$$

where b is half the slab thickness.

Within an accuracy of about 10 to 20 percent, the following approximation can be made for the temperature profile at the center of the slab:

$$\frac{T_1-T}{T_1-T_0} \simeq \frac{\alpha t}{b^2} \qquad (7\text{-}97)$$

This approximation can be used to obtain a corrected reference temperature T_r^*, where T_r^* increases with axial distance along the screw:

$$T_r^* = T_r + \left(\frac{4\alpha_s z}{v_{sz}H_s^2}\right)\Delta T_r \qquad (7\text{-}98)$$

where the bracketed term represents the Fourier number of the solid bed.

This corrected reference temperature T_r^* can be used in equation 7-95 instead of T_r.

The effect of this correction will be an accelerated melting towards the end of the melting zone. The heat flux from the interface into the solid bed $\dot{q}_{out}$ is again determined from Fourier's law:

$$-\dot{q}_{out} = -\frac{\dot{Q}_{out}}{A} = k_s \frac{dT(o)}{dy} = \rho_s C_s v_{sy}\Delta T_r \qquad (7\text{-}99)$$

A more correct determination of the heat flux into the solid bed could be made by using equation 7-96 and differentiating the temperature with respect to the normal distance y.

An additional complication is that the solid bed is assumed to be freely deformable. This means that the heat transfer situation is no longer determinate. Equations $7-94$ through $7-99$ can only be used if the rate of deformation of the solid bed is small relative to the rate of heat conduction into the solid bed. A typical solid bed melt velocity (v_{sy}) is 0.2 mm/sec. The thermal penetration thickness ($4\sqrt{\alpha t}$), see equation $6-99a$, is about 1.3 mm in one second when the thermal diffusivity (α) is $10^{-7} m^2/s$. Thus, the rate of heat conduction into the solid bed is about one order of magnitude higher than the rate of deformation of the solid bed. Thus, the rate of deformation of the solid bed, in most cases, is relatively small compared to the rate of heat conduction into the solid bed.

The temperature profiles in the melt film and solid bed are shown in Figure 7-24.

From a heat balance of the interface, the melting rate can be determined. The heat used to melt the polymer at the interface is determined by the heat flux into the interface minus the heat flux out of the interface:

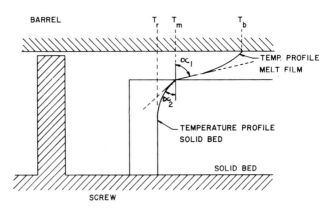

BARREL

T_i T_m T_b

α_1

TEMP. PROFILE
MELT FILM

α_2

TEMPERATURE PROFILE
SOLID BED

SOLID BED

SCREW

Figure 7-24. Temperature
Profiles and Heat Fluxes in
the Melting Model

$$v_{sy}\rho_s\Delta H_f \;=\; \frac{2k_m\Delta T_b+\mu\Delta v^2}{2H_m} \;-\; \rho_s C_s v_{sy}\Delta T_r \tag{7-100}$$

The melting velocity v_{sy} now becomes:

$$v_{sy} \;=\; \frac{2k_m\Delta T_b+\mu\Delta v^2}{2\rho_s H_m \Delta H} \tag{7-101}$$

where $\Delta H = \Delta H_f + C_s \Delta T_r$.

The factor ΔH can be considered the heat sink; it is the enthalpy difference between T_m and T_r. As melting occurs along the direction of relative motion of the solid bed, more molten polymer will accumulate in the melt film. Along a length of dx', the increase in mass flow as a result of melting is:

$$\Delta\dot M'_1 \;=\; \rho_s v_{sy}dx' \tag{7-102}$$

The prime with the mass flow indicates mass flow per unit length. The prime with the x-coordinate indicates the direction determined by the vectorial difference between the barrel and solid bed, see Figure 7-23 a.

This increase in mass flow will require an increase in melt film thickness. The corresponding increase in drag flow rate is:

$$\Delta\dot M'_1 \;=\; \frac{1}{2}\,\rho_m\Delta vdH_m \tag{7-103}$$

It can be assumed that the changes in melt film thickness occur only in direction x' of solid bed relative motion. In that case, by using equations 7-101 through 7-103, a differential equation describing the melt film thickness can be formulated:

$$H_m dH_m \;=\; \left(\frac{2k_m\Delta T_b+\mu\Delta v^2}{\rho_m\Delta v\,\Delta H}\right)dx' \tag{7-104}$$

If the melt film thickness at $x'=0$ is taken as the local radial clearance δ between flight and barrel, then the melt film thickness can be expressed as:

$$H_m(x') = \left[\frac{(4k_m\Delta T_b + 2\mu\Delta v^2)}{\rho_m\Delta v\Delta H}x' + \delta^2\right]^{1/2} \qquad (7\text{--}105)$$

The amount of polymer melting over the entire solid bed width can be found by:

$$\frac{d\dot{M}_p}{dz'} = \int_0^{W_s'}\rho_s v_{sy}dx' = \int_{H_m(0)}^{H_m(W_s')}\frac{1}{2}\rho_m\Delta vdH_m \qquad (7\text{--}106)$$

where W_s' is the solid bed width in direction x'.

This represents the melting rate per unit length in direction z' (see Figure 7-23a); it can be written as:

$$\frac{d\dot{M}_p}{dz'} = \frac{1}{2}\rho_m\Delta vH_m \left|_{H_m(0)}^{H_m(W_s')}\right. \qquad (7\text{--}107)$$

If the clearance is considered to be negligible, $H_m(0)=0$, the melting rate is:

$$\frac{d\dot{M}_p}{dz'} = \left[\frac{(2k_m\Delta T_b + \mu\Delta v^2)}{2\Delta H}\rho_m\Delta vW_s'\right] = \Omega\sqrt{W_s'} \qquad (7\text{--}108)$$

Equation 7-108 expresses the melting per unit length in direction z'. The direction x' is determined by the relative velocity Δv. The relationship between x' and the cross channel coordinate x is: $x = x'\sin(\theta+\varphi)$. The relationship between z' and the down channel coordinate z is: $z' = z\sin(\theta+\varphi)$. Thus, the melting rate per unit down channel distance is:

$$\frac{d\dot{M}_p}{dz} = \Omega\sqrt{W_s'}\ \sin(\theta+\varphi) \qquad (7\text{--}109a)$$

The relationship between z' and axial coordinate l is $l = z'\cos\theta$. Thus, the melting rate per unit axial length is:

$$\frac{d\dot{M}_p}{dl} = \frac{\Omega\sqrt{W_s'}}{\cos\theta} \qquad (7\text{--}109b)$$

The relationship between the solid bed width W_s' in x'-direction and W_s in cross-channel direction is $W_s = W_s'\sin(\theta+\varphi)$. With this relationship, the melting rate per unit down channel length can be written as:

$$\frac{d\dot{M}_p}{dz} = \Omega_1\sqrt{W_s} \qquad (7\text{--}109c)$$

where:

$$\Omega_1 = \left[\frac{(2k_m\Delta T_b + \mu\Delta v^2)\rho_m v_b \sin\varphi}{2\Delta H} \right]^{1/2} \qquad (7-109d)$$

This result will become obvious when it is realized that $\Delta v\sin(\theta + \varphi) = v_b\sin\varphi$, see equation 7-61.

From equation 7-107, it can be seen that the effect of a nonzero radial flight clearance is to reduce the local melting rate. This means that screw or barrel wear in the plasticating zone will reduce the melting performance of the extruder, see also Section 8.2.2.3. From equations 7-108 and 7-109, the contribution from heat conduction $2k_m\Delta T_b$ and the contribution from the viscous heat generation $\mu\Delta v^2$ can be clearly distinguished. As mentioned earlier, the viscous heat generation term is generally larger than the heat conduction term. These equations can also explain why sometimes an increase in barrel temperature does not result in improved melting performance. When the barrel temperature T_b is increased, the heat conduction term increases; however, the viscous heat generation term will decrease because the viscosity in the melt film will decrease with increasing temperature of the melt film. If the reduction in the viscous heat generation is larger than the increase in heat conduction, the net result will be a reduced melting rate. This can occur in polymers whose melt viscosity is very sensitive to temperature, such as PMMA, PVA, PVC, etc. For temperature dependence of melt viscosity, see Section 6.2.4 and Table 6.1.

Another interesting observation is that the local melting rate is directly determined by the width of the solid bed. Obviously, the maximum solid bed width is the channel width. The solid bed width in the very early part of melting can be assumed to be equal to the channel width. As melting proceeds, the solid bed width will generally reduce and, consequently, the local melting rate will reduce with it. In most cases, the highest melting rate is achieved at the start of melting, the melting rate then drops monotically with axial distance as the solid bed width reduces. This is an important consideration in screw design. For good melting performance, one would like to maintain a relatively wide solid bed over a substantial length in order to maintain the highest possible melting rate. This point will be discussed in more detail in Chapter 8 on screw design.

It should be noted that equations 7-102 through 7-109 are slightly different from the equations developed by Tadmor (62), see also reference 5 of chapter 1. The reason is that Tadmor assumed a constant melt film thickness. However, it is clear that the melt film thickness has to increase with cross-channel distance to accommodate the increased amount of melt. Both Shapiro (63) and Vermeulen (64, 65) have analyzed this point in detail and demonstrated that in a consistent model, the melt film thickness must vary with cross channel distance. If the melt film thickness is assumed to be constant, very high cross channel pressure gradients must occur to allow additional material in the melt film. These pressure gradients are of such high magnitude as to be unrealistic. If the thickness of the melt film is assumed constant across the width of the solid bed, the predicted melting rate will be lower by a factor of $\sqrt{2}$ compared to the case where the melt film thickness varies across the width of the solid bed.

The one unknown left at this point is the width of the solid bed. A relationship for the change in solid bed width with down channel distance can be obtained from a mass balance of the solid bed in down channel direction. This can be written as:

$$d\dot{M}_p = -\rho_s v_{sz} d(H_s W_s) \qquad (7-110)$$

In the melting zone, the channel is generally tapered; thus, the channel depth varies linearly with down channel distance:

$$H = H_f - A_z z \qquad (7-111)$$

where H_f is the channel depth of the feed section and A_z the degree of taper in down channel direction.

From equations 7-109 and 7-110, a differential equation is obtained describing the change in solid bed:

$$-\frac{\Omega_1 \sqrt{W_S}}{\rho_S v_{SZ}} = H_S \frac{dW_S}{dz} + W_S \frac{dH_S}{dz} \qquad (7-112)$$

The thickness of the solid bed H_S is primarily determined by the channel depth. Thus, it can be assumed that the change in H_S with distance equals the change in channel depth with distance:

$$\frac{dH_S}{dz} = \frac{dH}{dz} = -A_z \qquad (7-113)$$

This is an important point because this means that without melting, the compression in the channel would cause an increase in solid bed width. In this case:

$$dW_S = \frac{A_z}{H_S} dz \qquad (7-113a)$$

In the absence of melting the solid bed width would increase directly proportional to distance. Thus, there are two mechanisms affecting the width of the solid bed. Melting will cause a reduction of the solid bed width, but at the same time, channel taper will cause an increase in solid bed width. The reduction in W_S from melting should always exceed the increase in W_S from compression. If the compression is too rapid, large A_z, the melting cannot reduce the solid bed width fast enough. As a result, the solid bed width will grow and will plug the channel as it reaches the width of the screw channel. This puts an upper bound on the maximum compression ratio that can be applied in the plasticating section of an extruder screw, more on this in Section 8.2.2.

From equations 7-112 and 7-113, the following differential equation for the solid bed width is obtained if it is assumed that $H \simeq H_S$:

$$-\frac{\Omega_1}{\rho_S v_{SZ}} \sqrt{W_S} = H \frac{dW_S}{dz} - W_S A_z \qquad (7-114)$$

The solution of this equation is:

$$W_S = W_1 \left[\frac{\Omega_1}{A_z v_{SZ} \rho_S \sqrt{W_1}} - \left(\frac{\Omega_1}{A_z v_{SZ} \rho_S \sqrt{W_1}} - 1 \right) \left(\frac{H_f}{H_f - A_z z} \right)^{1/2} \right]^2$$

$$(7-115)$$

where W_1 is the solid bed width at $z = 0$.

The total length for melting can be obtained by setting $W_s = 0$:

$$Z_T = \frac{H_f v_{sz} \rho_s \sqrt{W_1}}{\Omega_1} \left(2 - \frac{A_z v_{sz} \rho_s \sqrt{W_1}}{\Omega_1} \right) \qquad (7-116)$$

From equation 7-115, it can be seen that an extreme condition is reached when the term $\Omega_1/(A_z v_{sz} \rho_s \sqrt{W_1})$ becomes unity. In this case, the solid bed width becomes independent of down channel distance and the shortest possible melting length is obtained $Z_T = H_f/A_z$. However, this condition cannot be achieved in practice because there would be no room for the polymer melt. Thus, in practical extrusion operations, the term $\Omega_1/(A_z v_{sz} \rho_s \sqrt{W_1})$ will be larger than unity to ensure a continuous reduction in solid bed width with distance.

7.3.1.1   NON-NEWTONIAN, NON-ISOTHERMAL CASE

A more realistic prediction of the melting performance can be obtained if the polymer melt is considered non-Newtonian and non-isothermal. However, this extension of the analysis results in coupled energy and momentum equations. Such problems generally do not allow analytical solutions. One approach to this problem, as suggested by Tadmor (61), is to assume a certain temperature profile and solve the equations. If a quadratic temperature profile is assumed, the solution becomes quite elaborate containing many error functions. Evaluation of the solutions requires substantial numerical analysis and number crunching; for details, the reader is referred to reference 5 of chapter 1. If a linear temperature profile is assumed in the melt film, the solution becomes much more manageable. The constitutive equation is:

$$\tau_{x'y} = m_r a_T \left(\frac{dv_{x'}}{dy} \right)^n \qquad (7-117)$$

where $a_T = \exp[\alpha_T(T_m - T)]$, see also equation 6-40.

The equation of motion (equation 5-3) becomes:

$$\frac{d\,\tau_{x'y}}{dy} = 0 \qquad (7-118)$$

This equation can be integrated to give:

$$\tau_{x'y} = K_1 \qquad (7-119)$$

where K_1 is an integration constant.

The assumed temperature profile can be written as:

$$T(y) = \Delta T_b \frac{y}{H_m} + T_m \qquad (7-120)$$

When this expression is substituted in equation 7-117, the velocity gradient will become dependent on normal distance y. Equation 7-117 becomes:

$$\frac{dv_{x'}}{dy} = \left(\frac{K_1}{m_r}\right)^S \exp\left(\frac{K_2}{H_m} y\right) \tag{7-121}$$

where $K_2 = s\alpha_T \Delta T_b$ and s is the reciprocal power law index $s = 1/n$.

With boundary conditions $v_{x'}(0) = 0$ and $v_{x'}(H_m) = \Delta v$, the solution becomes:

$$v_{x'}(y) = \Delta v \, \frac{1 - \exp\left(\frac{K_2}{H_m} y\right)}{1 - \exp K_2} \tag{7-122}$$

The shear rate distribution is determined by taking the first derivative of $v_{x'}(y)$ with respect to y:

$$\frac{dv_{x'}}{dy} = -\frac{\Delta v K_2 \exp\left(\frac{K_2}{H_m} y\right)}{H_m(1 - \exp K_2)} \tag{7-123}$$

The velocity gradient as a function of y can now be inserted into the energy equation 7-86. By taking boundary conditions $T(0) = T_m$ and $T(H_m) = T_b$, the solution becomes:

$$T(y) = T_m + \frac{\Delta T_b y}{H_m} + \frac{B_2 H_m^2}{K_2^2}\left[\exp\left(\frac{K_2 y}{H_m}\right) - 1 - (\exp K_2 - 1)\frac{y}{H_m}\right] \tag{7-124a}$$

where:

$$B_2 = -\frac{m_r}{k_m}\left[\frac{\Delta v K_2}{H_m(\exp K_2 - 1)}\right]^{n+1} \tag{7-124b}$$

For a temperature independent fluid ($\alpha_T = 0$), the factor K_2 becomes zero. When K_2 approaches zero:

$$B_2 \to -\frac{m_r}{k_m}\left(\frac{\Delta v}{H_m}\right)^{n+1} \tag{7-124c}$$

and

$$\left[\exp\left(\frac{K_2 y}{H_m}\right) - 1 - (\exp K_2 - 1)\frac{y}{H_m}\right] \to \frac{1}{2}K_2^2\left(\frac{y^2}{H_m^2} - \frac{y}{H_m}\right) \tag{7-124d}$$

Thus, for a temperature independent power law fluid, the temperature profile is:

$$T(y) = T_m + \frac{\Delta T_b}{H_m} y + \frac{m_r H_m^2}{2k_m} \left(\frac{\Delta v}{\Pi_m}\right)^{n+1} \left(\frac{y}{H_m} - \frac{y^2}{H_m^2}\right) \qquad (7-124e)$$

For a temperature independent Newtonian fluid $(n=1)$, the temperature profile becomes:

$$T(y) = T_m + \frac{\Delta T_b}{H_m} y + \frac{\mu \Delta v^2}{2k_m} \left(\frac{y}{H_m} - \frac{y^2}{H_m^2}\right) \qquad (7-124f)$$

Equation 7-124f corresponds, of course, to equation 7-92 derived earlier and also to equation 5-62. From equation 7-124a, the heat flux from the melt film into the solid-melt interface can be calculated. The first derivative of $T(y)$ with respect to y is:

$$\frac{dT}{dy} = \frac{\Delta T_b}{H_m} + \frac{B_2 H_m}{K_2^2} \left(K_2 \exp\frac{K_2 y}{H_m} - \exp K_2 + 1\right) \qquad (7-125)$$

The heat flux from the melt film into the interface is:

$$-\dot{q}_m = k_m \frac{dT(0)}{dy} = \frac{k_m \Delta T_b}{H_m} + \frac{k_m B_2 H_m}{K_2^2} (K_2 - \exp K_2 + 1) \qquad (7-126)$$

The heat balance for the interface now becomes:

$$v_{sy} \rho_s \Delta H_f = \frac{k_m \Delta T_b + B_3}{H_m} - \rho_s C_s v_{sy} \Delta T_r \qquad (7-127)$$

where:

$$B_3 = \frac{k_m B_2 H_m^2}{K_2^2} (K_2 - \exp K_2 + 1) \qquad (7-128)$$

Equation 7-127 corresponds to equation 7-100 describing the Newtonian, temperature-independent fluid. From equation 7-127, the solid bed melting velocity can be obtained:

$$v_{sy} = \frac{k_m \Delta T_b + H_m B_3}{\rho_s H_m \Delta H} \qquad (7-129)$$

Following the same procedure that was used to derive equation 7-105, the melt film thickness can be expressed as:

$$H_m(x') = \left[\frac{(4k_m\Delta T_b + 4B_3)x'}{\rho_m\Delta v\Delta H} + \delta^2\right]^{1/2} \qquad (7\text{-}130)$$

For a temperature-independent power law fluid, factor B_3 becomes:

$$B_3 = \frac{m_r H_m^2}{2}\left(\frac{\Delta v}{H_m}\right)^{n+1} \qquad (7\text{-}128\,a)$$

and the melt film thickness becomes:

$$H_m(x') = \left[\frac{4k_m\Delta T_b + 2m_r H_m^2(\Delta v/H_m)^{n+1}}{\rho_m\Delta v\Delta H}x' + \delta^2\right]^{1/2} \qquad (7\text{-}130\,a)$$

Equation 7-130a becomes equal to equation 7-105 when the power law index n is set to unity.

The melting rate per unit down channel length is still described by equation 7-107. If the clearance between flight and barrel is assumed zero, the melting rate is:

$$\frac{d\dot{M}_p}{dz'} = \left[\frac{k_m\Delta T_b + B_3}{\Delta H}\rho_m\Delta v W_s'\right]^{1/2} = \Omega * \sqrt{W_s'} \qquad (7\text{-}131)$$

By the same transformations as used before, the melting rate per unit length in z' direction can be rewritten as the melting rate per unit down channel lengthz:

$$\frac{d\dot{M}_p}{dz} = \Omega_2\sqrt{W_s'}\sin(\theta+\varphi) = \Omega_1 * \sqrt{W_s} \qquad (7\text{-}132)$$

where:

$$\Omega_1 * = \left[\frac{(k_m\Delta T_b + B_3)\rho_m v_b\sin\varphi}{\Delta H}\right]^{1/2} \qquad (7\text{-}132\,a)$$

This value Ω_1* can be used in the equations describing the solid bed width profile along the melting zone.

7.3.2   OTHER MELTING MODELS

After Tadmor's first publications on melting, many other workers started to study melting in single screw extruders. As a result, many publications appeared in the technical literature in the seventies. This section will review the various melting models proposed and analyze their advantages and disadvantages.

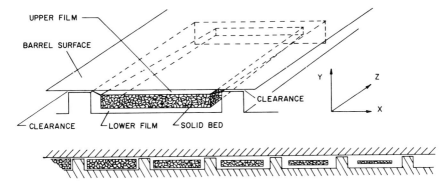

Figure 7-25. The Dekker/Lindt Melting Model

Many workers have repeated the screw extraction experiments as done by Maddock and later by Tadmor and coworkers; the majority of the workers confirmed the basic Tadmor melting model. Two exceptions are the findings of Klenk (66–68) and Dekker (69). Klenk observed the melting of PVC and noticed the melt pool located against the passive flight flank. This melting behavior was also observed by Gale (81) and Mennig (82) in PVC extrusion. Klenk attributed the unusual melting behavior to the slippage of the PVC melt along the wall. It should be mentioned that other workers have found the normal melting model with PVC (61, 70). Yet other workers (71) have found the melt pool location shifting from the passive flight to the active flight with increasing screw speed in PVC extrusion. Cox, et al. (80) observed the unusual melting behavior with the melt pool at the passive flight in extrusion of LDPE powder as well as the normal melting behavior. The unusual behavior occurred particularly at low screw speeds, with shallow channels, and with high barrel temperatures. Thus, the Tadmor melting model may not be valid in all cases.

Dekker (69) made observations of the melting behavior of polypropylene. He did not detect a clear melt pool at any side of the channel, but the solid bed was more or less suspended in a melt film. Lindt (72,73) developed a mathematical model describing this melting behavior, which is shown in Figure 7-25.

Some of the important assumptions in this model are:

1. The solid bed does not deform.

2. The solid bed is completely surrounded by a melt film and melting occurs at the entire circumference of the solid bed.

3. The melt film is constant in cross channel direction.

Assumption 3 causes some difficulties because a constant melt film thickness requires the presence of excessively large pressure gradients to accommodate the increased amount of polymer melt. This conflicts directly with assumption 1. A detailed analysis of the Lindt melting model was made by Gieskes (74) and Meijer (75). It appears that the Lindt melting model is unlikely to occur in the early stages of melting, but it may possibly occur in the later stages of melting.

A melting model in which the solid bed is considered to be subjected to a limited amount of deformation was proposed by Edmondson and Fenner (76–79). However, their analysis contains certain inconsistencies as pointed out by Meijer (75). Donovan (83, 84) extended the Tadmor analysis by removing the assumption of constant solid bed velocity and by incorporating the gradual heating of the solid bed along the melting zone. As mentioned earli-

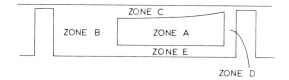

Figure 7-26. The 5-Zone Melting Model

er, Shapiro (63) and Vermeulen (64, 65) included the increasing melt film thickness with cross-channel distance. Shapiro, Pearson and coworkers have developed one of the most elaborate extensions of the Tadmor analysis (63, 85–87). Their model is a 5-zone model as shown in Figure 7-26.

The assumptions are more realistic and broader than most other analyses of melting. This will hopefully improve the accuracy of the model, but, unfortunately, it also increases the complexity of the computations quite substantially.

Hinrichs and Lilleleht (88) included the effect of flight clearance and used a helical coordinate system to account for the channel curvature. Sundstrom and Young (89) examined the effect of convective heat transfer and found that it can play an important role in the melting process, resulting in a melting rate about 20 percent higher than the predictions based on conductive heat transfer only. Sundstrom and Lo (89a) studied melting of amorphous polymers. They assumed the polymer-melt interface to be at the glass transition temperature and used a modified WLF equation to determine the shift factor for the temperature dependence of the viscosity. Chung (90,91) analyzed the effect of finite solid bed thickness, varying solid bed density, and varying solid bed velocity. Later Chung developed a screw simulator to study melting behavior of polymers. A substantial amount of experimental and theoretical work (94–97) on melting was done using the screw simulator.

A new approach to the analysis of melting in single screw extruders was taken by Viriyayuthakorn and Kassahun (234). They developed a new plasticating extrusion analysis program based on a finite element technique which can simulate three-dimensional flow with phase change. This program contains a number of new features that had not been incorporated in earlier work (prior to 1984) on melting in single screw extruders. A major new feature is the way the solid-melt phase change is handled. The conventional approach is to determine the location of the interface and use an energy balance of the interface to calculate the melting rate, as exemplified by equation 7-100. This method can introduce errors if the location of the interface is not accurately known and if the material does not have a sharp melting point, as is the case for most polymers. The technique used by Viriyayuthakorn and Kassahun is to absorb the latent heat of fusion term in the specific heat capacity and formulate the problem as though there is no phase change. Thus, a functional dependence of the specific heat capacity on temperature is used. This eliminates the need to first assume a certain melting model. The melting model is actually predicted from the calculations, something earlier workers have not been able to do.

Results from computer simulations for a HDPE polymer show that the Maddock/Tadmor melting mechanism occurs in the early stages of melting. At the beginning of the compression section, the solid bed tends to become totally encapsulated, while towards the end of the melting, the solid bed breaks apart into several pieces. At the time of publication of the paper by Viriyayuthakorn and Kassahun, no direct comparison was available between theoretical predictions and experimental results. Therefore, no statement can be made about the accuracy of the predictions. However, irregardless of the accuracy of the predictions, this model provides new capabilities that no doubt will prove very useful in future work on the analysis of plasticating extrusion. A drawback of the program is the need for very large computational capability. The simulations were performed on a Cray-1 computer, a computer not readily available to most process engineers. Even on this extremely powerful computer, one simulation can take as long as several hours. Work is underway to simplify the program to reduce the simulation time and to allow the program to be used on less powerful computers. However, considering the dramatic improvement in the number-crunching capability of new computers, the limitations in computational capabilities of computers will be less of a concern as more powerful and less expensive computers are introduced to the market.

A commendable effort has been made by Chung and coworkers to develop simple analytical expressions to predict the melting behavior of polymers (95). They developed analytical expressions valid for non-Newtonian fluids with temperature dependent viscosity, following an approach very similar to Pearson's (87).

The melting rate per unit area is:

$$\frac{d\dot{M}_p}{dA} = \dot{M}_p^0 \rho_m \Delta v \tag{7-133}$$

where $\dot{M}_p^0$ is the dimensionless melting efficiency; it represents the melting capacity per unit melting area per unit sliding distance. Various functional forms of $\dot{M}_p^0$ are given in (95). A simple expression that yields accurate results is:

$$\dot{M}_p^0 = \left(\frac{k_m \Delta T_b F_1(K_2)}{\rho_m \Delta H \Delta v W_s'}\right)^{1/2} \tag{7-134}$$

where K_2 is defined in equation 7–121 and $F_1(K_2)$ is:

$$F_1(K_2) = \frac{2(K_2 + 1 - expK_2)}{K_2(1 - expK_2)} \tag{7-135}$$

By curve fitting, the function $F_1(K_2)$ can be approximated by:

$$F_1(K_2) \approx \frac{20}{19 + (1 + K_2)^2} \tag{7-136}$$

If the melt viscosity is temperature independent, $K_2 = 0$ and $F_1(K_2) = 1$. By inserting equation 7–134 in equation 7–133, the melting rate per unit melt area becomes:

$$\frac{d\dot{M}_p}{dA} = \left(\frac{\rho_m k_m \Delta v \Delta T_b F_1(K_2)}{\Delta H W_s'}\right)^{1/2} \tag{7-137}$$

The melting rate per unit length normal to the sliding direction is obtained by multiplying expression 7–137 with W_s':

$$\frac{d\dot{M}_p}{dz'} = \left(\frac{\rho_m k_m \Delta v \Delta T_b F_1(K_2)}{\Delta H}\right)^{1/2} \sqrt{W_s'} \tag{7-138}$$

Equation 7-138 is comparable to equation 7-131; both equations are closed form analytical solutions. However, equation 7-138 is considerably more compact and easier to use. A word of caution is in order. Equation 7-138 has been experimentally verified with the screw simulator. The results of the screw simulator may not fully apply to actual melting in a single screw extruder. For instance, the screw simulator uses a molded solid polymer sample of one cubic inch. The solid bed in the extruder consists of compressed, partially sintered, polymeric particles. It is clear that an actual solid bed as occurs in the melting zone of an extruder may have different characteristics in terms of heat transfer properties and deformation behavior as compared to a molded solid block of the same material.

Equation 7-138 can be written in terms of melting rate per unit down channel by making the same coordinate transformation as discussed earlier in equation 7-109:

$$\frac{d\dot{M}_p}{dz} = \left(\frac{\rho_m k_m v_b \sin\varphi \Delta T_b F_1(K_2)}{\Delta H} \right)^{1/2} \sqrt{W_s} \qquad (7\text{-}139)$$

7.3.3   POWER CONSUMPTION IN THE MELTING ZONE

The mechanical power consumption in the melting zone can be determined by breaking down the power consumption in three parts: the power consumed in the melt film dZ_{mf}, the power consumed in the melt pool dZ_{mp}, and the power consumed in the clearance between flight and barrel dZ_{cl}. The power consumed in shearing the melt film is described by:

$$dZ_{mf} = v_b dz' \int_0^{W_s'} (\tau_{yz'} \sin\theta + \tau_{yx'} \cos\theta) dx' \qquad (7\text{-}140)$$

where $\tau_{yx'}$ is the shear stress in the direction x' of the relative velocity Δv between the solid bed and the barrel, see also Figure 7-23a. If the material can be described as a power law fluid, the shear stresses can be written as:

$$\tau_{yx'} = m \left(\frac{dv_{x'}}{dy} \right)^n \qquad (7\text{-}141a)$$

and

$$\tau_{yz'} = m \left(\frac{dv_{z'}}{dy} \right)^n \qquad (7\text{-}141b)$$

It is assumed that the flow in the melt film is a drag flow; i.e., pressure gradients are neglected. In this case:

$$\tau_{yx'} = m \left(\frac{\Delta v}{H_m} \right)^n \qquad (7\text{-}142a)$$

and

$$\tau_{yz}' = 0 \tag{7-142b}$$

The melt film thickness can be written according to equation 7-130:

$$H_m(x') = (K_4 x' + \delta^2)^{1/2} \tag{7-143}$$

where:

$$K_4 = \frac{4k_m \Delta T_b + 4B_3}{\rho_m \Delta v \Delta H} \tag{7-143a}$$

Factor B_3 is given by equation 7-128. The power consumption in the melt film can now be expressed as:

$$dZ_{mf} = m v_b dz' \Delta v^n \cos\theta \int_0^{W_s'} \frac{dx'}{(K_4 x' + \delta^2)^{n/2}} \tag{7-144}$$

By using equation 7-143, the integral can be rewritten as:

$$dZ_{mf} = m v_b dz' \Delta v^n \cos\theta \int_\delta^{H_{ms}} \frac{2H_m^{1-n}}{K_4} dH_m \tag{7-145}$$

where $H_{ms} = H_m(W_s')$, which is the maximum melt film thickness. The solution to equation 7-145 can be written as:

$$dZ_{mf} = m v_b dz' \Delta v^n \cos\theta \frac{2}{K_4(2-n)} \left(H_{ms}^{2-n} - \delta^{2-n} \right) \tag{7-146}$$

With $z' = z\sin(\theta + \varphi)$, the power consumption in the melt film can be written as:

$$dZ_{mf} = m v_b dz\sin(\theta + \varphi)\cos\theta \frac{2\Delta v^n}{K_4(2-n)} \left(H_{ms}^{2-n} - \delta^{2-n} \right) \tag{7-147}$$

From equations 7-146 and 7-147, it can be seen clearly that the power consumption in the melt film reduces with increasing clearance.

The power consumption in the melt pool will be relatively small compared to the other two terms. Because of the relatively complicated flow pattern in the melt pool, the derivation of the power consumption for a power law fluid will be rather involved. To simplify matters considerably, the power consumption in the melt pool can be approximated with the power consumption in a screw channel of width $W_m = W - W_s$ with a Newtonian fluid:

$$dZ_{mp} = (1+3r_d+4\tan^2\varphi) \, \frac{p\mu W_m v_{bz}^2 dz}{H} \qquad (7-148)$$

The derivation of this expression will be discussed in Section 7.4.1.3 on melt conveying. A similar expression for the melt conveying zone is equation 7-178.

The power consumption in the clearance can be easily determined if it is assumed that the velocity profile in the clearance is dominated by drag flow. In that case, the power consumption in the clearance can be written as:

$$dZ_{cl} = \frac{pv_b^{1+n} mwdz}{\delta^n} \qquad (7-149)$$

The total mechanical power consumption in the melting zone now becomes:

$$dZ = dZ_{mf} + dZ_{mp} + dZ_{cl} \qquad (7-150)$$

7.3.4   COMPUTER SIMULATION

One of the general problems of the melting theories is that the most realistic models are also the most complex. However, if the complexity goes beyond the level of the equations developed earlier, analytical solutions become very difficult to obtain, if not impossible. In this case, one has to use numerical techniques and computer simulation to find solutions to the equations. The main question is whether the more complex analysis will result in improved predictive ability. Relaxing certain assumptions quite often results in an improvement in accuracy that is relatively insignificant compared to the inherent uncertainties in most melting models. These are, among others, the actual location of the solid bed and melt pool, the shear strength and tensile strength of the solid bed as a function of time, temperature, and pressure, the actual melt film temperature, the actual temperature at the screw surface, the contribution of melting at the sides of the solid bed and at the screw surface, etc. The actual temperature in the melt film is most likely not fully developed and convective heat transport should be considered in determining the actual temperature profile.

Therefore, one has to strike a balance between the degree of sophistication of the analysis and the practical usefulness of the analysis. This balance will depend on the particular interests of the individual. For industrial applications, the author feels that the degree of sophistication of the analysis of melting as described in section 7.3.1 will usually be sufficient to analyze most practical extrusion problems. However, in some specific instances, one may want to go into much more detail on certain aspects of the melting process.

Analyses that require numerical techniques to arrive at solutions tend to be quite time consuming and require skilled personnel to develop the computer programs and to interpret the results of the computer simulations. Many people in the extrusion industry do not have the time or inclination to work through elaborate and complex analyses of melting. In this case, the preferred action is to use a less complicated analysis that yields analytical results. In most cases, actual predictions of melting performance can be made with a relatively simple programmable calculator. Another approach is to use a computer program developed else-where to analyze the problem. The danger is that if a person using such a program is not in-timately familiar with the theory behind the program, the assumptions, and the validity of the assumptions, it is quite likely that improper conclusions will be drawn from the predict-ed results. At this point in time, predictive computer programs to simulate the extrusion process are not truly fully predictive or 100 percent accurate, despite claims to the contrary by organizations providing such programs.

7.4   MELT CONVEYING

The melt conveying zone of the extruder starts at the point where the melting process is just completed. The melt conveying zone is also referred to as a pumping zone because, in most cases, the polymer melt has to be transported towards the die against a considerable head pressure. The melt conveying in an extruder was the subject of engineering studies as early as 1922 (98). The early work on melt conveying in extruders dealt with Newtonian fluids with temperature independent viscosity. This is a very convenient case to analyze because it is simple and yields clean, straightforward analytical solutions. The reason is that the cross channel flow can be analyzed independent of the down channel flow for a Newtonian fluid with temperature independent viscosity. Also, the pressure flow can be analyzed separate from the drag flow and the results can be superimposed.

These simplifications cannot be made when the fluid is non-Newtonian or when the viscosi-ty is temperature dependent. Most analyses on melt conveying can be categorized by the five most important assumptions made about the process:

1. Shear stress-shear rate relationship
   a. Linear (Newtonian)
   b. Non-linear (non-Newtonian)

2. Flow situation
   a. One-dimensional (down channel only)
   b. Two-dimensional (down and cross channel)
   c. Three-dimensional (down, cross channel and radial)

3. Effect of the flight flanks
   a. Negligible (infinite channel width)
   b. Not negligible (finite channel width)

4. Temperature effects
   a. Melt viscosity temperature independent
   b. Melt viscosity temperature dependent

5. Flight clearance
   a. Negligible (zero flight clearance)
   b. Not negligible (finite flight clearance)

The infinite channel width assumption applies to shallow channels, channels with a width to depth ratio higher than 10 (W/H > 10). If the depth of the channel is large relative to the width of the channel, the effect of the flight flanks on the down channel velocity profile has to be taken into account. Several reviews of the work on melt conveying in extruders have been written (101–106).

In addition to the five main assumptions of the analysis of melt conveying, there are a few more that are sometimes considered, such as curvature of the screw channel, the influence of an oblique channel end, etc. After the Newtonian analysis had been pretty much worked out, workers started to analyze non-Newtonian fluids in the early sixties. This adds significantly to the complexity of the analysis and generally the equations cannot be solved analytically, but have to be solved by numerical techniques. The simplest non-Newtonian case that can be considered is the one-dimensional, isothermal flow of the power law fluid in a channel of infinite width. This problem has been studied by various workers (107–113), but a closed-form solution was not obtained until 1983 (114). An exact analytical solution to even this simple problem has not been found to this date.

Relaxing other assumptions; i.e., temperature dependent viscosity, two- or three-dimensional flow, finite channel width, etc., substantially increases the degree of difficulty. Such analyses have been the subject of several Ph.D. studies (102, 115–121). The level of complexity can be increased almost at will by further relaxing the assumptions. However, this work can reach such a high level of complexity and sophistication that the usefulness to practicing polymer processing engineers becomes questionable. One should always try to balance the increased generality of the predictions against the time and effort spent on solving increasingly more complex problems. Somewhere, one can reach a point of diminishing return. To the practicing process engineer, the most important question should be whether the more general analysis will be applicable; e.g., are all the boundary conditions well known, and will it result in more accurate predictions. For instance, it probably does not make much sense to use a sophisticated non-isothermal melt conveying analysis to predict the melt temperature at the end of the screw if the actual screw temperature in the process is unknown. These practical considerations do not necessarily apply to the academic. On the other hand, it would make sense to perform a non-isothermal analysis to predict the effect of barrel temperature fluctuations on melt temperature or conveying rate. Besides the complications already discussed, there is the additional complication that the polymer melt is not a pure, inelastic power law fluid and significant time dependent effects can occur (e.g. 122–128).

Considering that in most extrusion operations the proper thermal boundary conditions are not known and further considering that analyses including the temperature dependence of viscosity are rather complex, the non-isothermal analysis of melt conveying will not be considered in this text. For such analyses the reader is referred to the literature; e.g., (102, 129). In the next section, melt conveying of Newtonian fluids and non-Newtonian fluids will be analyzed. The non-Newtonian fluids will be described with the power law equation (equation 6–23). The effect of the flight flank will be discussed and the difference between a one-dimensional and two-dimensional analysis will be demonstrated with particular emphasis on the implications for actual extruder performance.

7.4.1   NEWTONIAN FLUIDS

A very simple analysis of melt conveying can be made if the following assumptions are made:

1. The fluid is Newtonian.

2. The flow is steady.

3. The viscosity is temperature-independent.

4. There is no slip at the wall.

5. Body and inertia forces are negligible.

6. The channel width can be considered infinite.

7. The channel curvature is negligible (flat plate model).

The geometry is now simplified to Figure 7-27.

A flat plate is moving at velocity v_b over a flat rectangular channel with angle φ between v_b and the flights of the channel.

The equation of motion in down channel direction for this problem can be written as:

$$\frac{\partial P}{\partial z} = g_z = \frac{\partial \tau_{yz}}{\partial y} \tag{7-151}$$

The pressure is a function of down channel coordinate z only. Therefore, equation 7-151 can be integrated to give the shear stress profile in radial direction (y):

$$\tau_{yz} = \tau_0 + g_z y \tag{7-152}$$

where τ_0 is the shear stress at the screw surface, as yet unknown.

For a Newtonian fluid, equation 7-152 can be written as:

$$\mu \frac{dv_z}{dy} = \tau_0 + g_z y \tag{7-153}$$

By integration of equation 7-153, the down channel velocity as a function of normal distance y is obtained. By using boundary conditions $v_z(0)=0$ and $v_z(H)=v_{bz}$, the following expression is obtained:

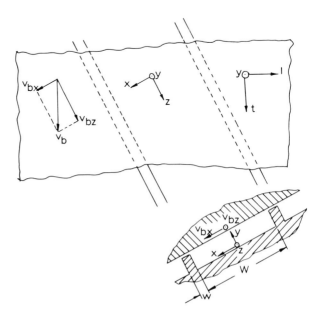

Figure 7-27. Flat Plate Model
for Melt Conveying

$$v_z(y) = \left(\frac{v_{bz}}{H} - \frac{g_z H}{2\mu}\right)y + \frac{g_z}{2\mu}y^2 \qquad\qquad (7\text{-}154)$$

The flow rate in down channel direction is obtained by integrating the down channel velocity over the cross-sectional area of the screw channel:

$$\dot{V} = \int_0^H pWv_z dy = \frac{1}{2} pWHv_{bz} - \frac{pWH^3 g_z}{12\mu} \qquad\qquad (7\text{-}155)$$

where p is the number of parallel flights and W the perpendicular channel width.

This represents the volumetric throughput of the melt conveying zone. The first term of equation 7–155 is the drag flow term. It represents the flow rate in pure drag flow; i.e., without a pressure gradient in down channel direction:

$$\dot{V}_d = \frac{1}{2} pWHv_{bz} \qquad\qquad (7\text{-}155\,a)$$

The second term of equation 7–155 is the pressure flow term. It represents the flow rate in pure pressure flow; i.e., without relative motion between the screw and barrel (zero screw speed):

$$\dot{V}_p = \frac{pWH^3 g_z}{12\mu} \qquad\qquad (7\text{-}155\,b)$$

Note that the pressure flow reduces the output when the pressure gradient is positive, which is often the case in actual extrusion operations.

Equation 7–155 is very useful because the output of the extruder can be determined from this equation. The effect of changes in screw geometry becomes quite obvious from equation 7–155. Particularly, the effect of channel depth is interesting. The drag flow rate is directly proportional to the channel depth but the pressure flow rate increases with the channel depth cubed. Thus, the pressure flow increases much faster with channel depth than drag flow. For this reason, the channel depth in the metering section is usually made quite shallow. The optimum channel depth that will give the highest output at a given screw speed and pressure gradient can be determined directly from equation 7–155, see also Chapter 8, Section 8.2.1.

The optimum channel depth can be determined by setting:

$$\frac{\partial \dot{V}}{\partial H} = 0 \qquad\qquad (7\text{-}156)$$

This results in the following optimum channel depth H*:

$$H^* = \left(\frac{2\mu v_{bz}}{g_z}\right)^{1/2} \qquad\qquad (7\text{-}156\,a)$$

The optimum helix angle can be determined from:

$$\frac{\partial \dot{V}}{\partial \varphi} = 0 \tag{7-156b}$$

This results in the following optimum helix angle φ^*:

$$\tan 2\varphi^* = \frac{6\,\mu v_b}{g_a H^2} \tag{7-156c}$$

where g_a is the axial pressure gradient ($g_a = g_z/\sin\varphi$). At the optimum channel depth the output becomes:

$$\dot{V}(H^*) = \frac{1}{3}\,WHv_{bz} = \frac{2}{3}\,\dot{V}_d \tag{7-157}$$

At the optimum channel depth H^*, the optimum helix angle becomes:

$$\varphi^* = 30° \tag{7-157a}$$

When the channel depth and helix angle are both at their optimum value, the output becomes:

$$\dot{V}(H^*, \varphi^*) = \frac{BHv_b}{4} \simeq 1.4\ HND^2 \tag{7-157b}$$

where B is the axial channel width and N the screw speed ($B \simeq \pi D \tan\varphi$ and $v_b = \pi DN$).

Another interesting observation is that the volumetric drag flow rate is independent of any fluid property for Newtonian fluids. Thus, the drag flow rate for water will be the same as oil, molten nylon, etc. The drag flow rate is directly proportional to screw speed because $v_{bz} = \pi DN\cos\varphi$, where N is the screw speed.

Many extruder screws are designed such that the drag flow is considerably larger than the pressure flow. In these cases, the approximate output of the extruder is determined from equation 7-155a. If the mass flow rate is required, the volumetric flow rate is simply multiplied with the polymer melt density. The reason that many screws are designed to have a relatively small pressure flow is that the extrusion operation tends to become less sensitive to diehead pressure fluctuations. If the pressure flow is only 10 percent of the total throughput, a 50 percent change in diehead pressure will cause only a 5 percent change in output. On the other hand, if the pressure flow is 50 percent of the total throughput, a 50 percent change in diehead pressure will cause a 25 percent change in output.

There are two ways to achieve a relatively low pressure sensitivity. One way is to reduce the depth of the channel in the metering section of the screw. The other way is to reduce the pressure gradient in the melt conveying zone by building up pressure in earlier zones, the plasticating and/or the solids conveying zone. The most common way to achieve the latter situation is by employing a grooved barrel section, as discussed in Section 7.2.2.2. Down channel velocity profiles for various values of the pressure flow (pressure gradient) are shown in Figure 7-28a.

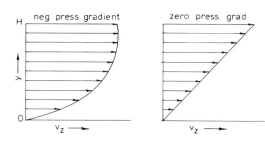

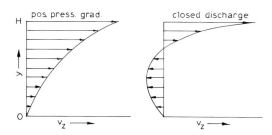

Figure 7-28a. Down Channel Velocity Profiles

By the same procedure employed to derive the down channel velocity, the cross channel velocity can be determined:

$$v_x(y) = \left(\frac{v_{bx}}{H} - \frac{g_x H}{2\mu}\right) y + \frac{g_x}{2\mu} y^2 \qquad (7-158)$$

In this case, g_x can be determined from the condition that there is no net flow in the cross channel direction:

$$\int_0^H v_x dy = 0 \qquad (7-159)$$

The cross channel pressure gradient is:

$$g_x = \frac{6\mu v_{bx}}{H^2} \qquad (7-160)$$

The resulting cross channel velocity profile is:

$$v_x = -\frac{2v_{bx}}{H} y + \frac{3v_{bx}}{H^2} y^2 \qquad (7-161)$$

The cross channel velocity profile is shown in Figure 7-28b.

It can be seen that the cross channel velocity at $y = 2H/3$ is zero. Thus, the material in the top 1/3 of the channel moves towards the active flight flank and the material in the bottom two-thirds of the channel moves towards the passive flight flank. It is clear that in reality the

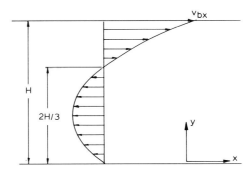

Figure 7-28 b. Cross Channel Velocity Profiles

situation becomes more complex at the flight flanks because normal velocity components must exist to achieve the circulatory flow patterns in cross channel direction. However, these normal velocity components will be neglected in this analysis. Normal velocity components were analyzed by Perwadtschuk and Jankow (129). The actual motion of the fluid is the combined effect of the cross and down channel velocity profiles. This is shown in Figure 7-29 at various levels of the pressure gradient.

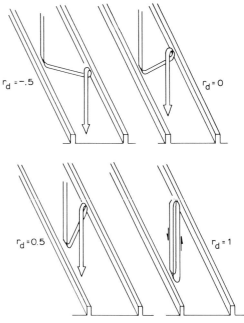

Figure 7-29. Fluid Motion in the Melt Conveying Zone

throttle ratio r_d is defined by equation 7-178a

7.4.1.1.   EFFECT OF FLIGHT FLANKS

If the channel width cannot be considered infinite, the equation of motion in the down channel direction becomes:

$$\frac{\partial P}{\partial z} = g_z = \frac{\partial \tau_{yz}}{\partial y} + \frac{\partial \tau_{xz}}{\partial x} \qquad\qquad (7\text{-}162)$$

Equation 7–162 is an expanded form of equation 7–151.

Considering the fluid Newtonian allows equation 7–162 to be written as:

$$\frac{\partial P}{\partial z} = \mu\left(\frac{\partial^2 v_z}{\partial y^2} + \frac{\partial^2 v_z}{\partial x^2}\right) \tag{7-163}$$

The solution to this equation is more difficult than the solution to equation 7–151. The case of pure pressure flow was first solved by Boussinesq (130) in 1868. The solution to the combined drag and pressure flow was first published in 1922 (98); the authorship of this publication remains a question. Since the 1922 publication, numerous workers have presented solutions to this problem. Meskat (131) reviewed various solutions and demonstrated that they were equivalent. The velocity profile resulting from the drag flow can be written as:

$$v_{zd}(x,y) = \frac{4v_{bz}}{\pi} \sum_{1,3,5\ldots}^{\infty} \frac{\sinh(i\pi y/W)}{i\sinh(i\pi H/W)} \sin(i\pi x/W) \tag{7-164}$$

The velocity profile resulting from the pressure flow can be written as:

$$v_{zp}(x,y) = \frac{g_z}{\mu}\left[\frac{y^2}{2} - \frac{Hy}{2} + \right.$$

$$\left. \frac{4H^2}{\pi^3} \sum_{1,3,5\ldots}^{\infty} \frac{\cosh[i\pi W/H(x/W-1/2)]}{i^3\cosh(i\pi W/2H)} \sin(i\pi y/H)\right] \tag{7-165}$$

The combined velocity is simply:

$$v_z(x,y) = v_{zd} + v_{zp} \tag{7-166}$$

The volumetric output can be conveniently expressed in the following form:

$$\dot{V} = \frac{1}{2}pWHv_{bz}F_d - \frac{pWH^3g_z}{12\mu}F_p \tag{7-167}$$

where the shape factor for drag flow F_d is:

$$F_d = \frac{16W}{\pi^3H} \sum_{1,3,5\ldots}^{\infty} \frac{1}{i^3}\tanh\left(\frac{i\pi H}{2W}\right) \tag{7-168}$$

and the shape factor for pressure flow F_p is:

$$F_p = 1 - \frac{192H}{\pi^5W} \sum_{1,3,5\ldots}^{\infty} \frac{1}{i^5}\tanh\left(\frac{i\pi W}{2H}\right) \tag{7-169}$$

In the range of most extruder screws (H/W < 0.6), the shape factors F_d and F_p can be quite accurately approximated with the following expressions:

$$F_d = 1 - 0.571 \frac{H}{W} \tag{7-170}$$

and

$$F_p = 1 - 0.625 \frac{H}{W} \tag{7-171}$$

Equations 7-170 and 7-171 are considerably easier to evaluate than the exact expressions 7-168 and 7-169. With the approximate expressions for the shape factor, the volumetric output can be expressed as:

$$\dot{V} = \frac{1}{2}p(W - 0.571H)Hv_{bz} - \frac{(W - 0.625H)pH^3 g_z}{12\mu} \tag{7-172}$$

In most extruder screws, the ratio of channel depth to channel width H/W will range from 0.10 to 0.03, with the latter value being more common than the former. This means that the correction as a result of the shape factors is usually less than 2 percent and essentially always less than 5 percent.

7.4.1.2   EFFECT OF CLEARANCE

Another source of error that can become quite important is the leakage flow through the clearance between the flight and the barrel. A normal design clearance (radial) is 0.001 D, where D is the diameter of the screw. When the clearance is normal, the flow through the clearance will be quite small. However, if the screw and/or barrel is subject to wear, the actual clearance can increase substantially beyond the normal design clearance. This can cause a considerable reduction in output and it is important to know how to evaluate the effect of clearance flow.

The clearance reduces the drag flow rate. The drag flow rate is reduced by a factor δ/H. The corrected drag flow becomes:

$$\dot{V}_d = \frac{1}{2}W(H-\delta)v_{bz} \tag{7-173}$$

where δ is the radial clearance.

The proper derivation of the pressure induced leakage flow is rather involved. For details of the derivation, the reader is referred to the publication of Mohr and Mallouk (228) or Tadmor (103). The total volumetric output including the effect of leakage can be written as:

$$\dot{V} = \frac{pW(H-\delta)v_{bz}}{2} - \frac{pWH^3 g_z}{12\mu}(1+f_L) \tag{7-174a}$$

The correction factor for pressure induced leakage through the flight clearance can be written as:

$$f_L = \frac{\mu w \delta^3}{\mu_{cl} W H^3} + \frac{(W+w)\mu \delta^3}{\mu W \delta^3 + \mu_{cl} w H^3} \left[\frac{6\mu v_{bz}(H-\delta)}{H^3 g_z} + \frac{W+w}{W\tan^2\varphi} \right] \quad (7\text{–}174\,b)$$

where:

$$W = \frac{\pi D \sin\varphi}{p} - w \qquad\qquad (7\text{–}174\,c)$$

and

$$v_{bz} = \pi D N \cos\varphi \qquad\qquad (7\text{–}174\,d)$$

The viscosity in the clearance μ_{cl} is differentiated from the viscosity in the channel μ because, in reality, the viscosity in the clearance will be substantially different from the viscosity in the channel as a result of differences in local temperature and shear rate. When the flight clearance is close to the normal design clearance, the value of f_L will be very close to zero and can be neglected unless extreme accuracy is required. However, when the radial clearance is considerably larger than the normal design clearance, for instance, as a result of wear, the actual value of f_L should be used in the expression for the total volumetric output, equation 7–174a.

It is interesting to note that even when the down channel pressure gradient g_z is zero, there is a pressure induced leakage flow. This results from the drag induced cross channel pressure gradient.

A significantly simpler expression for the pressure induced leakage flow is obtained by taking the following approach. The pressure induced leakage flow through the flight clearance can be approximated by considering the flight clearance as a rectangular slit of width

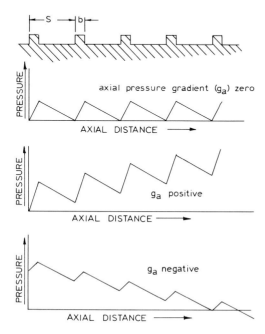

Figure 7-30a. Effect of Down Channel and Cross Channel Pressure Gradients on Pressure Differential Across the Flight

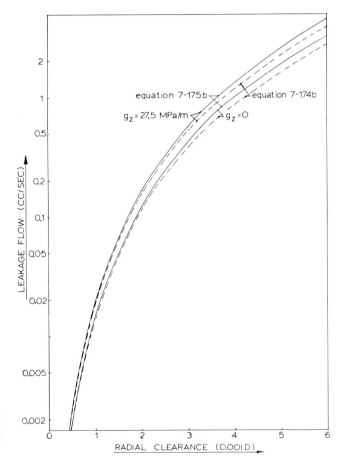

Figure 7-30b. Comparison of Leakage Flow Calculated With Equation 7-174b and Equation 7-175b

$\pi D\cos\varphi$, height δ, and depth w, with a pressure differential across the flight of ΔP_f. By using the equations in Table 7-1, section 7.5.1, the leakage flow can be written as:

$$\dot{V}_1 = \frac{\pi D\cos\varphi\,\delta^3}{12\,\mu_{cl}}\,\frac{\Delta P_f}{w} + \frac{pw\,\delta^3}{12\,\mu_{cl}}\,g_z \qquad (7\text{-}175)$$

The pressure differential across the flight results from both the down channel pressure gradient g_z and the cross channel gradient g_x. The latter is a drag induced pressure gradient; thus, it is present even in pure drag flow. When the actual leakage flow is very small, the cross channel pressure gradient is given by equation 7-160. The pressure differential across the flight can now be written as:

$$\Delta P_f = g_z\,\frac{\pi D\cos\varphi}{p} + g_x W \qquad (7\text{-}175a)$$

where W is the cross channel width.

The effect of down channel and cross channel pressure gradient of the pressure differential across the flight is illustrated in Figure 7-30a.

The leakage flow can now be written as:

$$\dot{V}_1 = \frac{\pi D \delta^3 \cos\varphi}{12 \mu_{cl} w} \left(\frac{\pi D g_z \cos\varphi}{p} + \frac{6 \mu v_{bx} W}{H^2} \right) + \frac{p w \delta^3}{12 \mu_{cl}} g_z \qquad (7\text{--}175\,\text{b})$$

Comparison of the leakage flow according to equation 7–174b to the leakage flow according to the simplified equation 7–175b is shown in Figure 7–30b.

The predictions are for a 114.3-mm (4.5-inch) extruder, screw speed 100 rpm, channel depth 5 mm, flight width 11.4 mm, helix angle 17.67 degrees, radial clearance 0.127 mm, and viscosity 6897 Pa-sec. The difference between the two equations is less than 5 percent at normal values of the clearance and about 10 percent at clearance values of about four times the normal value.

Equation 7–175b can be made more accurate by using an improved equation for the cross channel pressure gradient. Leakage through the flight clearance will reduce the cross channel pressure gradient to a value lower than the one given by equation 7–160. In Section 10.5, a more accurate expression for the drag induced pressure gradient is derived, taking into account the leakage flow over the flight clearance, see equations 10–108 through 10–112. If the viscosity in the clearance μ_{cl} is differentiated from the viscosity in the channel μ, the same approach leads to the following drag induced cross channel pressure gradient:

$$g_x = \frac{6 \mu \mu_{cl} (H-\delta) w v_{bx}}{\mu_{cl} w H^3 + \mu W \delta^3} \qquad (7\text{--}175\,\text{c})$$

Thus, the more accurate expression of the leakage flow can be written as:

$$\dot{V}_1 = \frac{\pi D \delta^3 \cos\varphi}{12 \mu_{cl} w} \left(\frac{\pi D g_z \cos\varphi}{p} + \frac{6 \mu \mu_{cl} (H-\delta) w W v_{bx}}{\mu_{cl} w H^3 + \mu W \delta^3} \right) + \frac{p w \delta^3}{12 \mu_{cl}} g_z \quad (7\text{--}175\,\text{d})$$

For normal values of the flight clearance, however, equation 7–175b will generally give sufficiently accurate results.

Figure 7–30b shows the leakage flow at two values of the down channel pressure gradient, 0 and 27.5 MPa/m. It is interesting to note that the leakage flow at a pressure gradient of 27.5 MPa/m is not much larger than the leakage flow at zero pressure gradient. This indicates that the drag induced pressure differential across the flight plays an important role and cannot be neglected. The value of the leakage flow at normal clearance values ($\delta = 0.001\,D$) is about 0.01 percent of the drag flow rate; at four times the normal clearance, the leakage flow is about 1 percent of the drag flow rate. Thus, the leakage flow becomes significant only when the clearance is larger than four times its normal value.

The total volumetric output, including the effect of leakage, equation 7–175b, can be written as:

$$\dot{V} = \frac{p W (H-\delta) v_{bz}}{2} - \frac{p W H^3 g_z}{12 \mu} - \frac{\pi D \delta^3 \cos\varphi}{12 \mu_{cl} w} \left(\frac{\pi D g_z \cos\varphi}{p} + \frac{6 \mu v_{bx} W}{H^2} \right) \quad (7\text{--}175\,\text{e})$$

$$- \frac{p w \delta^3}{12 \mu_{cl}} g_z$$

7.4.1.3   POWER CONSUMPTION IN MELT CONVEYING

The power consumption in the melt conveying zone is an important parameter to consider in screw design and in the analysis of actual extrusion operations. The power consumed for pumping in the channel is:

$$dZ_{ch} = pv_{bz}dz \int_0^W \tau_{yz}(H)dx + pv_{bx}dz \int_0^W \tau_{yx}(H)dx \qquad (7\text{--}176)$$

where the shear stresses are related to the barrel surface.

If the material is Newtonian:

$$\tau_{yx}(H) = \mu \left. \frac{dv_x}{dy} \right|_H \qquad (7\text{--}177\,a)$$

and

$$\tau_{yz}(H) = \mu \left. \frac{dv_z}{dy} \right|_H \qquad (7\text{--}177\,b)$$

,

The shear stresses can be evaluated from equation 7-177 and the equations for down channel velocity profile, equation 7-154 or 7-166, and the cross channel velocity profile, equation 7-161. The power consumption in the screw channel can be written as:

$$dZ_{ch} = (1+3r_d+4\tan^2\varphi) \frac{p\mu Wv_{bz}^2 dz}{H} \qquad (7\text{--}178)$$

where r_d is ratio of pressure flow to drag flow:

$$r_d = \frac{H^2 g_z}{6\mu v_{bz}} \qquad (7\text{--}178\,a)$$

Equation 7-178a is valid if the clearance is negligible. The ratio of pressure flow to drag flow is often referred to as the throttle ratio; Drossel quotient in German. It enables a brief expression for the combined drag and pressure flow:

$$\dot{V} = \frac{pWHv_{bz}}{2} (1 - r_d) \qquad (7\text{--}179)$$

The specific energy consumption in the channel can be determined by dividing the power consumption by the throughput:

$$\hat{Z}_{ch} = \frac{dZ_{ch}}{\dot{V}} = \frac{2\mu v_{bz}dz}{H^2} \frac{1+3r_d+4\tan^2\varphi}{1-r_d} \qquad (7\text{--}179\,a)$$

The pumping efficiency is the ratio of the theoretical energy requirement to develop pressure $\Delta P (= \dot{V}\Delta P)$ divided by the actual energy requirement $(= dZ_{ch})$. Thus, the pumping efficiency in the channel can be written as:

$$\varepsilon_{ch} = \frac{\dot{V}\Delta P}{dZ_{ch}} = \frac{3r_d - 3r_d^2}{1 + 3r_d + 4\tan^2\varphi} \tag{7-179b}$$

The optimum pumping efficiency can be determined by setting the first derivative of the pumping efficiency with respect to the throttle ratio equal to zero:

$$\frac{d\varepsilon_{ch}}{dr_d} = 0 \tag{7-179c}$$

This yields the following expression for the optimum throttle ratio r_d^*:

$$r_d^* = -\frac{1 + 4\tan^2\varphi}{3} \pm \frac{2(1 + 5\tan^2\varphi + 4\tan^4\varphi)}{3}^{1/2} \tag{7-180}$$

By inserting the optimum throttle ratio from equation 7-180 into equation 7-179b for the pumping efficiency, one can determine the optimum pumping efficiency in the channel. When the helix angle is zero, the optimum throttle ratio is 1/3 or -1 and the optimum pumping efficiency is also 1/3. A throttle ratio of -1 represents a large negative pressure gradient; this is not a situation that is likely to occur in practice. When the helix angle increases, the optimum throttle ratio increases, but the optimum pumping efficiency decreases. This is shown in Figure 7-31.

Thus, the highest pumping efficiency of the channel that can possibly be obtained is only 33.33 percent. The other 66.67 percent of the energy is dissipated in the fluid as heat. In practice, the actual pumping efficiency will be around 10 percent or less. The screw pump, therefore, is rather inefficient in developing pressure. Other types of pumps, such as a gear pump, can be more efficient in generating pressure. However, in many extrusion operations, the energy dissipated in the fluid is not wasted but effectively used to bring the polymer melt to the required melt temperature. Heating the polymer melt by viscous heat generation is more effective than heating by external barrel heaters.

The power consumption in the clearance can be determined quite easily if it is assumed that the velocity profile in the clearance is dominated by drag flow. The power consumption in the clearance can then be written as:

$$dZ_{cl} = \frac{pv_b^2 \mu_{cl} w dz}{\delta} \tag{7-181}$$

where w is the perpendicular flight width $(w = b\cos\varphi)$.

The power consumption in the clearance is directly proportional to the number of parallel flights, the local viscosity, the flight width, and inversely proportional to the radial clearance. It will be shown later that a substantial portion (in some cases 50 percent or more) of the total power consumption is consumed in the clearance. Therefore, the geometry of the flight and clearance become important geometrical variables when it comes to minimizing power consumption.

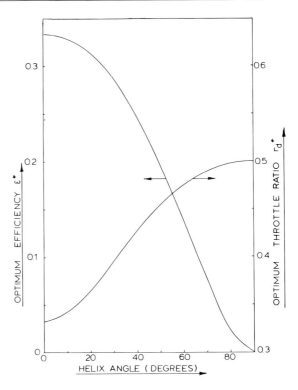

Figure 7-31. Optimum $r_d{}^*$ and $\varepsilon_{ch}{}^*$ as a Function of Helix Angle

The power consumption necessary to build up pressure in the polymer is determined by multiplying the volumetric flow rate with the pressure rise:

$$dZ_p = \dot{V}g_z dz \qquad\qquad (7\text{--}182)$$

The total pumping efficiency can now be determined from:

$$\varepsilon = \frac{dZ_p}{dZ_{ch}+dZ_{cl}} \qquad\qquad (7\text{--}182a)$$

The total pumping efficiency will usually be about 10 percent or less. The amount of power that is dissipated in the fluid as heat is:

$$dZ_{diss} = dZ_{ch} + dZ_{cl} - dZ_p \qquad\qquad (7\text{--}183)$$

This power should be used to calculate the temperature increase in the polymer melt. Since the pumping efficiency is usually less than 10 percent, the amount of power dissipated in the screw channel will generally be more than 90 percent of the total power input.

7.4.2   POWER LAW FLUIDS

In this section, the effect of the pseudoplastic behavior of the polymer melt on the conveying characteristics will be analyzed by describing the polymer melt as a power law fluid. As before, the flow is considered to be steady, fully developed, isothermal, and leakage flow is neglected. The effect of the flight flanks on the down channel velocity profile will be neglected. At first, the analysis will be one-dimensional, considering only the down channel velocity. Thus, the effect of cross channel flow will initially be neglected. Later, the analysis will be extended to a two-dimensional case, considering both the down channel and cross channel velocity.

7.4.2.1   ONE-DIMENSIONAL FLOW

For the one-dimensional analysis, the equation of motion is described by equations 7–151 and 7–152. The constitutive relationship, the power law equation, is written as:

$$\tau_{yz} = m \left| \frac{dv_z}{dy} \right|^{n-1} \frac{dv_z}{dy} \tag{7–184}$$

Equation 7–184 combined with equation 7–152 describes the basic problem. It is convenient to write the resulting equation in dimensionless form. For this purpose, the following dimensionless quantities are defined: the dimensionless depth $\xi = y/H$, the dimensionless down channel velocity $v_z^0 = v_z/v_{bz}$, and a reduced pressure gradient Γ_R. The reduced pressure gradient is defined as:

$$\Gamma_R = \left(\frac{Hg_z}{m} \right)^s \frac{H}{v_{bz}} \tag{7–185}$$

where s is the reciprocal power law index $(s = 1/n)$.

Equation 7–184 can be integrated to give:

$$\left| \frac{dv_z^0}{d\xi} \right|^{n-1} \frac{dv_z^0}{d\xi} = \Gamma_R^{\,n} (\xi - \lambda) \tag{7–186}$$

Variable λ in equation 7–186 represents the location where the shear rate is zero, which is also the location of the extremum in the velocity profile. This value needs to be known to eliminate the absolute value in equation 7–186. For the time being, only positive pressure gradients g_z will be considered. If the extremum occurs in the screw channel, then $0 \leqslant \lambda \leqslant 1$. When $\xi \geqslant \lambda$, equation 7–186 can be written as:

$$\frac{dv_z^0}{d\xi} = \Gamma_R (\xi - \lambda)^s \tag{7–187}$$

By integration and by using the appropriate boundary condition $v_z^0(1)=1$, the following expression is obtained:

$$v_z^0 = 1 - \frac{\Gamma_R}{s+1} [(1 - \lambda)^{s+1} - (\xi - \lambda)^{s+1}] \tag{7-188}$$

Similarly, for $\xi \leqslant \lambda$, the expression becomes:

$$v_z^0 = \frac{\Gamma_R}{s+1} [(\lambda - \xi)^{s+1} - \lambda^{s+1}] \tag{7-189}$$

At $\xi = \lambda$, the two velocities from equation 7-188 and 7-189 should be the same. From this equality, the following equation for λ results:

$$\lambda^{s+1} - (1 - \lambda)^{s+1} + \frac{s+1}{\Gamma_R} = 0 \tag{7-190}$$

From this equation, the value of λ can be determined. The condition for the existence of an extremum within the actual flow region is: $\Gamma_R \geqslant s+1$. The maximum velocity can be written as:

$$v_z^0\text{-max} = \frac{\Gamma_R}{s+1} \tag{7-191}$$

The extremum falls outside of the actual flow regime when $\Gamma_R < s+1$. By using the same procedure, the equation for λ when $\Gamma_R < s+1$ becomes:

$$(-\lambda)^{s+1} - (1 - \lambda)^{s+1} + \frac{s+1}{\Gamma_R} = 0 \tag{7-192}$$

Both the dimensionless velocity v_z^0 and the dimensionless flow rate $\dot{V}^0$ can be written into single expressions:

$$v_z^0 = \frac{|\xi - \lambda|^{s+1} - |\lambda|^{s+1}}{|1 - \lambda|^{s+1} - |\lambda|^{s+1}} \tag{7-193}$$

and

$$\dot{V}^0 = \frac{2}{s+2} (1 - \lambda - \Gamma_R |\lambda|^{s+1}) \tag{7-194}$$

The dimensionless flow rate is the actual flow rate divided by the drag flow rate, thus:

$$\dot{V}^0 = \frac{2\dot{V}}{WHv_{bz}} \tag{7-195}$$

Later in Chapter 8, Section 8.2.1, it will be shown that if the channel depth is optimized to give the highest output, the corresponding value of $\lambda = 0$. Thus, the corresponding optimum dimensionless flow rate is:

$$\dot{V}^0_{opt} = \frac{2}{s+2} = \frac{2n}{1+2n} \qquad (7\text{--}196)$$

And thus the optimum actual flow rate is:

$$\dot{V}_{opt} = \frac{n}{1+2n} WHv_{bz} \qquad (7\text{--}197)$$

For a Newtonian fluid $(n=1)$, equation 7-197 becomes equal to equation 7-157.

In order to determine the velocity profile and the flow rate, the value of λ has to be known. This involves solving equation 7-190 or 7-192. This is normally done by using a numerical technique; e.g., Newton-Raphson. Exact analytical solutions are only possible for the special case when s is a positive integer. However, by rewriting the equations and by performing a series expansion, a closed form solution can be obtained. If a new variable x is introduced in equation 7-190 with $x = \lambda - 0.5$, then the equation can be rewritten as:

$$(0.5 + x)^{s+1} - (0.5 - x)^{s+1} + \frac{s+1}{\Gamma_R} = 0 \qquad (7\text{--}198)$$

By performing a series expansion of the first two terms of equation 7-198 and neglecting the x terms of order four and higher, the following equation is obtained:

$$\frac{1}{3}s(s-1)2^{2-s} x^3 + 2^{1-s} x + \frac{1}{\Gamma_R} = 0 \qquad (7\text{--}199)$$

This is a standard third order (cubic) equation that can be readily solved. There is only one root that is real; the solution for λ when $\Gamma_R \geqslant s+1$ is:

$$\lambda = \frac{1}{2} + (q_1 + q_2)^{1/3} + (q_1 - q_2)^{1/3} \qquad (7\text{--}200)$$

where:

$$q_1 = -3 \cdot \frac{2^{s-3}}{s(s-1)\Gamma_R} \qquad (7\text{--}200\,a)$$

and

$$q_2 = [q_1{}^2 + (2s^2 - 2s)^{-1/2}]^{1/3} \qquad (7\text{--}200\,b)$$

This solution will be accurate when x is close to zero, which will generally be the case. In the region for which equation 7-198 applies, the values of λ will range between 0 and 0.5, thus x will range between -0.5 and 0.

Equation 7-192 can be rewritten by introducing a new variable $x = \lambda + 1$. After series expansion and neglecting x terms of order three and higher, a quadratic equation is obtained. The solution for λ when $\Gamma_R < s+1$ is:

$$\lambda = \frac{-C_2 \pm (C_2^{\,2} - 4C_1 C_3)^{1/2}}{2C_1} - 1 \qquad (7-201)$$

where:

$$C_1 = \frac{1}{2} s(s+1)(1-2^{s-1}) \qquad (7-201a)$$

$$C_2 = (s+1)(2^s-1) \qquad (7-201b)$$

$$C_3 = 1 - 2^{s+1} + \frac{s+1}{\Gamma_R} \qquad (7-201c)$$

Again, this solution will be accurate when x is close to zero. In the region where equation 7-192 applies, the value of λ will range from 0 (when $\Gamma_R = s+1$) to $-\infty$ when the pressure gradient is zero. Therefore, equation 7-201 cannot give accurate results at very small pressure gradients. However, at larger pressure gradients, equation 7-201 will give accurate results. The limited accuracy of the solution at small pressure gradients does not have much practical significance. In the analysis of real extruders, one is primarily concerned about the effect of large pressure gradients, not about the effect of small pressure gradients. The flow rate at small pressure gradients will be nearly equal to the drag flow rate. The predictions of the analytical solutions can now be plotted in the often used dimensionless form.

This is shown in Figure 7-32a, where the dimensionless output is plotted as a function of the dimensionless pressure gradient g_z^0, where:

$$g_z^{\,0} = \frac{g_z H}{6m} \left(\frac{H}{v_{bz}}\right)^n = \frac{\Gamma_R^n}{6} \qquad (7-202)$$

For a Newtonian fluid ($n = 1$), the familiar linear output-pressure relationship is found. However, when the power law index is less than unity, substantial deviations from the Newtonian characteristic occur. The deviations increase as the material becomes more pseudoplastic (more strongly non-Newtonian). The result is that for a pseudoplastic fluid, the pressure generating capability is drastically reduced compared to a Newtonian fluid. Or, at the same pressure gradient, the output is drastically reduced. For a fluid with a power law index less than 0.8, the use of the equations for Newtonian fluids will result in large errors!

Comparisons of Figure 7-32a to similar figures determined by other workers from numerical techniques; e.g., (132) or (106), reveals essentially indistinguishable results. This is a first indication that the analytical solutions for λ are quite accurate. Up to this point, only positive pressure gradients have been considered. From equation 7-194, it can be demonstrated rather easily that the output values for negative pressure gradients can be obtained from the following relationship:

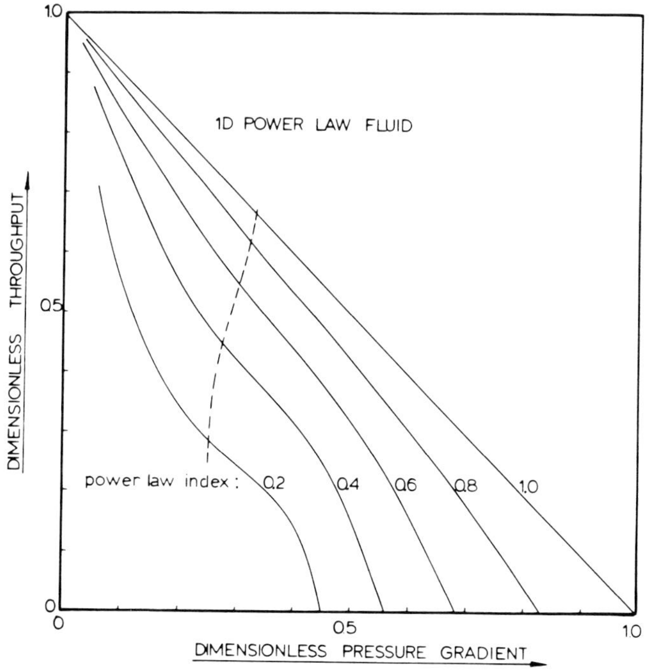

1D POWER LAW FLUID

power law index : 0.2   0.4 0.6 0.8 1.0

DIMENSIONLESS THROUGHPUT

DIMENSIONLESS PRESSURE GRADIENT

Figure 7-32a. Dimension-
less Troughput versus Pressure Gradient from Closed
Form Solution

$$\dot{V}^0(-g_z{}^0) \; = \; 2 \; - \; \dot{V}^0(+g_z{}^0) \tag{7-203}$$

The dashed line in Figure 7-32a shows at what point along each curve λ becomes zero. When λ is zero, the extremum in the velocity profile occurs right at the screw surface. The dimensionless throughput in this case is:

$$\dot{V}^0 \; = \; \frac{2}{2+s} \; = \; \frac{2n}{1+2n} \tag{7-204}$$

This is the same value as the optimum dimensionless flow rate described by equation 7-196. It is interesting to note that this throughput is determined only by the power law index. The data above the dashed line have been determined with equation 7-201. In this case, $\Gamma_R <$ $s+1$ and no extremum occurs in the velocity profile. The data below the dashed line have been determined with equation 7-200. In this case, $\Gamma_R \geqslant s+1$ and an extremum does occur in the velocity profile.

7.4.2.2   TWO-DIMENSIONAL FLOW

In this section, the previous analysis will be extended to include the effect of cross channel flow. The cross channel flow does not directly affect the conveying rate, but it does affect the total shear rate that the polymer is exposed to. Therefore, the viscosity will be affected and thus, the actual flow rate is affected as well. If the helix angle reduces to zero, there will be no cross channel flow and the results of the two-dimensional analysis will be identical to the one-dimensional analysis. This limiting case has been studied by Tadmor (133) and Dyer (134). More general two-dimensional analyses can be found in (132-145).

The difference between the one-dimensional and the two-dimensional analysis will increase with increasing helix angle and reducing power law index. From a practical point of view, the use of a two-dimensional analysis becomes important when large helix angles and strongly non-Newtonian fluids are analyzed. The equation of motion will again be the same as used before, see equations 7-151 and 7-152. The shear stress profiles can be written as:

$$\tau_{yz} = \tau_{zo} + g_z y \tag{7-205}$$

$$\tau_{yx} = \tau_{xo} + g_x y \tag{7-206}$$

The magnitude of the total shear stress is obtained from:

$$\tau = (\tau_{yx}^2 + \tau_{yz}^2)^{1/2} \tag{7-207}$$

The power law equation for the two-dimensional flow situation can be written as:

$$\tau = m \left[\left(\frac{dv_x}{dy} \right)^2 + \left(\frac{dv_z}{dy} \right)^2 \right]^{\frac{n-1}{2}} \frac{dv}{dy} \tag{7-208}$$

The direction of shear stress τ and velocity v is determined by τ_{yz} and τ_{yx}. At this point, there are three unknowns: the cross channel pressure gradient g_x, the cross channel shear stress at the screw surface τ_{xo}, and the down channel shear stress at the screw surface τ_{zo}. At the time of writing, no analytical solutions to this problem are known. In fact, an analytical solution does not seem possible. Therefore, some numerical scheme has to be used in order to determine the unknowns. Because this problem is of considerable importance to the proper analysis of melt conveying, it will be discussed in some detail.

Some initial values of g_x, τ_{xo} and τ_{zo} can be selected, for instance, by calculating the values for the Newtonian case. From those initial values, the corresponding velocity profiles and the flow rates in cross channel direction can be determined. The velocity profile in x and z direction can be determined from:

$$\frac{dv_x}{dy} = \frac{\tau_{yx}}{m^S} (\tau_{yx}^2 + \tau_{yz}^2)^{\frac{1-n}{2n}} \tag{7-209a}$$

and

$$\frac{dv_z}{dy} = \frac{\tau_{yz}}{m^S} (\tau_{yx}^2 + \tau_{yz}^2)^{\frac{1-n}{2n}} \tag{7-209b}$$

The net flow rate in cross channel direction should be zero if it is assumed the leakage over the flight is negligible. The cross channel flow rate is:

$$\dot{V}_x = \int_0^H v_x dy = 0 \tag{7-210}$$

The accuracy of the initial guesses of g_X, τ_{XO}, and τ_{ZO} can be evaluated by calculating $v_X(H)$, $v_Z(H)$ and $\dot{V}_X$. This can be done by using a standard numerical technique to integrate equation 7-209 and 7-210; e.g., Simpson's rule. The calculated values are then compared to the actual values: v_{bx}, v_{bz}, and 0 respectively. Unless the initial values are perfect, there will be residuals:

$$\Delta v_X = \int_0^H \frac{dv_X}{dy} \, dy - v_{bx} \tag{7-211a}$$

$$\Delta v_Z = \int_0^H \frac{dv_Z}{dy} \, dy - v_{bz} \tag{7-211b}$$

$$\Delta \dot{V}_X = \int_0^H v_X \, dy - 0 \tag{7-211c}$$

New values of g_X, τ_{XO}, and τ_{ZO} can be obtained by using a Newton-Raphson scheme. The residuals can be expressed as:

$$\Delta v_X = \frac{\partial v_X}{\partial g_X} \Delta g_X + \frac{\partial v_X}{\partial \tau_{XO}} \Delta \tau_{XO} + \frac{\partial v_X}{\partial \tau_{ZO}} \Delta \tau_{ZO} \tag{7-212a}$$

$$\Delta v_Z = \frac{\partial v_Z}{\partial g_X} \Delta g_X + \frac{\partial v_Z}{\partial \tau_{XO}} \Delta \tau_{XO} + \frac{\partial v_Z}{\partial \tau_{ZO}} \Delta \tau_{ZO} \tag{7-212b}$$

$$\Delta \dot{V}_X = \frac{\partial \dot{V}_X}{\partial g_X} \Delta g_X + \frac{\partial \dot{V}_X}{\partial \tau_{XO}} \Delta \tau_{XO} + \frac{\partial \dot{V}_X}{\partial \tau_{ZO}} \Delta \tau_{ZO} \tag{7-212c}$$

The partial derivatives can be determined by selecting a second set of data for g_X, τ_{XO}, and τ_{ZO} with values very close to the first values. When this is done, Δg_X, $\Delta \tau_{XO}$, and $\Delta \tau_{ZO}$ can be calculated by solving the three linear equations, which is a straightforward operation. The new values of g_X, τ_{XO}, and τ_{ZO} can now be determined from:

$$g_X\text{-new} = g_X\text{-old} + \Delta g_X \tag{7-213a}$$

$$\tau_{XO}\text{-new} = \tau_{XO}\text{-old} + \Delta \tau_{XO} \tag{7-213b}$$

$$\tau_{ZO}\text{-new} = \tau_{ZO}\text{-old} + \Delta \tau_{ZO} \tag{7-213c}$$

The iteration is repeated until the relative difference between the new and old value is less than a certain value, depending on the accuracy required. The throughput is determined from:

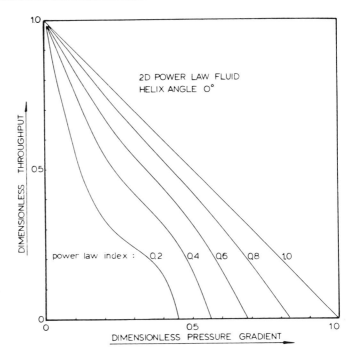

Figure 7-32b. Dimensionless Throughput versus Pressure Gradient from Numerical Calculations, Helix Angle is Zero

$$\dot{V}_Z = W \int_0^H v_Z dy \qquad\qquad (7-214)$$

A short Fortran program to perform this numerical procedure is given in Appendix 7-4. The program converges rapidly; the solution is usually obtained in five iterations. Figure 7-32b shows the dimensionless throughput-pressure gradient relationship from the two-dimensional analysis for a helix angle of zero degrees.

As mentioned before, these results should be the same as the results from the one-dimensional analysis as shown in Figure 7-32a. By comparing the two figures, it is clear that the results are virtually identical. This confirms the accuracy of the analytical solution for the one-dimensional flow of a power law fluid. The difference is generally less than one percent, except at small values of the dimensionless pressure gradient ($g_Z^0 < 0.1$). The loss of accuracy at low pressure gradients was predicted earlier and should not pose a serious problem in the analysis of real extrusion problems.

Figure 7-33 shows the dimensionless throughput versus pressure gradient for a helix angle of 17.66 degrees (square pitch screw).

The dimensionless throughput at 17.66 degrees helix angle is considerably lower than a zero helix angle. The difference is about 10 percent at small pressure gradients and about 40 percent at large pressure gradients when the power law index is less than one-half. At larger values of the power law index, the difference is generally less than 10 percent. This means that application of the equations for the one-dimensional case will lead to substantial errors for standard helix angles when the power law index is less than one-half! When the power law index is larger than one-half, these equations will be reasonably accurate.

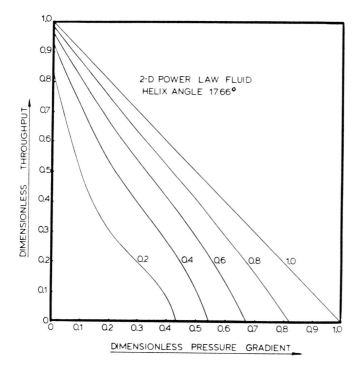

Figure 7–33. Dimensionless Throughput versus Pressure Gradient from Numerical Calculations, Helix Angle is 17.66 Degrees

It should be noted that a reduction in the dimensionless throughput does not necessarily mean that the actual throughput reduces as well. Obviously, when the helix angle is zero, the actual throughput will be zero. The dimensionless throughput is determined from:

$$\dot{V}^0 = \frac{2}{Hv_{bz}} \int_0^H v_z dy \qquad (7\text{–}215)$$

The actual volumetric throughput is related to the dimensionless throughput by:

$$\dot{V} = \frac{1}{2} HWv_{bz}\dot{V}^0 \approx \frac{1}{2} \pi^2 D^2 NHsin\varphi cos\varphi \dot{V}^0 \qquad (7\text{–}216)$$

The second equality of equation 7–216 is correct if the flight width is negligible. Figure 7–34 shows the dimensionless throughput versus pressure gradient for five helix angles when the power law index is one-half.

The dimensionless throughput reduces with increasing helix angle over the entire pressure gradient range. This demonstrates again that the equation for the one-dimensional case should not be used for large helix angles and/or small values of the power law exponent.

From Figures 7–31 through 7–34, it becomes clear that the Newtonian output-pressure gradient relationship is inacceptably inaccurate when the power law index of the polymer melt is less than 0.8. The one-dimensional power law (1-DPL) output-pressure gradient relationship is accurate only for small helix angles. Thus, for accurate results, a two-dimensional power law (2-DPL) analysis should be used. However, the 2-DPL analysis does not yield

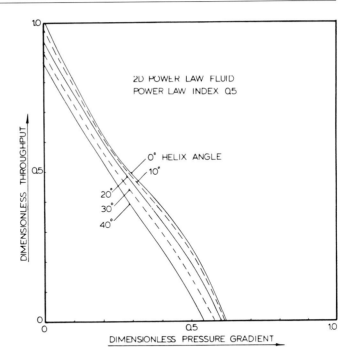

Figure 7-34. Dimensionless Throughput versus Pressure Gradient for Several Helix Angles

`

analytical solutions; numerical techniques have to be used to obtain results. One way to avoid complex calculations is to use the Newtonian output-pressure gradient relationship with correction factors for the non-Newtonian behavior of the polymer melt. Figure 7-35 shows the dimensionless output versus pressure gradient for an expression of $\dot{V}^0$ that incorporates such correction factors.

The correction factors apply to helix angles in the range of $15°$ to $25°$ and are determined to minimize the difference with a 2-DPL analysis. The dimensionless output is written as:

$$\dot{V}^0 = \frac{8+2n}{10} - \frac{3}{1+2n} \, g_z^0 \tag{7-217}$$

By using equation 7-202 for g_z^0 and equation 7-216 for $\dot{V}^0$, the Newtonian output-pressure gradient relationship with the correction factors for non-Newtonian behavior can be written as:

$$\dot{V} = \left(\frac{4+n}{10}\right) pWHv_{bz} - \left(\frac{1}{1+2n}\right) \frac{pWH^3 g_z}{4\mu} \tag{7-218}$$

where μ is calculated at the Couette shear rate in the channel ($\dot{\gamma} = v_{bz}/H$); thus

$$\mu = m\left(\frac{v_{bz}}{H}\right)^{n-1} \tag{7-218a}$$

With this equation, the difference with results from a 2-DPL analysis will generally be less than 10 percent when the power law index lies in the range of 0.3 to 1.0. Essentially all poly-

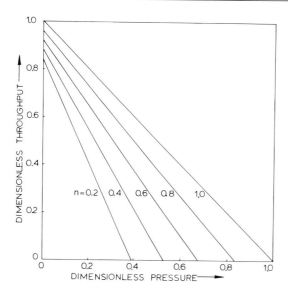

Figure 7-35. Dimensionless Through-put versus Pressure Gradient by Linear Approximation

mers fall in this range of power law indici. Thus, equation 7-218 should provide a useful relationship between output and pressure gradient for use in the analysis of practical extrusion problems. Of course, equation 7-218 is 100 percent accurate when the power law index is unity; i.e., for a Newtonian fluid. The correction factors in equation 7-217 and 7-218 have been determined for helix angles normally used in the extrusion industry, with the helix angle usually ranging from about 15° to 25°. If helix angles considerably below or above this range are used, then other correction factors will be more appropriate.

Another method to avoid complex calculations is to use an effective viscosity to be used in the Newtonian equation in order to predict output for a non-Newtonian fluid. This method was proposed by Booy (132). A drawback of this method is that it only corrects the pressure flow term. As a result, the method does not work at small pressure gradients. The effective viscosity ratio has to be obtained from a graph. The ratio can be approximated by taking the effective viscosity ratio equal to the power law index. However, the latter approximation will be considerably less accurate than equation 7-218.

Potente (235) also proposed approximate equations to predict the output as a function of pressure gradient for power law fluids. However, Potente's equations are only valid in a limited range of the dimensionless output $\dot{V}^0$, $0.6 \leqslant \dot{V}^0 \leqslant 1$. Thus, for a power law fluid with index $n = 0.3$, the equations are valid only when the dimensionless pressure gradient g_z^0 is less than about 0.1. In practical extrusion problems, one is generally more concerned about the effect of large pressure gradients than about the effect of small pressure gradients. Therefore, the equations developed by Potente have limited usefulness.

7.5   DIE FORMING

In this functional zone, the shaping of the polymer takes place. In many respects, the die forming zone is the most important functional zone because the actual shape of the final product develops in this zone. The die forming zone is essentially always a pressure consuming zone. The pressure built up in the preceding functional zones is used in the die forming zone. The diehead pressure is the pressure required to force the polymer melt through the die. The diehead pressure is not determined by the extruder but by the extruder die. The variables that affect the diehead pressure are:

1. The geometry of the flow channel in the die.

2. The flow properties of the polymer melt.

3. The temperature distribution in the polymer melt.

4. The flow rate through the die.

When these variables remain the same, the diehead pressure will be the same whether a single screw extruder or a twin screw extruder is used. The main function of the extruder is to supply a homogeneous polymer melt to the die at the required rate and diehead pressure. The rate and diehead pressure should be steady and the polymer melt should be homogeneous in terms of temperature and consistency. Die design and analysis of flow in dies are two of the most complicated elements of polymer process engineering. One of the reasons is that in a proper analysis of the die forming process, the polymer melt can no longer be assumed to be purely viscous because some important die flow phenomena such as extrudate swell (often erroneously referred to as die swell) cannot be explained with this simplifying assumption. Therefore, the polymer melt has to be analyzed as a viscoelastic fluid and this complicates the analysis of the die forming process considerably. Even if the polymer melt is assumed to be purely viscous, the analysis will generally be quite complicated because many dies have flow channels of complex shape. There are only two basic flow channel geometries that are rather easy to analyze: the circular flow channel and the slit flow channel (rectangular cross section with W >> H). Most other geometries are quite difficult to analyze.

7.5.1   VELOCITY AND TEMPERATURE PROFILES

Velocity profiles and temperature profiles in extruder dies are intimately related because of the high polymer melt viscosity and because the melt viscosity is temperature dependent. It is important to understand and appreciate this interrelationship to understand the die forming process and the variables that influence this process. The relationship between velocity and temperature profiles can be illustrated by considering the down channel velocity profile in a slit die. This is shown in Figure 7–36a. The more or less parabolic velocity profile is observed which is typical of pressure driven flow (pipe flow). The solid line represents a Newtonian fluid ($n = 1$) and the dashed line represents a non-Newtonian fluid ($n = 0.5$). The velocity curve for the Newtonian fluid is an exact parabole; the curve for the nonNewtonian fluid is not purely parabolic, it has a flattened center region and a larger gradient at the wall. From the velocity profile, one can obtain the shear rate profile by determining the local gradients of the velocity profile. This is shown in Figure 7–36b, again showing both the Newtonian and non-Newtonian fluid. For both fluids, the shear rate in the center of the flow channel is zero and the highest shear rate occurs at the wall. The wall shear rate for the non-Newtonian fluid, however, is considerably higher than for the Newtonian fluid.

As a result of the velocity gradients, there will be heat generation in the fluid from the viscous dissipation of energy. In rectilinear flow, the rate of energy dissipation per unit volume is given by:

$$\dot{E}_d = \tau_{yz} \frac{dv_z}{dy} \qquad (7-219)$$

This is a simplified version of the general expression for energy dissipation, equation 5–5 d. If the fluid can be described by the power law equation, then equation 7–219 becomes:

$$\dot{E}_d = m \left| \frac{dv_z}{dy} \right|^{n+1} = m \dot{\gamma}^{n+1} \qquad (7\text{--}220)$$

Thus, the local viscous dissipation is determined by the local shear rate raised to the power $n+1$. Since the highest shear rate occurs at the wall, it is clear that the highest viscous dissipation will also occur at the wall. As a result of the non-uniform shear rate distribution in the flow channel, there will be a non-uniform viscous heat generation in the flow channel. The largest amount of viscous heat generation occurs at the wall. As a result of the viscous heat generation, the temperature of the polymer melt will increase. But since the viscous heat generation is non-uniform across the flow channel, the temperature rise of the polymer melt will also be non-uniform across the flow channel. In the hypothetical case that the wall is perfectly insulated, i.e., no heat flux through the wall, the highest stock temperature will occur at the wall. This corresponds to adiabatic conditions at the wall; since the wall is usually a metal with high thermal conductivity, truly adiabatic conditions are rarely achieved. The other extreme condition would be if the wall was maintained at constant temperature. In this case, the heat flux through the wall would be such as to maintain a perfectly constant temperature along the wall. This is referred to as an isothermal wall boundary condition. Because of the high thermal conductivity of the wall, the isothermal boundary condition is more likely to occur than the adiabatic boundary condition. Adiabatic conditions can be approached if the die is very well insulated. In most actual cases, the true thermal boundary condition will be somewhere between isothermal and adiabatic depending on the design of the die and external conditions around the die. A typical temperature profile resulting from the velocity profiles shown in Figure 7–36a is shown in Figure 7–36c.

Initially, the maximum temperature will occur close to the wall; later, this maximum will shift towards the center. A quantitative method of evaluating temperature profiles will be discussed next.

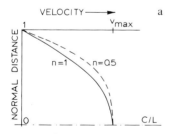

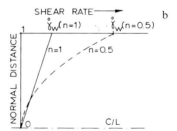

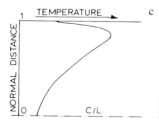

Figure 7–36. Velocity, Shear Rates, and Temperature Profiles in Flow Channels

Important relationships for shear stress, shear rate, velocity and flow rate for the pressure flow of power law fluids in a slit flow channel are given in Table 7-1. Table 7-2 shows the same relationships for a circular flow channel. These relationships are valid for fluids with temperature independent viscosity.

TABLE 7-1. PRESSURE FLOW OF A POWER LAW FLUID THROUGH A SLIT

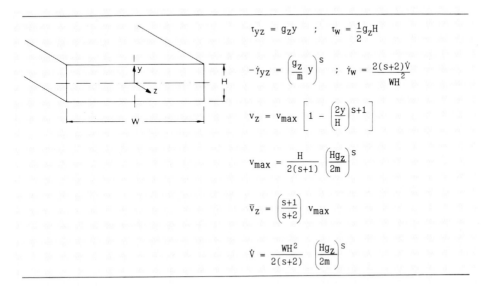

$$\tau_{yz} = g_z y \quad ; \quad \tau_w = \frac{1}{2} g_z H$$

$$-\dot\gamma_{yz} = \left(\frac{g_z}{m} y\right)^s \quad ; \quad \dot\gamma_w = \frac{2(s+2)\dot V}{WH^2}$$

$$v_z = v_{max}\left[1 - \left(\frac{2y}{H}\right)^{s+1}\right]$$

$$v_{max} = \frac{H}{2(s+1)}\left(\frac{Hg_z}{2m}\right)^s$$

$$\bar v_z = \left(\frac{s+1}{s+2}\right) v_{max}$$

$$\dot V = \frac{WH^2}{2(s+2)}\left(\frac{Hg_z}{2m}\right)^s$$

TABLE 7-2. PRESSURE FLOW OF A POWER LAW FLUID THROUGH A CIRCULAR CHANNEL

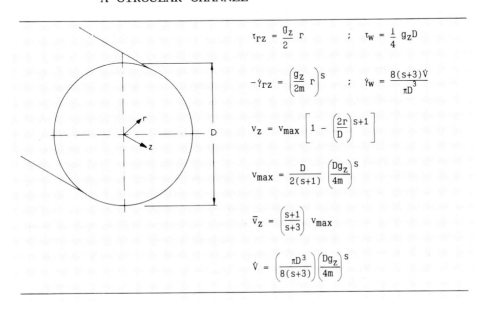

$$\tau_{rz} = \frac{g_z}{2} r \quad ; \quad \tau_w = \frac{1}{4} g_z D$$

$$-\dot\gamma_{rz} = \left(\frac{g_z}{2m} r\right)^s \quad ; \quad \dot\gamma_w = \frac{8(s+3)\dot V}{\pi D^3}$$

$$v_z = v_{max}\left[1 - \left(\frac{2r}{D}\right)^{s+1}\right]$$

$$v_{max} = \frac{D}{2(s+1)}\left(\frac{Dg_z}{4m}\right)^s$$

$$\bar v_z = \left(\frac{s+1}{s+3}\right) v_{max}$$

$$\dot V = \left(\frac{\pi D^3}{8(s+3)}\right)\left(\frac{Dg_z}{4m}\right)^s$$

Dinh and Armstrong (146) have developed general analytical solutions for the local temperature change due to viscous dissipation for non-Newtonian fluids. Their results apply to fluids with flow properties that are insensitive to temperature. The dimensionless normal distance is defined as $y^0 = 2y/H$ for a slit or $y^0 = 2y/D$ for a circular flow channel. The dimensionless velocity is defined as $v_z^0 = v_z/v_{max}$; it can be expressed in terms of dimensionless normal distance as follows:

$$v^0 = 1 - (y^0)^{S+1} \qquad\qquad (7\text{-}221)$$

Since the analysis deals with rectilinear flow, the subscript of the velocity will be deleted. The dimensionless viscosity is defined as:

$$\eta_r = \frac{\eta}{\eta_0} = \left| \frac{dv^0}{dy^0} \right|^{n-1} \qquad\qquad (7\text{-}222)$$

where:

$$\eta_0 = m \left(\frac{v_{max}}{H} \right)^{n-1} \qquad\qquad (7\text{-}222\,a)$$

The energy equation expressed in terms of dimensionless quantities can be written as:

$$v^0 \frac{\partial \theta}{\partial z^0} = \frac{\partial^2 \theta}{\partial (y^0)^2} + \eta_r \left(\frac{dv^0}{dy^0} \right)^2 \qquad\qquad (7\text{-}223)$$

where:

$$\theta = \frac{k(T - T_0)}{\eta_0 v_{max}^2} \qquad\qquad (7\text{-}223\,a)$$

$$z^0 = \frac{kz}{\rho C_p v_{max} H^2} \qquad\qquad (7\text{-}223\,b)$$

The following boundary conditions will be considered:

$$\text{at } z^0 = 0, \quad \theta = 0 \qquad\qquad (7\text{-}223\,c)$$

$$\text{at } y^0 = 0, \quad \frac{\partial \theta}{\partial y^0} = 0 \qquad\qquad (7\text{-}223\,d)$$

$$\text{at } y^0 = 1, \quad \frac{\partial \theta}{\partial y^0} = -N\theta \qquad\qquad (7\text{-}223\,e)$$

where N is the Biot number as defined in equation 5–74. When N is zero, there is no exchange of heat; adiabatic conditions prevail. When N is infinitely large, the wall temperature equals the temperature of the polymer melt; this corresponds to isothermal conditions. Normal values for the Biot number in extruder dies range from 1 to 100. As long as the Biot number is non-zero, there will be a fully developed temperature profile. However, when the Biot number is zero, the temperature in the fluid will continue to rise without limit.

The fully developed temperature profile θ_1 is the solution to the following differential equation:

$$\frac{d^2\theta_1}{d(y^0)^2} = -\eta_r \left(\frac{dv^0}{dy^0}\right)^2 \qquad (7\text{–}224)$$

with the following boundary condition:

$$\frac{d\theta_1(0)}{dy^0} = 0 \text{ and } \frac{d\theta_1(1)}{dy^0} = -N\theta_1(1) \qquad (7\text{–}224\,a)$$

The solution to this equation is:

$$\theta_1(y^0) = \left(\frac{n+1}{n}\right)^{n+1} \left[\frac{1}{N} + \frac{n^2}{(2n+1)(3n+1)} - \frac{(y^0)^{3+s}n^2}{(2n+1)(3n+1)}\right] \qquad (7\text{–}225)$$

where s is the reciprocal power law index (s = 1/n).

Equation 7–225 is a very useful relationship to determine fully developed temperature profiles in pipe flow of power law fluids. The maximum fully developed temperature always occurs at the center of the flow channel as can be seen from equation 7–225 and from Figure 7–37, which shows the fully developed temperature profile under isothermal wall conditions at various values of the power law index.

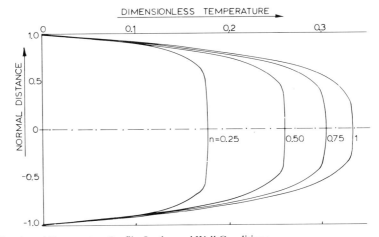

Figure 7–37. Fully Developed Temperature Profile, Isothermal Wall Conditions

As the fluid becomes more pseudoplastic, the fully developed temperature profile becomes more flattened in the center. The temperature remains almost constant in a center region which extends for about half the channel height. The larger temperature gradients occur in a relatively thin region at the walls.

The temperature profile as a function of normal distance and down channel distance is postulated to be of the form:

$$\theta\,(y^0, z^0) = \theta_1 - \sum_i c_i x_i (y^0) \exp(-a_i z^0) \tag{7-226}$$

For the details of the determination of eigenfunctions x_i and eigenvalues a_i, the reader is referred to the paper by Dinh and Armstrong (146). The eigenvalues are given by:

$$a_j = \left[\frac{\left(j - \frac{1}{4} \right)\pi - \alpha}{\int_0^1 \sqrt{v^0} \, dy^0} \right]^2 \tag{7-227}$$

where $j = 1,2,3 \ldots$ and α is given by:

$$\alpha = \arctan \frac{-3\left[3^{1/3}\Gamma\left(\frac{2}{3}\right) + (\dot{\gamma}_w a)^{-1/3} N\Gamma\left(\frac{1}{3}\right) \right]}{\left[3^{5/6}\Gamma\left(\frac{2}{3}\right) - 3^{1/2}(\dot{\gamma}_w a)^{-1/3} N\Gamma\left(\frac{1}{3}\right) \right]} \tag{7-228}$$

where $\pi/3 \leqslant \alpha \leqslant 2\pi/3$, $\dot{\gamma}_w$ is the dimensionless shear rate at the wall, and $\Gamma(p)$ is the gamma function. The gamma function is defined as:

$$\Gamma(p) = \int_0^\infty x^{p-1} \exp(-p) \, dx \tag{7-229a}$$

where:

$$\Gamma(p+1) = p\Gamma(p) \text{ if } p > 0$$
$$\Gamma(p+1) = p! \quad \text{ if } p \text{ is a positive integer}$$
$$\Gamma(1) = 1$$
$$\Gamma\left(\frac{1}{2}\right) = \pi^{1/2}$$

In evaluating the eigenvalues using equation 7-227, a convenient relationship involving gamma functions is:

$$\int_0^1 x^{c_1-1}(1-x)^{c_2-1}\,dx = \int_0^\infty \frac{x^{c_1-1}}{(1+x)^{c_1+c_2}}\,dx = \frac{\Gamma(c_1)\Gamma(c_2)}{\Gamma(c_1+c_2)}$$

$$(7-229b)$$

The eigenfunctions x_j are given by:

$$x_j \simeq (v^0)^{-1/4}\left(\sqrt{a_j}\int_{y^0}^1 \sqrt{v^0}\,dy^0\right)^{1/2} F$$

$$(7-229c)$$

where:

$$F = \left[3^{1/3}\Gamma\left(\frac{2}{3}\right)J_{-1/3}\left(\sqrt{a_j}\int_{y^0}^1 \sqrt{v^0}\,dy^0\right) + \right.$$

$$\left. + (\dot{\gamma}_w a)^{-1/3}N\Gamma\left(\frac{1}{3}\right)J_{1/3}\left(\sqrt{a_j}\int_{y^0}^1 \sqrt{v^0}\,dy^0\right)\right]$$

$$(7-230)$$

where J_v is the Bessel function of the first kind of order v.
The Bessel function J_v is given by:

$$J_v(x) = \left(\frac{1}{2}x\right)^v \sum_{i=0}^\infty \frac{\left(-\frac{1}{4}x^2\right)^i}{i!\,\Gamma(i+v+1)}$$

$$(7-231)$$

The expansion coefficients c_i in equation 7-226 for the slit problem are given by:

$$c_i = \frac{\int_0^1 \theta_1 v^0 x_i\,dy^0}{\int_0^1 v^0 x_i^2\,dy^0}$$

$$(7-232)$$

If the fluid is a power law fluid and the wall condition is isothermal, the eigenvalues become:

$$a_j = \left[2\sqrt{\pi}\,\frac{\Gamma\left(\frac{3}{2}+\frac{1}{1+s}\right)}{\Gamma\left(1+\frac{1}{1+s}\right)}\left(j-\frac{7}{12}\right)\right]^2$$

$$(7-233)$$

The dimensionless temperature as a function of dimensionless normal distance for a Newtonian fluid is shown in Figure 7-38.

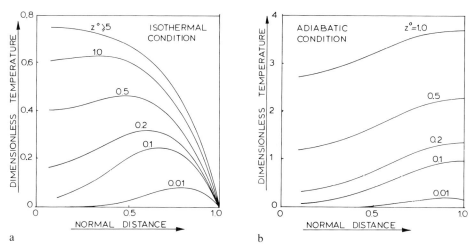

Figure 7–38. Temperature versus Distance for Newtonian Fluid

Figure 7–38a shows the temperature profiles under isothermal wall conditions and Figure 7–38b under adiabatic wall conditions. Figure 7–39 shows the temperature profiles for a power law fluid with power law index n = 0.5. Figure 7–39a shows the temperature profiles under isothermal wall conditions and 7–39b under adiabatic wall conditions.

The dimensionless temperature and down channel distance in Figures 7–38 and 7–39 are related to the average fluid velocity. In comparing Figure 7–38 to 7–39, it is seen that increased pseudoplasticity reduces the temperature build-up in the polymer melt at equal volumetric flow rates. It is also quite apparent that the temperature build-up under adiabatic conditions is substantially higher than under isothermal conditions.

The fully developed temperature profile, equation 7–225, is rather easy to calculate. However, the developing temperature profile, equation 7–226 through 7–233, involves some rather lengthy and complex calculations. Although the solutions to the developing tempera-

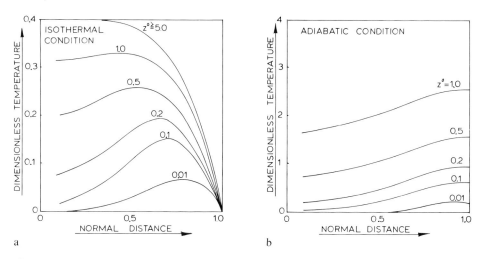

Figure 7–39. Temperature versus Distance for Power Law Fluid With n = 0.5

ture profile are analytical, obtaining actual results still requires computations that go beyond the capabilities of most pocket calculators. In many practical cases, one would like to
know to what extent the actual temperature profile approaches the fully developed temperature profile. This can be determined by using a dimensionless axial distance Z_{Gz}, defined
as:

$$Z_{Gz} = \frac{Z}{LN_{Gz}} \qquad (7-234)$$

where L is the length of the channel and N_{Gz} the Graetz number as defined in equation
5-68. If $Z_{Gz} \geqslant 1$, the temperature profile will be essentially fully developed in most practical heat transfer situations as analyzed, for instance, by Winter (147). For a slit flow channel,
the axial length Z_1 at which the temperature profile will be fully developed can now be expressed as:

$$Z_1 = \frac{\dot{V}H}{\alpha W} \qquad (7-235)$$

In most practical extrusion operations, the length of the die land is much too short to reach
a fully developed temperature profile. Thus, in order to determine the actual stock temperatures, one must evaluate the developing temperature profile.

The total amount of power dissipated in the flow channel of a die is simply determined by
the product of flow rate and pressure drop along the flow channel. Thus:

$$Z_d = \dot{V}\Delta P \qquad (7-236)$$

If it is assumed that all this power is used to raise the temperature of the polymer melt, i.e.,
adiabatic conditions, then the volume average rise in melt temperature can be determined
from:

$$\Delta\bar{T} = \frac{\Delta P}{\rho C_p} \qquad (7-237)$$

Thus, the volume average adiabatic temperature rise is directly proportional to the pressure
drop. If the pressure drop is 30 MPa ($=4350$ psi), a typical volume average temperature rise
is about $10°$ C. In most cases, however, the actual volume average temperature rise will be
less because heat transfer will take place at the die wall. In other words, adiabatic conditions will not occur in actual extrusion operation. It is important to realize, though, that local temperatures can be considerably higher than the volume average temperature. The
highest shear rates occur at the wall and consequently the highest viscous heat generation
occurs at the wall. Therefore, the polymer close to the wall will heat up much faster than the
polymer in the center region of the channel. Thus, it is quite possible that a local temperature rise close to the wall can be many times higher than the volume average temperature
rise.

In extrusion, one is always concerned about temperature uniformity. One of the important
requirements of the extruder is to deliver to the die a polymer melt of uniform consistency
and temperature. However, it should be realized that, even if the polymer melt entering the
die is uniform in temperature, non-uniformities in temperature will develop in the die as a

result of the non-uniform velocity gradients. This is inherent to die flow. The temperature build-up and non-uniformities can be reduced by reducing the shear rate. This can be achieved by lowering the flow rate through the die or by opening up the die flow channel. Another possibility is to use coextrusion with the outer layer being thin and of low viscosity. One can also use an external lubricant in the polymer to reduce a die flow problem; however, this may well introduce other problems as well (e. g., solids conveying problems, loss of mechanical properties, etc.).

7.5.2 EXTRUDATE SWELL

A well-known and typical phenomenon in polymer melt extrusion is the swelling of the extrudate as it leaves the die. This is sometimes referred to as die swell; however, it is not the die but the polymer that swells. The elasticity of the polymer melt is largely responsible for the swelling of the extrudate upon leaving the die. This is primarily due to the elastic recovery of the deformation that the polymer was exposed to in the die. The elastic recovery is time-dependent. A die with a short land length will have a large amount of swelling, while a long land length will reduce the amount of swelling. The polymer has what is often called a "fading memory". A deformation can be recovered to a large extent shortly after the occurrence of the deformation; however, after longer times the recoverable deformation reduces. Thus, a certain amount of relaxation occurs in the die depending on the geometry of the flow channel.

It should be noted that extrudate swelling is not unique to viscoelastic fluids. It can also occur in an inelastic or purely viscous fluid; this has been demonstrated experimentally and theoretically. Obviously, in a inelastic fluid, the mechanism of extrudate swell is not an elastic recovery of prior deformation. The swelling is caused by a significant rearrangement of the velocity profile as the polymer leaves the die, this is shown in Figure 7–40.

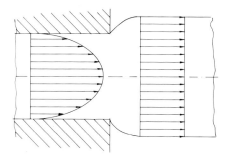

Figure 7–40. Change in the Velocity Profile in the Die Exit Region

The velocity profile changes from an approximately parabolic velocity profile in the die to a straight velocity profile (plug flow) a short distance away from the die. In a Newtonian fluid, this causes a small amount of extrudate swelling (about 10 percent) at low Reynolds numbers ($N_{Re} < 16$). Viscoelastic fluids exhibit about the same amount of swelling at low shear rates, but much larger amounts of swelling can occur at high shear rates (over 200 percent!).

One of the main problems with extrudate swell is that it is generally not uniformly distributed over the extrudate. This means that some areas of the extrudate swell more than others. If the geometry of the exit flow channel of the die is made to match the geometry of the re-

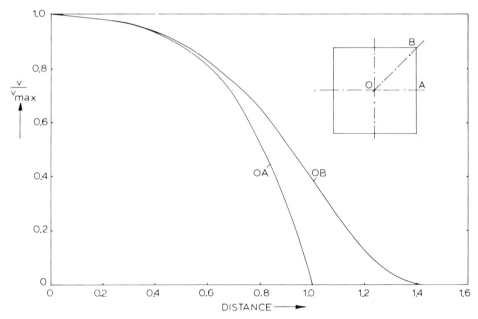

Figure 7-41. Velocity Profiles in a Square Channel

quired product, the uneven swelling will cause a distortion of the extrudate and the required product geometry cannot be obtained. Draw down cannot cure this problem! Therefore, the geometry of the exit flow channel must generally be different from the required product geometry. This can be understood by analyzing the velocity profiles in the flow channel. Figure 7-41 shows the velocity profile in a flow channel with a square cross section.

It can be seen that the shear rates at the wall vary significantly. The wall shear rate in the corner is relatively low, while the highest shear rate occurs at the middle of the wall. Therefore, the elastic recovery in the middle will be larger than the elastic recovery at the corners. This results in a "bulged" extrudate. It is not possible to obtain a perfectly square extrudate

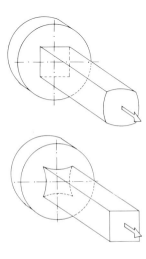

Figure 7-42. Uneven Swelling of Extrudate and Possible Correction

with a perfectly square flow channel. To eliminate this problem one has to modify the shape of the flow channel to compensate for the uneven swelling of the extrudate. A good die designer must anticipate the amount of uneven swelling and design the flow channel accordingly. This is a very difficult task and the determination of the flow channel geometry is often done by a "trial and error" process. Accurate mathematical prediction of the die swell profile is quite difficult and, therefore, determination of the proper flow channel geometry to minimize uneven swelling by engineering calculations is generally not practical. The non-uniform extrudate swell and the correction of the flow channel geometry is illustrated in Figure 7-42.

The amount of swelling is very much dependent on the nature of the material. Some polymers exhibit considerable swell (100 to 300 percent), e. g., polyethylenes; other polymers exhibit lower swell, e. g., polyvinylchloride. When PVC is extruded at relatively low temperatures (165° to 175° C), the swell ranges from 10 to 20 percent only. This is one of the reasons that PVC is such a popular material in profile extrusion; it conforms quite well to the geometry of the die flow channel and has good melt strength.

7.5.3   DIE FLOW INSTABILITIES

In extrusion, certain die flow instabilities can occur that may seriously affect the entire extrusion process and render the extruded product unacceptable. Two very important die flow instabilities are shark skin and melt fracture.

7.5.3.1   SHARK SKIN

Shark skin manifests iteself as a regular ridged surface distortion, with the ridges running perpendicular to the extrusion direction. A less severe form of shark skin is the occurrence of matness of the surface, where the glossy surface cannot be maintained. Shark skin is generally thought to be formed in the die land or at the exit. It is dependent primarily on the temperature and the linear extrusion speed. Factors such as shear rates, die dimensions, approach angle, surface roughness, L/D ratio, material of construction, seem to have little or no effect on shark skin.

The mechanism of shark skin is postulated to be caused by the rapid acceleration of the surface layers of the extrudate when the polymer leaves the die; this is illustrated in Figure 7-40. If the stretching rate is too high, the surface layer of the polymer can actually fail and form the characteristic ridges of the shark skin surface (148). High viscosity polymers with narrow molecular weight distribution (MWD) seem to be most susceptible to shark skin instability (149, 150).

The shark skin problem can generally be reduced by reducing the extrusion velocity and increasing the die temperature, particularly at the land section. There is some evidence that running at very low temperatures can also reduce the problem (151). Selection of a polymer with a broad MWD will also be beneficial in reducing shark skin. Using an external lubricant can also reduce the problem. This can be done by using an additive to the polymer or by coextruding a thin, low viscosity outer layer.

7.5.3.2   MELT FRACTURE

Melt fracture is a severe distortion of the extrudate which can take many different forms: spiraling, bambooing, regular ripple, random fracture, etc., see Figure 7-43.

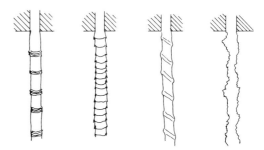

Figure 7-43. Various Forms of Melt Fracture

It is not a surface defect like shark skin, but is associated with the entire body of the molten extrudate. However, many workers do not distinguish between shark skin and melt fracture, but lump all these flow instabilities together under the term melt fracture. There is a large amount of literature on the subject of melt fracture (e. g., 152–164). Despite the large number of studies on melt fracture, there is no clear agreement as to the exact cause and mechanism of melt fracture. It is quite possible that the mechanism is not the same for different polymers and/or different flow channel geometries (169). Linear polymers tend to develop an instability of the shear flow in the die land, while branched polymers tend to develop instabilities in the converging region of the die flow channel.

However, there is relatively uniform agreement that melt fracture is triggered when a critical wall shear stress is exceeded in the die. This critical stress is in the order of 0.1 to 0.4 MPa (15 to 60 psi). A number of mechanisms have been proposed to explain melt fracture. Some of the more popular ones are:

    i)   critical elastic deformation in the entry zone
   ii)   critical elastic strain
  iii)   slip-stick flow in the die

The effect of the entry zone has been demonstrated by many workers. In general, the smaller the entry angle, the higher the deformation rate at which instability occurs. Gleiszle (230) has proposed a critical elastic strain as measured by recoverable strain. Based on measurements with 11 fluids, he proposed the existence of a critical value of the ratio of first normal stress difference to the shear stress; the average value being 4.63 for 11 widely different fluids with a standard deviation of about 5 percent. Much larger differences were found in the critical shear stress; the average being 3.7E5 Pa for the polymer with a standard deviation of about 55 percent. In 1961, Benbow, Charley and Lamb (232, 233) introduced the slip-stick mechanism to explain flow instability and extrudate distortion. Above a certain critical stress, the polymer melt is believed to experience intermittent slipping due to lack of adhesion between the melt and the die wall in order to relieve excessive deformation energy absorbed as a result of flow through a die. A large number of workers have observed slippage by various techniques.

More recent work by Utracki and Gendron on pressure oscillations in extrusion of polyethylenes (231) led them to conclude that the pressure oscillation does not seem to be related to elasticity or slip. They conclude that the parameter responsible for pressure oscillations is the critical strain (Hencky) value ε_c of the melt. For LLDPE, $\varepsilon_c < 3$, for HDPE, $\varepsilon_c < 2$, while for LDPE, $\varepsilon_c > 3.5$. The instability seems to be based on the inability of the polymer melt to sustain levels of strain larger than the critical strain.

Streamlining the flow channel geometry has been found to reduce the tendency for melt fracture in branched polymers. Increased temperatures, particularly at the wall of the die land, enable higher extrusion rates before melt fracture appears. The critical wall shear stress appears to be relatively independent of the die length, radius, and temperature. The critical stress seems to vary inversely with molecular weight, but seems to be independent of MWD. Certain polymers exhibit a superextrusion region, above the melt fracture range, where the extrudate is not distorted (165). This process is particularly advantageous with polymers that melt fracture at relatively low rates, such as FEP. In superextrusion, the polymer melt is believed to slip relatively uniformly along the die wall. The occurrence of slip in extruder dies has been studied by a number of investigators (e. g., 166, 168). However, it is still not clear whether the slip is actual loss of contact of polymer melt and metal wall or whether it is failure of a thin polymeric layer very close to the metal surface.

The melt fracture problem can be reduced by streamlining the die, increasing the temperatures at the die land, running at lower rates, reducing the MW or the polymer melt viscosity, increasing the cross sectional area of the exit flow channel, or by using an external lubricant. In some instances, the melt fracture problem can be solved by going to superextrusion; this process is used particularly often in the wire coating industry where high line speeds are quite important for economic production.

7.5.3.3   DRAW RESONANCE

Draw resonance occurs in processes where the extrudate is exposed to a free surface stretching flow, such as blown film extrusion, fiber spinning, and blow molding. It manifests itself in a regular cyclic variation of the dimensions of the extrudate. An extensive review (169) and an analysis (170) of draw resonance was done by Petrie and Denn. Draw resonance occurs above a certain critical draw ratio while the polymer is still in the molten state when it is taken up and rapidly quenched after take-up.

Draw resonance will occur when the resistance to extensional deformation decreases as the stress level increases. The total amount of mass between die and take-up may vary with time because the take-up velocity is constant but not necessarily the extrudate dimensions. If the extrudate dimensions reduce just before the take-up, the extrudate dimensions above it have to increase. As the larger extrudate section is taken up, a thin extrudate section can form above it; this can go on and on. Thus, a cyclic variation of the extrudate dimensions can occur. Draw resonance does not occur when the extrudate is solidified at the point of take-up because the extrudate dimensions at the take-up are then fixed (171, 172). Isothermal draw resonance is found to be independent of the flow rate. The critical draw ratio for almost Newtonian fluids such as nylon, polyester, polysiloxane, etc. is approximately 20. The critical draw ratio for strongly non-Newtonian fluids such as polyethylene, polypropylene, polystyrene, etc. can be as low as 3 (173). The amplitude of the dimensional variation increases with draw ratio and draw down length.

Various workers have performed theoretical studies of the draw resonance problem by linear stability analysis. Pearson and Shah (174, 175) studied inelastic fluids and predicted a critical draw ratio of 20.2 for Newtonian fluids. Fisher and Denn (176) confirmed the critical draw ratio for Newtonian fluids. Using a linearized stability analysis for fluids that follow a White-Metzner equation, they found that the critical draw ratio depends on the power law index n and a viscoelastic dimensionless number. The dimensionless number is a function of the die take-up distance, the tensile modulus, and the velocity at the die. Through their analysis, Fisher and Denn were able to determine stable and unstable operating regions. In some instances, draw resonance instability can be eliminated by increasing the draw ratio, although under most operating conditions draw resonance is eliminated by reducing the draw ratio.

White and Ide (177-180) demonstrated experimentally and theoretically that polymers whose elongational viscosity increases with time or strain do not exhibit draw resonance, but undergo cohesive failure at high draw ratios. A polymer that behaves in such a fashion is LDPE. Polymers whose elongational viscosity decreases with time or strain do exhibit draw resonance at low draw ratios and fail in a ductile fashion at high draw ratios. Examples of polymers that behave in such a fashion are HDPE and PP. Lenk (181) proposed a unified concept of melt flow instability. His main conclusions are that all flow instabilities originate at the die entrance and that melt fracture and draw resonance are not distinct and separate flow phenomena; both are caused by elastic effects that have their origin at the die entrance. Lenk's analysis, however, is purely qualitative and does not offer much help in the engineering design of extrusion equipment or in determining how to optimize process conditions to minimize instabilities.

7.6   DEVOLATILIZATION

Devolatilization is a function that is not performed on all extruders, as opposed to the other functions: solids conveying, melting, melt conveying, and mixing. Therefore, devolatilizing extruders are relatively specialized. However, as polymer processing operations become more sophisticated, it is becoming less unusual for the extruder to be used for continuous devolatilization. There have been relatively few engineering analyses of the devolatilization process in extruders. The first major effort seems to have been the work by Latinen (182). Other analytical studies of devolatilization in extruders have been made by Coughlin and Canevari (183), Roberts (184), Biesenberger (185, 186), and Denson (236). The physical model of devolatilization in a single screw extruder is shown in Figure 7-44.

The exposure time λ_f of the wiped film as it travels with the barrel surface from the flight clearance to the melt pool is:

$$\lambda_f = \frac{1 - Y}{N} \qquad\qquad (7\ 238)$$

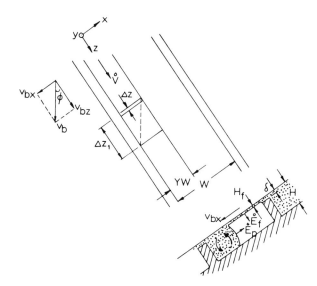

Figure 7-44. Model for Devolatilization in Screw Extruders

where N is the screw speed and Y the fraction of the channel width occupied by the melt pool. If the channel is partially filled, then $Y < 1$ and Y represents the degree of fill in the extraction section.

The total volumetric flow rate of the wiped film is:

$$\dot{V}_f = v_b H_m L_b \tag{7-239}$$

where H_m is the melt film thickness and L_b is the total axial length of the devolatilizing zone.

The feedback ratio n_f is:

$$n_f = \dot{V}_f / \dot{V} \tag{7-240}$$

where $\dot{V}$ is the total volumetric flow rate through the extruder.

The feedback ratio is the part of the material that splits off of the main material flow to form a film through which volatiles are escaping. The feedback ratio n_f can be interpreted as the extent of the surface renewal of the melt film. As the melt film enters the melt pool, again a certain amount of back mixing will take place, depending on the width of the melt pool and the helix angle. The melt pool will also loose volatile because of the exposed surface. The exposure time of the exposed surface of the melt pool is limited because of the circulatory motion in the melt pool. Roberts (184) approximated the exposure time of the melt pool by:

$$\lambda_p = \frac{H}{v_{bx}} \tag{7-241}$$

where H is the channel depth and v_{bx} the cross channel component of the barrel velocity.

The local bulk evaporation rate $\dot{E}_p(z)$ can be expressed as:

$$\dot{E}_p(z) = 2 \left(\frac{D'}{\pi \lambda_p} \right)^{1/2} H \Delta z [C(z) - C_e] \tag{7-242}$$

where D' is the diffusion coefficient, C(z) the local bulk average concentration of the melt pool, and C_e the equilibrium concentration at the vapor-liquid interface. For a derivation of equation 7-242, see Section 5.4.2.

The local film evaporation $\dot{E}_f(z)$ can be expressed as:

$$\dot{E}_f(z) = v_{bx} H_f \Delta z [C(z) - C'(z)] \tag{7-243}$$

where C'(z) is the concentration of the film reentering the melt pool at point z, having left the melt pool a distance Δz_1, downstream, see Figure 7-44. The formulation of the model can be completed by taking a steady state material balance on the volatile component over a differential element of volume of a thickness Δz. The balance simply states that the reduction in convective transport of the volatile component equals the film evaporation plus the bulk evaporation:

$$\dot{V}[C(z) - C(z+\Delta z_1)] = \dot{E}_p(z) + \dot{E}_f(z) \tag{7-244}$$

Now the value $C(z+\Delta z_1)$ has to be determined to obtain $C'(z)$. Concentration $C(z+\Delta z_1)$ is the initial concentration of the melt film. This melt film reenters the melt pool at location z with concentration $C'(z)$. As suggested by Roberts (184), a Taylor's series expansion about z can be used:

$$C(z+\Delta z_1) = C(z) + \frac{dC(z)}{dz} \Delta z_1 + \frac{d^2C(z)}{dz^2} \frac{\Delta z_1^2}{2} + \ldots \qquad (7-245)$$

Terms of order three and higher will be assumed to be negligible. The concentration $C'(z)$ at which the polymer reenters the melt pool is the initial concentration $C(z+\Delta z_1)$ of the film as it leaves the melt pool minus the amount of volatile lost through evaporation in the film during film exposure time λ_f. If X_f is the stage efficiency of the diffusing film, the concentrations $C'(z)$ and $C(z+\Delta z_1)$ can be related by:

$$\frac{C'(z) - C_e}{C(z+\Delta z_1) - C_e} = 1 - X_f \qquad (7-246)$$

By using equations 7–245 and 7–246, the mass balance equation can be rewritten as:

$$v_{bx}H_f(1-X_f) \frac{\Delta z_1^2}{2} \frac{d^2C}{dz^2} + [v_{bx}H_f(1-X_f)\Delta z_1 - \dot{V}] \frac{dC}{dz}$$

$$- [X_f v_{bx}H_f + k_pH](C-C_e) = 0 \qquad (7-247)$$

where:

$$k_p = \left(\frac{4D'}{\pi\lambda_p}\right)^{1/2} \qquad (7-247a)$$

Equation 7–247 can be written in dimensionless form:

$$\frac{1}{N_{Pe}} \frac{d^2\hat{C}}{d\hat{z}^2} - \frac{d\hat{C}}{d\hat{z}} - E_x\hat{C} = 0 \qquad (7-248)$$

where:

$$\hat{z} = \frac{z}{L} \qquad (7-248a)$$

$$\Delta\hat{z}_1 = \frac{\Delta z_1}{L} \qquad (7-248b)$$

$$N_{Pe} = \left(\frac{2}{\Delta\hat{z}_1}\right)\left(\frac{1}{n_f(1-X_f)\Delta\hat{z}} - 1\right) \qquad (7-248c)$$

$$E_x = \frac{n_f X_f + k_p HL/\dot{V}}{1 - n_f(1-X_f)\Delta\hat{z}_1} \qquad (7-248d)$$

$$n_f = \frac{v_{bx} H_f L}{\dot{V}} \qquad (7-248e)$$

$$\hat{C} = \frac{C - C_e}{C_0 - C_e} \qquad (7-248f)$$

The Peclet number N_{Pe} represents the effect of longitudinal backmixing. Backmixing can be neglected when the Peclet number is very large ($N_{Pe} \gg 1$). In the extreme case of pure plug flow, the Peclet number $N_{Pe} = \infty$; when $N_{Pe} = 0$, the entire flow channel acts as an ideal mixer. The extraction number E_x is a measure of the overall devolatilization efficiency. The backmixing term disappears when the melt film thickness is zero. This is to be expected because the backmixing is caused by the transport in the melt film. If the film thickness is zero, there can be no transport in the melt film and consequently no backmixing. If the film stage efficiency $X_f = 1$, the backmixing term will disappear as well. This is also easy to understand because in this case the concentration $C'(z)$ at which the melt film reenters the melt pool is known and will equal C_e. Thus, the Taylor's series expansion is no longer necessary and the first order differential equation can be determined directly from the mass balance, equation 7-244.

The magnitude of the backmixing effect will be directly determined by the distance Δz_1. This distance is directly related to the degree of fill Y, and the channel width W, and the helix angle φ:

$$\Delta z_1 = \frac{YW}{\tan\varphi} \qquad (7-249)$$

The channel width W, however, is also a function of the helix angle. If the screw is single flighted and if the flight width is negligible, distance Δz_1, can be written as:

$$\Delta z_1 = Y\pi D\cos\varphi \qquad (7-250)$$

From equation 7-250, it can be seen that for a certain size extruder the backmixing effect will increase with the degree of fill in the extraction section. Thus, the channel depth in the extraction section should be significantly larger than the channel depth of the preceeding screw section, the metering section. The backmixing effect will also increase when the helix angle becomes smaller. Thus, one would like to have a relatively small helix angle in the extraction section to increase the backmixing effect.

If the backmixing effect can be neglected ($N_{Pe} \gg 1$), equation 7-248 becomes simply:

$$\frac{d\hat{C}}{d\hat{z}} = - E_x\hat{C} \qquad (7-251)$$

The concentration thus becomes an exponential function of distance. If $\hat{C}(0) = 1$, the concentration profile becomes:

$$\hat{C} = \exp(-E_x\hat{z}) \tag{7-252}$$

where:

$$E_x = n_f X_f + k_p HL/\dot{V} \tag{7-252a}$$

The devolatilization efficiency of the machine X_T is a function of the individual stage efficiency X and the extent of surface renewal. In continuous equipment, such as screw extruders, the extent of surface renewal is described by the factor n_f (equation 7-240 or 7-248e); in batch devolatilizers, the extent of surface renewal is described by n, the discrete number of surface renewals. The film stage efficiency X_f is a function of the surface-to-volume ratio and the exposure time λ_f. It can generally be written as a single function of the ratio λ_f/λ_D. Thus, the overall efficiency X_T can be described as a function of the surface renewal factor n_f and the ratio λ_f/λ_D:

$$X_T = X_T (n_f, \; \lambda_f/\lambda_D) \tag{7-253}$$

The effectiveness of the devolatilization operation is strongly dependent on the actual length of the devolatilization zone L_B. In actual extrusion, the actual length L_B will often be considerably longer than the length of the extraction section of the screw L_e. Thus, the length L_B can extend into the pump section of the extruder. This is determined by the filled length of the pump section L_{pf}.

If the length of the pump section is L_p, the actual devolatilization length L_B can be determined from:

$$L_B = L_e + L_p - L_{pf} \tag{7-254}$$

where L_e is the length extraction section, L_p the length pump section, and L_{pf} the filled length of the pump section. The latter can be calculated from melt conveying theory if the following parameters are known:

i)   the geometry of the pump section
ii)   the throughput
iii)   the flow properties of the polymer
iv)   the diehead pressure

From the melt conveying theory of Newtonian fluids, the length L_{pf} can be determined from:

$$L_{pf} = \frac{pWH^3 P\sin\varphi}{6\mu pWHv_{bz} - 12\mu\dot{V}} \tag{7-255}$$

where P is the diehead pressure.

This equation was derived using equation 7-155; it does not take into account the leakage flow or the effect of the flight flanks. The diehead pressure P is related to the total volumetric flow rate $\dot{V}$ by the die constant K:

$$P = K\dot{V} \tag{7-256}$$

In most two-stage devolatilizing extruder screws, the through put $\dot{V}$ is determined by the metering section as a result of its shallow channel depth H_m. If the throughput is determined by the drag flow rate of the metering section and if the helix angle and the number of parallel flights is constant, the expression for L_{pf} can be simplified to:

$$L_{pf} = \frac{pWH^3 H_m K \sin\varphi}{12\mu(H - H_m)} \qquad (7-257)$$

Thus, in order to keep L_{pf} short and L_B long, the depth of the metering section H_m should be small compared to the depth of the pump section H. Also the restriction of the die K should be made as small as possible.

7.7   MIXING

Mixing can be broadly defined as a process to reduce the non-uniformity of a composition. The basic mechanism of mixing is to induce physical motion of the ingredients. The types of motion that can occur are molecular diffusion, turbulent motion, and convective motion. The first two types of motion are essentially limited to gasses and low viscosity liquids. Convective motion is the predominant motion in high viscosity liquids, such as polymer melts. As discussed in Section 5.3.3, polymer melts are not capable of turbulent motion as a result of their high viscosity; motion in polymer melts is always by laminar flow.

Convective mixing by laminar flow is referred to as laminar mixing. This is the type of mixing that occurs in polymer melt extrusion. The mixing action generally occurs by shear flow and elongational flow. If the components to be mixed are fluids and do not exhibit a yield point, the mixing is distributive. This is sometimes referred to as extensive mixing. The process of distributive mixing can be described by the extent of deformation or strain to which the fluid elements are exposed. The actual stresses involved in this process are irrelevant in the description of the distributive mixing. If the mixture contains a component that exhibits a yield stress, then the actual stresses involved in the process become very important. If the component exhibiting a yield point is a solid, this type of mixing is referred to as dispersive mixing, sometimes as intensive mixing. In dispersive mixing, a solid component needs to be broken down, but the breakdown only occurs after a certain minimum stress (yield stress) has been exceeded. If the component exhibiting a yield point is a liquid, the mixing process is referred to as homogenization. An example of dispersive mixing is the manufacture of a color concentrate where the breakdown of the pigment agglomerates below a certain critical size is of crucial importance. An example of distributive mixing is the manufacture of a polymer blend, where none of the components exhibit a yield point.

Distributive mixing and dispersive mixing are not physically separated. In dispersive mixing, there will always be distributive mixing. However, the reverse is not always true. In distributive mixing, there can be dispersive mixing only if there is a solid component with a yield stress and if the stresses acting on this component exceed the yield stress.

A very important aspect of the study of mixing is the characterization of the mixture. A complete characterization requires the specification of the size, shape, orientation, and spatial location of every discrete element of the minor component. This, of course, is generally impossible. Various theories and techniques have been devised to describe and measure the goodness of mixing (187–200). Some of the characterization techniques are quite sophisticated and most are very time consuming. Quantitative characterization is very important to workers doing research on mixing. However, such techniques are often impractical in actual polymer processing operations. Visual observation, although qualitative, is often sufficiently accurate to determine whether a product is acceptable or not. Therefore, the various characterization theories and techniques will not be discussed here. For more information on this subject, the reader is referred to the literature (187–200). In this section, the primary emphasis will be on the description of the mixing process in a screw extruder.

7.7.1   MIXING IN SCREW EXTRUDERS

Mixing is an essential function of the screw extruder. It occurs in all screw extruders as opposed to devolatilization, which is done only on specialized machines. The mixing zone in the extruder extends from the start of the plasticating zone to the end of the die, assuming that significant mixing only takes place when the polymer is in the molten state. The fact that mixing starts at the beginning of the melting zone presents a practical problem and an analytical problem. It means that at the end of the melting zone there will be a considerable non-uniformity in the mixing history of the polymer.

A polymer element that melts early will be exposed to a significant mixing history by the time it reaches the end of the melting zone. On the other hand, a polymer element that melts at the very end of the melting zone will have hardly any mixing history at the end of the melting zone.

Similar problems occur in the melt conveying zone. A polymer element at about 2/3 of the height of the channel will have no cross-channel velocity component and as a result will have a short residence time in the melt conveying section and little mixing history. A polymer element at about 1/3 of the channel will have considerable cross channel velocity and relatively low down channel velocity. As a result, this element will have a long residence time in the melt conveying zone and a large mixing history. This can be verified by analyzing the velocity profiles in the metering section as discussed in Section 7.4. It is clear, therefore, that the mixing action is not uniformly applied to all elements of the polymer melt. As a result of the inherent transport process in a screw extruder there will be considerable non-uniformities in the intensity of the mixing action and the duration of the mixing action. This is also true for the extruder die. Fluid elements in the center of the flow channel are exposed to very low shear rate and their residence is short because the velocities are highest in the center. Fluid elements at the wall are exposed to high shear rates and their residence time is long because of the low velocities at the wall. Thus, even if a perfectly mixed fluid enters a die, non-uniformities can be expected as the fluid leaves the die.

The mixing process in extruders is generally analyzed by determining the velocity profiles occurring in the screw channel. From the velocity profiles, the deformation at various locations in the fluid can be determined. In most analyses, the fluid is considered Newtonian, the components have the same flow properties (i.e., a rheologically homogeneous fluid mixture), and the flow through the flight clearance is neglected. Another common assumption is a two-dimensional flow pattern in the screw channel; only flow in down-channel and cross channel direction.

When two viscous liquids are mixed, the interfacial area increases and the striation thickness decreases. Spencer and Wiley (201) have proposed to use the interfacial area as a quantitative measure of the goodness of mixing. Mohr et al. (189) used the striation thickness to describe the mixing process. If a surface element with arbitrary orientation is located in a simple shear flow field, the surface area A after a total shear strain of γ can be demonstrated to be (201):

$$A = A_0(1 - 2\cos\alpha_x \cos\alpha_y \gamma + \cos^2\alpha_x \gamma^2)^{1/2} \qquad (7-258)$$

where A_0 is the original surface area, α_x the angle of the vector normal to A_0 with the x axis, and α_y the angle of the vector normal to A_0 with the y axis.

Angles α_x, α_y, and α_z determine the initial orientation of the surface element under consideration. The three angles are related by:

$$\cos^2\alpha_x + \cos^2\alpha_y + \cos^2\alpha_z = 1 \qquad (7-259)$$

If the total shear strain is very large ($\gamma \gg 1$), then equation 7-258 becomes:

$$A = A_0|\cos\alpha_x|\gamma \qquad (7-260)$$

Equation 7-260 indicates that the increase in interfacial area is directly proportional to the total shear strain and $\cos\alpha_x$. Thus, the total shear strain is an important variable in the description of the mixing process in a shear flow field. The initial orientation α_x is also very important. If the initial surface is oriented parallel to the flow field ($\alpha_x = 90°$), then the increase in interfacial area is zero. However, if the initial surface orientation is perpendicular to the flow field ($\alpha_x = 0$), then the increase in interfacial area is maximum. At low strains, it can be seen from equation 7-258 that the interfacial area can increase or decrease with strain, depending on the initial orientation.

If the interfaces are initially randomly oriented, the mean change in interfacial area becomes (202):

$$A = \frac{1}{2} A_0 \gamma \qquad (7-261)$$

Equation 7-261 is valid when the total strain is very large ($\gamma >> 1$). The striation thickness is defined as the total volume divided by half the total interfacial surface:

$$r = \frac{V}{A/2} \qquad (7-262)$$

If the minor component is initially introduced as randomly oriented cubes of height H and with a volume fraction φ, the striation thickness can be expressed as:

$$r = \frac{2H}{3\varphi\gamma} \qquad (7-263)$$

Thus, the striation thickness is directly proportional to initial domain size of the minor component and inversely proportional to the volume fraction and total shear strain. This indicates that a small striation thickness is achieved more easily when the initial domain size of the minor component is small and the volume fraction large.

If the simple shear field is disrupted by a short mixing section that produces a randomly oriented minor component, the interfacial area at the outlet of the mixing section is:

$$A_1 = \frac{1}{2} A_0 \gamma_1 \qquad\qquad (7\text{-}264)$$

where γ_1 is the total shear strain the fluid is exposed to before the mixing section. It is assumed that the shear strain in the mixing section itself is insignificant. If the simple shear field is restored after the mixing section, then the total interfacial area after another exposure to shear strain γ_1 becomes:

$$A_2 = \frac{1}{2} A_1 \gamma_1 = A_0 \left(\frac{1}{2} \gamma_1\right)^2 \qquad\qquad (7\text{-}265)$$

Similarly, after n mixing sections and n shear strain exposures of the same magnitude γ_1, the total interfacial area will be:

$$A_n = A_0 \left(\frac{1}{2} \gamma_1\right)^n \qquad\qquad (7\text{-}266)$$

From equation 7-266, it can be seen that the generation of interfacial area can be increased substantially by inclusion of mixing sections that randomize the minor component. The improvement can be evaluated by comparing n mixing sections and $n\gamma_1$ strain exposures to simple shear mixing without mixing sections but the same total strain exposure. The ratio of the interfacial areas is:

$$\frac{A_n(\gamma_1)}{A_1(n\gamma_1)} = \frac{\left(\frac{1}{2} \gamma_1\right)^{n-1}}{n} \qquad\qquad (7\text{-}267)$$

This ratio is shown in Figure 7-45.

It is obvious that randomizing mixing sections greatly improve the generation of interfacial area, and thus the mixing performance. Erwin and Ng (205) have constructed an experimental mixing apparatus by which the results of equations 7-266 and 7-267 and Figure 7-45 have been experimentally verified. However, the mixing apparatus does not lend itself to practical mixing operations.

If the mixing section is capable of orienting the minor component in the most favorable direction, i.e., perpendicular to the velocity, the total interfacial area after n mixing sections and n shear strain exposures of magnitude γ_1 will be:

$$A_n = A_0 (C\gamma_1)^n \qquad\qquad (7\text{-}268)$$

where $C = 1/2$ if the initial orientation is random and $C = 1$ if the initial orientation is most favorable.

Equations 7-266 and 7-267 demonstrate, at least qualitatively, that incorporation of mixing devices can substantially improve laminar mixing performance. In a dynamic mixing device, such as an extruder, it may be difficult to design a mixing section that will orient the minor component in the most favorable orientation. However, random orientation may be

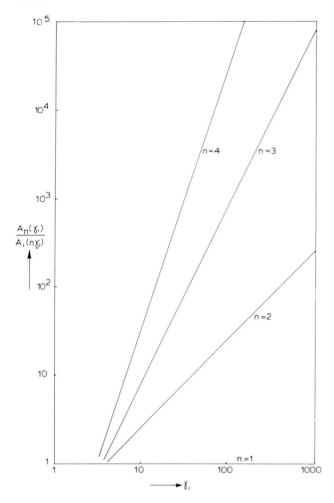

Figure 7–45. Mixing Efficiency Increase with Mixing Section

more feasible. In a static mixing device, it is easier to control the orientation of the minor component and effective laminar mixing can occur in such mixing devices, see Section 7.7.2.

The energy requirement to achieve a certain amount of increase in interfacial area was studied by Erwin (204). In uniaxial extension flow, the energy per unit volume is related to the surface area increase by:

$$E_{ue} = \frac{12\eta}{t} \left(\ln\frac{5A}{4A_0} \right)^2 \qquad (7-269)$$

where η is the viscosity of the fluid and t the duration of the extensional deformation.

In biaxial extensional flow, the energy per unit volume is:

$$E_{be} = \frac{3\eta}{t} \left(\ln\frac{5A}{4A_0} \right)^2 \qquad (7-270)$$

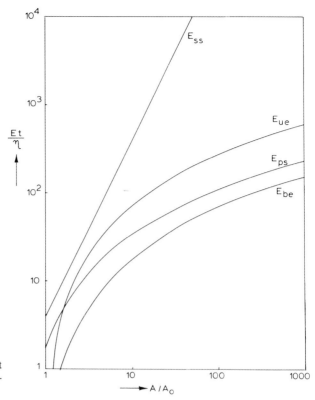

Figure 7-46. Energy Requirement for Mixing in Various Flow Situations

In plane strain elongation (two-dimensional elongation), the energy per unit volume is:

$$E_{ps} = \frac{4\eta}{t} \left(\ln\frac{2A}{A_0} \right)^2 \qquad\qquad (7\text{--}271)$$

In simple shear, the energy per unit volume is:

$$E_{ss} = \frac{4\eta}{t} \left(\frac{A}{A_0} \right)^2 \qquad\qquad (7\text{--}271\,a)$$

Figure 7-46 shows the normalized energy per unit volume Et/η versus the ratio of interfacial areas.

It is obvious that simple shear is the most inefficient flow. When the area ratio becomes larger than 100, the energy requirement in simple shear is several orders of magnitude higher than extensional flow or pure shear flow.

In continuous mixers, different fluid elements will invariably experience different amounts of strain, as discussed earlier for the screw extruder. Tadmor and Lidor (206) proposed the use of strain distribution functions (SDF), similar to residence time distribution functions (RTD). The SDF for a continuous mixer $f(\gamma)d\gamma$ is defined as the fraction of exiting flow rate that experienced a strain between γ and $d\gamma$. It is also the probability of an entering fluid element to exit with that strain. The cumulative SDF, $F(\gamma)$, is defined by the following expression:

$$F(\gamma) = \int_{\gamma_0}^{\gamma} f(\gamma)d\gamma \qquad (7-272)$$

where γ_0 is the minimum strain.

$F(\gamma)$ represents the fraction of exiting flow rate with strain less or equal to γ. The mean strain of the exiting stream is:

$$\overline{\gamma} = \int_{\gamma_0}^{\infty} \gamma f(\gamma)d\gamma \qquad (7-273)$$

Tadmor and Pinto (207) used the weighted average total strain (WATS) to describe mixing performance in the non-homogeneous flow field of a single screw extruder. This is defined as:

$$WATS = \int_{0}^{\infty} \gamma(t)f(t)dt \qquad (7-274)$$

where $f(t)$ is the RTD function and $\gamma(t)$ the strain undergone by a fluid element at time t. The WATS does not produce an experimentally measurable quantity describing mixing, as discussed by Ottino and Chella (208) in an extensive review of laminar mixing of polymeric liquids. Another limitation is that the initial orientation of the minor component is not considered and changes in orientation are not taken into account. Ottino (209, 210) developed a description of laminar mixing using the mathematical structure of continuum mechanics. This description allows evaluation of the role of initial orientation and the definition of mixing efficiency. This approach was applied to mixing in single screw extruders (211). The mixing achieved was expressed in terms of mixing cup average area stretch "η". This factor was found to depend on down channel distance, channel width to height ratio, helix angle, throttle ratio, and the initial orientation. The effect of helix angle and throttle ratio (ratio of pressure flow to drag flow) is shown in Figure 7–47.

The mixing in single screw extruders was also studied by Tadmor and Klein (103) who used the mean strain to evaluate the mixing performance of the extruder; their result is shown in Figure 7–48.

By comparing Figure 7–47 to Figure 7–48, it is clear that Tadmor's results correspond reasonably well with Ottino's results. The mixing performance improves as the throttle ratio increases. This is to be expected since the output per revolution will decrease, thus the mean residence time will increase with the throttle ratio, while the local shear rates will remain approximately the same. The mixing performance reduces as the helix angle increases from $10°$ to $30°$. As the helix angle increases, the output per revolution increases as well, resulting in a reduced residence time. The residence time seems to play an overriding role because the cross channel mixing improves with increasing helix angle, but the overall mixing performance reduces. Thus, the reduction in mean residence time overrides the effect of improved cross channel mixing as the helix angle increases from $10°$ to $30°$. It will be shown later, see Chapter 8, that $30°$ happens to be the optimum helix angle for Newtonian fluids with respect to output. This means that at a helix angle of $30°$, the shortest residence time is achieved, provided the depth of the channel is optimum as well.

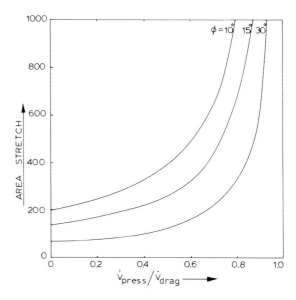

Figure 7-47. Mixing Performance as a function of Helix Angle and Throttle Ratio

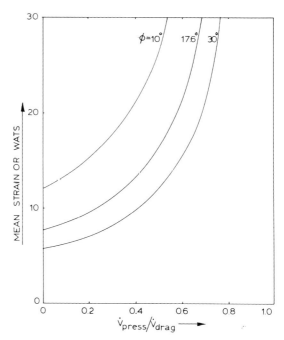

Figure 7-48. Mixing Performance as a function of Helix Angle and Throttle Ratio

Analyses of laminar mixing in screw extruders generally deal with highly simplified problems. The leakage flow through the flight is usually neglected, as well as the normal velocity components close to the flight flanks. Just these two simplifications constitute a severe limitation on the validity and applicability of results of any mixing analysis. The normal velocity components achieve a reorientation of the minor component. When the normal velocity components are neglected, the reorientation of the interfacial area cannot be properly accounted for. This is a particularly severe problem in the analysis of multiflighted mixing sections, such as the Dulmage mixing section, see Section 8.7.2.

The laminar mixing action can only be analyzed if the exact flow patterns are known. This can be done reasonably well for simple rectangular flow channels in an extruder screw. However, when mixing sections are incorporated into the extruder screw, the flow patterns generally become quite complex. In one respect, this is desirable because complex velocity profiles tend to improve the mixing effectiveness. However, on the other hand, the mathematical description of the velocity profiles can become very involved. This is particularly true when the minor fluid component has flow properties that are different from the major fluid component; i.e., when the fluid mixture is rheologically non-homogeneous. As a result, quantitative analysis of mixing sections in extruders is quite difficult. Therefore, development of mixing devices has been mostly empirical. Even the latest developments in mixing theory provide little guidance for the actual design of mixing sections in screw extruders.

7.7.2   STATIC MIXING DEVICES

Static mixers are incorporated in a flow channel through which the flow is a pure pressure driven flow. The liquid is mixed by solid plates of various shapes, designed to improve the laminar mixing of the liquid mixture. The static mixer does not contain any moving solid parts; it is sometimes referred to as motionless mixer. The static mixer is used for laminar distributive mixing and is often incorporated between the extruder barrel and the die. Qualitative discussions on the use of static mixers in the polymer processing industry have been presented by Skoblar (212) and Smoluk (213).

More than 30 different static mixing devices are commercially available today. Three geometries that are reasonably well known in the U.S. are shown in Figure 7-49.

Figure 7-49a shows the Kenics mixer geometry. The fluid is split in two and reoriented 180° over the length of one element and then split again. Figure 7-49b shows the Ross ISG mixer. The liquid is divided into four circular channels that are reoriented in one element. Figure 7-49c shows the Koch element (license from Sulzer, Switzerland). The liquid is divided into many separate streams between corrugated plates, with the channel of neighboring plates oriented under an angle of 90°. This geometry allows substantial mixing within one element. The next element is oriented under 90° with the previous element. Thus, reorientation occurs at various levels. The mixing performance of static mixers is often related to the number of striations N_s formed with a certain number of mixing elements n_e. For the Kenics mixer, this relationship is:

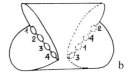

a

b

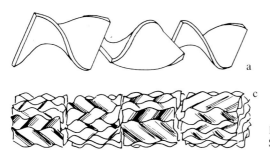

c

Figure 7-49. Three Types of Static Mixing Sections

$$N_S = 2^{n_e} \qquad\qquad (7\text{--}275)$$

For the Ross ISG mixer, N_S is:

$$N_S = 4^{n_e} \qquad\qquad (7\text{--}276)$$

For the Koch mixer, the following relationship can be used:

$$N_S = \frac{p}{2}(2p)^{n_e-1} \qquad\qquad (7\text{--}277)$$

where p is the number of plates per element.

The mixing thus occurs primarily by stream splitting and reorientation in an ordered fashion. The reorientation can be easily controlled and optimized. Schott et al. (214) have reviewed the Ross and the Kenics mixer. The Kenics mixer was used by Han and Kim (215) in studies on two-phase polymer blends. The improvement of heat transfer and thermal homogeneity with static mixers was studied by Nauman (216).

The pressure drop in a static mixing unit with laminar flow is:

$$\Delta P = N_{sm}\,\frac{4\eta\dot{V}L}{\pi D^4} \qquad\qquad (7\text{--}278)$$

where N_{sm} is a constant depending only on the mixer geometry.

Thus, for a particular static mixer geometry the pressure drop can be predicted if the value of N_{sm} is known. The Swiss Federal Institute performed mixing experiments on a large number of static mixers (217). The homogeneity at the outlet of the mixer was evaluated

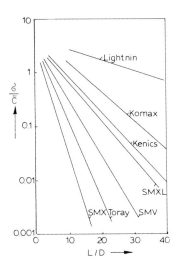

Figure 7–50. Ratio σ/C versus L/D for Various Static Mixers

TABLE 7-3.   COMPARISON OF VARIOUS COMMERCIAL STATIC
MIXERS COURTESY OF REFERENCE (218)

Mixer	*L/D	N_{sm}	Vol_{rel}	$Holdup_{rel}$	D_{rel}	L_{rel}	ΔP_{rel}
Koch SMX	9	1237	1.0	1.0	1.0	1.0	1.0
Koch SMXL	26	245	1.8	1.8	0.84	2.4	0.6
Koch SMV	18	1430	4.6	4.5	1.3	2.7	2.3
Kenics	29	220	1.9	1.8	0.84	2.7	0.6
Etoflo HV	32	190	2.0	2.0	0.84	2.7	0.6
Komax	38	620	8.9	8.2	1.3	5.4	2.1
Lightnin	100	290	29.0	27.0	1.4	15.3	2.6
PMR	320	500	511.0	460.0	2.4	86.0	14.5
Toray	13	1150	1.9	0.9	1.1	1.6	1.35
N-Form	29	544	4.5	3.8	1.1	3.6	1.4
Ross ISG	10	9600	9.6	3.4	2.1	2.3	8.6

*The L/D ratio refers to the ratio required to achieve $\sigma/\overline{C} \leqslant 0.05$.

with a conductive tracer method. The local tracer concentrations were determined by moving a conductivity cell across the outlet of the mixer. The mean tracer concentration $\overline{C}$ and the standard deviation σ were evaluated according to standard mathematical techniques. The pressure drop was measured with a "Kistler" piezoresistive pressure indicator. Figure 7-50 shows the ratio $\sigma/\overline{C}$ versus the L/D ratio for various commercial static mixers.

Other comparative data of commercial static mixers is shown in Table 7-3.

Ottino (219) analyzed the mechanical mixing efficiency for static mixers. He derived expressions for the increase in interfacial area and the efficiency index as a function of the power law index of the liquid. By taking the Kenics and Ross mixers as examples, Ottino concludes that the Ross mixer is more effective for highly pseudoplastic fluids, whereas the Kenics mixer is more effective for relatively Newtonian fluids.

7.7.3   DISPERSIVE MIXING

The discussion so far has primarily been concerned with distributive laminar mixing. However, in actual extrusion operations the requirement for good dispersive mixing is often more critical than the distributive mixing. This is particularly true in extrusion of compounds with pigments or in small gage extrusion (e.g., low denier fiber spinning, thin film extrusion, etc.). In dispersive mixing, the actual stresses acting on the agglomerates determine whether or not the agglomerate will break down. The breakdown stress will depend on the size, shape and nature of the agglomerate. The stresses acting on the agglomerate will depend on the flow field and the rheological properties of the fluid.

One of the first engineering analyses of dispersive mixing was made by Bolen and Colwell (220). They assumed that the agglomerates break when the internal stresses, induced by viscous drag on the particles, exceed a certain threshold value. Bird et al. (221) analyzed the forces acting on a single agglomerate in the form of a rigid dumbbell, consisting of two spheres of radii r_1 and r_2. The centers of the spheres are separated by a distance L and the dumbbell is located in a homogeneous flow field of an incompressible Newtonian fluid. As a result of the viscous drag acting on the spheres, a force will develop in the connector. The force depends on the drag on each sphere and on the orientation of the dumbbell. Tadmor (222) adopted Bird's approach and extended the analysis to spheres of different radius and to include the effect of Brownian motion.

In a steady simple shear flow, the maximum connector force develops when the dumbbell is oriented at a 45° angle to the direction of shear; this force is:

$$F_{max} = 3\pi\mu\dot{\gamma}L\frac{r_1 r_2}{r_1 + r_2} \qquad (7\text{-}279)$$

When the spheres are in contact with each other, equation 7–29 reduces to:

$$F_{max} = 3\pi\mu\dot{\gamma}r_1 r_2 = 3\pi\tau r_1 r_2 \qquad (7\text{-}280)$$

From equations 7–279 and 7–280, it can be seen that the force in the connector is directly proportional to the shear stress and to the product of the radii. Thus, in the same flow field, the force in the connector (breakdown force) reduces as the size of the spheres reduce. If the breakdown force of the agglomerate remains about constant, the breakdown will proceed until a certain minimum agglomerate size is reached. Further breakdown will not occur because the flow field will be unable to generate sufficient breakdown force in the agglomerate. For a steady elongational flow, the maximum force in the connector is obtained when the dumbbell is aligned in the direction of flow. When the spheres are in contact with each other, the maximum force is:

$$F_{max} = 6\pi\mu\dot{\varepsilon}r_1 r_2 \qquad (7\text{-}281)$$

where $\dot{\varepsilon}$ is the rate of elongation.

Equations 7–280 and 7–281 assume that the spheres do not affect the flow field and that dumbbell interaction can be neglected.

By comparing equations 7–280 and 7–281, it can be seen that the breakdown force in elongational flow is twice as large as in simple shear at the same rate of deformation. In most commercial mixers used for dispersive mixing, the actual flow patterns in the mixer will be a combination of shear flow and elongational flow. Unfortunately, the flow patterns are often too complex to allow accurate mathematical description. As a result, the mixer geometry is generally not determined from an engineering analysis but from experience, empirical knowledge, and intuition.

From equations 7–280 and 7–281, it can be seen that the breakdown force is directly proportional to the viscosity of the matrix. This has some important practical implications. Dispersive mixing should be done at as low a temperature as possible to increase the viscosity and thus the breakdown force. If both a dispersive and a distributive mixing element are required in a single screw extruder, the dispersive element should be placed upstream of the distributive element. This placement is more likely to result in a relative low stock temperature at the inlet of the dispersive mixing element, while the stock temperature at the inlet of the distributive mixing element will be relatively high as a result of the viscous heat generation in the dispersive mixing element. The low stock temperature in the dispersive mixing element will enhance dispersive mixing, while the high stock temperature in the distributive mixing element will improve the energy efficiency of the distributive mixing step. The need for high viscosity in dispersive mixing explains why it is often easier to produce a master-batch of a high filler loading and let it down later, than to produce a compound with relatively low filler loading. The viscosity of the masterbatch will be much higher and, therefore, the dispersive mixing action will be much more effective. Master batching is a very common technique in the mixing and compounding industry. It should be remembered that in many instances, agglomerate breakdown can be achieved more effectively in high speed solid-solid mixing than in liquid-solid mixing (225-227). The forces that can be transmitted in solid-solid mixing are generally much higher than in liquid-solid mixing and agglomerate breakdown will occur faster and to a larger extent, resulting in very finely dispersed particles.

Dispersive mixing in single screw extruders was studied in detail by Martin (223, 224). An interesting finding of this study was that dispersive mixing is determined not only by the shear stress acting on an agglomerate but also by the exposure time. It was found that in dispersive mixing of carbon blacks, a certain minimum shear stress exposure time was necessary to accomplish breakdown; this minimum exposure time was about 0.2 second. Below this minimum exposure time, no breakdown occurred no matter how high the shear stress. This finding runs counter to common belief that dispersive mixing is only determined by the actual level of stress acting on the agglomerate. However, the minimum stress exposure time is a very important consideration in the design of dispersive mixing equipment, such as a dispersive mixing section on an extruder screw. It was also found that the mixing performance of an extruder is very much dependent on the length of the plasticating zone, which depends on the operating conditions. Incorporation of mixing sections can substantially improve extruder mixing performance and enable deepening of the metering section as compared to screws without mixing sections.

APPENDIX 7 – 1

CONSTANTS OF EQUATION 7 – 30

$$a_{11} = 1 + \frac{\sin\beta_e \sin 2\psi_w}{2\psi_w - \pi} \tag{1}$$

$$a_{12} = \frac{a^2 \sin\beta_e}{2\psi_w - \pi} \left\{ \frac{2\cos 2\psi_w}{2\psi_w - \pi} + \sin 2\psi_w \left[1 - \frac{2}{(2\psi_w - \pi)^2} \right] \right\} + \frac{a^2}{3} \tag{2}$$

$$a_{13} = \sin\beta_e \sin 2\psi_w \left[\frac{2}{2\psi_w - \pi} + \frac{1}{a} \right] \tag{3}$$

$$\alpha_{14} = \alpha\sin\beta_e\sin2\psi_W + \frac{2\alpha^2\sin\beta_e}{2\psi_W - \pi}\left\{\frac{2\cos2\psi_W}{2\psi_W - \pi} + 2\sin2\psi_W\left[1 - \frac{2}{(2\psi_W - \pi)^2}\right]\right\} \quad (4)$$

$$\alpha_{21} = \frac{-\sin\beta_e}{2\psi_W - \pi}(1 + \cos2\psi_W) \quad (5)$$

$$\alpha_{22} = \frac{\alpha^2\sin\beta_e}{2\psi_W - \pi}\left[\frac{2\sin2\psi_W}{2\psi_W - \pi} + \frac{2(1+\cos2\psi_W)}{(2\psi_W - \pi)^2} - \cos2\psi_W\right] \quad (6)$$

$$\alpha_{23} = -\sin\beta_e(1 + \cos2\psi_W)\left[\frac{2}{2\psi_W - \pi} + \frac{1}{\alpha}\right] \quad (7)$$

$$\alpha_{24} = \alpha(1 - \sin\beta_e\cos2\psi_W) + \frac{2\alpha^2\sin\beta_e}{2\psi_W - \pi}\left[\frac{2\sin2\psi_W}{2\psi_W - \pi} + \frac{2(1+\cos2\psi_W)}{(2\psi_W - \pi)^2} - \cos2\psi_W\right] \quad (8)$$

$$\beta_{11} = \frac{-\rho g\sin\alpha}{\alpha} \quad (9)$$

$$\beta_{12} = \rho c_m\lambda \quad (10)$$

$$\beta_{21} = \frac{-\rho g(\cos\alpha - 1)}{\alpha} \quad (11)$$

$$A = \alpha_{11}\alpha_{22} - \alpha_{21}\alpha_{12} \quad (12)$$

$$B = \alpha_{11}\alpha_{24} + \alpha_{13}\alpha_{22} - \alpha_{21}\alpha_{14} - \alpha_{23}\alpha_{12} \quad (13)$$

$$C = \alpha_{13}\alpha_{24} - \alpha_{23}\alpha_{14} \quad (14)$$

$$E = \beta_{11}(\alpha_{22} + \alpha_{24}) - \beta_{21}(\alpha_{12} + \alpha_{14}) \quad (15)$$

$$F = \rho(\alpha_{24} - 2\alpha_{22}) \quad (16)$$

$$G = \beta_{21}(\alpha_{11} + \alpha_{13}) - \beta_{11}(\alpha_{21} + \alpha_{23}) \quad (17)$$

$$H = -\rho(\alpha_{23} - 2\alpha_{21}) \quad (18)$$

where β_e is the effective angle of internal friction and ψ is the angle between the major principle stress and the radial r-axis. Further:

$$\psi_w = \frac{\pi}{2} + \frac{1}{2}\left(\arcsin\frac{\sin\beta_w}{\sin\beta_e} + \beta_w\right) \tag{19}$$

$$c_m = \frac{1}{\alpha\,\bar{V}}\int_0^\alpha v^2 d\alpha \quad (c_m = 1 \text{ for uniform velocity profile}) \tag{20}$$

$$\lambda = r\bar{v} = \frac{r}{\alpha}\int_0^\alpha v\,d\alpha \tag{21}$$

where v is the radial velocity component.

APPENDIX 7 – 2

CONSTANTS OF THE ENGSTAD EQUATION 7 – 36

σ_{co} is the constant part of the failure function; the unconfined yield strength is written as:

$$\sigma_c = k\sigma_1 + \sigma_{co} \tag{1}$$

The term Y_b is determined from:

$$Y_b = \frac{(\beta_w+\alpha)\sin\alpha}{\sin^2(\beta_w+\alpha)} + \frac{\sin\beta_w}{\sin(\beta_w+\alpha)} - \frac{kY}{2\,(X-1)}\frac{\sin\alpha + \sin(2\beta_w+\alpha)}{\sin\alpha}(1+\sin\beta_e) \tag{2}$$

For a passive stress state:

$$X = \frac{\sin\beta_e}{1-\sin\beta_e}\left[\frac{\sin\alpha + \sin(2\beta_p+\alpha)}{\sin\alpha}\right] \tag{3}$$

$$Y = \frac{(\beta_p+\alpha)\sin\alpha + \sin\beta_p\sin(\beta_p+\alpha)}{(1-\sin\beta_e)\sin^2(\beta_p+\alpha)} \tag{4}$$

$$\beta_p = \frac{1}{2}\left(\beta_w + \arcsin\frac{\sin\beta_w}{\sin\beta_e}\right) \quad \text{with } 0 < \arcsin \text{ value} < \frac{\pi}{2} \tag{5}$$

For an active stress state:

$$X = \frac{\sin\beta_e}{1+\sin\beta_e} \left[\frac{\sin(2\beta_a+\alpha) - \sin\alpha}{\sin\alpha} \right] \tag{6}$$

$$Y = \frac{\left(\alpha + \beta_a - \frac{\pi}{2}\right)\sin\alpha + \cos\beta_a\cos(\alpha+\beta_a)}{(1+\sin\beta_e)\cos^2(\alpha+\beta_a)} \tag{7}$$

$$\beta_a = \frac{1}{2}\left(\beta_w + \arcsin\frac{\sin\beta_w}{\sin\beta_e}\right) \quad \text{with } \frac{\pi}{2} < \text{arcsin value} < \pi \tag{8}$$

APPENDIX 7-3

The solids conveying angle as determined by using the Darnell and Mol solution procedure:

$$\theta = \arcsin\left[\frac{(1 + K^2 - M^2)^{1/2} - KM}{1 + K^2} \right] \tag{1}$$

where:

$$K = \frac{D_m(\sin\varphi_m + f_s\cos\varphi_m)}{D_b(\cos\varphi_m - f_s\sin\varphi_m)} \tag{2}$$

and:

$$M = M_1 + M_2 + M_3 \tag{3}$$

$$M_1 = \frac{2Hf_s}{W_b f_b}\sin\varphi_b\left(K+\frac{D_m}{D_b}\cot\varphi_m\right) \tag{4}$$

$$M_2 = \frac{W_s f_s}{W_b f_b}\sin\varphi_b\left(K+\frac{D_s}{D_b}\cot\varphi_s\right) \tag{5}$$

$$M_3 = \frac{W_m H}{W_b z_b f_b}\sin\varphi_m\left(K+\frac{D_m}{D_b}\cot\varphi_m\right)\ln\frac{P_1}{P_0} \tag{6}$$

APPENDIX 7–4

FORTRAN PROGRAM TO ANALYZE TWO-DIMENSIONAL FLOW OF A POWER LAW FLUID

```
          PROGRAM EXP
          REAL H
          DIMENSION C(200),Z(201),TX(201),TY(201),T(201),S(201),SX(201)
          DIMENSION SY(201),VX(201),VY(201),DX(201),DY(201),A(4,4),B(3)
          DIMENSION GRA(3),SOX(3),SOY(3),SOQ(3)
          READ (6,20) M,NPRVEL
20        FORMAT (2I10)
          READ (6,30) REV,RK,D
          READ (6,30) HMIN,DELH,HMAX
          READ (6,30) GMIN,DELG,GMAX
          READ (6,30) RNMIN,DELRN,RNMAX
          READ (6,30) PHIMIN,DELPHI,PHIMAX
          READ (6,30) PGMIN,DELPG,PGMAX
          WRITE (7,29) M,NPRVEL
29        FORMAT (2I10)
30        FORMAT (3F10.4)
          WRITE (7,31) REV
31        FORMAT (' SCREW SPEED IN RPM IS',F10.2)
          WRITE (7,32) HMIN
32        FORMAT (' CHANNEL DEPTH IN CM IS',F10.4)
          WRITE (7,320) DELH
320       FORMAT (' INCREMENTAL DEPTH IN CM IS',F10.4)
          WRITE (7,321) HMAX
321       FORMAT (' MAXIMUM DEPTH IN CM IS',F10.4)
          WRITE (7,33) D
33        FORMAT (' SCREW DIAMETER IN CM IS',F10.4)
          WRITE (7,34) RK
34        FORMAT (' POWER LAW CONST. IN POISE IS',F14.4)
          WRITE (7,36) GMIN
36        FORMAT (' INITIAL RED. PRESS. GRADIENT IS',F10.6)
          WRITE (7,37) DELG
37        FORMAT (' INCREMENTAL PRESS. GRADIENT IS',F10.6)
          WRITE (7,38) GMAX
38        FORMAT (' FINAL PRESSURE GRADIENT IS',F10.6)
          WRITE (7,39) RNMIN
39        FORMAT (' INITIAL POWER LAW INDEX IS',F10.6)
          WRITE (7,40) DELRN
40        FORMAT (' INCREMENTAL POWER LAW INDEX IS',F10.6)
          WRITE (7,41) RNMAX
41        FORMAT (' FINAL POWER LAW INDEX IS',F10.6)
          WRITE (7,42) PHIMIN
42        FORMAT (' INITIAL HELIX ANGLE IN DEGR. IS',F10.4)
          WRITE (7,43) DELPHI
43        FORMAT (' INCREMENTAL HELIX ANGLE IS',F10.4)
          WRITE (7,44) PHIMAX
44        FORMAT (' FINAL HELIX ANGLE IN DEGREES IS',F10.4)
          WRITE (7,440) PGMIN
440       FORMAT (' INITIAL AXIAL PRESS. GRADIENT IN PSI/INCH',F10.4)
          WRITE (7,441) DELPG
441       FORMAT (' INCREMENTAL AXIAL PRESS. GRADIENT IS',F10.4)
          WRITE (7,442) PGMAX
442       FORMAT (' MAXIMUM AXIAL PRESS. GRADIENT IN PSI/INCH',F10.4)
          WRITE (7,444)
444       FORMAT (' ','REDTHRUPUT PRESS-GRAD HELIXANGLE AC-THRUPUT DIM',
       X  'THRUPUT DOWNCHA-TX CROSSCH-TY CONSTANTCY')
          H=HMIN
          G=GMIN
```

```
        RN=RNMIN
        PG=PGMIN
        TXO=10000.0
        TYO=10000.0
        CY=10000.0
        REV=REV/60.0
        PHIMIN=PHIMIN*17.4533/1000.0
        DELPHI=DELPHI*17.4533/1000.0
        PHIMAX=PHIMAX*17.4533/1000.0
        PHI=PHIMIN
45      UO=3.14159*D*REV*COS(PHI)
        VO=UO*TAN(PHI)
        IF (PG .EQ. 0) GO TO 48
        CX=27145.67*SIN(PHI)*PG
C       ABOVE LINE USED WITH CONSTANT AXIAL PRESSURE GRADIENT
        GO TO 49
48      CX=6.0*RK*G/H*(UO/H)**RN
49      NR=0
50      DO 70 I=1,((4*M)+1)
        Z(I)=(FLOAT(I-1))*H/(4.*(FLOAT(M)))
        TX(I)=TXO+(CX*Z(I))
        TY(I)=TYO+(CY*Z(I))
        T(I)=((TX(I)**2)+(TY(I)**2))**.5
        S(I)=(T(I)/RK)**(1./RN)
        SX(I)=S(I)*TX(I)/T(I)
        SY(I)=S(I)*TY(I)/T(I)
70      CONTINUE
        VX(1)=0.0
        VY(1)=0.0
        DO 80 I=2,M+1
        K=((FLOAT(I-2))*4)
        DO 90 J=1,5
        L=K+J
        DX(J)=SX(L)
        DY(J)=SY(L)
90      CONTINUE
        CALL SIMP(4,DX,UMX)
        CALL SIMP(4,DY,UMY)
        VX(I)=VX(I-1)+(UMX*H/(4.*(FLOAT(M))))
        VY(I)=VY(I-1)+(UMY*H/(4.*(FLOAT(M))))
80      CONTINUE
        CALL SIMP(M,VX,XOX)
        CALL SIMP(M,VY,YOY)
        QX=XOX*H/(FLOAT(M))
        QY=YOY*H/(FLOAT(M))
250     DIVX=VX(M+1)-UO
        DIVY=VY(M+1)-VO
        DIVQ=QY
        IF (NR-1) 180,185,185
180     STVX=DIVX
        STVY=DIVY
        STVQ=DIVQ
C       WRITE(7,251) TXO,TYO,CY,STVX,STVY,STVQ
251     FORMAT (6E12.3)
183     IF ((AMAX1(STVX,STVY,STVQ))-1E-5) 500,500,187
187     LAL=1
        NR=NR+1
        GRA(1)=0.01*TXO
        TXO=TXO+GRA(1)
        GO TO 50
185     SOX(LAL)=DIVX-STVX
        SOY(LAL)=DIVY-STVY
        SOQ(LAL)=DIVQ-STVQ
        IF (LAL-2) 200,210,220
200     LAL=2
        TXO=TXO-GRA(1)
        GRA(2)=0.01*TYO
```

```
              TYO=TYO+GRA(2)
              GO TO 50
210           LAL=3
              TYO=TYO-GRA(2)
              GRA(3)=0.01*CY
              CY=CY+GRA(3)
              GO TO 50
220           LAL=1
              CY=CY-GRA(3)
              DO 270 MP=1,3
              A(1,MP)=SOX(MP)/GRA(MP)
              A(2,MP)=SOY(MP)/GRA(MP)
              A(3,MP)=SOQ(MP)/GRA(MP)
270           CONTINUE
              B(1)=-STVX
              B(2)=-STVY
              B(3)=-STVQ
              OK=A(2,1)/A(1,1)
              A(2,2)=A(2,2)-(A(1,2)*OK)
              A(2,3)=A(2,3)-(A(1,3)*OK)
              B(2)=B(2)-(B(1)*OK)
              OK=A(3,1)/A(1,1)
              A(3,2)=A(3,2)-(A(1,2)*OK)
              A(3,3)=A(3,3)-(A(1,3)*OK)
              B(3)=B(3)-(B(1)*OK)
              OK=A(3,2)/A(2,2)
              A(3,3)=A(3,3)-(A(2,3)*OK)
              B(3)=B(3)-(B(2)*OK)
              DLCY=B(3)/A(3,3)
              DLTY=(B(2)-(A(2,3)*DLCY))/A(2,2)
              DLTX=B(1)-(A(1,3)*DLCY)
              DLTX=(DLTX-(A(1,2)*DLTY))/A(1,1)
              TXO=TXO+DLTX
              TYO=TYO+DLTY
              CY=CY+DLCY
              NR=0
              GOTO 50
500           QRED = QX*2.0/(H*UO)
              QREDA=QRED*SIN(2.0*PHI)
              QA=QX*3.14159*D*SIN(PHI)
              GA=CX*H/(6.0*RK)*(H/UO)**RN
              PHIDGR=PHI*1000.0/17.4533
              WRITE(7,501) QREDA,GA,PHIDGR,QA,QRED,TXO,TYO,CY
501           FORMAT (8E11.4)
              IF (QX .LT. 0) GO TO 560
              CONTINUE
              IF (NPRVEL .NE. 1) GO TO 556
              WRITE (7,10000)
10000         FORMAT ('1')
554           WRITE(7,555)  (Z((4*I)+1),VX(I),VY(I),I=1,M)
555           FORMAT (3E15.4)
556           IF (G .GE. GMAX) GO TO 560
              G=G+DELG
              GO TO 45
558           IF (PG .GE. PGMAX) GO TO 600
              PG=PG+DELPG
              PHI=PHIMIN
              H=HMIN
              WRITE (7,559) PG
559           FORMAT (' NEW AXIAL PRESSURE GRADIENT IS ',F10.4)
              GO TO 45
560           IF (RN .GE. RNMAX) GO TO 570
              RN=RN+DELRN
              G=GMIN
              WRITE (7,565) RN
565           FORMAT (' NEW POWER LAW EXPONENT IS',F6.3)
              TXO=10000.0
```

```
          TYO=10000.0
          CY=10000.0
          GO TO 45
570       IF (PHI .GE. PHIMAX) GO TO 580
          PHI=PHI+DELPHI
          G=GMIN
          TXO=10000.0
          TYO=10000.0
          CY=10000.0
          GO TO 45
580       IF (H .GE. HMAX) GO TO 558
          H=H+DELH
          PHI=PHIMIN
          WRITE (7,585) H
585       FORMAT (' NEW CHANNEL DEPTH IS',F10.4)
          TXO=10000.0
          TYO=10000.0
          GO TO 45
600       STOP
          END

          SUBROUTINE SIMP(I,C,S)
          DIMENSION C(200)
          S=0.
          DO 55 L=1,(I-1),2
          S=S+(2.*C(L))+(4.*C(L+1))
55        CONTINUE        .
          S=((S+(C(I+1)))-C(1))/3.
          RETURN
          END
          40            0
SCREW SPEED IN RPM IS     100.00
CHANNEL DEPTH IN CM IS     0.2500
INCREMENTAL DEPTH IN CM IS      0.0200
MAXIMUM DEPTH IN CM IS     0.2500
SCREW DIAMETER IN CM IS     6.3500
POWER LAW CONST. IN POISE IS     10000.0000
INITIAL RED. PRESS. GRADIENT IS   0.300000
INCREMENTAL PRESS. GRADIENT IS    0.050000
FINAL PRESSURE GRADIENT IS    0.300000
INITIAL POWER LAW INDEX IS    1.000000
INCREMENTAL POWER LAW INDEX IS    0.200000
FINAL POWER LAW INDEX IS    1.000000
INITIAL HELIX ANGLE IN DEGR. IS    10.0000
INCREMENTAL HELIX ANGLE IS     2.5000
FINAL HELIX ANGLE IN DEGREES IS    50.0000
INITIAL AXIAL PRESS. GRADIENT IN PSI/INCH 3000.0000
INCREMENTAL AXIAL PRESS. GRADIENT IS 2000.0000
MAXIMUM AXIAL PRESS. GRADIENT IN PSI/INCH15000.0000
```

REDTHRUPUT	PRESS-GRAD	HELIXANGLE	AC-THRUPUT	DIMTHRUPUT	DOWNCHA-TX	CROSSCH-TY	CONSTANTCY
0.1882E+00	0.4499E+00	0.1000E+02	0.7800E+01	0.5501E+00	-0.4579E+06	-0.4619E+06	0.5543E+07
0.1836E+00	0.5656E+00	0.1250E+02	0.7610E+01	0.4344E+00	-0.9049E+06	-0.5757E+06	0.6908E+07
0.1582E+00	0.6836E+00	0.1500E+02	0.6557E+01	0.3164E+00	-0.1350E+07	-0.6884E+06	0.8261E+07
0.1122E+00	0.8045E+00	0.1750E+02	0.4650E+01	0.1955E+00	-0.1793E+07	-0.7998E+06	0.9598E+07
0.4587E-01	0.9286E+00	0.2000E+02	0.1902E+01	0.7137E-01	-0.2232E+07	-0.9097E+06	0.1092E+08
0.4018E-01	0.1057E+01	0.2250E+02	-0.1666E+01	-0.5683E-01	-0.2667E+07	-0.1018E+07	0.1221E+08
0.1453E+00	0.1190E+01	0.2500E+02	-0.6025E+01	-0.1897E+00	-0.3097E+07	-0.1124E+07	0.1349E+08
0.2688E+00	0.1328E+01	0.2750E+02	-0.1114E+02	-0.3282E+00	-0.3521E+07	-0.1228E+07	0.1474E+08
0.4097E+00	0.1473E+01	0.3000E+02	-0.1698E+02	-0.4731E+00	-0.3938E+07	-0.1330E+07	0.1596E+08
0.5668E+00	0.1625E+01	0.3250E+02	-0.2350E+02	-0.6254E+00	-0.4348E+07	-0.1429E+07	0.1715E+08
0.7391E+00	0.1787E+01	0.3500E+02	-0.3064E+02	-0.7865E+00	-0.4749E+07	-0.1526E+07	0.1831E+08
0.9251E+00	0.1958E+01	0.3750E+02	-0.3835E+02	-0.9578E+00	-0.5142E+07	-0.1619E+07	0.1943E+08
0.1124E+01	0.2141E+01	0.4000E+02	-0.4658E+02	-0.1141E+01	-0.5525E+07	-0.1710E+07	0.2052E+08
0.1333E+01	0.2338E+01	0.4250E+02	-0.5525E+02	-0.1338E+01	-0.5897E+07	-0.1797E+07	0.2156E+08
0.1551E+01	0.2551E+01	0.4500E+02	-0.6431E+02	-0.1551E+01	-0.6258E+07	-0.1881E+07	0.2257E+08
0.1778E+01	0.2784E+01	0.4750E+02	-0.7369E+02	-0.1784E+01	-0.6607E+07	-0.1961E+07	0.2353E+08
0.2010E+01	0.3041E+01	0.5000E+02	-0.8331E+02	-0.2041E+01	-0.6943E+07	-0.2038E+07	0.2445E+08

```
NEW AXIAL PRESSURE GRADIENT IS  5000.0000
0.8557E-01 0.7498E+00 0.1000E+02 0.3547E+01 0.2502E+00-0.1636E+07-0.4619E+06 0.5543E+07
0.2421E-01 0.9427E+00 0.1250E+02 0.1004E+01 0.5728E-01-0.2374E+07-0.5757E+06 0.6908E+07
0.6971E-01 0.1139E+01 0.1500E+02-0.2890E+01-0.1394E+00-0.3107E+07-0.6884E+06 0.8261E+07
0.1955E+00 0.1341E+01 0.1750E+02-0.8102E+01-0.3408E+00-0.3833E+07-0.7998E+06 0.9598E+07
0.3521E+00 0.1548E+01 0.2000E+02-0.1460E+02-0.5477E+00-0.4553E+07-0.9097E+06 0.1092E+08
0.5384E+00 0.1761E+01 0.2250E+02-0.2232E+02-0.7614E+00-0.5264E+07-0.1018E+07 0.1221E+08
0.7529E+00 0.1983E+01 0.2500E+02-0.3121E+02-0.9829E+00-0.5965E+07-0.1124E+07 0.1349E+08
0.9941E+00 0.2214E+01 0.2750E+02-0.4121E+02-0.1214E+01-0.6654E+07-0.1228E+07 0.1474E+08
0.1260E+01 0.2455E+01 0.3000E+02-0.5224E+02-0.1455E+01-0.7331E+07-0.1330E+07 0.1596E+08
0.1549E+01 0.2709E+01 0.3250E+02-0.6421E+02-0.1709E+01-0.7994E+07-0.1429E+07 0.1715E+08
```

REFERENCES - CHAPTER 7

1. "Mechanics Applied to the Transport of Bulk Materials, " S. C. Cowin, editor, ASME, New York (1979).
2. "Continuum Mechanical and Statistical Approaches in the Mechanics of Granular Materials", S.C. Cowin and M.Satake, editors. Proceedings U.S.-Japan Seminar, Sendai, Japan, Gakujutsu Bunken Fukyu-Kai (1978).
3. C. Orr, Jr., "Particulate Technology", Macmillan, New York (1966).
4. R. L.Brown and J.C. Richards, "Principles of Powder Mechanics", Pergamon Press, Oxford (1966).
5. J. E.Richards, editor, "The Storage and Recovery of Particulate Solids", Inst. Chem. Eng., Working Party Rep. (1966).
6. K. Wieghardt, Ann. Rev. Fluid Mech., 7, 89-114 (1975).
7. S. B.Savage, J.Fluid Mech., 92, 53-96 (1979).
8. A. W.Jenike, "Gravity Flow of Bulk Solids", Bulletin No.108 of the Utah Engineering Experimental Station, Univ. of Utah, Salt Lake City (1961).
9. A. W.Jenike, "Storage and Flow of Solids", Bulletin No. 123 of the Utah Engineering Experimental Station, Univ. of Utah, Salt Lake City (1964).
10. M. Shahinpoor and J.S.S. Siah, J. Non-Newtonian Fluid Mech., 9, 147-156 (1981).
11. M. Shahinpoor, J. Non-Newtonian Fluid Mech., 12, 31-38 (1983).
12. K. I. Kanatani, Int. J.Eng. Sci., 17, 419-432 (1979).
13. J. T.Jenkins and S.C. Cowin, in reference 1, p.79-89.
14. R. L.Brown and P.G. Hawksley, Fuel, 26, 171 (1947).
15. J. Lee, S.C. Cowin, and J.S. Templeton, Trans. Soc. Rheol., 18, 247-269 (1974).
16. P. M.Blair-Fish and P.L. Bransby, Trans. A.S.M.E., J.Eng. Ind., 95, 17-26 (1973).
17. G. C.Gardner, Chem. Eng. Sci., 21, 261-273 (1966).
18. M. Levinson, B.Shmutter, and W.L. Resnick, Powder Technol., 16, 29-43 (1977).
19. L. Bates and D.Kershan, Chem. Eng. Progress, 71, 66-68 (1975).
20. L. M.Connelly, in reference 1, pp.35-59.
21. R. Butterfield, R.M. Harkness, and K.Z. Andrews, Geotechnique, 8, 308 (1970).
22. H. A.Janssen, Zeitschrift VDI, 1045 (1895).
23. D. M.Walker, Chem. Eng. Sci., 21, 975-997 (1966).
24. J. K.Walters and P.M. Nedderman, Chem. Eng. Sci., 28, 1907-1908 (1973).
25. P. L.Bransby and P.M. Blair-Fish, Chem. Eng. Sci., 29, 1061-1074 (1974).
26. G. Hagen, Berliner Monatsberichte Akad. d.Wiss., 35-42 (1852).
27. R. L.Brown, Nature, 191, 458-461 (1961).

28. J. R. Johanson, Trans. S. M. E., 232, 69–80 (1965).
29. S. B. Savage, British J. Appl. Phys., 16, 1885–1888 (1965).
30. S. B. Savage, Ph. D. thesis, McGill Univ., Montreal (1967).
31. S. B. Savage and M. Sayed, in reference 1, pp. 1–24.
32. J. M. Rausch, Ph. D. thesis, Princeton Univ. (1948).
33. J. Bosley, C. Schofield, and C. A. Shook, Trans. Inst. Chem. Eng., 47, T147-T153 (1969).
34. W. N. Sullivan, Ph. D. thesis, California Institute of Technology, Pasadena (1972).
35. D. M. Walker, Powder Technol., 1, 228 (1967).
36. R. K. Eckhoff and P. G. Leversen, Powder Technol., 10, 51 (1974).
37. G. Engstad, Chem. Eng. Sci., 30, 1273–1283 (1975).
38. Y. Lee, Combustion, 32, 20–27 (1960).
39. O. Richmond, Mechanical Engineering, 85, 46–49 (1963).
40. G. C. Gardner, Chem. Eng. Sci., 18, 35–39 (1963).
41. O. Richmond and H. L. Morrison, in reference 1, pp. 103–111.
42. J. F. Ingen Housz in "Fortschritte beim Extrudieren", Carl Hanser Verlag, Munich (1976).
43. W. H. Darnell and E. A. J. Mol, SPE J., 12, 20 (1956).
44. J. G. A. Lovegrove, Ph. D. thesis, London Univ. (1972).
45. E. Broyer and Z. Tadmor, Polym. Eng. Sci., 12, 12–24 (1972).
46. Z. Tadmor and E. Broyer, Polym. Eng. Sci., 12, 378–386 (1972).
47. H. B. Kessler, R. M. Bonner, P. H. Squires, and C. F. W. Wolf, SPE J., 16, 267 (1960).
48. H. Decker, "Die Spritzmaschine", P. Troester, Hannover, W. Germany (1941).
49. A. Schneiders, Plastverarbeiter, 19, 797–799 (1968).
50. G. Menges, W. Predoehl, R. Hegele, R. Kosel, and W. Elbe, Plastverarbeiter, 20, 79–88 and 188–190 (1969).
51. G. Fuchs, Plastverarbeiter, 19, 765–771 (1968) and 20, 237–244 (1969).
52. W. L. Krueger, SPE ANTEC, Boston, 676–678 (1981).
53. R. Rautenbach and E. Goldacker, Kunststoffe, 61, 104–107 (1971).
54. G. Menges and R. Hegele, Plastverarbeiter, 23, 332–338 (1975).
55. B. Franzkoch and G. Menges, SPE ANTEC, Washington, D. C., 512–515 (1978).
56. E. Langecker, G. R. Langecker, and W. Fillman, Plastverarbeiter, 28, 531–535 (1977).
57. E. Langecker, German patent application DDS 22 05426, filed February 2, 1972.
58. C. Maillefer, U. S. 4,154,535 (also CH 612375, FR 2,385,517, DE 2,813,585 and SE 7,80,3477).
59. E. Gruenschlosz, SPE ANTEC, New Orleans, 160–165 (1979).
60. B. H. Maddock, SPE ANTEC, New York, 383 (1959).
61. Z. Tadmor, I. Duvdevani, and I. Klein, Polym. Eng. Sci., 7, 198 (1967).
62. Z. Tadmor, Polym. Eng. Sci., 6, 185 (1966).
63. J. Shapiro, Ph. D. thesis, Cambridge Univ., England (1971).
64. J. R. Vermeulen, P. M. Gerson, and W. J. Beek, Chem. Eng. Sci., 26, 1445–1455 (1971).
65. J. R. Vermeulen, P. G. Scargo, and W. J. Beek, Chem. Eng. Sci., 26, 1457–1465 (1971).
66. G. Menges and P. Klenk, Kunststoffe, 57, 590 (1967).
67. P. Klenk, Rheologica Acta., 7, 75–78 (1968).
68. P. Klenk, Plastverarbeiter, 21, 537 (1968).
69. J. Dekker, Kunststoffe, 66, 130 (1976).
70. T. E. Fahey, M. Sc. Thesis, Case Western Reserve Univ., September (1981).
71. F. R. Kulas and N. P. Thorshaug, J. Appl. Polym. Sci., 23, 1781–1794 (1979).
72. J. T. Lindt, Polym. Eng. Sci., 16, 284–291 (1976).
73. J. T. Lindt, Polym. Eng. Sci., 21, 1162–1166 (1981).
74. K. A. Gieskes, Internal Report, Twente Univ. of Technology, Mech. Eng. Dept., the Netherlands (1979).
75. H. E. H. Meijer, Ph. D. thesis, Twente Univ. of Technology, the Netherlands (1980).

76. I. R. Edmondson, Ph. D. thesis, Univ. of London (1972).
77. I. R. Edmondson and R. T. Fenner, Polymer, 16, 49 (1975).
78. R. T. Fenner, Polymer, 18, 617 (1977).
79. A. P. D. Cox and R. T. Fenner, Polym. Eng. Sci., 20, 562–571 (1980).
80. A. P. D. Cox, J. G. Williams, and D. P. Isherwood, Polym. Eng. Sci., 21, 86–92 (1981).
81. G. M. Gale, Plastics and Polymer, 6, 183 (1970).
82. G. Mennig, Kunststoffe, 71, 359–362 (1981).
83. R. C. Donovan, Polym. Eng. Sci., 11, 247–257 (1971).
84. R. C. Donovan, Polym. Eng. Sci., 11, 485–491 (1971).
85. J. Shapiro, A. L. Halmos, and J. R. A. Pearson, Polymer, 17, 905–918 (1976).
86. A. L. Halmos, J. R. A. Pearson, and R. Trottnow, Polymer, 19, 1199–1216 (1978).
87. J. R. A. Pearson, Int. J. Heat Mass Transfer, 19, 405–511 (1976).
88. D. R. Hinrichs and L. U. Lilleleht, Polym. Eng. Sci., 10, 268–278 (1970).
89. D. W. Sundstrom and C. C. Young, Polym. Eng. Sci., 12, 59–63 (1972).
89a. D. W. Sundstrom and J. R. Lo, Polym. Eng. Sci., 18, 422–426 (1978).
90. C. I. Chung, Modern Plastics 45, 178 (1968) and 45, 110 (1968).
91. C. I. Chung, SPE J., 26, 32 (1970).
92. C. I. Chung, Polym. Eng. Sci., 11, 93–98 (1971).
93. C. I. Chung, R. J. Nichols, and G. A. Kruder, Polym. Eng. Sci., 14, 29 (1974).
94. E. M. Mount and C. I. Chung, Polym. Eng. Sci., 18, 711–720 (1978).
95. E. M. Mount, J. G. Watson, and C. I. Chung, Polym. Eng. Sci., 22, 729–737 (1982).
96. D. E. McClelland and C. I. Chung, Polym. Eng. Sci., 23, 100–104 (1983).
97. K. H. Chung and C. I. Chung, Polym. Eng. Sci., 23, 191–196 (1983).
98. N. N., Engineering, 114, 606 (1922). This publication is often attributed to H. S. Rowell and D. Finlayson; however, this is disputed by Holmes (99). Rowell and Finlayson did publish on the subject later in 1928 (100).
99. D. B. Holmes, Ph. D. thesis, Delft University of Technology, the Netherlands (1967).
100. H. S. Rowell and D. Finlayson, Engineering, 126, 249 (1928).
101. H. A. A. Helmy, Ph. D. thesis, Univ. of Bradford, England (1975).
102. R. T. Fenner, "Extruder Screw Design", Illiffe, London (1970).
103. Z. Tadmor and I. Klein, "Engineering Principles of Plasticating Extrusion", van Nostrand Reinhold Comp., New York, (1970).
104. J. Nebrensky, J. F. T. Pittman, and J. M. Smith, Polym. Eng. Sci., 13, 209 (1973).
105. R. T. Fenner, Polymer, 18, 617 (1977).
106. Z. Tadmor and C. E. Gogos, "Principles of Polymer Processing", John Wiley & Sons, New York (1979).
107. Y. Mori and T. K. Matsumoto, Rheol. Acta, 1, 240 (1958).
108. R. E. Colwell and K. R. Nicholls, Ind. Eng. Chem., 51, 841 (1959).
109. W. E. Ball and R. E. Colwell, 43rd Nat'l. Mtg. AIChE, Tulsa, OK, Sept. (1960).
110. D. J. Weeks and W. J. Allen, J. Mech. Eng. Sci., 4, 380 (1962).
111. B. S. Clyde and W. A. Holmes-Walker, Int. Plast. Eng., 2, 338 (1962).
112. F. W. Kroesser and S. Middleman, Polym. Eng. Sci., 5, 231 (1965).
113. R. W. Flumerfelt, M. W. Pierick, S. L. Cooper, and R. B. Bird, Ind. Eng. Chem., 8, 354 (1969).
114. C. J. Rauwendaal, SPE ANTEC, Chicago, 186–199 (1983).
115. H. J. Zamodits, Ph. D. thesis, Univ. of Cambridge, England (1964).
116. B. Yates, Ph. D. thesis, Univ. of Cambridge, England (1968).
117. B. Martin, Ph. D. thesis, London Univ., England (1972).
118. K. Palit, Ph. D. thesis, London Univ., England (1972).
119. N. R. Neelakantan, Ph. D. thesis, Univ. of Wales, England (1974).
120. K. P. Choo, Ph. D. thesis, Univ. of Wales, England (1977).
121. H. Kuehnle, Kunststoffe, 5, 267 (1982).

122. D. E. Hanson, Polym. Eng. Sci., 9, 405 (1967).
123. Z. K. Walczak, J. Appl. Polym. Sci., 17, 153 (1973).
124. M. Rokuda, J. Appl. Polym. Sci., 23, 463 (1979).
125. J. M. Dealy and W. K. W. Tsang, J. Appl. Polym. Sci., 26, 1149 (1981).
126. T. Y. Liu, D. S. Soong, and M. C. Williams, Polym. Eng. Sci., 21, 675 (1981).
127. J. W. White and W. Minoshima, Polym. Eng. Sci., 21, 1113 (1981).
128. B. Maxwell and A. Breckwoldt, J. Rheol., 25, 55 (1981).
129. W. P. Perwadtschuk and V. I. Jankow, Plaste und Kautschuk, 28, 36–44 (1981).
130. J. Boussinesg, J. Mathematique Pures et Appliquees, series 2, 13, 377–424 (1868).
131. W. Meskat, Kunststoffe, 45, 87 (1955).
132. M. L. Booy, Polym. Eng. Sci., 21, 93 (1981).
133. Z. Tadmor, Polym. Eng. Sci., 6, 203 (1966).
134. D. F. Dyer, AIChE J., 15, 823 (1969).
135. R. M. Griffith, Ind. Eng. Chem., 1, 180 (1962).
136. K. Hayashida, Rheol. Acta, 2, 261 (1962).
137. H. Kruger, Kunststoffe, 10, 711 (1963).
138. D. R. Rea and W. R. Schowalter, Trans. Soc. Rheol., 11, 125 (1967).
139. H. J. Zamodits and J. R. A. Pearson, Trans. Soc. Rheol., 13, 357 (1969).
140. K. Palit and R. T. Fenner, AIChE J., 18, 628 (1972).
141. H. H. Winter, Rheol. Acta., 1, 1 (1978).
142. M. L. Hami and J. F. T. Pittman, Polym. Eng. Sci., 20, 339 (1980).
143. K. P. Choo, N. R. Neelakantan, and J. F. T. Pittman, Polym. Eng. Sci., 20, 349 (1980).
144. K. P. Choo, M. L. Hami, and J. F. T. Pittman, Polym. Eng. Sci., 21, 100 (1981).
145. J. Nebrensky, J. F. T. Pittman, and J. M. Smith, Polym. Eng. Sci., 13, 209 (1973).
146. S. M. Dinh and R. C. Armstrong, AIChE J., 28, 294–301 (1982).
147. H. H. Winter, Polym. Eng. Sci., 15, 84–89 (1975).
148. F. N. Cogswell, J. Non-Newtonian Fluid Mech., 2, 37–47 (1977).
149. M. T. Dennison, Trans. J. Plastics Inst., 35, 803–808 (1967).
150. J. J. Benbow and E. R. Howells, Trans. J. Plastics Inst., 30, 240–254 (1960).
151. British Patent 32559/72.
152. J. P. Tordella, in "Rheology", Vol. 4, F. R. Eirich, ed., Academic Press, New York (1969), Chapter 3.
153. J. L. White, Appl. Polym. Symp., No. 20, 155 (1973).
154. J. M. Lupton and J. W. Regester, Polym. Eng. Sci., 5, 235 (1965).
155. L. L. Blyler and A. C. Hart, Polym. Eng. Sci., 10, 193 (1970).
156. S. M. Barnett, Polym. Eng. Sci., 7, 168 (1967).
157. E. Boudreaux and J. A. Cuculo, J. Macromol. Sci. – Rev. Macromol Chem., C16, 39–77 (1977–1978).
158. C. D. Han and R. R. Lamonte, Polym. Eng. Sci., 11, 385 (1971).
159. J. L. den Otter, Rheol. Acta, 10, 200–207 (1971).
160. T. W. Huseby, Trans. Soc. Rheol., 10, 181–190 (1966).
161. A. P. Metzger and C. W. Hamilton, SPE Trans., 4, 107–112 (1964).
162. G. V. Vinogradov et al., Polym. Eng. Sci., 12, 323–334 (1972).
163. J. J. Benbow and P. Lamb, SPE Trans, 3, 7–17 (1963).
164. W. Phillippoff and F. H. Gaskins, Trans. Soc. Rheol., I, 263–284 (1958).
165. U. S. Patent 2,991,508 by R. T. Fields and C. F. W. Wolf to E. I. du Pont de Nemours and Company, issued July 11, 1961.
166. A. M. Kraynik and W. R. Schowalter, J. Rheol., 25, 95–114 (1981).
167. R. F. Westover, Polym. Eng. Sci., 6, 83 (1966).
168. R. A. Worth, J. Parnaby, and H. A. A. Helmy, Polym. Eng. Sci., 17, 257 (1977).
169. C. J. S. Petrie and M. M. Denn, AIChE J., 22, 109–236 (1976).
170. C. J. S. Petrie and M. M. Denn, AIChE J., 22, 236–246 (1976).

171. J. C. Miller, SPE Trans., 3, 134 (1963).
172. S. Kase, J. Appl. Polym. Sci., 18, 3279 (1974).
173. G. F. Cruz-Saenz, G. J. Donnelly, and C. B. Weinberger, AIChE J., 22, 441 (1976).
174. J. R. A. Pearson and Y. T. Shah, Trans. Soc. Rheol., 16, 519 (1972).
175. J. R. A. Pearson and Y. T. Shah, Ind. Eng. Chem. Fundam., 13, 134 (1979).
176. R. J. Fischer and M. M. Denn, Chem. Eng. Sci., 30, 1129 (1975).
177. Y. Ide and J. L. White, J. Appl. Polym. Sci., 20, 2511-2531 (1976).
178. Y. Ide and J. L. White, J. Non-Newt. Fluid Mech., 2, 281-298 (1977).
179. Y. Ide and J. L. White, J. Appl. Polym. Sci., 22, 1061-1079 (1978).
180. J. L. White and Y. Ide, J. Appl. Polym. Sci., 22, 3058-3074 (1978).
181. R. S. Lenk, J. Appl. Polym. Sci., 22, 1781-1785 (1970).
182. G. A. Latinen, ACS Adv. in Chem. Series, 34, 235 (1962).
183. R. W. Coughlin and G. P. Canevari, AIChE J., 15, 560 (1969).
184. G. W. Roberts, AIChE J., 16, 878 (1970).
185. J. A. Biesenberger, Polym. Eng. Sci., 20, 1015 (1980).
186. J. A. Biesenberger and G. Kessides, Polym. Eng. Sci., 22, 832 (1982).
187. J. T. Bergen, G. W. Carrier, and J. A. Krumbansh, SPE NATEC, Detroit (1958).
188. P. V. Danckwerts, Appl. Sci. Res., Sec. A3, 279-296 (1952).
189. W. D. Mohr, R. L. Saxton, and C. H. Jepson, Ind. Eng. Chem., 49, 1855 (1957).
190. N. Nadav and Z. Tadmor, Chem. Eng. Sci., 28, 2115 (1973).
191. W. M. Hess, V. E. Chirico, and P. C. Vegvari, Elastomerics, Jan., 24-34 (1980).
192. D. H. Sebastian and J. A. Biesenberger, SPE ANTEC, Chicago, 121-123 (1983).
193. H. J. Suchanek, SPE ANTEC, Chicago, 117-120 (1983).
194. T. Dobroth, G. Druhak, and L. Erwin, SPE ANTEC, Chicago, 124-126 (1983).
195. G. Krassowski and G. Mennig, Kunststoffe, 73, 127-131 (1983).
196. Z. Tadmor and C. Gogos, "Principles of Polymer Processing", Wiley, New York (1979), Chapter 7.
197. R. J. Cembrole, Rubber Chem. Techn., 56, 233-243 (1983).
198. K. K. Mohanty, J. M. Ottino, and H. T. Davis, Chem. Eng. Sci., 37, 905 (1982).
199. J. W. Hiby, Int. Chem. Eng., 21, 197 (1981).
200. C. L. Tucker, Chem. Eng. Sci., 36, 1829 (1982).
201. R. S. Spencer and R. M. Wiley, J. Colloid. Sci., 6, 133 (1951).
202. L. Erwin, Polym. Eng. Sci., 18, 572-576 (1978).
203. L. Erwin, Polym. Eng. Sci., 18, 738-740 (1978).
204. L. Erwin, Polym. Eng. Sci., 18, 1044-1047 (1978).
205. K. Y. Ng and L. Erwin, Polym. Eng. Sci., 21, 212-217 (1981).
206. Z. Tadmor and G. Lidor, Polym. Eng. Sci., 16, 450-461 (1976).
207. G. Pinto and Z. Tadmor, Polym. Eng. Sci., 10, 279-288 (1970).
208. J. M. Ottino and R. Chella, Polym. Eng. Sci., 23, 357-379 (1983).
209. J. M. Ottino, Ph. D. thesis, Univ. of Minnesota, Minneapolis, MN (1979).
210. J. M. Ottino, W. E. Ranz, and C. W. Macosko, AIChE J., 27, 565-577 (1981).
211. R. Chella and J. M. Ottino, Ind. Eng. Chem. Fundam., 24, 170-180 (1985).
212. S. M. Skoblar, Plastics Technology, Oct., 37-43 (1974).
213. G. Smoluk, Plastics World, May, 40-43 (1978).
214. N. R. Schott, B. Weinstein, and D. LaBombard, Chem. Eng. Prog., 71, 54-58 (1975).
215. C. D. Han and Y. W. Kim, J. Appl. Polym. Sci., 19, 2831-2843 (1975).
216. E. B. Nauman, AIChE J., 25, 246-258 (1979).
217. P. T. Allocca, paper presented at the 73 rd Annual AIChE Meeting, Chicago, Nov. (1980).
218. P. T. Allocca, Fiber Producer, April, 12-19 (1982).
219. J. M. Ottino, AIChE J., 29, 159-161 (1983).
220. W. R. Bolen and R. E. Colwell, SPE J., 14, 24-28 (1958).

221. R. B. Bird, H. R. Warner, Jr., and D. C. Evans, "Kinetic Theory and Rheology of Dumbbell Suspension with Brownian Motion", Fortschritte Hochpolymerenforschung, Springer Verlag, 8, 1–90 (1971).
222. Z. Tadmor, Ind. Eng. Chem. Fundam., 15, 346–348 (1976).
223. G. Martin, Industrie-Anzeiger, 93, 2651–2656 (1971).
224. G. Martin, Ph. D. thesis, Univ. of Stuttgart, W. Germany (1972).
225. B. Miller, Plastics World, Jan., 58–64 (1982).
226. T. B. Reeve and W. L. Dills, J. Color & Appearance, 1, 25–29 (1971).
227. N. N., Plastics Compounding, Jan./Febr., 20–32 (1981).
228. W. D. Mohr and R. S. Mallouk, Ind. Eng. Chem., 51, 765 (1959).
229. H. A. A. Helmy, SPE ANTEC, Chicago, 146–150 (1983).
230. W. Gleiszle, Rheol. Acta, 21, 484 (1982).
231. L. A. Utracki and R. Gendron, J. Rheol., 5, 28, 601–623 (1984).
232. J. J. Benbow, R. V. Charley, and P. Lamb, Nature, 192, 223 (1961).
233. J. J. Benbow and P. Lamb, SPE Trans., Jan., 7–17 (1963).
234. M. Viriyayuthakorn and B. Kassahun, SPE ANTEC, New Orleans, 81–84 (1984).
235. H. Potente, Rheol. Acta, 22, 387–395 (1983).
236. C. D. Denson, Adv. Chem. Eng., 12, 61–104 (1983).
237. G. Menges, W. Feistkorn, and G. Fischbach, Kunststoffe, 74, 695–699 (1984).

PART III
PRACTICAL APPLICATIONS

8   EXTRUDER SCREW DESIGN

The single most important mechanical element of a screw extruder is the screw. The proper design of the geometry of the extruder screw is of crucial importance to the proper functioning of the extruder. If material transport instabilities occur as a result of improper screw geometry, even the most sophisticated computerized control system cannot solve the problem. Screw design is often still considered to be more of an art than a science. As a result, misconceptions about certain aspects of screw design still abound today. Since the theory of single screw extrusion is now well-developed, see Chapter 7, the design of screws for single screw extruders can now be based on solid engineering principles. Thus, screw design for single screw extruders should no longer be an art, but a science based on the principles of polymer processing engineering. Unfortunately, people involved in screw design are not always up to date on extrusion theory. As a result, many extruder screws in use today perform considerably below maximum possible performance, solely because of improper screw design. An example is the still-common use of the square pitch extruder screw. This is a screw with constant pitch with the pitch being equal to the diameter of the screw; this pitch corresponds to a helix angle of 17.66°. It can be demonstrated quite easily that the square pitch is far from optimum with respect to melting and melt conveying. The latter fact has been known since the early 1950's, yet most extruder screws in use today still use the constant square pitch design.

A factor that may have contributed to the state of affairs in screw design is that there has not been a comprehensive text dealing with screw design. The objective of this chapter is to demonstrate how extrusion theory can be used to properly design extruder screws. Hopefully, this will provide a solid foundation, based on engineering principles, from which better and more effective screw designs can be developed in the future. The principles of screw design are not only important in designing new extruder screws, but also in the analysis of processing problems of an existing extrusion line. It is important to be able to recognize whether a problem is related to poor screw design or to another part of the process. Thus, knowledge of the basic principles of screw design is important to essentially every person involved with extruders.

8.1   MECHANICAL CONSIDERATIONS

Irregardless of the details of the screw geometry, it is important that the screw have sufficient mechanical strength to withstand the stresses imposed by the conveying process in the extruder.

8.1.1   TORSIONAL STRENGTH OF THE SCREW ROOT

An important requirement for the extruder screw is the ability to transmit the torque required to turn the screw. The most critical area of the screw in this respect is the feed section. In the feed section, the cross-sectional area of the root of the screw is generally the smallest and thus the torsional strength the lowest. Also, in the feed section, the entire torque has to be transmitted, while further downstream only a fraction of the total torque has to be transmitted. The torque that is transmitted can be determined from the power to the screw, Z_{screw}, and the screw rpm, N.

$$T = C \frac{Z_{screw}}{N} \qquad (8-1)$$

where Z_{screw} is the power to the screw, N the rotational speed of the screw, and C a conversion constant.

If Z is expressed in horsepower and N in rev/min., the constant C must equal 7124 to give a torque expressed in Newtonmeter. This formula is based on the relationship between torque, angular frequency ω, and power:

$$Z = T\omega = \frac{2\pi TN}{60} \qquad (8-2)$$

where ω is expressed in radians/sec and N in rpm.

From equation 8-1, it can be seen that the transmitted torque is directly proportional to the horsepower and inversely proportional to the screw speed. The power to the screw is related to the motor power by:

$$Z_{screw} = Z_{motor} \, \varepsilon_{motor} \, \varepsilon_{transmission} \qquad (8-3)$$

where ε_{motor} is the efficiency of the motor and $\varepsilon_{transmission}$ is the efficiency of the transmission.

The total efficiency is generally around 0.75. Thus, the power to the screw will be about 75 percent of the motor power. The stresses in the screw shaft as a result of the torque on the screw are shown in Figure 8-1.

The shear stress in the shaft resulting from torque T can be expressed as:

$$\tau = \frac{2Tr}{\pi R^4} = \frac{Tr}{J} \qquad (8-4)$$

where J is the polar moment of inertia.

The maximum stress occurs at the circumference ($r = R$), and the maximum stress is:

$$\tau_{max} = \frac{2T}{\pi R^3} = \frac{TR}{J} \qquad (8-5)$$

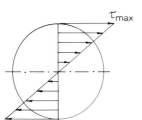

Figure 8-1. Stresses in the Screw Shaft

In order to avoid failure in the screw shaft, the maximum stress should be less than the allowable stress τ_a of the metal of the screw. The allowable stress of metal ranges from about 50 MPa to 100 MPa. Thus, in order to have sufficient torsional strength, the following inequality should hold:

$$R > \left(\frac{2T}{\pi \tau_a}\right)^{1/3} \tag{8-6}$$

From inequality 8-6, the maximum channel depth in the feed section can be calculated:

$$H_{max} = \frac{1}{2} D - \left(\frac{2T}{\pi \tau_a}\right)^{1/3} \tag{8-7}$$

where D is the OD of the extruder screw.

Consider, as an example, a 150-mm extruder, running at 80 rpm and consuming 200 hp at the screw. The torque to the screw, according to equation 8-1, is T = 17810 Nm. If τ_a = 100 MPa, then the maximum channel depth in the feed section becomes:

$$H_{max} = 0.0750 - 0.0484 = 0.0266 \text{ [m]} = 26.6 \text{ [mm]} \tag{8-8}$$

It should also be realized that equations 8-6 and 8-7 are valid for solid screws only. If the screw is cored for cooling or heating, the torsional strength of the screw shaft will be reduced. In this case, the polar moment of inertia of the screw shaft becomes:

$$J = \frac{\pi(R_s^4 - R_c^4)}{2} \tag{8-9}$$

where R_s is the radius of the root of the screw and R_c is the radius of the core channel in the screw.

8.1.2 STRENGTH OF THE SCREW FLIGIIT

The primary loading of the screw flights occurs by the pressure differential ΔP between the leading and trailing side of the flight. In addition, there will be loading from the shear stress acting on the tip of the flight as a result of the leakage flow through the clearance. This situation is shown in Figure 8-2.

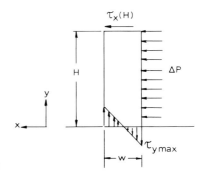

Figure 8-2. Stresses Acting on a Screw Flight

As a result of the stresses acting on the screw flight, there will be a distribution of normal stresses in the flight that reaches a maximum value at the screw surface. The torque resulting from this normal stress distribution counteracts the torque resulting from τ_x and ΔP. If the maximum normal stress at the wall is τ_{ymax}, then a torque balance per unit flight length yields:

$$\frac{1}{6} w^2 \tau_{ymax} = \frac{1}{2} H^2 \Delta P + Hw\, \tau_x(H) \qquad (8-10)$$

Thus, the maximum normal stress is:

$$\tau_{ymax} = \frac{3H^2 \Delta P}{w^2} + \frac{6H\, \tau_x(H)}{w} \qquad (8-11)$$

The cross flight shear stress τ_x acting in the screw flight also reaches a maximum at the screw surface. The maximum shear stress τ_{max} is obtained from a simple force balance in x-direction:

$$w\, \tau_{xmax} = w\, \tau_x(H) + H\Delta P \qquad (8-12)$$

Thus, the maximum shear stress is:

$$\tau_{xmax} = \tau_x(H) + \frac{H}{w} \Delta P \qquad (8-13)$$

For the combined shear and normal stress, the following criterion can be used:

$$\tau_{comb} = (\tau_{ymax}^2 + 3\, \tau_{xmax}^2)^{1/2} < \tau_a \qquad (8-14)$$

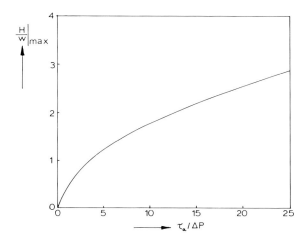

Figure 8-3. Slenderness Ratio versus $\tau_a/\Delta P$

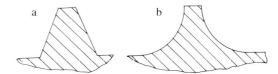

Figure 8–4. Flights with Varying Flight Width

From equations 8–11 through 8–14, the ratio of flight height to flight width can be determined such that $\tau_{comb} < \tau_a$. If the shear stress $\tau_x(H)$ at the tip of the flight can be neglected, the ratio of flight height to flight width should be:

$$\frac{H}{W} < \left[\frac{1}{3}\left(\frac{1}{4} + \frac{\tau_a^{\,2}}{\Delta P^2}\right)^{1/2} - \frac{1}{6}\right]^{1/2} \qquad (8-15)$$

This relationship is shown in Figure 8–3.

The ratio of flight height to flight width can be thought of as a slenderness ratio. In general, the ratio of the allowable stress to pressure difference, $\tau_a/\Delta P$, will be larger than 25. Thus, a slenderness ratio of 2 (a common value used in screw design) will yield more than adequate strength if the shear stress at the flight tip is negligible. In some instances, the shear stress at the flight tip cannot be neglected. For instance, in a situation where considerable lateral forces are acting on the screw, metal-to-metal contact can occur between screw flight and barrel. In such a case, very large shear stresses can act on the flight tip and these stresses must then be taken into account. However, this is a rather abnormal situation and should be avoided if at all possible.

It should be noted that the stress concentration in the base of the flight can be reduced significantly if the flight width is varied with flight height, see Figure 8–4.

An additional advantage of the flight geometries shown in Figure 8–4 is that the flight width at the tip can be reduced considerably compared to the standard flight geometry without increasing the stresses in the base of the flight. It will be shown later that there are substantial functional benefits that can be derived from the flight geometries as shown in Figure 8–4.

8.1.3   LATERAL DEFLECTION OF THE SCREW

There are various causes that will tend to deflect the screw. The most obvious cause is the force of gravity acting on the screw. If the drive support of the screw is considered rigid and the supporting function of the polymer and barrel is neglected, then the sagging of the screw by its own weight can be represented by Figure 8–5.

The amount of sag at the end of the screw is:

$$Y(L) = \frac{qL^4}{8EI} \qquad (8-16)$$

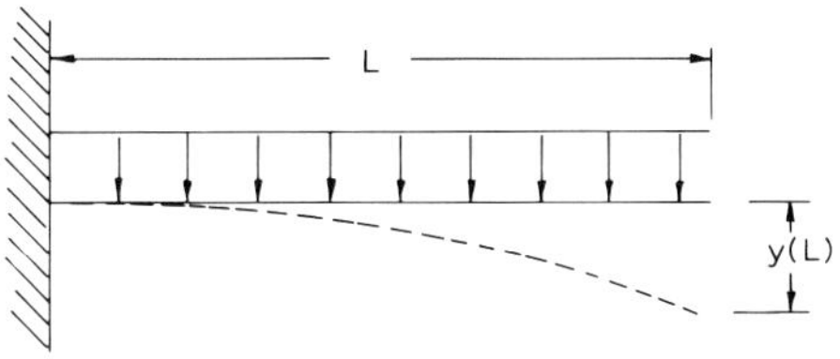

Figure 8–5. Screw as a Cantilever

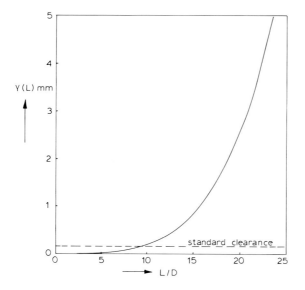

Figure 8-6. Unrestrained Sag versus Screw Length

where q is the force per unit length from the weight of the screw, E is the elastic modulus, and I is the moment of inertia. If the presence of the screw channel is neglected, then equation 8-16 can be written as:

$$Y(L) = \frac{2a_g \rho L^4}{ED^2} \qquad (8-17)$$

where a_g is the gravitational acceleration and ρ the density of the screw material.

Consider a 150-mm extruder screw with a density of 7850 kg/m^3 and an elastic modulus of 210 GPa ($=$ 210E9 Pa); the sag for this example is:

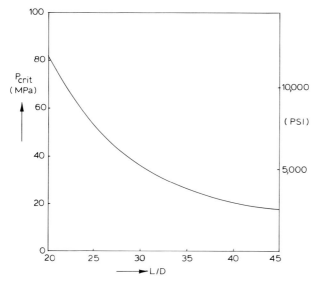

Figure 8-7. Critical Head Pressure versus L/D Ratio

$$Y(L) = 3.256E\text{-}5 \; L^4 \qquad\qquad (8\text{--}18)$$

The unrestrained sag as a function of screw length is shown in Figure 8-6.

It is clear that the unrestrained sag starts to exceed the standard radial clearance (≈ 0.2 mm) when the L/D ratio exceeds 10. At normal L/D ratios of 20 to 30, the unrestrained sag is about one to two orders of magnitude larger than the standard radial clearance. From these simple considerations, it is obvious that the polymer between the screw and barrel must play a considerable support function to prevent contact between screw and barrel. The supporting force necessary to counteract the sagging by the weight of the screw has to increase very strongly when the L/D is increased.

Another mechanism that can cause lateral movement of the screw is buckling. The collapse force required to cause buckling of a uniform cantilever is:

$$F = \frac{\pi^2 EI}{4L^2} \qquad\qquad (8\text{--}19)$$

Thus, the head pressure necessary to cause buckling is:

$$P = \frac{\pi^2 ED^2}{64L^2} \qquad\qquad (8\text{--}20)$$

because the moment of inertia $I = \dfrac{\pi D^4}{64}$

The critical head pressure for buckling as a function of the L/D ratio is shown in Figure 8-7.

The elastic modulus is taken as 210 GPa. Considering that normal head pressures range from 20 MPa to 60 MPa, it is clear that the occurrence of buckling is a distinct possibility when the L/D is greater than 20. The equations above do not take into account the weakening of the screw as a result of the screw channel. Fenner and Williams (1) analyzed the effect of channel depth and compression ratio. They found that the critical head pressure for buckling reduces when the channel depth is taken into account. At a ratio of H/D=0.05 and a compression ratio C=3, their predicted critical head pressure values are less than half of those predicted by equation 8-20 and shown in Figure 8-7. This simply indicates that buckling is more likely to occur when the L/D is greater than 20. Thus, the polymer will have to support the screw in the center region of the extruder to prevent screw to barrel contact as a result of buckling. However, the lateral force resulting from buckling F_b is of the order:

$$F_b \approx 4F\delta/L \approx \pi D^2 P\delta/L \qquad\qquad (8\text{--}21)$$

where F is the force acting on the end of the screw.

Since the ratio δ/L is about 0.00005, the lateral force resulting from buckling will be quite small.

Another possible deflection mechanism is the occurrence of whirling. Whirling occurs when a shaft reaches a critical speed and becomes dynamically unstable with large lateral amplitudes. This phenomenon is due to the resonance frequency when the rotational speed corresponds to the natural frequencies of lateral vibration of the shaft. For uniform beams vibrating in flexure, the natural frequencies can be expressed as:

$$f_W = C_n \left(\frac{a_g E I}{w_1 L^4} \right)^{1/2} \tag{8-22}$$

where a_g is the gravitational acceleration and w_1 is the weight per unit length.

The value C_n for a beam with one end clamped and one end free is $C_n = 0.560$. For a circular beam, the critical rotational speed becomes ($N = 2\pi f$):

$$N_W = 0.88 \frac{D}{L^2} \left(\frac{E}{\rho} \right)^{1/2} \tag{8-23}$$

When $E = 210$ GPa and $\rho = 7850$ kg/m^3, the critical rotational speed can be written as:

$$N_W = \frac{4549.5}{D(L/D)^2} \tag{8-24}$$

For a 150 – mm extruder with $L = 30\,D$, the critical whirling speed $N_W = 33.7$ rev/sec = 2022 rpm. It is clear from these numbers that whirling is not likely to occur in normal extrusion operations. Fenner and Williams (1) also calculated the critical rotational speed for whirling in extruders; however, they used an incorrect equation for the whirling speed. Instead of weight per unit length, they used mass per unit length in equation 8 – 22. As a result, their values of the whirling speed are lower by a factor of $\sqrt{a_g}$.

Another possible cause of lateral deflection of the screw is non-uniform pressure distribution around the circumference of the screw. Figure 8-8 shows a possible pressure distribution that will result in a considerable lateral force on the screw.

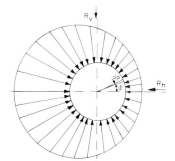

Figure 8-8. Circumferential Pressure Distribution That Can Cause Screw Deflection

The horizontal reaction component can be determined from:

$$R_h = \int_0^{2\pi} P(\varphi) \cos \varphi R d\varphi \tag{8-25}$$

The vertical reaction component can be determined from:

$$R_v = \int_0^{2\pi} P(\varphi) \sin \varphi R d\varphi \tag{8-26}$$

If the pressure is on the average ΔP larger on the left side than on the right side, the horizontal reaction force over length L is:

$$R_h = \Delta PDL \qquad\qquad\qquad (8-27)$$

If $D=L=150$ mm and $\Delta P=1$ MPa ($\simeq 145$ psi), the reaction force will be 22.5kN ($\simeq 5000$ lbf.). In reality, the situation is more complicated because the pressure varies both in down channel and cross channel direction. However, from this simple example, it is clear that even a small pressure differential of only 1 MPa can cause a significant lateral reaction force on the screw. The reaction force is of such magnitude that it can easily deflect the screw into the barrel and cause substantial wear. This type of pressure induced deflection is most likely to occur when the pressure reaches a sharp maximum or minimum somewhere along the screw. It is also more likely to occur with large helix angles as compared to small helix angles.

A likely location for sharp pressure peaks is the end of the compression section. If the compression ratio is high, there is a large chance of plugging of the solid bed. This can cause local pressure peaks that can deflect the screw into the barrel. It is interesting to note that generally the most severe wear does indeed occur towards the end of the compression section. This indicates that quite often the compression ratio is too high, causing sharp pressure peaks towards the end of the compression section.

If one considers all the possible causes of lateral screw deflection: gravity, buckling, whirling, and pressure induced deflection, it seems that pressure induced deflection is the most likely cause of lateral deflection. The wear caused by pressure induced deflection is most likely located towards the end of the compression section. This type of wear can be quite dramatic. The author experienced a case where a 150-mm extruder was wearing at a rate of about 1 mm (0.040 inch) in about 24 hours. The cause turned out to be installation of a relatively long grooved barrel section without a corresponding change in screw design. A reduction of the length of the grooved barrel section solved the problem; a change in screw design (lower compression ratio) probably would have solved the problem as well. However, this example illustrates that pressure induced screw deflection can cause severe wear problems. Unfortunately, this type of problem is not easy to diagnose because the exact pressure profile along the length of the extruder is generally not known. Another complicating factor is that this type of wear may take several weeks or even a few months to become really significant. This can make it difficult to relate the wear to the event that triggered it because there can be a considerable time lag.

8.2   OPTIMIZING FOR OUTPUT

Generally, the objective of a screw design is to deliver the largest amount of output at acceptable melt quality. Unfortunately, high output and mixing quality are, to some extent, conflicting requirements. As output goes up, residence time goes down and so does the mixing quality. However, the mixing quality can generally be restored by incorporating mixing sections; either a mixing section along the screw or a static mixing section.

It is also important to realize that all functional zones of the extruder are interdependent. It makes little sense to drastically increase the pumping capacity if the melting rate is the real bottleneck in the process. Thus, before designing a new extruder screw to replace an existing screw, one should determine what part of the extruder is limiting the rate. This process will be covered in Chapter 11 on troubleshooting extruders.

8.2.1   OPTIMIZING FOR MELT CONVEYING

The melt conveying theory discussed in Section 7.4 can be used to determine the optimum screw geometry for melt conveying. This optimum geometry will normally be used in the metering section of the extruder screw. The optimum channel depth for output can be determined from:

$$\frac{\partial \dot{V}}{\partial H} = 0 \tag{8-28}$$

The volumetric melt conveying rate for a Newtonian fluid, neglecting the effect of the flight flanks, is given by equation 7-155. Considering that the channel width $W = \pi D \sin\varphi/p - w$, down channel barrel velocity $v_{bz} = \pi D N \cos\varphi$, and down channel pressure gradient $g_z = g_a \sin\varphi$, where g_a is the axial pressure gradient, equation 7-155 can be written as:

$$\dot{V} = 0.5\,\pi D N H (\pi D \sin\varphi \cos\varphi - pw\cos\varphi) - \frac{H^3 P}{12\,\eta L}(\pi D \sin^2\varphi - pw\sin\varphi) \tag{8-29}$$

where L is the axial distance over which pressure P is built up; thus, the axial pressure gradient $g_a = P/L$.

The polymer melt viscosity η will depend on the local shear rate in the screw channel. If it is assumed that the representative shear rate in the channel is the Couette shear rate and that the polymer melt behaves as a power law fluid, that melt viscosity can be represented by:

$$\eta = m\left(\frac{\pi D N}{H}\right)^{n-1} \tag{8-30}$$

where m is the consistency index and n the power law index, see equation 6-23.

The optimum channel depth for output rate H_r^* can now be determined from equations 8-28 through 8-30.

$$H_r^* = \left[\frac{6(\pi D N)^n\,mL}{(n+2)P\tan\varphi}\right]^{\frac{1}{n+1}} \tag{8-31}$$

The optimum depth depends on the diameter, screw speed, power law index, consistency index, pressure gradient, and helix angle. The optimum helix angle for output can be determined from:

$$\frac{\partial \dot{V}}{\partial \varphi} = 0 \tag{8-32}$$

By using the same equations for the melt conveying rate, the optimum helix angle for output rate φ_r^* has to be determined from the following equation:

$$\frac{\pi D \cos^2 \varphi_r^* - \pi D \sin^2 \varphi_r^* + pw \sin \varphi_r^*}{2\pi D \sin \varphi_r^* \cos \varphi_r^* - pw \cos \varphi_r^*} = \frac{PH^{n+1}}{6mL(\pi DN)^n}$$ (8-33)

Equation 8-33 cannot be easily solved in the form it is in. However, the equation can be simplified considerably if it is assumed that $w = 0$ and by defining a dimensionless down channel pressure gradient g_z^0. Equation $8-33$ can now be written as:

$$\frac{\cos^2 \varphi_r^* - \sin^2 \varphi_r^*}{2\cos \varphi_r^*} = g_z^0 (\cos \varphi_r^*)^n$$ (8-34)

where:

$$g_z^0 = \frac{PH^{n+1} \sin \varphi}{6mL(\pi DN \cos \varphi)^n}$$ (8-35)

The optimum helix angle now has to be determined from:

$$2\cos^2 \varphi_r^* - 2g_z^0 (\cos \varphi_r^*)^{n+1} = 1$$ (8-36)

For the Newtonian case ($n = 1$), the solution becomes:

$$\cos \varphi_r^* = \left(\frac{1}{2 - 2g_z^0}\right)^{1/2}$$ (8-37)

For the extreme non-Newtonian case ($n = 0$), the solution is:

$$\cos \varphi_r^* = \frac{1}{2} g_z^0 + \frac{1}{2} [(g_z^0)^2 + 2]^{1/2}$$ (8-38)

Figure 8-9 shows the optimum helix angle as a function of the dimensionless down channel pressure gradient for $n = 1$ and $n = 0$.

It can be seen that the two curves are relatively close; thus, the effect of the pseudoplastic behavior is not very pronounced.

It may be more interesting to optimize the channel depth and helix angle simultaneously. This can be done by inserting equation 8-31 into equation 8-33. After some calculations, the optimum helix angle can be determined to be:

$$\sin \varphi_r^* = \left(\frac{n}{2n+2}\right)^{1/2} + \hat{w}\left(\frac{n+2}{4n}\right)$$ (8-39)

where $\hat{w}$ is a reduced flight width:

$$\hat{w} = \frac{pw}{\pi D}$$ (8-39a)

where p is the number of flights, w the perpendicular flight width, and D the screw OD.

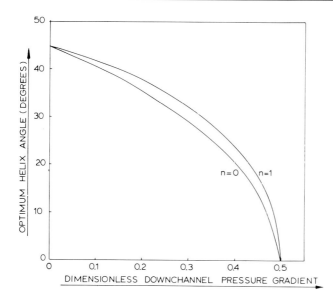

Figure 8-9. Optimum Helix Angle versus Dimensionless Down Channel Pressure Gradient

The details of the derivation can be found in an article on screw design of two-stage extruder screws (2). The optimum helix angle is only a function of the power law index and the reduced flight width. Figure 8-10 shows the optimum helix angle as a function of the power law index at various values of the reduced flight width.

The simplicity of equation 8-39 makes it a useful and convenient expression for optimizing the geometry of the melt conveying zone of an extruder. The corresponding optimum channel depth can be found by inserting equation 8-39 into equation 8-31. The optimum channel depth resulting from simultaneous optimization is simply:

$$H_r^* = \left[\frac{6mv_b^{\,n}}{(n+2)g_a\tan\varphi_r^*} \right]^{\frac{1}{n+1}} \tag{8-40}$$

Unfortunately, the optimum channel depth is dependent on many more variables then the optimum helix angle. The latter depends only on the power law index and reduced flight width. In addition to these variables, the optimum channel depth also depends on the screw speed, screw diameter, consistency index, and pressure gradient. This means that it is not possible to design a universally optimum screw geometry. Thus, one has to determine the most likely operating parameters that the screw is likely to encounter and design for those parameters.

It should be noted that fundamentally it is not entirely correct to take an expression derived for a Newtonian fluid and insert a power law viscosity form into it. However, if this simplification is not made, the analysis becomes much more complex and analytical solutions much more difficult to obtain, if not impossible. Results of the analytical solutions have been compared to results of numerical computations for a two-dimensional flow of a power law fluid. In most cases, the results are within 10 to 20 percent (2). It should be noted that the results are exact when the power law index is unity; i.e., for Newtonian fluids. However, if the optimum depth and helix angle for a pseudoplastic fluid are calculated using expressions valid for Newtonian fluids only, very large errors can result, particularly when the

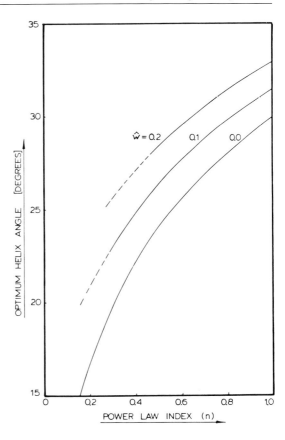

Figure 8–10. Optimum Helix Angle versus the Power Law Index in Simultaneous Optimization

power law index is about one-half or less. It is very important, therefore, to take the pseudoplastic behavior into account, since the large majority of polymers are strongly non-Newtonian.

Instead of optimizing the screw geometry for output, it may be desirable to optimize for pressure generating capability.

The optimum depth H_p^* and helix angle φ_p^* for pressure generation is found by setting:

$$\frac{\partial P}{\partial H} = \frac{\partial P}{\partial \varphi} = 0 \tag{8–41}$$

An expression for pressure P can be found by rewriting equation 8-29. The optimum channel depth is:

$$H_p^* = \frac{n+2}{n+1} \frac{2\dot{V}}{\pi^2 D^2 N \cos\varphi(\sin\varphi - \hat{w})} \tag{8–42}$$

The optimum helix angle has to be determined from:

$$\frac{\cos\varphi_p^* \, (2\sin\varphi_p^* - \hat{w})}{(\sin\varphi_p^* - \hat{w})^2} = \frac{\pi^2 D^2 NH}{2\dot{V}} \tag{8–43}$$

When the depth and helix angle are optimized simultaneously, equation 8–42 is inserted into equation 8–43. The resulting optimum helix angle for simultaneous optimization of the pressure generating capacity is:

$$\sin\varphi_p^* = \left(\frac{n}{2n+2}\right)^{1/2} + \hat{w}\left(\frac{n+2}{4n}\right) \tag{8-44}$$

Comparing this result to the optimum helix angle for output φ_r^*, equation 8–39, it is clear that the two expressions are r the same, thus:

$$\varphi_p^* = \varphi_r^* = \varphi^* \tag{8-45}$$

The optimum channel depth for simultaneous optimization for pressure generation is found by inserting equation 8–39 into equation 8–37. The resulting expression can be shown to be identical to equation 8–35, thus:

$$H_p^* = H_r^* = H^* \tag{8-46}$$

Therefore, simultaneous optimization of depth and helix angle for output yields the same results as simultaneous optimization for pressure generating capability.

It should be noted that the expressions for optimum channel depth are only valid when the pressure gradient is positive. When the pressure gradient is negative, the pressure flow will be in the forward direction and the output increases with channel depth without reaching a maximum value. In this situation, the depth of the metering section will be determined by the requirements for complete melting and good mixing.

The optimum channel depth and helix angle can also be determined from the melt conveying theory of power law fluids when the flow is considered to be one-dimensional, see also Section 7.4.2. By combining equations 7–194 and 7–195, the output can be written as:

$$\dot{V} = \frac{WHv_{bz}}{s+2}\,(1 - \lambda - \Gamma_R|\lambda|^{s+1}) \tag{8-47}$$

where:

$$\Gamma_R = \frac{(g_a\sin\varphi)^s H^{1+s}}{m^s v_b \cos\varphi} \tag{8-47a}$$

The optimum channel depth can again be determined by taking the partial derivative of output $\dot{V}$ and setting the result equal to zero. The resulting expression is:

$$1 - \lambda - (s+2)\,\Gamma_R|\lambda|^{s+1} - H\,\frac{\partial\lambda}{\partial H}\,[1 + \Gamma_R\,(s+1)|\lambda|^s] = 0 \tag{8-48}$$

For large positive pressure gradients, equation 7–190 should be used to evaluate the first derivative of λ with respect to channel depth H. This results in the following expression:

$$\frac{\partial \lambda}{\partial H} = \frac{(s+1)}{H\Gamma_R} \frac{1}{[\lambda^S + (1-\lambda)^S]} = \frac{1}{H} \frac{(1-\lambda)^{S+1} - \lambda^{S+1}}{\lambda^S + (1-\lambda)^S} \qquad (8\text{-}49)$$

Inserting equation 8–49 into equation 8–48 yields an expression with only λ and s:

$$1 - \lambda + \frac{(s+1)(s+2)|\lambda|^{S+1}}{\lambda^{S+1} - (1-\lambda)^{S+1}} + \frac{\lambda^{S+1} - (1-\lambda)^{S+1} - (s+1)^2|\lambda|^S}{\lambda^S + (1-\lambda)^S} = 0$$

$$(8\text{-}49\,a)$$

A complete solution of equation 8–49 a may be rather involved; however, it can be seen quite easily that $\lambda = 0$ is a solution of equation 8–49 a. Thus, the channel depth is optimized when $\lambda = 0$; i.e., when the velocity gradient becomes zero at the screw surface. The optimum helix angle is obtained by taking the first derivative of output $\dot{V}$ and setting the result equal to zero. This results in the following expression:

$$2\cos 2\varphi^* (1-\lambda-\Gamma_R|\lambda|^{S+1}) + \sin 2\varphi^* \left[-\frac{\partial \lambda}{\partial \varphi} - \frac{\partial \Gamma_R}{\partial \varphi}|\lambda|^{S+1} - \Gamma_R(s+1)|\lambda|^S \frac{\partial \lambda}{\partial \varphi} \right] = 0$$

$$(8\text{-}49\,b)$$

For simultaneous optimization of channel depth and helix angle, $\lambda = 0$ and equation 8–49 b becomes:

$$\frac{\partial \lambda}{\partial \varphi} = 2\cot 2\varphi^* \qquad (8\text{-}49\,c)$$

When $\lambda = 0$, the first derivative of λ with respect to helix angle φ, determined from equation 7–190, becomes:

$$\frac{\partial \lambda}{\partial \varphi} = \frac{1}{(s+1)^2} \frac{\partial \Gamma_R}{\partial \varphi} = \frac{s+\tan^2\varphi}{(s+1)\tan\varphi} \qquad (8\text{-}49\,d)$$

Inserting equation 8–49 d into equation 8–49 c yields the solution of the optimum helix angle:

$$\tan\varphi^* = \left(\frac{1}{s+2}\right)^{1/2} = \left(\frac{n}{1+2n}\right)^{1/2} \qquad (8\text{-}49\,e)$$

To compare this result with the result from the modified Newtonian analysis, equation 8–44, the optimum helix angle can be written as:

$$\sin\varphi^* = \left(\frac{n}{1+3n}\right)^{1/2} \qquad (8\text{-}50)$$

Equation 8–50 obviously is different from equation 8–44 when the flight width is neglected ($\hat{w} = 0$).

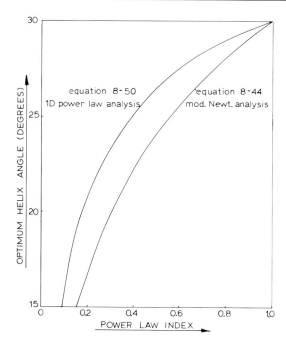

Figure 8–11. Optimum Helix Angle versus Power Law Index

Figure 8–11 compares the results of simultaneous optimization from the modified Newtonian analysis and the one-dimensional power law analysis.

It is clear that the one-dimensional power law analysis yields higher values of the optimum helix angle than the modified Newtonian analysis, except when the power law index is unity.

Figure 8–12a compares the results of simultaneous optimization from the modified Newtonian analysis and the two-dimensional power law analysis, obtained by numerical computations (2).

It can be seen that the values of the optimum helix angle from the modified Newtonian analysis are about 10 to 20 percent above those from the two-dimensional power law analysis. This indicates that the results from the modified Newtonian analysis, equation $8-44$, may be more appropriate than the results from the one-dimensional power law analysis. The use of the one-dimensional power law analysis leads to errors when the helix angle is substantially above zero, as discussed in Section 7.4.2. Therefore, one would like to use a two-dimensional power law analysis. However, there are no analytical solutions for this case. From numerical computations, it is possible to develop a plot of optimum helix angle as a function of a dimensionless axial pressure gradient g_a^0, where g_a^0 is:

$$g_a^0 = \frac{HP}{6mL} \left(\frac{H}{v_{bz}}\right)^n \qquad\qquad (8-51)$$

The plot of the optimum helix angle versus dimensionless axial pressure gradient is shown in Figure 8–12b.

It should be noted that the optimum helix angle in Figure 8–12b is not the result of simultaneous optimization of channel depth and helix angle, but optimization of the helix angle

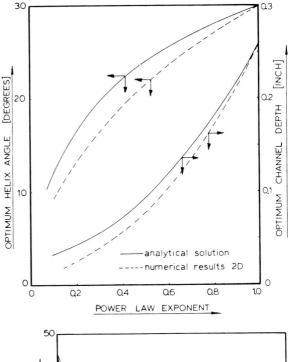

Figure 8 – 12a. Optimum Helix Angle versus Power Law Index

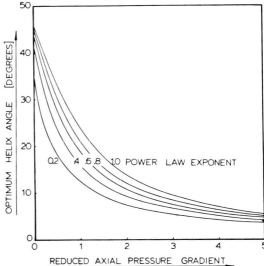

Figure 8-12b. Optimum Helix Angle versus Reduced Axial Pressure Gradient

only. Figure 8-13 shows the optimum helix angle versus dimensionless down channel pressure gradient as again determined from a two-dimensional analysis of a power law fluid.

Figure 8-13 can be compared directly to the results of the modified Newtonian analysis shown in Figure 8-9. It is quite clear that the modified Newtonian analysis underestimates the effect of the non-Newtonian behavior.

Another approach to optimization of the screw geometry for melt conveying can be made by using the Newtonian flow rate equation with correction factors for pseudoplastic behavior, see equation 7-218. If the flight width is neglected, equation 7-218 can be written as:

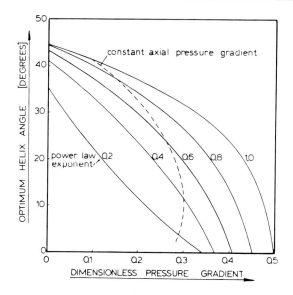

Figure 8-13. Optimum Helix Angle versus Dimensionless Down Channel Pressure Gradient

$$\dot{V} = \left(\frac{4+n}{10}\right) \pi^2 D^2 HN \sin\varphi\cos\varphi - \left(\frac{1}{1+2n}\right) \frac{\pi DH^3 g_a \sin^2\varphi}{4\mu} \qquad (8-52)$$

The optimum helix angle can be determined by taking the first derivative of output $\dot{V}$ with respect to the helix angle and setting the result equal to zero. This results in the following expression for the optimum helix angle φ^*:

$$\frac{\cos^2\varphi^* - \sin^2\varphi^*}{\cos\varphi^*} = \left(\frac{1}{1+2n}\right) \left(\frac{10}{4+n}\right) 3g_z^0 \cos\varphi^* \qquad (8-53)$$

The solution is:

$$\cos\varphi^* = \left[2 - \frac{30g_z^0}{(1+2n)(4+n)}\right]^{-1/2} \qquad (8-53a)$$

where the dimensionless down channel pressure gradient is given by:

$$g_z^0 = \frac{g_z H^2}{6\mu v_{bz}} \qquad (8-53b)$$

Figure 8-14a shows the optimum helix angle as determined from equation 8-53a as a function of the dimensionless pressure gradient at various values of the power law index.

It can be seen that the optimum helix angle is strongly dependent on the power law index. Comparison with results from the two-dimensional power law analysis, shown in Figure 8-13, indicates a reasonable agreement when the power law index is larger than one-half

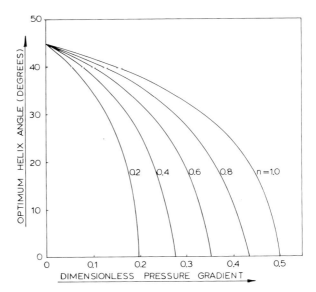

Figure 8-14a. Optimum Helix
Angle versus Dimensionless
Down Channel Pressure Gradient

($n > 0.5$), but relatively large differences when the power law index is smaller than one-half
($n < 0.5$).

If the optimum helix angle is expressed as a function of the axial pressure gradient, the optimum helix angle becomes:

$$\tan 2\varphi^* = \frac{2(1+2n)(4+n)\mu v_b}{5g_a H^2} \qquad (8-54)$$

If the reduced axial pressure gradient g_a^0 is defined as:

$$g_a^0 = \frac{H^2 g_a}{6\mu v_b} \qquad (8-54a)$$

then the optimum helix angle can be expressed as:

$$\varphi^* = 0.5 \arctan\left[\frac{(1+2n)(4+n)}{15 g_a^0}\right] \qquad (8-54b)$$

Similarly, the optimum channel depth can be determined to be:

$$H^* = \left[\frac{(4+n)(1+2n)2\mu v_{bz}}{15g_z}\right]^{1/2} \qquad (8-54c)$$

The optimum helix angle as a function of reduced axial pressure gradient is shown in Figure 8-14b.

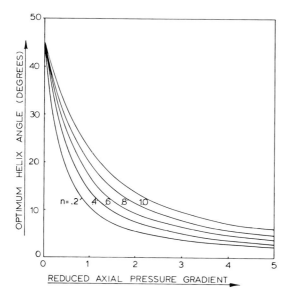

Figure 8–14b. Optimum Helix Angle versus Reduced Axial Pressure Gradient

Figure 8–14b can be compared directly to Figure 8–12b showing results from a two-dimensional power law analysis. The agreement between the two sets of results is quite reasonable.

Summarizing, it can be concluded that for simultaneous optimization of the channel depth and helix angle, the modified Newtonian analysis yields reasonably accurate results when compared to the two-dimensional power law analysis. The important equations are 8–39 for the optimum helix angle and 8–40 for the optimum channel depth. The results of simultaneous optimization from a one-dimensional power law analysis are less accurate than the modified Newtonian analysis.

For optimization of just the helix angle or the channel depth, the results from the Newtonian flow rate equation with correction factors for pseudoplastic behavior, equation 8–54, are more accurate than the results from the modified Newtonian analysis, equation 8–36. It should again be noted that simultaneous optimization only makes sense for relatively large positive pressure gradients. When the pressure gradient is negative, the output increases monotonically with channel depth. Thus, in this case, there is no optimum channel depth.

8.2.2   OPTIMIZING FOR PLASTICATING

The geometry for optimum melting performance can be determined from the equations developed in Section 7.3. A convenient relationship to use is the total axial melting length. It is convenient because the effect of varying the helix angle can be evaluated directly, whereas the total down channel melting length has to be converted to axial melting length for a one-to-one comparison. The total axial melting length can be determined from equation 7–116:

$$L_T = \frac{H_f v_{sz} \rho_s \sqrt{W_1} \sin\varphi}{\Omega_1} \left(2 - \frac{A_1 v_{sz} \rho_s \sqrt{W_1} \sin\varphi}{\Omega_1} \right) \qquad (8-55)$$

The initial solid bed width W_1 is:

$$W_1 = \frac{\pi D \sin\varphi}{p} - w \qquad (8\text{-}56)$$

The solid bed velocity can be expressed as:

$$v_{sz} = \frac{\dot{M}}{\rho_s H_f(\pi D \sin\varphi - pw)} \qquad (8\text{-}57)$$

The term Ω_1 is given by equation 7-109d; it can be written as:

$$\Omega_1 = \left[\frac{[2k_m\Delta T_b + \mu(v_b^2 + v_{sz}^2 - 2v_b v_{sz}\cos\varphi)]\rho_m v_b \sin\varphi}{2\Delta H}\right]^{1/2} \qquad (8\text{-}58)$$

The term A_l in equation 8-55 is the compression in axial direction. A_l is related to the compression in down channel direction A_z by:

$$A_z = A_l \sin\varphi \qquad (8\text{-}59)$$

With these equations, the effect of various geometrical variables can be determined. Figure 8-15 shows the total axial melting length as a function of the helix angle at various values of the flight width.

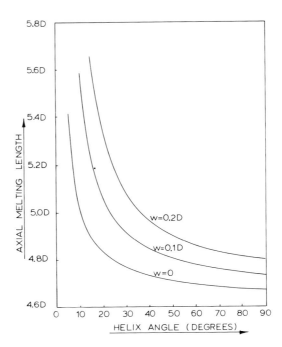

Figure 8-15. Axial Melting Length versus Helix Angle

The results shown in Figure 8-15 are predictions for a 50-mm extruder running at a screw speed of 100 rpm with an output of 100 kg/hr.; the barrel temperature is set 50°C above the melting point of the polymer. The channel depth in the feed section is 5 mm, the axial compression $A_l = 0.008$, and the number of flights $p = 1$. The following polymer properties were used:

Melt density	$\rho = 7800 \text{ kg/m}^3$
Thermal conductivity	$k_m = 0.25 \text{ J/ms}°\text{K}$
Melt viscosity	$\mu = 1500 \text{ Ns/m}^2$
Enthalpy change	$\Delta H = 4.5\text{E5 J/kg}$

These properties are typical of polyethylene, see Table 6-1.

8.2.2.1   EFFECT OF HELIX ANGLE

From Figure 8-15, it can be seen that the melting length reduces strongly with helix angle at relatively small angles. At larger helix angles, the melting length reduces weakly with helix angle. Very little improvement is obtained by increasing the helix angle beyond 30°. The melting length increases as the flight width increases, particularly at small helix angles. It is clear from Figure 8-15 that there is no optimum helix angle for which the axial melting length reaches a minimum. The shortest melting length is obtained at a helix angle of 90°; however, the melting length at 90° is only about 3 percent less than the melting length at 30° and about 8 percent less than the melting length at 17.66° (square pitch). Considering that a helix angle of 90° does not produce any forward drag transport, it makes little sense to apply such an extreme helix angle for the sake of melting.

8.2.2.2   EFFECT OF MULTIPLE FLIGHTS

From equations 8-55 through 8-58, the effect of the number of parallel flights can be evaluated directly. Figure 8-16 shows the axial melting length versus helix angle for a single flighted, double flighted, and triple flighted screw design.

The predictions are for the same example used earlier in this section. It is clear from Figure 8-16 that the melting length can be reduced substantially when the number of flights is increased, provided the helix angle is sufficiently large. It can be easily shown that the melting length with p parallel flights L(p) is related to the melting length with one flight L(1) by:

$$L(p) = \frac{L(1)}{\sqrt{p}} \qquad\qquad (8-60)$$

Expression 8-60 is valid for zero compression screws ($A_l = 0$) when the flight width w is negligible; however, it is reasonably accurate even if $A_l \neq 0$. Figure 8-16 shows curves with flight width $w = 0$ and $w = D/10$ for each number of flights. This is shown to demonstrate that the effect of flight width becomes more pronounced as the number of flights is increased. Substantial errors can be made if the flight width is neglected in a multi-flighted screw geometry. Figure 8-16 also shows that the effect of the helix angle on melting length becomes stronger when the number of flights is increased. With three parallel flights, the melting length at 90° is about 20 percent less than at 17.66° (square pitch), while in the single flighted design the difference is only about 8 percent.

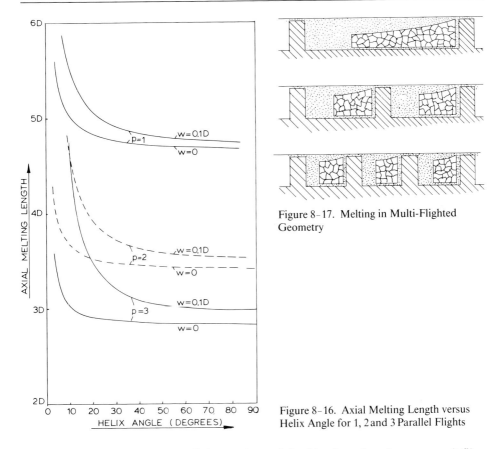

Figure 8–17. Melting in Multi-Flighted Geometry

Figure 8–16. Axial Melting Length versus Helix Angle for 1, 2 and 3 Parallel Flights

The beneficial effect of multiple flights can be explained by the reduced average melt film, see Figure 8–17.

As the number of flights is increased, the width of the individual solid beds becomes smaller. As a result, the average melt film thickness reduces, because the melt film thickness increases with the solid bed width. The thinner melt film will increase the viscous heat generation in the melt film, causing improved melting performance.

For optimum melting performance, the most favorable geometry is a multi-flighted design with relatively narrow flights and relatively large helix angle, around 25° to 30°. This combination of screw geometry variables is important because a multi-flighted geometry with a small helix angle may actually reduce the melting performance.

The effect of helix angle, flight width, and number of flights can also be analyzed in terms of the volumetric efficiency of the extruder screw. This is defined simply as the channel volume divided by the total volume of channel and flight. The volumetric efficiency ε_V can be expressed as:

$$\varepsilon_V = 1 - \frac{pw}{\pi D \sin\varphi} \tag{8–61}$$

Figure 8–18 shows the volumetric efficiency as a function of the helix angle at various numbers of flights.

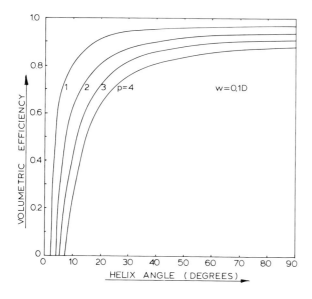

Figure 8–18. Volumetric Efficiency versus Helix Angle at Various Numbers of Flights (w = 0.1 D)

The volumetric efficiency rises sharply at small helix angles and more slowly at large helix angles. The highest ε_v is obtained when the helix angle is 90°. When the number of parallel flights is increased at a small helix angle, the volumetric efficiency drops considerably. When $\varphi = 17.66°$ (square pitch), the volumetric efficiency is about 85 percent at p = 1, but it drops to about 50 percent at p = 4. When $\varphi = 30°$, the volumetric efficiency is about 95 percent at p = 1 and drops to about 75 percent at p = 4. Thus, the adverse effect of multiple flights on volumetric efficiency is much less severe at large helix angles than it is at small helix angles.

8.2.2.3   EFFECT OF FLIGHT CLEARANCE

The melting rate for a non-zero clearance can be determined from equation 7–107; it can be written as:

$$\frac{d\dot{M}_p}{dz'} = \frac{1}{2}\rho_m \Delta v H_m (W_s') - \frac{1}{2}\rho_m \Delta v \, \delta \tag{8–62}$$

Equation 8–62 expresses the melting rate per unit length z', where z' is the direction of the relative velocity between the solid bed and barrel. The melting rate per unit down channel distance z can be written as:

$$\frac{d\dot{M}_p}{dz} = \left[\Omega_1^2 \, W_s + \left(\frac{1}{2}\rho_m \, \delta v_b \sin\varphi\right)^2\right]^{1/2} - \frac{1}{2}\rho_m \, \delta v_b \sin\varphi \tag{8–63}$$

From equation 8–63, it is clear that the melting rate reduces with increasing clearance. Figure 8–19a shows the melting rate as a function of radial clearance for the example used earlier, when $\varphi = 17.66$ degrees and $W_s = 1/2\,D$.

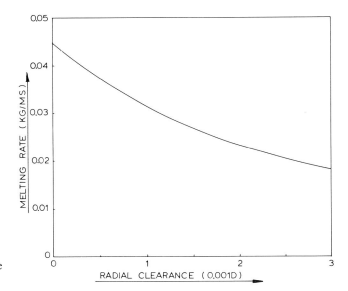

Figure 8-19a. Melting Rate versus Radial Clearance

The melting rate drops monotonically with increasing radial clearance. Considering that the standard clearance is 0.001 D; a doubling of the standard clearance causes a reduction in melting rate of about 25 percent. A tripling of the standard clearance causes a reduction in melting rate of about 35 percent. It is clear that wear in the plasticating zone of the extruder has a detrimental effect on the melting performance. It is important, therefore, to make sure that the radial clearance in this region is within reasonable limits. Unfortunately, screw and barrel wear often occurs in the plasticating zone of the extruder, as discussed in Section 8.1.3. This type of wear will adversely affect the melting performance and thus reduce over-all extruder performance. Symptoms of this can be temperature non-uniformities and throughput and pressure fluctuations. If these problems occur, it is good practice to check for screw and barrel wear. If the clearance is more than two or three times the standard clearance, the screw and/or barrel should be replaced.

Wear in the compression section of the screw is often caused by large compression ratios. This will be discussed next.

8.2.2.4   EFFECT OF COMPRESSION RATIO

The compression of the channel depth tends to widen the solid bed, as discussed in Section 7.3. Melting tends to reduce the width of the solid bed. If compression is too rapid, the melting may be insufficient and the solid bed can grow in width. This will generally cause plugging of the channel by the solid bed and should be avoided if at all possible. Plugging will cause output fluctuations, but it may also cause wear in the compression section of the screw, as discussed in Section 8.1.3. It was discussed in Section 7.3 that plugging can be avoided if:

$$\frac{\Omega_1}{A_z v_{sz} \rho_s \sqrt{W_1}} > 1 \qquad\qquad (8\text{--}64)$$

The compression ratio X_c is the channel depth in the feed section divided by the channel depth in the metering section. The axial length of the compression section is L_c. Thus, in or-

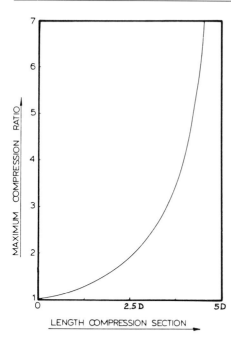

Figure 8–19 b. Minimum Length of the Compression Section versus Compression Ratio

der to avoid plugging, the length of the compression section should obey the following inequality:

$$L_C > \frac{v_{sz}\rho_s\sqrt{W_1}\, H_f(X_C - 1)\sin\varphi}{\Omega_1 X_C} \qquad (8-65)$$

From expression 8–65, the minimum allowable length of the compression section can be determined if the compression ratio is known. Figure 8–19 b shows the minimum L_C versus the compression ratio for the example used earlier.

If a large compression ratio is used, the length of the compression section must be quite long to avoid plugging.

The most dangerous combination in design of the compression section is a large compression ratio and a short compression section length. This will lead very easily to plugging. Rapid compression screws should be avoided for this very reason. One of the myths in screw design is that certain polymers, e. g., nylon, require a very rapid compression screw in order to extrude properly. Many screws have been designed with a compression length of less than one diameter. Such a screw can only work if the majority of the melting occurs before the compression section. However, this defeats the purpose of having a compression section in the first place. Thus, from a functional point of view, rapid compression screws do not make much sense, as they are susceptible to surging and wear. It is important to realize that there is no polymer that requires a very rapid compression screw in order to extrude properly. The benefits of very rapid compression screws are imaginary and based on a serious misconception. Obviously, a very rapid compression is quite possible in the second stage of a two-stage extruder screw because essentially all the melting will have taken place in the first stage, see Section 8.5.2.

8.2.3   OPTIMIZING FOR SOLIDS CONVEYING

Optimization of the solids conveying process is very important because solids conveying is the basis of the entire plasticating extrusion process. If instabilities occur in the solids conveying zone, these instabilities will transmit to the downstream zones and cause fluctuations in output and pressure. It was already discussed in Section 7.2.2.2 that grooved barrel sections provide a powerful tool to improve solids conveying rate and stability. There are, however, some important considerations with respect to screw design to optimize the solids conveying process.

8.2.3.1   EFFECT OF CHANNEL DEPTH

It was discussed in Section 7.2.2 that there appears to be an optimum channel depth for which the solids conveying rate reaches a maximum. At low values of the pressure increase over the solids conveying section, this optimum channel depth is indeed apparent because this optimum channel depth does not occur when the channel curvature is taken into account. At higher values of the pressure increase, however, there is an actual optimum channel depth even when the channel curvature is taken into account. This is shown in Figure 8-20 for a 114-mm (4.5-inch) extruder running at 60 rpm; the coefficient of friction is 0.5 on the barrel and 0.3 on the screw. When the pressure gradient increases, the optimum channel depth decreases.

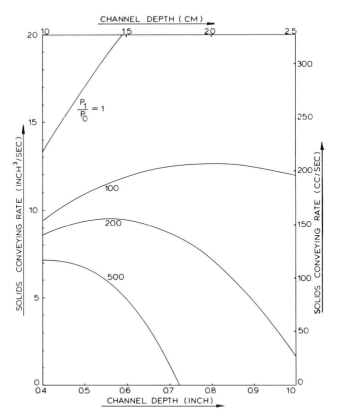

Figure 8-20. Solids Conveying Rate versus Channel Depth

The optimum channel depth can be obtained by taking the first derivative of the solids conveying rate $\dot{M}_S$ with respect to the channel depth H and setting the result equal to zero:

$$\frac{\partial \dot{M}_S}{\partial H} = 0 \qquad (8-66)$$

Equation 8-66 does not have a simple closed form solution. The optimum channel depth can be evaluated by using a numerical or graphical method. The optimum channel depth will increase with the coefficient of friction on the barrel. It will decrease with the coefficient of friction on the screw, the number of flights, and the pressure gradient. Unfortunately, the actual coefficients of friction are generally not known; thus, actual prediction of the optimum channel depth is usually not possible. The channel depth in the feed section is often about 0.10D to 0.15 D.

8.2.3.2   EFFECT OF HELIX ANGLE

The helix angle in the feed section will also have an optimum value for which the solids conveying rate reaches a maximum. This is obvious if one realizes that a zero degree helix angle results in zero rate and a ninety degree helix angle also results in zero rate. Thus, somewhere between zero and ninety degrees, the solids conveying rate will reach a maximum. The optimum helix angle can be determined from:

$$\frac{\partial \theta}{\partial \varphi} + \frac{\tan \theta}{\tan \varphi} = \frac{\tan \theta}{\tan(\theta+\varphi)} \left(1 + \frac{\partial \theta}{\partial \varphi} \right) \qquad (8-67)$$

where:

$$\frac{\partial \theta}{\partial \varphi} = \frac{1}{\cos(\theta+\varphi)} \frac{\partial}{\partial \varphi} \left[\frac{(1+f_S^2-k^2)^{1/2}-f_S k}{1+f_S^2} \right] -1 \qquad (8-68)$$

Equation 8-67 does not have a convenient analytical solution. Thus, the optimum helix angle can be determined by using a numerical or graphical method. Again, the coefficients of friction need to be known to determine the optimum helix angle. Therefore, accurate prediction of the optimum helix angle is usually not possible. In most extruder screws, the helix angle in the feed section ranges from 15° to 25°, with the most common angle being 17.66°.

8.2.3.3   EFFECT OF NUMBER OF FLIGHTS

The effect of the number of parallel flights in the feed section has already been discussed in Section 7.2.2. Increasing the number of parallel flights reduces the open cross-sectional area of the channel, see Figure 8-18, and increases the area of contact between the solid bed and screw. Both of these factors have a negative impact on the solids conveying performance, particularly when the helix angle is relatively small. The volumetric efficiency shown in Figure 8-18 is the open cross-sectional area relative to the total area between the root of the screw and the barrel.

8.2.3.4   EFFECT OF FLIGHT CLEARANCE

If the solid polymeric particles are compacted into a solid bed, there will be practically no leak flow through the flight clearance. In many cases, the flight clearance is smaller than the particle size of the polymer. Thus, even if the polymer particles are not fully compacted, the actual radial clearance will generally not be too critical.

8.2.3.5   EFFECT OF FLIGHT GEOMETRY

Dekker (3) studied the effect of various flight geometries on solids conveying performance. He proposes that many extrusion instabilities may be due to internal deformation of the solid bed. Internal deformation is more likely to occur when the internal coefficient of friction of the polymer particles is low. Spherical particles tend to have a lower internal coefficient of friction than non-spherical (e. g. cylindrical) particles and are, therefore, more susceptible to internal solid bed deformation. This may explain the often observed difference in extrusion behavior between strand pelletized and die-face pelletized material.

Dekker compared a trapezoidal flight geometry to a standard rectangular flight geometry. He found that the trapezoidal geometry resulted in higher throughput and more stable extrusion performance, particularly at a high screw speed. The trapezoidal flight geometry is also better from a stress distribution point of view as discussed in Section 8.1.2, see Figure 8–4a. It is expected that the curved flight geometry shown in Figure 8–4b will give similar benefits in terms of solids conveying as the trapezoidal flight geometry shown in Figure 8–4a, see also Section 7.2.2. Another important benefit of these flight geometries is that the contact area between the solid bed and the screw is reduced. This should have an additional positive effect on solids conveying performance.

8.3   OPTIMIZING FOR POWER CONSUMPTION

In some cases, optimizing for power consumption may be more important than optimizing the output. This is the case if the power consumption is excessive, causing high stock temperatures and increasing the chance of degradation. Some polymers have inherent properties that result in high power consumption. A typical example is linear low density polyethylene, LLDPE. The cause of its high power consumption is generally high polymer melt viscosity. The problem can be particularly severe if the viscosity remains high at high shear rates, i. e., if the polymer is not very shear thinning (large power law index). The most difficult polymers with respect to power consumption are low melt index polymers (fractional M. I.) with relatively large power law indici. Such polymers require special attention to the screw design in order to minimize the power consumption.

The screw geometry that will result in the least amount of power consumption can be determined from the expressions of power consumption developed in Chapter 7. In this section, attention will be focused on the melt conveying zone. However, other functional zones can be analyzed by the same procedure. Equations for power consumption in melt conveying are given in Section 7.4.1.3, equations 7–178 through 7–183. When optimizing for power consumption, one should be concerned about the power consumption at a certain level of throughput. Minimizing power consumption without considering the throughput does not make much sense. The power consumption is optimum when the ratio of power consumption to throughput reaches a minimum. This ratio is the specific energy consumption, SEC. It is the mechanical energy expended per mass of material and is usually expresed in kWhr/kg or hphr/lb. A high level of SEC translates into a large amount of energy expended per mass of polymer; this will result in large temperature increases in the polymer and possibly degradation. Thus, the optimum screw geometry for power consumption is that geometry for which the ratio of power consumption to output reaches a minimum.

8.3.1   OPTIMUM HELIX ANGLE

The optimum helix angle φ^* for power consumption can be determined by setting:

$$\frac{\partial Z}{\partial \varphi} \dot{V} = \frac{\partial \dot{V}}{\partial \varphi} Z \tag{8-69}$$

To evaluate the first derivative of power consumption Z with respect to helix angle φ, the power consumption has to be written as an explicit function of the helix angle. Using equations 7–178 and 7–181, the following expressions can be derived:

$$Z = a_1 + a_2 \sin^2 \varphi + a_3 \sin\varphi\cos\varphi + \frac{a_4}{\sin\varphi} \tag{8-70}$$

where:

$$a_1 = \frac{p\mu\pi Dv_b^2 dl}{H} \tag{8-71a}$$

$$a_2 = \frac{3p\mu\pi Dv_b^2 dl}{H} \tag{8-71b}$$

$$a_3 = \frac{p\pi Dv_b Hg_a dl}{2} \tag{8-71c}$$

$$a_4 = \frac{p\mu_{cl}wv_b^2 dl}{\delta} \tag{8-71d}$$

To evaluate the first derivative of output $\dot{V}$ with respect to the helix angle φ, the output has to be written as an explicit function of the helix angle. By using equation 7–175, the output can be written as:

$$\dot{V} = b_1 \sin\varphi\cos\varphi + b_2 \sin^2\varphi \tag{8-72}$$

where:

$$b_1 = \frac{1}{2}p\pi Dv_b (H-\delta) - \frac{(\pi D)^2 + (pb)^2}{12\mu_{cl}bp} g_a \delta^3 \tag{8-73a}$$

$$b_2 = -\frac{p\,\pi DH^3 g_a}{12\mu} - \frac{\pi^2 D^2 \delta^3 \mu V b}{2H^2 \mu_{c1}b} \qquad (8\text{-}73b)$$

The optimum helix angle now has to be found from the following equation:

$$\left(a_2\sin2\varphi + a_3\cos2\varphi - a_4\frac{\cos\varphi}{\sin^2\varphi}\right)\left(\frac{1}{2}\,b_1\sin2\varphi + b_2\sin^2\varphi\right) =$$

$$\left(b_1\cos2\varphi + b_2\sin2\varphi\right)\left(a_1 + a_2\sin^2\varphi + a_3\sin\varphi\cos\varphi + \frac{a_4}{\sin\varphi}\right)$$

$$(8\text{-}74)$$

This equation does not have an obvious simple, analytical solution. One can find the solution either graphically or by using a numerical technique, such as the Newton-Raphson method. Figure 8-21 shows the optimum helix angle as a function of channel depth for a 50-mm extruder, running at 100 rpm.

The flight width is 0.1 D, the radial clearance 0.001 D, and the number of flights is one. The power law index of the polymer melt is 0.5, the consistency index $m = 1E4$ Pa-sec^n, and the

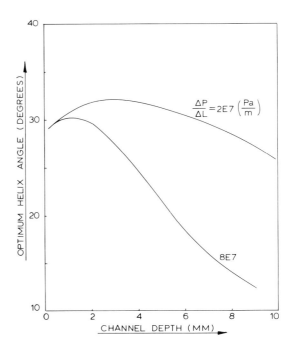

Figure 8-21. Optimum Helix Angle versus Channel Depth

melt density 0.8 gr/cc. It can be seen from Figure 8-21 that the optimum is relatively insensitive to the changes in the channel depth when the pressure gradient is moderate. However, at relatively high pressure gradients, the optimum helix angle depends strongly on the channel depth; at large channel depth values, the optimum helix angle becomes quite small.

Figure 8-22 shows the actual specific energy consumption, SEC, at the optimum helix angle as a function of channel depth.

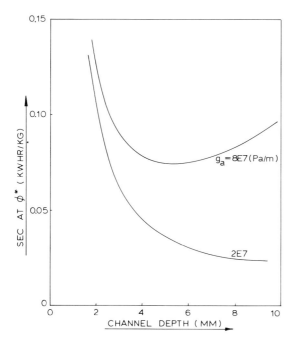

Figure 8-22. SEC at Optimum Helix Angle versus Channel Depth

The SEC at a small channel depth is quite high, but reduces with increasing channel depth. When the pressure gradient is high, the SEC reaches a minimum and starts to increase when the channel depth is further increased. Figure 8-22 thus illustrates a graphical method to determine the optimum channel depth for power consumption when both the channel depth and the helix angle are optimized simultaneously. It can further be seen in Figure 8-22 that the higher pressure gradient causes a substantially higher specific energy consumption. In the normal channel depth range, $H = 0.10D - 0.05D$, the difference in SEC is about 50 percent. This will improve the mixing efficiency of the metering section, but it will also cause more viscous heat generation in the melt and may overheat the polymer.

The optimum channel depth H^* can be calculated from:

$$\frac{\partial Z}{\partial H} \dot{V} = \frac{\partial \dot{V}}{\partial H} Z \tag{8-75}$$

This results in another lengthy expression similar to equation 8-74. This expression is difficult to solve analytically and is generally solved numerically or graphically, as shown in Figure 8-22.

8.3.2   EFFECT OF FLIGHT CLEARANCE

The effect of the radial clearance on the optimum helix angle is shown in Figure 8-23 for two values of the pressure gradient.

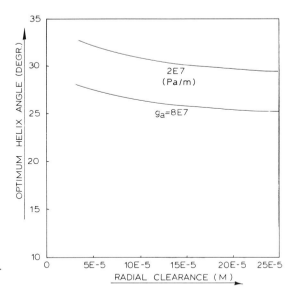

Figure 8-23. Optimum Helix Angle versus Radial Clearance

It can be seen that the radial clearance has relatively little effect on the optimum helix angle. The corresponding SEC as a function of clearance is shown in Figure 8-24.

It can be seen that there is an optimum value of the clearance δ^* for which the SEC reaches a minimum value. For the large pressure gradient, the optimum clearance $\delta^* \simeq 20$ E-5 m and for the small pressure gradient, $\delta^* \simeq 30$ E-5 m. This is interesting when one considers that the standard radial clearance is $\delta = 5$E-5 m (0.001 D). Thus, the standard radial clearance is not necessarily the best clearance in the metering section of the extruder. However, as discussed in Section 8.2.2, the radial clearance in the plasticating zone should be as small as possible to enhance the melting capacity. This indicates that a varying clearance along the length of the extruder may not be a bad idea. This can be achieved by varying the barrel O. D. and/or the screw diameter. Unfortunately, wear usually occurs right at a location where it should not happen: in the compression section. If wear would occur in the metering section, it could actually have a beneficial effect.

The optimum radial clearance can be determined from:

$$\frac{\partial Z}{\partial \delta} \, \dot{V} = \frac{\partial \dot{V}}{\partial \delta} \, Z \tag{8-76}$$

Again, the resulting equation cannot be easily solved analytically. Thus, the optimum clearance can be found by solving equation 8-76 numerically or graphically.

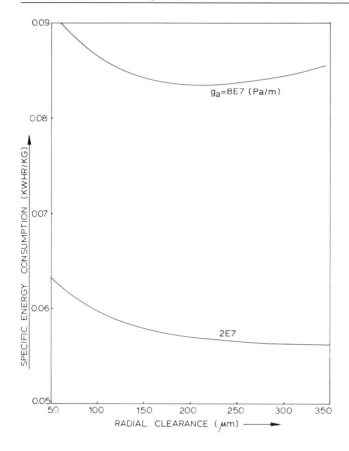

Figure 8-24. SEC at Optimum Helix Angle versus Radial Clearance

8.3.3 EFFECT OF FLIGHT WIDTH

If the various contributions to power consumption are carefully examined, it can be seen that a substantial portion is consumed in the clearance between the flight tip and the barrel; see, for instance, equation 8-71 d. The power consumption in the clearance is inversely proportional to the radial clearance and directly proportional to the total flight width pw. However, the viscosity in the clearance will generally be lower than the viscosity in the channel since the polymer melt is pseudoplastic. Figure 8-25 shows how the ratio of power consumption in the clearance Z_{cl} to the total power consumption Z_t depends on the ratio of flight width w to channel width plus flight width W + w for a polymer with a power law index n = 0.5.

Figure 8-25 shows that the relative contribution of the power consumption in the clearance increases strongly when the flight width increases. A typical ratio of w/(W + w) is about 0.1; in the example shown in Figure 8-25 this corresponds to a power consumption in the clearance of about 40 percent of the total power consumption! The power consumed in the clearance does not serve any useful purpose. It does not aid in transporting the polymer forward, but causes a viscous heating of the polymer. Therefore, one would like to make the flight width as narrow as possible to reduce the power consumption in the clearance. The power consumption in the clearance will be more pronounced when the material is more Newtoni-

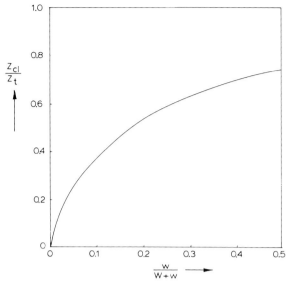

Figure 8-25. Z_{cl}/Z_{total} versus w/(W+w)

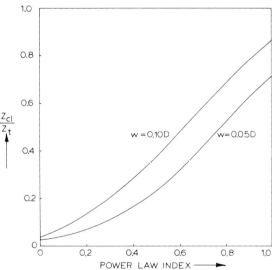

Figure 8-26. Z_{cl}/Z_{total} versus Power Law Index

an in flow behavior, i.e., less shear-thinning. This is shown in Figure 8-26, where Z_{cl}/Z_t is plotted against the power law index of the polymer melt.

The viscosity in the clearance is determined from:

$$\mu_{cl} = m \left(\frac{v_b}{\delta}\right)^{n-1} \tag{8-77}$$

The viscosity in the channel is determined from:

$$\mu = m \left(\frac{v_b}{H}\right)^{n-1} \tag{8-78}$$

The results shown in Figure 8-25 are for a standard radial clearance of 0.001 D. The contribution of the power consumption in the clearance rises dramatically when the power law index increases. When the fluid is Newtonian, around 80 percent of the total power is consumed in the clearance! Thus, problems with excessive power consumption are more likely to occur when the material of a certain melt index is less shear thinning. This appears to be the main reason behind the extrusion problems encountered with linear low density polyethylene, LLDPE (4, 5).

The most logical way to reduce power consumption in the metering section is to reduce the ratio $w/(W+w)$. This can be achieved in two ways. One is to increase the channel width W by increasing the helix angle, as discussed in Section 8.3.1. The other approach is to reduce the flight width itself. The combination of increasing the helix angle and decreasing the flight width is obviously most effective. This allows a reduction of the $w/(W+w)$ ratio from a typical value of 0.1 down to around 0.03. The minimum flight width is not determined by functional considerations but by mechanical considerations. Functionally, one would like the flight width to be almost infinitely thin. However, there must obviously be sufficient mechanical strength in the flight to withstand the forces acting on it. These mechanical considerations are discussed in detail in Section 8.1.2. The flight width in the metering section w_m can be considerably narrower than the flight width in the feed section w_f because the flight height or channel depth is much smaller in the metering section. In order to keep the mechanical stresses in the flight approximately the same, the following rule can be used to determine the flight width in the metering section:

$$w_m = w_f \left(\frac{H_m}{H_f}\right)^{1/2}$$

(8-79)

Considering that the flight width in the feed section is generally about 0.1 D, equation 8-79 can be written as:

$$w_m = \frac{w_f}{\sqrt{X_c}} = \frac{D}{10\sqrt{X_c}}$$

(8-80)

where X_c is the channel depth ratio H_f/H_m.

Based on these considerations, a new screw design was recently developed to reduce power consumption in materials like LLDPE; the screw is often referred to as the LL-screw (4, 5).

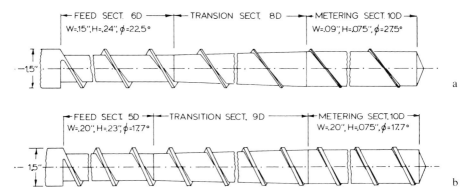

Figure 8-27. The LL-Extruder Screw (a) versus a Standard Screw (b)

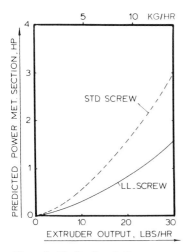

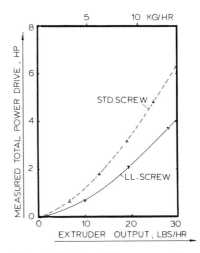

Figure 8-28. Predicted Output versus Power Consumption

Figure 8-29. Actual Output versus Power Consumption

Figure 8-27 shows a standard screw geometry and the LL-screw geometry. Both screws have a 38 mm (1.5 inch) diameter and a 24 L/D ratio.

Essentially, the only differences between the two screws is the helix angle and the flight width. Thus, the differences in performance between the two screws can be attributed solely to these two geometrical factors. Figure 8-28 shows the predicted output versus hp curves for the two screws shown in figure 8-27.

Figure 8-29 shows the actual output versus hp curves for the same two screws.

It is clear that the optimized geometry of the LL-screw results in considerably lower power consumption compared to the standard geometry. The drop in power consumption is around 35 to 40 percent! It is also interesting to note that the predicted power consumption, Figure 8-28, agrees quite well with the actual power consumption, Figure 8-29. The difference in the ordinates is due to the fact that Figure 8-28 is predicted power for the metering section only, while Figure 8-29 is actual total power consumption. A patent has been filed on the LL-screw (4) and the screw is manufactured and sold under license by Migrandy Corporation as the Migrandy LL-screw.

It should be noted that the benefits of a reduced flight width and increased helix angle are valid for the plasticating zone as well. By considering equation 7-149, it can be seen that the power consumption in the plasticating zone of the extruder is also reduced when the flight width is reduced. In Section 8.2.2, it was discussed that increasing the helix angle improves melting performance. Thus, the combination of increased helix angle and reduced flight width should have a beneficial effect not only on the melt conveying zone, but also on the melting zone of the extruder.

8.4   SINGLE FLIGHTED EXTRUDER SCREWS

In the previous sections of this chapter, screw design was analyzed by functional performance. By using the extrusion theory developed in Chapter 7, it was shown how the screw design can be determined quantitatively for optimum performance. In this section, screw design will be approached from another angle. Screw designs in use today will be described and their advantages and disadvantages will be discussed and analyzed.

8.4.1   THE STANDARD EXTRUDER SCREW

In many discussions on extrusion, reference is made to a so-called standard or conventional extruder screw. In order to define this term more quantitatively, the general characteristics of the standard extruder screw will be listed; see also Figure 8-30:

- total length 20D − 30D
- length of feed section 4D − 8D
- length of metering section 6D − 10D
- number of parallel flights 1
- flight pitch 1D (helix angle 17.66°)
- flight width 0.1D
- channel depth in feed section 0.10D − 0.15D
- channel depth ratio 2 − 4

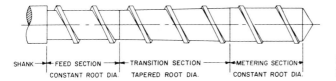

SHANK ─►│◄─ FEED SECTION ─►│◄─ TRANSITION SECTION ─────►│◄─METERING SECTION─►│
      │ CONSTANT ROOT DIA. │ TAPERED ROOT DIA. │ CONSTANT ROOT DIA.│

Figure 8-30. The Standard Extruder Screw

These dimensions are approximate, but it is interesting that the majority of the extruder screws in use today have the general characteristics listed above. For profile extrusion of PA, PC, and PBTB, Brinkschroeder and Johannaber (46) recommend a channel depth in the feed section $H_f \simeq 0.11\,(D+25)$ and a channel depth in the metering section $H_m \simeq 0.04$ $(D+25)$, where channel depth H and diameter D are expressed in mm. Based on these guidelines, the geometry of a standard extruder screw can be determined in a minimum length of time.

Based on the design methodology developed in Sections 8.2 and 8.3, it should be clear that the standard screw design is by no means an optimum screw design. It has developed over the last several decades mostly in an empirical fashion and works reasonably well with many polymers. However, significant improvements in performance can be made by functional optimization using extrusion theory. In this light, it is somewhat surprising that the standard extruder screw is still so popular today. It probably indicates a lack of awareness of the implications of extrusion theory on screw design and the improvements that can be realized from functional optimization of the screw geometry. Another interesting note is that several manufacturers of extruder screws claim to use sophisticated computer programs to optimize the screw geometry, but often still end up with a standard square pitch screw. It can be shown from an elementary analysis that the square pitch geometry is not optimum for melting or melt conveying. Thus, if the result of the screw optimization by computer is a square pitch geometry, this indicates that either the computer program is incorrect or the person using the program is not using it correctly.

8.4.2 MODIFICATIONS OF THE STANDARD EXTRUDER SCREW

There are a large number of modifications of the standard extruder screw in use today. It will not be possible to mention all of them, but an effort will be made to discuss the more significant ones. Figure 8–31 shows the standard screw with an additional flight in the feed section.

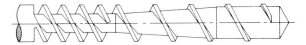

Figure 8–31. Standard Screw with Additional Flight in Feed Section

The additional flight is intended to smooth out the pressure fluctuation caused by the flight interrupting the in-flow of material from the feed hopper every revolution of the screw. An additional benefit of the double-flighted geometry is that the forces acting on the screw are balanced; thus, screw deflection is less likely to occur. On the negative side, the additional flight reduces the open cross-sectional channel area and increases the contact area between solid bed and screw. Thus, pressure surges may be reduced, but the actual solids conveying rate will be reduced as well.

Figure 8–32a shows a variable pitch extruder screw.

The varying pitch allows the use of the locally optimum helix angle, i.e., optimum helix angle for solids conveying in the feed section and optimum helix angle for melt conveying in the metering section of the screw. This design is covered by a U.S. patent (6) and described in a 1980 ANTEC paper (7). Figure 8–32b shows a variable pitch extruder screw as often used for rubber extrusion, see also Section 2.1.4.

In this design, the pitch decreases with axial distance as opposed to the screw shown in Figure 8–32a. The reducing pitch causes a lateral compression of the material in the screw channel; as a result, the normal compression from the reducing channel depth can be reduced or eliminated altogether. In fact, many of these variable reducing pitch screws maintain the same channel depth along the entire length of the screw.

It should be noted that the variable reducing pitch screw is not a high performance screw. It is designed primarily to exert minimal shear to the polymer; the L/D ratio is generally quite short, about 10. This screw is clearly not designed for high output.

A major supplier of LLDPE is recommending a variable reducing pitch (VRP) screw for extrusion with LLDPE (61). Considering the approach developed in Section 8.3, this screw design would seem inappropriate for LLDPE. As discussed in Section 8.3, power consumption can be reduced by increasing the helix angle and flight clearance and by reducing the flight width. The VRP screw recommended for LLDPE does not reduce the flight width and reduces the helix angle. This combination of screw design parameters should result in in-

Figure 8–32a. Variable Pitch Extruder Screw, Increasing Pitch

Figure 8–32b. Variable Pitch Extruder Screw, Reducing Pitch

creased power consumption instead of reduced power consumption. The reason, however, that this VRP screw works is that the clearance between flights and barrel is substantially larger than the normal design clearance, about double! This design feature is not much emphasized, however it is the key to the performance of the VRP screw for LLDPE because the power consumption in the flight clearance plays such an overriding role in LLDPE extrusion. Based on the arguments developed in Section 8.3, it is clear that a variable increasing pitch (VIP) screw with a larger flight clearance would be significantly better than the VRP screw. A disadvantage of the larger clearance is reduced melting capacity and reduced heat exchange between the polymer melt and barrel. As a result, the VRP screw with increased flight clearance will not be suitable for extrusion of polymers other than LLDPE.

Figure 8-33 shows an extruder screw without a metering section; the so-called zero-meter screw (8).

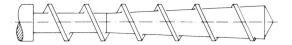

Figure 8-33. Zero-Meter Screw

This screw is more appropriate for a plasticating unit of an injection molding machine. The zero-meter screw is used to reduce the temperature build-up in the material by deepening the depth of the channel in the melt conveying zone of the extruder. The obvious drawback is that the pressure generating capability of the screw will be adversely affected, but this is not a major concern in injection molding applications. In other applications, however, the approach outlined in Section 8.3 is recommended. An extension of the zero-meter screw is the zero-feed zero-meter screw shown in Figure 8-34.

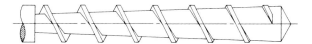

Figure 8-34. Zero-Feed, Zero-Meter Screw

This screw essentially consists of only a compression section. This allows a very gradual compression of the material. The screw has been in commercial use for many years and has been successfully used with many polymers; in particular, nylon.

The exact opposite of the zero-feed zero-meter screw is the very rapid compression screw, shown in Figure 8-35.

Figure 8-35. Rapid Compression Screw

The length of the compression section is generally less than 1 D in these screws. Unfortunately, this screw is often referred to as a nylon screw. This is unfortunate because it implies that nylon should be extruded on a rapid compression screw. However, this is a major misconception in screw design and the success of the zero-feed zero-meter screw with nylon should make that quite clear. Nylon has a relatively narrow melting range and turns into a relatively low viscosity melt quite readily. However, this does not mean that the compression should be very rapid. The maximum compression can be determined from equation 8-64. The relative width of the melting range of the polymer is totally immaterial to the determination of the maximum compression of the extruder screw. The low melt viscosity of nylon wil reduce the melting rate and it indicates that a gradual compression screw will be much more appropriate for nylon than a rapid compression screw. This was conclusively demonstrated as early as 1963 by Bonner (9), who found that the gradual compression screw reduced air entrapment, pressure and output fluctuations, and improved extrudate quality.

The zero-meter screw is used to reduce the viscous heat generation (power consumption) in the melt conveying zone of the extruder by having a relatively deep channel in this portion of the screw, with the depth reducing linearly with distance. Another similar approach is the decompression screw shown in Figure 8-36.

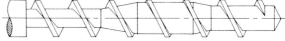

Figure 8-36. Decompression Screw

The final portion of the screw has a deep channel section following a decompression section. The channel depth is constant over the last screw section. Again, the deeper channel in the final screw section will reduce the pressure generating capability of the screw. A more effective power reduction can be obtained by changing not only the channel depth, but the channel depth, helix angle, flight width, and radial clearance in an optimum fashion as discussed in Section 8.3.

8.5   DEVOLATILIZING EXTRUDER SCREWS

Devolatilizing extruder screws are used to extract volatiles from the polymer in a continuous fashion. Such extruders have one or more vent ports along the length of the extruder through which volatiles escape. Some of the applications of vented extruders are:

• Removal of monomers and oligomers in the production of polymers (e.g., PS, HDPE, PP, etc.).

• Removal of reaction products of condensation polymerization (e.g., water, methanol, etc.) and oligomers from nylon and polyesters.

• Removal of air with filled polymers, particularly with glass fiber reinforced polymers.

• Removal of residual carrier fluid in emulsion and suspension polymerization (e.g., PS, PVC, etc.)

• Removal of water from hygroscopic polymers (e.g., ABS, PMMA, PA, PC, SAN, CA, PU, PPO, Polysulfone, etc.); all polymer particles can have surface moisture left from underwater pelletizing or surface condensation from storage at varying temperatures and relative humidity.

• Removal of solvent and unreacted monomers in solution polymerization (e.g., HDPE).

• Removal of volatile components in compounding of polymers with additives and other ingredients.

Removal of water from hygroscopic polymers is a common use of vented extruders. Most polymers require less than 0.2 percent moisture in order to properly extrude. In some polymers, this percentage is considerably lower, e.g., PMMA <0.1 percent, PA <0.1 percent, ABS <0.1 percent, CA <0.05 percent, PBTB <0.05 percent, and PC <0.02 percent. Many polymers have an equilibrium moisture content at room temperature and 50 percent R. H. (relative humidity) that is considerably higher than the maximum allowable percentage moisture content for extrusion. Some values of the equilibrium moisture content of hygroscopic polymers (10) are: ABS 1.5 percent, PMMA 0.8 percent, PBTP 0.2 percent, PC 0.2 percent, and PA 3 percent. Such polymers require significant drying or extrusion devolatilization to manufacture good products. In many cases, extrusion devolatilization is preferred over drying.

Conventional two-stage extruders can generally reduce the level of volatiles only a fraction of one percent. For example, for PP/Xylene with an initial solvent concentration of 0.3 to 1.0 percent, the amount of solvent removed by single vent extrusion is about 50 percent.

8.5.1 FUNCTIONAL DESIGN CONSIDERATIONS

Figure 8–37 shows a typical two-stage devolatilizing extruder screw.

The screw consists of at least five distinct geometrical sections. The first three sections: feed, compression, and metering, are the same as on a conventional screw. After the metering section there is a rapid decompression followed by the extraction section, which, in turn, is followed by a rapid compression and a pump section. Two important functional requirements for good devolatilization are zero pressure in the polymer under the vent port and completely molten polymer under the vent port.

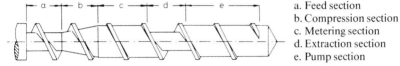

a. Feed section
b. Compression section
c. Metering section
d. Extraction section
e. Pump section

Figure 8–37. Typical Two-Stage Extruder Screw

The zero pressure requirement is to avoid vent flow, i. e., polymer melt escaping through the vent port. The complete fluxing requirement has several reasons. If the polymer is not completely molten in the metering section, there may not be a good seal between the vent port and the feed opening. This will limit the amount of vacuum that can be applied at the vent port. A good vacuum is generally quite important in order to obtain effective devolatilization. Another reason for the complete melting requirement has to do with diffusion coefficients. The devolatilization process in extruders is often controlled by diffusion (2), see also Section 7.6. Diffusion coefficients are very much temperature dependent. When the polymer is below the melting point, diffusion generally occurs at an extremely low rate. The polymer, therefore, should be above the melting point to increase the rate of diffusion and with it the devolatilization efficiency. Even when the polymer is in the molten state, the diffusion coefficients can often be increased substantially by increasing the temperature of the polymer melt (11). Further, when the polymer is in the molten state, surface renewal is possible. This greatly enhances the devolatilization process. The extent of surface renewal is a strong function of the screw design; a multiflighted, large pitch extraction section will be beneficial to the devolatilization efficiency. Thus, for the highest devolatilization effectiveness, the polymer should be completely molten and at relatively high temperature when it reaches the extrusion section. The complete melting requirement can be worked out by the procedure developed in Section 8.2.2.

The zero pressure requirement can be fulfilled by ensuring that the channel in the extraction section is only partially filled with polymer. There is no chance of pressure build-up, at least in the down channel direction, when the screw channel is not fully filled. In order to achieve this partial fill, the depth of the extraction section has to be considerably larger than the depth of the metering section, usually at least three times larger, and the transport capacity of the pump section must be larger than the transport capacity of the metering section. In other words, one has to make sure that the polymer can be transported away from the vent port at a rate at least as high as the rate with which it can be supplied to the vent section. If the transport capacity of the pump section is insufficient, the polymer melt will back up in the pump section and eventually escape through the vent port.

If the flight pitch is constant and the polymer melt viscosity can be described by equation 8–78, the maximum diehead pressure for effective devolatilization can be written as (2):

$$P_{max} = \frac{6v_b^n mL}{\tan\varphi} \left(\frac{H_p - H_m}{H_p^{n+2}}\right) \tag{8-81}$$

This equation was derived by assuming zero pressure gradient in the metering section and by using the Newtonian throughput-pressure relationship, equation 7–155. The optimum channel depth in the pump section H_p^* can be obtained by setting:

$$\frac{\partial P_{max}}{\partial H_p} = 0 \tag{8-82}$$

This results in the following expression for the optimum channel depth in the pump section:

$$H_p^* = \frac{(n+2)H_m}{n+1} \tag{8-83}$$

The ratio of depth in the pump section to depth in the metering section is often referred to as pump ratio X_p. The optimum pump ratio X_p^* according to equation 8–83 is only a function of the power law index; this is shown in Figure 8–38.

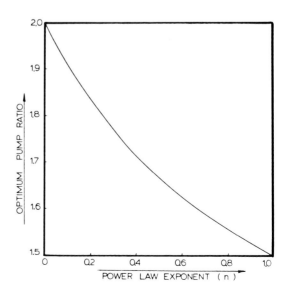

Figure 8–38. Optimum Pump Ratio versus Power Law Index

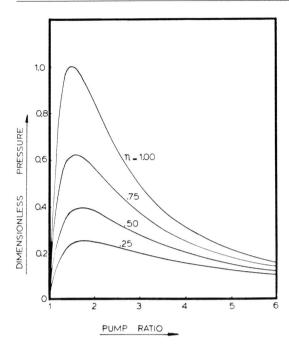

Figure 8-39. Dimensionless Maximum Pressure versus Pump Ratio

The pump ratio should increase when the power law index decreases. The practical lower limit is 1.5 and should be used for polymers with almost Newtonian flow characteristics. The upper limit of the pump ratio is 2.0 and should be used for polymers with very strong pseudoplastic flow behavior.

Strictly speaking, one cannot insert a power law melt viscosity in the Newtonian through-put-pressure relationship, as discussed earlier in Section 8.2.1. Thus, equations 8-81 and 8-83 are not 100 percent accurate. However, they are much more accurate than predictions based on pure Newtonian behavior, because the latter can cause substantial errors; see, for instance, Figures 7-32 through 7-34. The dimensionless maximum pressure is shown as a function of the pump ratio X_p in Figure 8-39.

The dimensionless maximum pressure is the actual maximum pressure divided by the peak maximum pressure for the Newtonian case ($X_p = 1.5$). The dimensionless maximum pressure can be expressed as:

$$P_{max}^0 = \frac{P_{max}}{P_{max}^*(n=1)} = \frac{27(X_p-1)}{4X_p^{n+2}} \left(\frac{V_b}{H_m}\right)^{1-n} \qquad (8-84)$$

Figure 8-39 shows clearly that the peak maximum pressure is highest for the Newtonian fluid and reduces steadily when the power law index reduces. This indicates that the pressure generating capacity reduces as the fluid becomes more pseudoplastic. At the same time, the optimum pump ratio increases with reducing power law index. From equation 8-84, it can be seen that the dimensionless maximum pressure also depends on the average shear rate in the screw channel. Figure 8-40 shows how the dimensionless maximum pressure varies with the pump ratio at several values of the average shear rate.

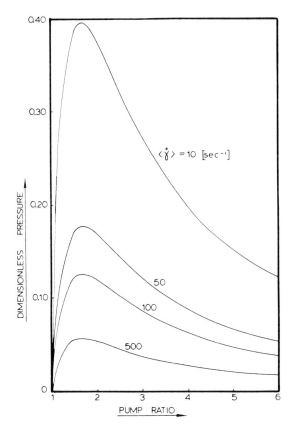

Figure 8-40. Dimensionless Pressure versus Pump Ratio When n = 0.5

It is evident from Figure 8-40 that increases in shear rate have a strong effect on the pressure generating capability. This can be seen clearly from the following relationship:

$$\frac{\overset{U}{P}_{max}(\dot{\gamma}_1)}{\overset{0}{P}_{max}(\dot{\gamma}_2)} = \left(\frac{\dot{\gamma}_1}{\dot{\gamma}_2}\right)^{1-n} = \left(\frac{N_1}{N_2}\right)^{1-n} \qquad (8-85)$$

where N is the screw speed.

In practical terms, this has important implications. It means that when the screw speed is increased, the pressure generating capability increases less than proportional to the screw speed. Thus, if the output increases approximately proportional to the screw speed, at some point the diehead pressure can exceed P_{max} causing vent flow. This will happen more readily when the material is more pseudoplastic, i.e., when the power law index is closer to zero.

8.5.2   VARIOUS VENTED EXTRUDER SCREW DESIGNS

There are many different designs of devolatilizing single screw extruders with widely differing devolatilization capacity. Some of the more common ones will be described and discussed next.

8.5.2.1   CONVENTIONAL VENTED EXTRUDER SCREW

The conventional vented extruder screw is shown in Figure 8–37. In many cases, mixing sections are incorporated into the metering section of the screw to improve the homogeneity of the melt entering the extraction section. The volatiles travel with the polymer up to the vent port. This type of venting is referred to as forward devolatilization. The length of the extraction zone is usually 2 D to 5 D. The channel depth in the extraction section is large, particularly if the polymer foams in the extraction section. The channel depth in the extraction section can be as large as 0.4D on large diameter extruders, 0.3D on smaller extruders. In order to achieve frequent surface renewal, the extraction section is often designed with multiple flights, see Figure 8–41 a.

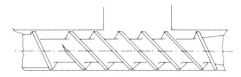

Figure 8–41 a. Extraction Section With Multiple Flights

For the same reason, the helix angle is often increased from the conventional 17.66° (square pitch) to as high as 40° (12).

As discussed in section 8.5.1, the optimum pump ratio ranges from 1.5 to 2.0. In practice, the pump ratio is often selected in the range from 1.2 to 1.4. However, these lower pump ratios make the extruder more susceptible to vent flow. Sometimes vent flow is avoided by starve feeding the extruder. However, if starve feeding is necessary to avoid vent flow, it indicates a deficiency in the screw design and it might be better to modify the screw or design a new one. Carley (13) recommends the use of rear valving to adjust the flow rate of the material entering the extraction section. This is shown schematically in Figure 8–41 b.

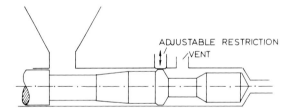

Figure 8–41 b. Rear Valving

Another approach was taken by Heidrich (56), who developed a vented extruder with axial adjustment capability of the screw. This allows variation of a conical gap at the end of the metering section; an example of this feature is shown in figure 8–45. In vented extruders without external adjustment capability of the first stage resistance, it is generally a good idea to incorporate a pressure consuming mixing element. This improves melt homogeneity and reduces the pressure of the melt entering the extraction section.

8.5.2.2   BYPASS VENTED EXTRUDER SCREW

Another method to control the flow rate into the extraction section is to use a bypass system. This system was proposed by Willert of Egan Machinery Company (14). Maddock and Matzuk (15) discussed the principles of the bypass vented extruder and described actual experiments with different screw geometries. The bypass vented extruder is shown schematically in Figure 8–42.

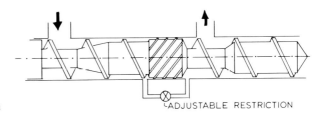

Figure 8-42. Bypass System Egan

The polymer melt is forced from the metering section into a bypass flow channel by incorporating a multiflighted screw section with reversed pitch and shallow channels between the metering section and the extraction section. The bypass channel has one or more adjustable restrictions to control the rate of flow into the extraction section. The polymer flows from the bypass channel into the beginning of the extraction section. Maddock and Matzuk concluded that bypass venting allows a wider range of operation, is less susceptible to instabilities, and is less operator-sensitive.

A different bypass venting system was described by Anders (16); it is shown in Figure 8-43.

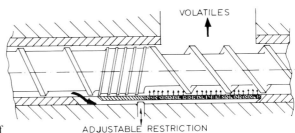

Figure 8-43. Bypass System Berstorff

The melt flows from the metering section into a bypass channel, which runs essentially parallel to the screw. The melt is forced into the bypass channel by a shallow multiflighted screw section of reversed pitch located just downstream of the bypass channel inlet. The bypass channel extends about 2 D into the extraction section. The polymer melt flows into the extraction section through a large number of holes. This increases the surface area generation and improves the devolatilization efficiency. The bypass channel has adjustable restrictions to control the flow of material to the extraction section. The devolatilization capability is further enhanced by using a multiflighted, large pitch design in the extraction section. The bypass system described by Anders is used in multi-stage vented extrusion.

8.5.2.3 REARWARD DEVOLATILIZATION

Rearward devolatilization is used on melt-fed extruders. In rearward devolatilization, the volatiles are extracted upstream of the feed opening of the extruder. This is shown schematically in Figure 8-44.

This machine is used in melt fed extrusion. The vent port is generally located at least 1 D from the feed opening to avoid polymer melt getting to the vent port. To improve the devolatilization capability, the melt is often forced into the extruder through numerous small holes, examples of this feature are shown in figure 8-45 and 8-48.

In order to avoid plugging of the vent port, a feedback control mechanism can be incorporated that controls the degree of fill of the extruder. This can be done by measuring the pres-

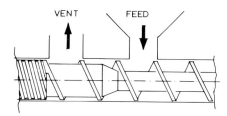

Figure 8-44. Rearward Devolatilization

sure at the beginning of the metering section and by using this reading to adjust the screw speed or feed rate to maintain the same pressure, and thus the same degree of fill.

8.5.2.4 MULTI-VENT DEVOLATILIZATION

Multi vent devolatilization is used when large amounts of volatiles need to be removed from the polymer. In a well-designed system, as much as 15 percent of volatiles can be removed in one extrusion operation. Figure 8-45 shows a schematic of a multi-stage system used for devolatilization of molten polystyrene.

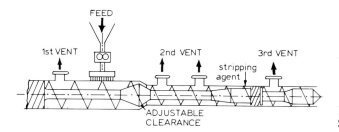

Figure 8-45. Efficient Multi-Stage Degassing System

This system incorporates rearward venting, stranding of the melt at the inlet to increase surface area, variable gap before the second vent port by axial screw adjustment, water injection, and bypass venting at the final vent port. Such a system can reduce the monomer level from 15 percent down to as low as 0.1 percent in one operation. Such devolatilization performance is quite comparable to that of twin screw devolatilization systems.

At high levels of volatiles, the initial devolatilization will be quite rapid because the process will occur primarily by a foam devolatilization, see Section 5.4.2. When the volatile level reduces, the devolatilization will occur by molecular diffusion, reducing the rate of devolatilization considerably. In this situation, the devolatilization can be greatly improved by injecting a stripping agent into the polymer. The stripping agent is generally introduced in or right before a mixing section. This causes foaming of the polymer at the extraction section, resulting in much improved devolatilization. A common stripping agent is water; also used are low boiling organics, or nitrogen. In some cases, the volatile and the stripping agent can form an azeotropic mixture which boils at a lower temperature than either of the components. An example is styrene monomers and water (17).

A more conventional three-stage extruder screw is shown in Figure 8-46.

This system has two vent ports and is used in applications where a single vent port cannot remove a sufficient amount of volatiles. It is used, for instance, in ABS extrusion as described by Brozenick and Kruder (18) and with acrylic, polycarbonate, polypropylene, etc.,

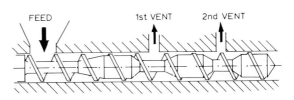

Figure 8-46. Conventional Three-
Stage Degassing System

as described by Nichols, Kruder, and Ridenour (19). This system can remove moisture levels as high as 5 to 7 percent.

8.5.2.5   CASCADE DEVOLATILIZATION

In many polymer devolatilization systems, two extruders are arranged in a cascade arrangement. The first extruder is primarily used for solids conveying, plasticating, and mixing. The second extruder is primarily used for melt conveying, i.e., for pumping. The first extruder is often a multi-screw extruder; the second extruder is generally a single screw extruder. Figure 8-47 shows a planetary gear extruder feeding a single screw extruder.

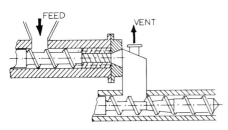

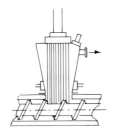

Figure 8-47. Cascade Devolatilization with a
Planetary Gear Extruder

Figure 8-48. Stranding for Improved Degassing

The venting takes place between the first and second extruder. The system shown in Figure 8-47 is often used for devolatilization of PVC. The devolatilization effectiveness can be improved by stranding the polymer melt as it enters the second extruder. This is shown in Figure 8-48.

The distance of the strand die to the second extruder is made reasonably long to improve devolatilization. The distance is limited by the fact that the strands cannot cool below the point where the intake of the second extruder is affected.

The major advantage of the cascade devolatilizing system is that the control of the output of the first stage to the pressure generating capability of the second stage is much better than in a single extruder devolatilization system. Obviously, the cost will be higher and the decision for one system or the other must be based on the importance of improved flexibility and controllability.

8.5.2.6   VENTING THROUGH THE SCREW

An interesting development in devolatilizing extrusion was described by Bernhardt in 1956 (20). In this extruder, the volatiles are removed through the screw instead of through a vent port in the barrel. The screw has a hollow core connecting with a lateral hole in the extraction section of the screw, see Figure 8-49a.

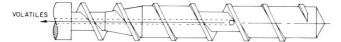

Figure 8-49a. Venting Through the Screw

The volatiles are withdrawn through a rotary union at the rear of the screw. This venting was tested in practice on acrylic and was found to perform reliably for extensive periods of time.

An obvious advantage of this approach is that venting can be done on an extruder not equipped with a vent port in the barrel. An equally obvious disadvantage is that plugging of the vent channel in the screw may cause a complete shut-down. The plug may be removed by a blast of high pressure air into the core of the screw. However, if this does not work, the screw has to be pulled and cleaned. Plugging of the vent channel in the screw, however, may not be as much of a problem as one might think. It should be remembered that the polymer has to adhere to the barrel in order to move forward; however, it does not have to adhere to the screw. Thus, a vent port in the barrel is much more likely to accumulate molten polymer than a vent port in the screw, particularly if the vent port is located close to the trailing flight flank. This type of venting, however, does not seem to have found widespread acceptance.

8.5.2.7   VENTING THROUGH A FLIGHTED BARREL

Kearney and Hold (62) proposed a new devolatilizer with helical flights in the barrel and a smooth screw section (rotor), see Figure 8-49b. This device is called a rotating drum devolatilizer (RDD).

The barrel has multiple helical flights as shown in Figure 8-49b. Each turn of the helical channels has an oblong opening following the helical path of the channel. The volatiles are removed through these openings, which are located in the same angular position. The vent openings of channels operating at the same vacuum level are covered by a single manifold. Upstream of the vent opening is a replaceable melt barrier, which is used to provide a hydraulic seal between the various stages. The material moves through the RDD by virtue of the contact between the polymer melt and the rotor. Therefore, there is little tendency of the

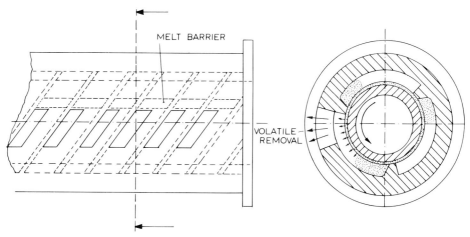

Figure 8-49b. Rotating Drum Devolatilizer

material to accumulate in the vent port. This situation is similar to the conditions existing in venting through the screw described in Section 8.5.2.6. The multiple flights in the housing provide for good mixing and surface renewal, resulting in effective devolatilization. The rotating drum devolatilizer can, in principle, be mounted on the end of an existing extruder. However, full exploitation of its potential benefits will probably require incorporation into new machinery, specially designed to take advantage of the benefits of the RDD.

8.5.3   VENT PORT CONFIGURATION

As discussed in Section 8.5.2.6, the polymer has to adhere to the barrel in order to be conveyed forward. This means that a vent port in the barrel is likely to pick up molten polymer. It is almost inevitable, simply by the nature of the conveying process in a screw extruder. For this reason, most vented single screw extruders tend to have a gradual accumulation of polymer in the vent port, requiring periodic cleaning of the vent port in order to maintain devolatilization efficiency. In order to minimize accumulation of polymer in the vent port, it is important that the shape of the vent port be such that there is a minimal chance of material to hang up. For this reason, the leading edge of the vent port is often undercut, with the undercut making a small angle with the O. D. of the screw. This is shown in Figure 8-50a.

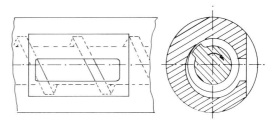

Figure 8-50a. Vent Port Geometry to Avoid Vent Flow

In addition to the undercut, the vent port is often offset from the vertical, again to minimize the chance of polymer scraping off at the leading edge of the vent port. The length of the vent port is usually 0.5D to 1.5D and the width about 0.1D to 0.25D.

Another well-designed vent port geometry is shown in Figure 8-50b.

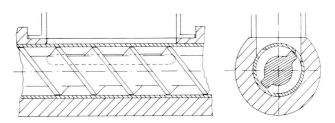

Figure 8-50b. Vent Port Geometry to Avoid Vent Flow

Again, this geometry minimizes the chance of polymer melt to be scraped off at the leading edge of the vent port. Another interesting vent port design is shown in Figure 8-51.

In this design, the feed port is combined with the vent port. The volatiles are removed through the annular space between the feed pipe and the feed port housing. The system is used in conjunction with a vacuum feed hopper as described by Franzkoch (21). Such a sys-

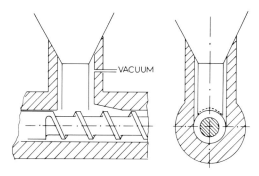

Figure 8–51. Combined Feed and Vent Port

tem allows devolatilization to occur in the feed hopper section of the extruder. This type of devolatilization is successfully used with powders, particularly when the material in the hopper can be heated. The vacuum feed hopper system and some of the problems associated with it are discussed in Section 3.4. Hopper devolatilization of pellets is generally unsuccessful because the larger particle size reduces the surface area and thus the devolatilization effectiveness, as discussed in Section 5.4.1.

8.6   MULTI-FLIGHTED EXTRUDER SCREWS

There are a large number of screw geometries using multiple flights. The additional flights can be the same as the main flight; this geometry will be referred to as the conventional multi-flighted screw. In other designs, the additional flight(s) is different in geometry and function from the main flight; this geometry will be referred to as the barrier flight screw geometry.

8.6.1   THE CONVENTIONAL MULTI-FLIGHTED EXTRUDER SCREW

The conventional multi-flighted extruder screw has a number of advantages and disadvantages. The basic geometry is shown in Figure 8–52.

Figure 8–52. Conventional Multi-Flighted Screw

The multi-flighted screw geometry adversely affects the solids conveying and melt conveying rate, as discussed in Chapter 7 and Sections 8.2.1 and 8.2.3. On the other hand, however, the multi-flighted screw geometry can significantly improve melting performance, as discussed in Section 8.2.2, provided the helix angle is sufficiently large. The multi-flighted screw geometry adversely affects power consumption, as discussed in Section 8.3, particularly when the polymer melt is low in melt index and relatively Newtonian in its flow behavior.

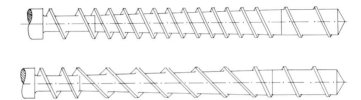

Figure 8-53. Multiple Flighted Screw for Improved Melting

An advantage of the double-flighted screw geometry is the symmetry of the screw flights. This can reduce the tendency of screw deflection by abrupt changes in pressure along the screw channel; this point was discussed in Section 8.1.3. Another possible advantage of a double-flighted screw geometry in the feed section is a more regular intake of material, as discussed in Section 8.4.2. Design considerations of double flighted extruder screws were discussed by Maddock (22) based on computer simulations. He predicts higher melting rate and reduced power consumption. The latter prediction seems incorrect and must have resulted from improper consideration of the power consumption in the flight clearance. However, the prediction of improved melting is correct and this seems to be the main benefit of a multi-flighted screw geometry. Since solids conveying and melt conveying rates are adversely affected by a multi-flighted geometry, it makes sense to incorporate multiple flights only along a particular section of the screw. This should be the screw section where melting will occur. Two possible screw designs are shown in Figure 8-53.

8.6.2   BARRIER FLIGHT EXTRUDER SCREWS

Barrier flight extruder screws have been around since the early 60's. Presently, barrier screws enjoy widespread popularity in the U.S. Every major U.S. extruder manufacturer offers at least one type of barrier flight extruder screw. There are various types of barrier screws and there is little agreement as to which type is better than the other. Even the advantages and disadvantages of barrier screws compared to regular single-flighted screws are not widely known or agreed upon. In this section, a functional analysis will be made of barrier screws based on extrusion theory. From this analysis, the advantages and disadvantages of barrier screws will become clear, also indicating the preferred geometrical configuration for certain applications.

The inventor of the barrier flight extruder screw is Maillefer, a pioneer in the field of extrusion. He first applied for a patent in Switzerland on December 31, 1959 (23) and later applied for patents in various other countries. Patents were granted, among other countries, in Germany (24) and in England (25). Maillefer filed a patent application in the U.S. on December 20, 1960. However, Maillefer did not obtain a patent in the U.S. because of a surprising situation. Geyer from Uniroyal filed a patent application for a barrier screw on April 5, 1961, several months after Maillefer's filing date (26). Geyer's patent application describes a barrier flight extruder screw almost identical to the one invented by Maillefer. As a result, an interference procedure developed which resulted in the granting of Geyer's patent and the rejection of Maillefer's claim to the patent. This may seem rather surprising because Maillefer filed his patent application several months before Geyer did. The explanation is that Maillefer, by U.S. patent law, was treated as a foreign national, while Geyer was treated as a U.S. national. If Uniroyal could demonstrate that Geyer conceived of his invention before Maillefer filed his application, which Uniroyal managed to do in court, Uniroyal would be legally entitled to the patent. Maillefer, being considered a foreign national, could not use the priority date based on the conception of the idea.

This little piece of history explains why some time elapsed, seven years, between the filing of Geyer's patent and the date of issue (April 2, 1968). It is an interesting situation because such an interference does not arise often at all. It is also interesting because Uniroyal has enforced its patent quite rigorously; various law suits have been filed as a result of alleged infringement of the Geyer patent. Uniroyal issued licenses to several companies, among others to Sterling Extruder Corporation who sells the screw under the name Sterlex High Performance Screw.

The principle of all barrier screws is very much the same. At the beginning of the barrier section a barrier flight is introduced into the screw channel. The clearance between the barrier flight and the barrel is generally larger than the clearance between the main flight and the barrel. The barrier clearance is large enough so that polymer melt can flow over the barrier, but it is too small for solid polymer particles to flow over the barrier. As a result, the solid bed will be located at the active side of the barrier flight and the polymer melt mostly at the passive side of the barrier flight. Thus, the barrier flight causes a phase separation, confining the solid bed to one side of the barrier flight while allowing the polymer melt to the other side. This is illustrated in Figure 8–54.

Thus, the barrier screw has a solids channel and a melt channel. In the down channel direction, the solids channel reduces in cross-sectional area, while the melt channel correspondingly increases in cross-sectional area. At the end of the barrier section, the solids channel reduces to zero, while the melt channel starts to occupy the full channel again.

This geometry ensures complete melting of the solids because the solids cannot travel beyond the barrier section, unless they are able to cross the barrier clearance. This is only possible if the particle has reduced to a size that will allow rapid melting after any possible crossing of the barrier. Another benefit of the barrier design is that all the polymer has to flow through the barrier clearance where it is briefly subjected to relatively high level of shear. This causes a certain amount of mixing, similar to the mixing in a fluted mixing element, see Section 8.7.1.

By the nature of the barrier geometry, the solid bed is confined to a channel that is considerably narrower than the total channel width. This has a possible benefit that the solid bed is less likely to break up, although this has not been conclusively demonstrated. On the other hand, it limits the space available to the solid bed. Thus, the screw design will have to be tailored more carefully to the melting profile of the solid bed. The solid bed is more likely to plug the solids channel. Melting must begin considerably before the start of the barrier section in order to allow the introduction of the barrier flight. If this were not the case, plugging would occur immediately.

In order to obtain maximum melting efficiency, the solids channel should be filled from flank to flank with solid polymer with only a thin melt film between the solid bed and the

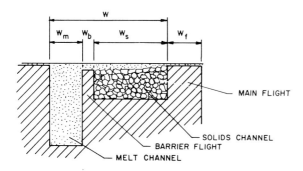

Figure 8–54. Phase Separation in Barrier Screw

barrel. Obviously, this also represents a situation that can easily develop into plugging because the melting has to match the reducing cross-section of the solids channel. In practice, therefore, there is likely to be more polymer melt in the solids channel than just in the melt film. The assumption of the solid bed occupying the full width of the solids channel can be used to determine the maximum possible melting rate. It should be realized, however, that this rate is not likely to be realized in practice. The comparable melting performance of a regular compression screw can be obtained by using the ideal compression A_1^*:

$$A_1^* = \frac{\Omega_1}{v_{sz}\rho_s\sqrt{W_1}\ \sin\varphi} \qquad\qquad (8\text{--}86)$$

This results in the following ideal axial melting length for a standard compression screw, see equation 8-55 and 7-116:

$$L_T^* = \frac{H_f v_{sz}\rho_s\sqrt{W_1}\ \sin\varphi}{\Omega_1} \qquad\qquad (8\text{--}87\,\text{a})$$

or as down channel melting length:

$$Z_T^* = \frac{H_f v_{sz}\rho_s\sqrt{W_1}}{\Omega_1} = \frac{H_f}{\psi} \qquad\qquad (8\text{--}87\,\text{b})$$

where ψ is given by equation 8-91 a.

The following analysis of the melting performance of various barrier screws is based on the analysis developed by Meijer and Ingen Housz (27). This analysis provides a clear and logical approach to the determination of the melting capacity as a function of the barrier section geometry.

8.6.2.1   THE MAILLEFER SCREW

The Maillefer extruder screw is shown in Figure 8-55.

The main feature of this barrier screw is that the helix angle of the barrier flight is larger than the helix angle of the main flight. As a result, there is a continuous reduction in the

Figure 8-55. Maillefer Screw

width of the solids channel and a corresponding increase in the width of the melt channel. This geometry allows a smooth and gradual change of the solids channel as well as the melt channel. A drawback of this geometry is that the solids channel reduces in width. This causes a corresponding reduction in melting rate, since this is directly determined by the width of the solid bed.

The melting performance of the Maillefer screw geometry can be analyzed by a down channel mass balance:

$$\frac{d\dot{M}_S}{dz} = -\frac{d\dot{M}_p}{dz} \qquad\qquad (8\text{--}88)$$

where:

$$\dot{M}_S = \rho_S v_{Sz} W_S H_S \qquad\qquad (8\text{--}88a)$$

The right-hand term, the melting rate per unit down channel distance, is given by equation 7-109c. In the analysis of the melting performance, a complication arises in that the solid bed velocity cannot be assumed constant. This can be appreciated by comparing the solid bed width profile of a standard screw with constant channel depth to the channel width profile of the solids channel in a Maillefer screw. The width of the solid bed W_S in a standard zero-compression screw as a function of down channel distance z can be written as:

$$\frac{W_S}{W_1} = \left(1 - \frac{z}{Z_T}\right)^2 \qquad\qquad (8\text{--}89)$$

where W_1 is the channel width and Z_T the total melting length. This expression is derived based on the assumption that the solid bed velocity v_{Sz} is constant along the melting zone. This relationship is shown in Figure 8-56.

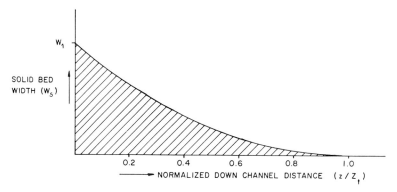

Figure 8-56. Solid Bed Width Profile with Constant Solid Bed Velocity

At the beginning of melting, the width of the solid bed reduces quite rapidly. As melting proceeds, the solid bed width continues to decline but the rate of change reduces with distance. At the final stage of melting, the rate of reduction of solid bed width approaches zero. This varying rate of change in the width of the solid bed is due to the fact that the melting rate reduces as the width of the solid bed reduces.

The width of the solids channel of the Maillefer screw can be written as:

$$W_{1S} = W_1 - z\tan\Delta\varphi = W_1\left(1 - \frac{z}{Z_T}\right) \qquad\qquad (8\text{--}90)$$

where $\Delta\varphi$ is the difference in helix angle between the main flight and barrier flight and Z_T is the total length of the barrier section in the down channel direction.

The relationship becomes clear from examination of a picture of the unwrapped geometry of the Maillefer barrier section, this is shown in Figure 8-57.

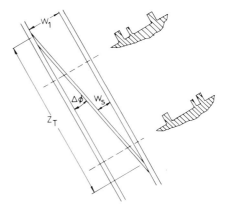

Figure 8-57. Unwrapped Maillefer Barrier Geometry

By comparing the channel width profile, Figure 8-57, to the solid bed width profile in a standard screw with constant solid bed velocity, Figure 8-56, it is clear that the two profiles are considerably different. If the solid bed is to occupy the full width of the channel W_{1s} in the Maillefer screw, the velocity of the solid bed will have to change along the barrier section. This may require substantial deformations in the solid bed, but it will be assumed that the solid bed is capable of undergoing the required deformations. In reality, the solid bed may not occupy the full width of the channel. However, this is not important at this point since the objective of this exercise is to determine the highest possible melting rate with the Maillefer screw geometry.

If it is assumed that the solid bed width equals the channel width W_{1s}, equation 8-88 can be written as:

$$\frac{dv_{sz}}{dz}\left(1-\frac{z}{Z_T}\right) - \frac{v_{sz}}{Z_T} = -\frac{v_{sz}\psi}{H}\left(1-\frac{z}{Z_T}\right)^{1/2} \tag{8-91}$$

where:

$$\psi = \frac{\Omega_1}{\rho_s v_{sz} W_1^{1/2}} \tag{8-91a}$$

This equation cannot be solved without making some simplifying assumptions. Meijer and Ingen Housz (27) solved this problem by initially taking the term $v_{sz}\psi$ constant and evaluating this term by its value at the beginning of melting, i.e., $v_{sz}\psi = v_{szo}\psi_o$. This results in the following expression:

$$\frac{v_{sz}}{v_{szo}}\left(1-\frac{z}{Z_T}\right) = 1 - \frac{2\psi_o Z_T}{3H}\left[1 - \left(1-\frac{z}{Z_T}\right)^{3/2}\right] \tag{8-92}$$

When $z = Z_T$, the left-hand term becomes zero; for the right-hand term to be zero as well, the melting length has to be:

$$Z_T = \frac{3H}{2\psi_0} = \frac{3}{2} Z_T^*$$

(8-93)

where Z_T^* is the shortest possible melt length in down channel direction for a standard compression screw, see equation 8-87b.

By substituting equation 8-93 into equation 8-92, an expression is obtained for the solid bed velocity as a function of down channel distance z:

$$\frac{v_{sz}}{v_{szo}} = \left(1 - \frac{z}{Z_T}\right)^{1/2}$$

(8-94)

Equation 8-94 can now be inserted into equation 8-91 to obtain a more accurate solution of equation 8-91. This can be done by writing Ω_1 as a function of v_{sz}, following Meijer and Ingen Housz (27):

$$\Omega_1 = \Omega_0 \left[1 + E\left(1 - \frac{v_{sz}}{v_{szo}}\right)\right]$$

(8-95)

where E is the coefficient relating Ω_1 to v_{sz}.

By ignoring higher order terms of E, the solution becomes:

$$Z\text{-Maillefer} = \frac{3}{2}\left(1 - \frac{E}{4}\right)\frac{H}{\psi}$$

(8-96)

This represents the shortest possible melting length with a Maillefer-type barrier screw. A reasonable maximum value of $E = 0.4$; this results in the following melting length:

$$Z\text{-Maillefer} \approx \frac{4}{3} Z_T^*$$

(8-97)

Thus, the best melting length in a Maillefer screw is about 30 percent longer than an ideal compression screw. For comparison, it is interesting to note that a non-ideal compression screw with a channel depth ratio of 4:1 has a total melting length of $Z_T = 3/2 \, Z_T^*$, as can be determined from equation 7-116. From a theoretical analysis of the melting performance of a Maillefer screw, it can be concluded that the best melting length is about 30 percent longer than an ideal compression screw and about 10 percent shorter than a standard compression screw with a compression ratio of four. Considering that the maximum melting performance of the Maillefer screw is only about 15 percent better than a standard compression (4:1) screw, it can be concluded that the Maillefer screw does not offer a significant benefit over a standard compression screw in terms of melting performance, particularly since the actual melting performance of the Maillefer screw is likely to be less than the maximum melting performance.

The advantage of the Maillefer screw is primarily the physical separation of the melt pool and the solid bed. As a result, there is less chance of formation of a melt film between the solid bed and the screw and, therefore, there is less chance of solid bed breakup. Thus, the melting process can occur in a more stable fashion but not necessarily at a higher rate.

8.6.2.2   THE BARR SCREW

The Barr screw is shown in Figure 8–58.

Figure 8–58. Barr Screw

The initial part of the barrier section is the same as the Maillefer screw. However, when the melt channel is sufficiently wide, the barrier flight starts to run parallel to the main flight.

The cross-sectional area of the solids channel is then reduced by reducing the channel depth while at the same time the channel depth of the melt channel is increasing. The advantage of this geometry is that the solids channel width is not continuously reducing, as in the Maillefer geometry, but remains constant at a relatively large value. This increases the solid-melt interfacial area and improves the melting performance. It has a few drawbacks as well, however. Since the melt channel is narrow and quite deep in the latter part of the barrier section, the melt conveying efficiency of the melt channel will be far less than in the Maillefer screw. Also, at the end of the barrier section the melt channel changes from a narrow, deep channel to a wide, shallow channel over a short length. This rapid change in channel geometry will not be conducive to stable flow conditions. These relatively abrupt changes in channel geometry do not occur in the Maillefer screw.

The Barr screw was developed by Barr and Chung when they worked at the Hartig Plastics Machinery Division of Midland-Ross Corporation. Barr applied for a patent in August 1971, the patent issued in October 1972 (28). Hartig is selling barrier screws covered by this patent under the names "MC3" and "MC4" screw. Later when Barr and Chung left Hartig, Chung applied for a patent on a modified Barr screw. The application was filed in July 1975 and the patent issued in January 1977 (29). The modification consists primarily of a difference in the transition from the final melt channel geometry to the metering section. Screws covered by this later patent are sold by Robert Barr, Inc. under the names Barr II, Barr III, and Barr ET screw. In 1982, Uniroyal brought suit against Robert Barr, Inc. for patent infringement. The suit was settled for an undisclosed amount of money.

A very similar barrier screw is described by Willert in a patent application dated August 26, 1981 (30). The difference in the barrier screw is that the main flight becomes a barrier flight at the beginning of the barrier section, while at the same time the main flight branches off at an angle and then starts to run parallel to the barrier flight. The barrier screw design is shown in Figure 8–59.

Interestingly enough, Willert's description is essentially identical to the patent of Hsu (32). This patent was filed in January 1973 and issued January 1975. Screws covered by Hsu's patent are sold under the name "Maxmelt Screw" by the Plastics Machinery Division of Hoover Universal. Coincidentally, Hsu used to work at Hartig at the time when the basic

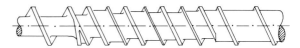

Figure 8–59. The Lacher/Hsu/Willert Barr Screw

concepts of the Barr screw were being developed. Another patent on a barrier screw was obtained by Lacher (57) from NRM Corporation in 1966. This geometry is also very similar to the one described by Hsu (32) and Willert (30). Both Maillefer and NRM have become licensees of Uniroyal.

The melting performance of the Barr screw can be analyzed by the same procedure followed for the Maillefer screw. The initial portion of the barrier section can be analyzed just as a Maillefer screw. If z_{T_1} is the length of the initial Maillefer portion of the barrier section, the solid bed velocity at $z = Z_{T_1}$ is approximately:

$$v_{sz} = v_{szo} \left(1 - \frac{Z_{T1}}{Z_T} \right)^{1/2} \tag{8-98}$$

The melting length of the Maillefer portion is:

$$Z_{T1} = \frac{3HZ_{T1}}{2\psi Z_T} \tag{8-99}$$

In the parallel portion of the barrier section, the width of the solids channel is constant. The highest melting performance will be reached if the width of the solid bed fills the entire width of the solids channel. If the solid bed velocity is assumed constant, the total melting length can be found by using the equations derived for the standard extruder screw, see equation 7-116. The total melting length for the parallel barrier portion is simply:

$$Z_{T2} = \frac{H}{\psi} \left(1 - \frac{Z_{T1}}{Z_T} \right) \tag{8-100}$$

The total melting of the barrier section is the sum of equations 8-99 and 8-100. This results in the following expression for the melting length of the Barr screw:

$$Z\text{-Barr} = \frac{H}{\psi} \left(1 + \frac{1}{2} \frac{Z_{T1}}{Z_T} \right) \tag{8-101}$$

In equation 8-101, the effect of the varying solid bed velocity on Ω_1 has been neglected. Since the solid bed varies only in the initial portion of the barrier section, this will not affect the accuracy very much as long as the initial Maillefer section is relatively short relative to the total length of the barrier section.

In comparison to the melting length Z_T^* of the ideal compression screw, the melting length of the Barr screw is:

$$Z\text{-Barr} = \left(1 + \frac{Z_{T1}}{2Z_T} \right) Z_T^* \tag{8-102}$$

A typical ratio of Z_{T1}/Z_T is about 0.2. Thus, the shortest possible melting length of the Barr screw is about 10 percent longer than the Z_T^* of the ideal compression screw and about 25 percent shorter than the melting length of the standard compression (4:1) screw. The Barr screw is slightly more efficient than the Maillefer screw in terms of melting performance, about 15 percent. Thus, the Barr screw is marginally better than the Maillefer screw in melting performance. However, the Maillefer screw is better in terms of melt conveying performance in the barrier section.

8.6.2.3 THE DRAY AND LAWRENCE SCREW

The Dray and Lawrence screw is shown in Figure 8-60.

The screw geometry is very similar to the Barr screw with the one major difference being an abrupt change in helix angle in the main flight at the point where the barrier flight is introduced. This allows the width of the solids channel to stay just as wide as the full channel width of the feed section. Obviously, this is done in an attempt to maintain the solids channel as wide as possible. However, this also causes an abrupt change in the direction of the solid bed velocity and this could lead to instabilities.

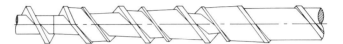

Figure 8-60. Dray and Lawrence Screw

If the effect of the change in helix angle on Ω_1 is neglected, the melting length can be shown to be (27):

$$Z\text{–DL} = \frac{H}{\psi}\left(\frac{\sin\varphi_f}{\sin\varphi_b}\right)^{1/2} = \left(\frac{\sin\varphi_f}{\sin\varphi_b}\right)^{1/2} Z_T^* \qquad (8\text{–}103)$$

where φ_f is the helix angle in the feed section and φ_b the helix angle in the barrier section.

In this case, however, the down channel melting length does not provide a good basis for comparison because the helix angle is different along the screw. The total axial melting length can be expressed as:

$$L\text{–DL} = \left(\frac{\sin\varphi_b}{\sin\varphi_f}\right)^{1/2} L_T^* \qquad (8\text{–}104)$$

If typical values are used for φ_b and φ_f, the melting length of the Dray and Lawrence screw will be about 10 to 20 percent longer than the ideal compression screw. The melting performance of the Dray and Lawrence screw is thus about the same as the Barr screw and slightly better than the Maillefer screw. A patent on this barrier screw was applied for by Dray of Feed Screw, Inc. and Lawrence of Owens-Illinois, Inc. (31). The patent was filed in May 1970 and was issued in March 1972. The screw based on this patent is sold by Feed Screw, Inc. under the name "Efficient Screw." At the end of the barrier section, the melt channel has to make a transition to the melt channel of the metering section. Since the solids channel is quite wide, this transition tends to be quite abrupt, even more so than with the Barr screw. Thus, the Dray and Lawrence screw has two abrupt changes in screw channel geometry, one at the beginning and one at the end of the barrier section. This will tend to make the screw more susceptible to surging types of instabilities than the Maillefer screw, which has much more gradual transitions in the screw channel geometry.

8.6.2.4   THE KIM SCREW

The Kim screw is basically an improvement of the Dray and Lawrence screw. A picture of the Kim screw is shown in Figure 8-61.

Figure 8-61. Kim Screw

In the Kim screw, the helix angle of the main flight and the barrier flight is changed gradually to obtain a smooth transition from feed section to barrier section. In the Kim screw, the width of the solids channel remains constant as with the Dray and Lawrence screw. A patent was filed in August 1972 and issued February 1975 (33). The patent was reissued in August 1975 (34) with the number of claims reduced from six to two. The assignee is B. F. Goodrich Company. The screw was licensed to Davis-Standard Division of Crompton & Knowles Corporation; it is sold under the name "VPB" screw. The VPB screw is covered both by the early Uniroyal patent (26) and the B. F. Goodrich patent (34).

Again, the melting performance can be analyzed by the procedure used for the Maillefer screw. Because of the continuously varying helix angle, the analysis is rather involved. Ingen Housz and Meijer (27) found for the total melting length of the Kim screw:

$$\text{L--Kim} = \frac{3\hat{W}_S(1+0.5\hat{W}_S)}{2[(1+\hat{W}_S)^{3/2}-1]} L_T^* \qquad (8-105)$$

where $\hat{W}_S$ is the ratio of the final melt channel width W_{1m} to the initial solids channel width W_{1S}:

$$\hat{W}_S = \frac{W_{1m}}{W_{1S}} \qquad (8-105a)$$

The value of $\hat{W}_S$ in the Kim screw is one, which results in the following value of the melting length:

$$\text{L--Kim} \approx \frac{5}{4} L_T^* \qquad (8-106)$$

This means that the melting performance of the Kim screw is slightly lower than the Dray and Lawrence screw and the Barr screw. It has an advantage over the Dray and Lawrence screw in that the transition from feed to barrier section occurs more smoothly. However, at the end of the barrier section the same difficulty arises as with the Dray and Lawrence screw.

8.6.2.5   THE INGEN HOUSZ SCREW

The Ingen Housz screw combines barrier geometry with multi-flighted geometry to obtain significant improvements in melting. A picture of the barrier section geometry is shown in Figure 8-62.

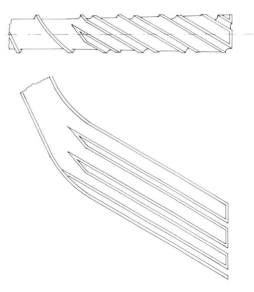

Figure 8 – 62. Ingen Housz Screw

The barrier section geometry is shown with the screw channel unrolled onto a flat plane. The solid bed is divided into several parallel solid channels. The melt is collected in several parallel melt channels. It is possible to achieve this multi-flighted geometry by a significant increase in the helix angle. The total melting length can be expressed as (27):

$$L-IH = \left(\frac{1+\hat{W}_S}{p}\right)^{1/2} L_T^*$$ (8–107)

With $\hat{W}_S = 0.25$ and and $p = 3$ the total melting length becomes:

$$L-IH \sim 0.65 L_T^*$$ (8–108)

This means that with this barrier screw geometry it is possible to obtain a minimum melting length that is shorter than the ideal compression screw. In fact, the multi-flighted barrier screw geometry is the only one that yields significant benefits in terms of melting performance compared to the standard compression screw. A U.S. patent on the Ingen Housz screw was issued August 19, 1980 (35).

The Ingen Housz barrier screw has a few drawbacks that may or may not be significant. At the start of the barrier section, the solid bed is sliced into several narrower solid beds. This requires easy deformability of the solid bed. If there is considerable resistance against this deformation in the solid bed, it could lead to instabilities. The melt conveying in the barrier section occurs in deep, narrow channels with a large helix angle. As a consequence, the melt conveying capacity will be poor. This essentially requires the use of a grooved barrel section in the feed section of the extruder to ensure a negative pressure gradient along the barrier section to reach sufficient melt conveying rate. Finally, the transition from the barrier section to the metering section will be difficult if the metering section is of conventional geometry. However, if a grooved barrel section is used, the metering section can be deleted altogether since pressure build-up in the melt conveying zone is no longer necessary.

Results of extensive experimental tests with the Ingen Housz screw are described by Ingen Housz and Meijer (36). Tests were run on a 60-mm extruder with LDPE, HDPE, and PP. The high melting capacity of the screw could only be utilized if sufficient solids conveying capacity was made available by the use of a grooved barrel section. Outputs as high as 200 kg/hr at 100 rpm were achieved while the total length of the screw was only 16 D. The output was found to be sensitive to the particle size of the polymer. High outputs were achieved with larger particles, while the output with smaller particle size polymer was considerably lower, sometimes as much as 50 percent. This demonstrates that the output can never be higher than the solids conveying rate in the feed section. Even with the grooved barrel section, the solids conveying rate for some polymers was insufficient to supply the melting section with enough material to utilize the full melting capacity.

8.6.2.6   SUMMARY OF BARRIER SCREWS

The characteristics of the various barrier screws are summarized in Table 8-1. The Maillefer screw has many desirable characteristics despite the fact that its melting performance is not quite as good as the other barrier screws. The Ingen Housz screw clearly has the best melting performance; however, that is at the expense of geometrical simplicity.

Figure 8-63. DFM Screw

From a functional analysis, a double flighted Maillefer (DFM) screw would seem to be a good compromise between considerations concerning geometry and output. With a double flighted geometry, the melting performance can be improved about 30 percent in the best case. This would make the DFM screw more efficient in melting capacity than the Barr screw, Dray and Lawrence screw, and the Kim screw. In order to minimize the adverse effect of the additional flight, the helix angle of the main flight should be relatively large. However, the helix angle should not be too large in order to maintain good melt conveying capability. A helix angle of about 25° would seem like a reasonable compromise. Figure 8-63 shows a possible configuration of the DFM screw. The characteristics of the DFM screw are included in table 8-1.

A double flighted compression screw (4:1 ratio) has a melting capacity only slightly less than the DFM screw, see table 8-1, and better than the Barr screw, Dray and Lawrence screw, and Kim screw. The double flighted compression screw will be easier to manufacture than any barrier screw. The advantage of barrier screws that they keep unmolten material from reaching the metering section can also be obtained by incorporating a fluted mixing section at the beginning of the metering section.

TABLE 8-1. CHARACTERISTICS OF VARIOUS BARRIER EXTRUDER SCREWS

	Transition Feed-Barrier	Transition Barrier-Metering	Minimum Melting Length	Melt Conveying Capability	Ease of Manufacture	Require Grooved Barrel Section
Maillefer	smooth	smooth	$1.3\ L_T^*$	good	good	no
Barr	smooth	abrupt	$1.1\ L_T^*$	moderate	fair	no
DL	abrupt	abrupt	$1.1\ L_T^*$	fair	fair	no
Kim	smooth	abrupt	$1.2\ L_T^*$	fair	difficult	no
Ingen Housz	abrupt	abrupt	$0.65\ L_T^*$	poor	difficult	yes
DFM	smooth	smooth	$0.9\ L_T^*$	good	good	no
Compression (4:1) Single Flighted	smooth	smooth	$1.5\ L_T^*$	good	excellent	no
Compression (4:1) Double Flighted	smooth	smooth	$1.1\ L_T^*$	good	excellent	no

8.7   MIXING SCREWS

The mixing capacity of standard extruder screws is relatively limited. As a result, many modifications have been made to the standard extruder screw in an effort to improve the mixing capacity. The number of mixing elements that have been used on extruder screws is nearly infinite. Therefore, it is not possible to discuss all mixing screws used in the industry. This discussion will be limited to the more common types of mixing elements. Before selecting a mixing element, it is important to determine whether distributive or dispersive mixing is required, see Section 7.7. Therefore, the mixing sections will be divided into distributive and dispersive mixing sections to indicate their preferred application.

8.7.1   DISPERSIVE MIXING ELEMENTS

Dispersive mixing elements are used when agglomerates or gels need to be broken down. This is particularly important in small or thin gauge extrusion, e.g., fiber spinning, thin film extrusion, etc. The most common dispersive mixing section is the fluted or splined mixing section. In this mixing section, one or more barrier flights are placed along the screw such that the material has to flow over the barrier flight(s). In the barrier clearance the material is subjected to a high shear rate; the corresponding shear stress should be large enough to break down the particles in the polymer melt. A well-known fluted mixing section is the Union Carbide (UC) mixing section invented by LeRoy (37), see Figure 8-64.

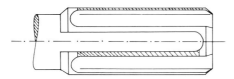

Figure 8-64. UC Mixing Section

Maddock from Union Carbide published results of experiments with this mixing section (38); since then the mixing section is often referred to as the Maddock mixing section.

The UC mixing section has longitudinal splines, i.e., a barrier flight helix angle of 90 degrees. All material has to flow over the barrier flight because the inlet channel is closed at the end of the mixing section. Thus, all of the material is forced over the barrier flight, yielding a uniform dispersive mixing action. A drawback of this mixing section is that it is a pressure consuming mixing section, i.e., it reduces the output of the extruder. Also, the longitudinal geometry with constant channel depth results in stagnating regions. Thus, the design will be less suitable with materials of limited thermal stability.

Another fluted mixing section is the Egan mixing section invented by Gregory and Street (47), see Figure 8-65.

Figure 8-65. Egan Mixing Section

In this mixing section, the splines run in a helical direction, i. e., the barrier flight helix angle is less than 90 degrees. The advantage of the helical splines is the fact that this enables forward drag transport in the inlet and outlet channel. As a result, the fluted mixing section with helical flutes will consume less pressure than the fluted mixing section with longitudinal flutes, see Figure 8-72. Thus, the helically fluted mixing will reduce the extruder output to a lesser extent. In fact, if the mixing section is properly designed it can even generate pressure, causing an improvement in extruder output. Another feature of the Egan mixing section is a gradual reduction of the depth of the inlet channel, leading to zero depth at the end of the mixing section. This channel depth profile is reversed in the outlet channel. The channel depth taper reduces the chance of hangup of material, and thus reduces the chance of degradation.

A similar mixing device was later patented by Gregory (48). The difference between this mixing device and the Egan mixing section is a constant depth in inlet and outlet channel and a concave channel geometry. Another similar mixing section was patented by Dray (49), see Figure 8–66.

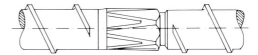

Figure 8-66. Dray Mixing Section

The main difference in this mixing section is that the outlet channel is open at the start of the mixing section. Thus, not all material is forced over the barrier clearance. Therefore, this mixing device will not result in a uniform shear history of the material. An extreme form of a fluted mixing section is the annular blister ring, see Figure 8-67.

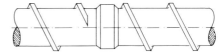

Figure 8-67. Blister Ring

It is simply a smooth cylindrical screw section with a small radial clearance. All the material has to pass through this clearance to exit from the extruder. Since no positive drag transport takes place over the barrier clearance, the pressure drop over the blister ring will be high compared to other fluted mixing sections. A blister ring with a small barrier clearance will generally cause a significant reduction in output. The pressure drop over the blister ring can be written by using the expressions in Table 7-1:

$$\Delta P = \frac{2m\Delta L}{\delta} \left[\frac{2\dot{V}(s+2)}{\pi D \delta^2} \right]^n \tag{8-109}$$

This expression is for a power law fluid and is valid if the effect of the screw rotation on the melt viscosity can be neglected. In reality, however, the effective viscosity in the clearance will be reduced as a result of the rotation of the screw. The shear stress is composed of a shear stress in tangential direction and a shear stress in axial direction. The tangential shear stress can be determined by evaluating the stress at the center of the channel where the axial shear stress is zero. This yields the following expression for the shear stress in tangential direction:

$$\tau_t = m \left(\frac{\pi D N}{\delta} \right)^n \tag{8-110}$$

The tangential shear stress is constant over the depth of the channel. The axial shear stress can be related to the axial pressure gradient by a simple force balance; this yields:

$$\tau_a = y \frac{\Delta P}{\Delta L} \tag{8-111}$$

The total shear stress is obtained by vectorial addition of the axial and tangential shear stress:

$$\tau = \left[m^2 \left(\frac{\pi D N}{\delta} \right)^{2n} + \left(y \frac{\Delta P}{\Delta L} \right)^2 \right]^{1/2} \tag{8-112}$$

The axial velocity gradient can be determined from:

$$\frac{dv}{dy} = \left(\frac{\tau}{m} \right)^S \frac{\tau_a}{\tau} = \frac{1}{m^S} \tau_a \tau^{S-1} \tag{8-113}$$

The axial velocity is obtained by integration of the axial velocity gradient:

$$v = \int \left(\frac{dv}{dy} \right) dy + C \tag{8-114}$$

And the volumetric flow rate is obtained by integration of the axial velocity:

$$\dot{V} = \int_0^{\delta/2} 2\pi D v \, dy \tag{8-115}$$

Closed form solutions of equations 8-114 and 8-115 are only possible for a few specific values of the power law index n, namely those for which $(1-n)/2n$ is an integer. Worth (50) derived a solution for a power law index value $n = 1/3$. This is a useful case because many of the high volume commodity polymers have a power law index close to $1/3$, see also Table 6-1. When $n = 1/3$, the throughput as a function of pressure can be written as:

$$\dot{V} = \frac{\pi D \Delta P}{m^3 \Delta L} \left[\frac{\delta^3 \tau_t^2}{12} + \frac{\delta^5}{80} \left(\frac{\Delta P}{\Delta L} \right)^2 \right] \tag{8-116}$$

The pressure drop now has to be found by solving a cubic equation. This solution can be written as:

$$\Delta P = (q+r)^{1/3} + (q-r)^{1/3} \tag{8-117}$$

where:

$$q = \frac{40m^3 \Delta L^3 \dot{V}}{\pi D \delta^5} \qquad (8\text{--}118\,a)$$

$$p = \frac{-20\,\tau_t^2 \Delta L^2}{9\,\delta^2} \qquad (8\text{--}118\,b)$$

$$r = (q^2 - p^3)^{1/2} \qquad (8\text{--}118\,c)$$

A comparison of the pressure drop predicted with equation 8-118 to the ΔP predicted with equation 8-109 is shown in Figure 8-68.

The prediction is for a 114-mm (4-1/2-inch) extruder running at 100 rpm with a blister length of 12.7 mm (1/2 inch) and a flow rate of 131 cc/sec (8 inch^3/sec). The reduction of the pressure drop as a result of the screw rotation is about 15 percent. This indicates that the simple equation 8-109 gives a reasonably accurate prediction of the pressure drop.

The pressure drop in a fluted mixing section can be calculated for a Newtonian fluid. The first theoretical analysis was performed by Tadmor and Klein (51). Their final equation for the pressure drop contains five dimensionless numbers, which makes determination of the effect of certain design variables rather indirect. A non-isothermal and non-Newtonian analysis was performed by Lindt et al. (52). This analysis requires numerical techniques to solve the equations. Therefore, this analysis can only be used if one develops the computer

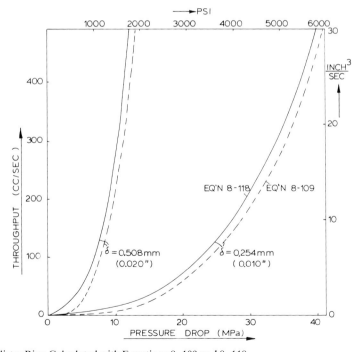

Figure 8-68. ΔP Over Blister Ring Calculated with Equations 8-109 and 8-118

software to perform the calculations. A simpler analysis was made by the author (53), leading to closed form analytical solutions from which the effect of the most important design variables can be easily evaluated.

To determine the pressure drop as a function of flow rate, one pair of inlet and outlet channels will be examined in detail, see Figure 8-69a.

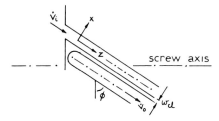

Figure 8-69a. Inlet and Outlet Channel Geometry

The volumetric flow rate at the entrance to the inlet channel is $\dot{V}_i(o)$, where the subscript i refers to the inlet channel. $\dot{V}_i(o)$ is the total volumetric output of the extruder divided by the number of inlet channels. The flow rate through the inlet channel decreases in down channel direction as a result of leakage over the barrier flight. At the same time there is a corresponding increase in the flow rate through the outlet channel. If the fluid is considered Newtonian and isothermal, the flow rate in the inlet channel as a function of down channel distance z can be written as:

$$\dot{V}_i(z) = \frac{1}{2} F_d WH(1-\delta/H)^2 v_b \cos\varphi - \frac{1}{12\mu} F_p WH^3 \frac{dP_i}{dz} \tag{8-119}$$

where δ is the radial clearance of the barrier flight.

The radial clearance of the non-barrier flight is taken to be zero. The initial pressure gradient G_1 in the inlet channel can be obtained from:

$$G_1 = \frac{dP_i(o)}{dz} = \frac{6\mu F_d (1-\delta/H)^2 v_b \cos\varphi}{F_p H^2} - \frac{12\mu \dot{V}_i(o)}{F_p WH^3} \tag{8-120}$$

The flow rate in the exit or outlet channel $\dot{V}_e(z)$ can be written as

$$\dot{V}_e(z) = \frac{1}{2} F_d WH v_b \cos\varphi - \frac{1}{12\mu} F_p WH^3 \frac{dP_e}{dz} \tag{8-121}$$

The subscript e refers to the exit channel. Considering that the flow rate at the exit of the outlet channel $\dot{V}_e(z_m)$ equals the flow rate at the entry to the inlet channel $\dot{V}_i(o)$, the pressure gradient at the exit of the outlet channel G_2 can be written as:

$$G_2 = \frac{6\mu F_d v_b \cos\varphi}{F_p H^2} - \frac{12\mu \dot{V}_i(o)}{F_p WH^3} \tag{8-122}$$

The leakage flow from the inlet channel to the outlet channel $\dot{V}_l$ is a combination of drag flow and pressure flow. The leakage flow per unit down channel distance $\dot{V}_l'$ can be written as:

$$\dot{V}_1' = \frac{1}{2} v_b \delta \sin\varphi + \frac{\delta^3}{12\mu_{cl}w_{cl}} (P_i - P_e) \qquad (8-123)$$

where w_{cl} is the perpendicular barrier flight width and μ_{cl} the polymer melt viscosity in the clearance.

From a mass balance, it follows that the local flow rate in the exit channel equals the total flow rate minus the local flow rate in the inlet channel:

$$\dot{V}_e(z) = \dot{V}_i(o) - \dot{V}_i(z) \qquad (8-124)$$

This equation leads to the following relationship between the two pressure gradients:

$$\frac{dP_i(z)}{dz} + \frac{dP_e(z)}{dz} = \frac{12\mu}{F_p WH^3} \left[F_d WH(1-\delta/H) v_b \cos\varphi - \dot{V}_i(o) \right] \qquad (8-125)$$

Equation 8-125 is valid when the local channel width W and channel depth H of the inlet channel are the same as those of the outlet channel. If the dimensions of both channels do not change with down channel distance, the sum of the pressure gradients will be constant along the length of the mixing section. In this case, equation 8-125 can be written as:

$$\frac{dP_i(z)}{dz} + \frac{dP_e(z)}{dz} = -A_1 \qquad (8-126)$$

Thus, the pressure in the exit channel can be related to the pressure in the inlet channel by:

$$P_e = -A_1 z + A_2 - P_i \qquad (8-127)$$

where:

$$A_1 = \frac{12\mu}{F_p WH^3} \left[\dot{V}_i(o) - F_d WH(1-\delta/H) v_b \cos\varphi \right] \qquad (8-127a)$$

Another important relationship can be obtained by considering that the local change in flow rate of the inlet channel over an incremental increase in down channel distance equals the local leakage flow:

$$\frac{d\dot{V}_i}{dz} = -\dot{V}_1' \qquad (8-128)$$

If the channel dimensions do not change in down channel direction, equation 8-128 can be written as:

$$\frac{F_p WH^3}{12\mu} \frac{d^2 P_i}{dz^2} = \frac{1}{2} v_b \delta \sin\varphi + \frac{\delta^3}{12\mu_{cl}w_{cl}} (P_i - P_e) \qquad (8-129)$$

With equation 8-127, the differential equation can be written as:

$$\frac{d^2P_i}{dz^2} - B_1P_i = \frac{1}{2} A_1B_1z + \frac{6\mu v_b \delta \sin\varphi}{F_pWH^3} - \frac{1}{2} A_2B_1 \qquad (8\text{-}130)$$

where:

$$B_1 = \frac{2\mu\delta^3}{F_pWH^3\mu_{cl}w_{cl}} \qquad (8\text{-}130\,\text{a})$$

Equation 8-130 is a nonhomogeneous equation of the second order. The solution to the homogeneous equation is:

$$P_i^* = C_1\exp(-B_2z) + C_2\exp(B_2z) \qquad (8\text{-}131)$$

where:

$$B_2 = \sqrt{B_1} \qquad (8\text{-}131\,\text{a})$$

An obvious particular solution is:

$$P_i^{**} = -\frac{1}{2} A_1z - \frac{3\mu_{cl}w_{cl}v_b\sin\varphi}{\delta^2} + \frac{1}{2} A_2 \qquad (8\text{-}132)$$

The general solution is the sum of the solution to the homogeneous equation plus the particular solution. Thus, the pressure profile in the inlet channel can be described by:

$$P_i = C_1\exp(-B_2z) + C_2\exp(B_2z) - \frac{1}{2} A_1z - \frac{3\mu_{cl}w_{cl}v_b\sin\varphi}{\delta^2} + \frac{A_2}{2} \qquad (8\text{-}133)$$

With equation 8-127, the pressure profile in the outlet channel can be written as:

$$P_e = -C_1\exp(-B_2z) - C_2\exp(B_2z) - \frac{1}{2} A_1z + \frac{3\mu_{cl}w_{cl}v_b\sin\varphi}{\delta^2} + \frac{A_2}{2} \qquad (8\text{-}134)$$

Two boundary conditions are necessary to evaluate constants C_1 and C_2. The following boundary conditions can be used:

$$\frac{dP_i(o)}{dz} \approx \frac{dP_e(z_m)}{dz} \approx \frac{6\mu F_d(1-\delta/H)v_b\cos\varphi}{F_pH^2} - \frac{12\mu\dot{V}_i(o)}{F_pWH^3} \qquad (8\text{-}135)$$

This results in the following expressions for C_1 and C_2:

$$C_1 = \frac{-6\mu\dot{V}_i(0)}{F_p WB_2 H^3[\exp(-B_2 z_m)-1]} \qquad (8-136)$$

$$C_2 = \frac{-6\mu\dot{V}_i(0)}{F_p WB_2 H^3[1-\exp(B_2 z_m)]} \qquad (8-137)$$

Figure 8-69b shows the pressure profile in the inlet channel and outlet channel for a 114-mm (4.5-inch) extruder running at 100 rpm with a throughput of 164 cc/sec (10 inch^3/sec).

The length of the mixing section is 2 D, the barrier clearance is 0.5 mm (0.020 inch), the helix angle is 45 degrees, and the number of inlet channels is 3. The local viscosity is evaluated with the power law equation by using the Couette shear rate; the consistency index is 13 800 Pa-sec^n (2 psi-sec^n) and the power law index is 0.5. The pressure in the inlet channel reduces initially, but later starts to increase again. This indicates some degree of pressure generating capability of the inlet channel. In the outlet channel, the pressure rises initially and drops in the later portion of the channel. Thus, both the inlet channel and outlet channel have some pressure generating capability. This is primarily achieved by the helical orientation of the flutes.

The importance of the helix angle is shown in Figure 8-70.

Two sets of pressure profiles are shown, one for a mixing section with a 90 degree helix angle and one for a mixing section with a 50 degree helix angle. In this case, the barrier

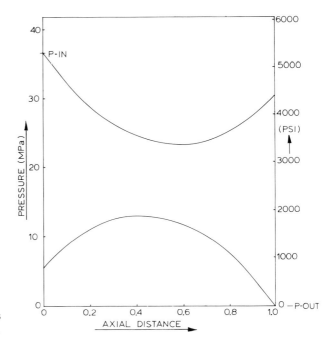

Figure 8-69b. Pressure Profiles in the Inlet and Outlet Channel

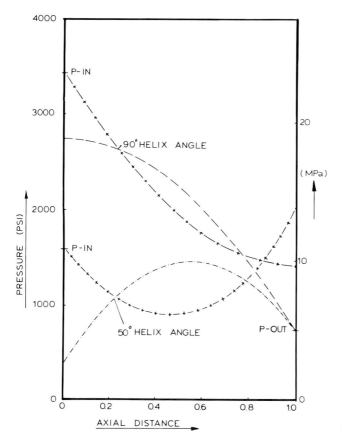

Figure 8-70. Pressure Profiles at Two Helix Angles

clearance is 0.635 mm (0.025 inch) and the throughput is 131 cc/sec (8 inch³/sec). The profiles are determined such that the final pressure has the same value (5 MPa = 725 psi). It is quite evident that the helix angle has a strong effect on the pressure profiles and the total pressure drop. There is a significant pressure generation in the inlet channel and outlet channel of the helical mixing section, resulting in a relatively small total pressure drop. On the other hand, there is no pressure generating capacity in the axially oriented mixing section as evidenced by the monotonic drop in pressure in both the inlet and outlet channel. This results in a rather large total pressure drop, about three times as high as the helically oriented mixing section!

The total pressure drop over the mixing section ΔP_m is simply:

$$\Delta P_m = P_i(o) - P_e(z_m) \qquad (8\text{-}138)$$

With equations 8-133 through 8-137, this results in the following expression for the pressure drop over the mixing section:

$$\Delta P_m = \frac{2T[1+\exp(B_2 z_m)]}{1-\exp(B_2 z_m)} - \frac{6\mu_{cl} w_{cl} v_b \sin\varphi}{\delta^2} + \frac{A_1 z_m}{2} \qquad (8\text{-}139)$$

where:

$$T = \frac{-6\,\mu\dot{V}_i(o)}{F_p W H^3 B_2} \tag{8-139a}$$

When $B_2 z_m$ ranges between 0 and 1, the following approximation can be made:

$$\frac{1 + \exp(B_2 z_m)}{1 - \exp(B_2 z_m)} \simeq \frac{-2}{B_2 z_m} \tag{8-140}$$

The total pressure drop can now be written as the sum of two terms:

$$\Delta P_m \simeq \Delta P_{cl} + \Delta P_{ch} \tag{8-141}$$

The pressure drop in the clearance ΔP_{cl} is given by the first two terms on the right-hand side of equation 8-139. The pressure drop inlet and outlet channel ΔP_{ch} is the last term of equation 8-139.

The first term ΔP_{cl} is the pressure drop caused by the clearance. When $B_2 z_m$ is less than unity, equation 8-140 can be used to express ΔP_{cl} as:

$$\Delta P_{cl} = \frac{12\,\mu_{cl} w_{cl}}{z_m \delta^3} \left[\dot{V}_i(o) - \frac{\delta L_m v_b}{2} \right] \tag{8-142a}$$

When $B_2 z_m$ is larger than unity, ΔP_{cl} can be expressed as:

$$\Delta P_{cl} = \frac{12\,\mu_{cl} w_{cl}}{z_m \delta^3} \left[\left(\frac{3}{4} + 0.3\ B_2 z_m \right) \dot{V}_i(o) - \frac{\delta L_m v_b}{2} \right] \tag{8-142b}$$

The second term ΔP_{ch} is the pressure drop in the inlet and outlet channel; this can be written as:

$$\Delta P_{ch} = \frac{6\,\mu z_m}{F_p W H^3} \left[\dot{V}_i(o) - F_d W(H-\delta) v_b \cos\varphi \right] \tag{8-143}$$

The first term ΔP_{cl} is inversely proportional to the cube of the barrier clearance. Thus, when the barrier clearance is small, the pressure drop over the clearance will increase very rapidly and will be the major component of the total pressure drop. The pressure drop over the clearance ΔP_{cl} can be made zero by making sure that the drag flow rate over the barrier clearance equals the flow rate at the entrance to the inlet channel. Thus, the mixing section should be designed such that:

$$\frac{1}{2}\ \delta L_m v_b \geq \dot{V}_i(o) \tag{8-144}$$

If the flow rate through the mixing section is assumed to be about two-thirds of the drag flow rate of the preceeding screw section and the axial length is about two diameters ($L_m \approx 2\,D$), then equation 8–144 can be simplified to the following form:

$$\delta \geq C\,\frac{D}{p} \qquad\qquad (8-145)$$

D is the screw diameter, p the number of inlet channels, and the constant C in many cases is about 0.01. When the pressure drop over the clearance is zero, the total pressure drop becomes simply:

$$\Delta P_m = \Delta P_{ch} = \frac{6\,\mu z_m}{F_p W H^3}\left[\dot{V}_i(0) - F_d W(H-\delta)v_b\cos\varphi\right] \qquad\qquad (8-146)$$

The pressure drop over the channel reaches its maximum value when the helix angle is 90 degrees. Thus, this corresponds to the most unfavorable geometry because it will result in the largest drop in output. If the channel depth and helix angle are optimized simultaneously, the pressure drop in the channel will reach a minimum when the helix angle is 52.24 degrees (63). The corresponding optimum channel depth is $H^* = 0.314\dot{V}/(F_p D^2 N)$.

The various factors that influence the pressure drop over the mixing section can now be easily analyzed. The pressure drop increases proportionally with the flow rate through the mixing section, as shown in Figure 8–71.

Therefore, the design of the mixing section has to be matched to the preceeding screw section in order to avoid excessive pressure drop. The effect of the helix angle is shown in Figure 8–72.

As discussed earlier, the helix angle has a strong effect on the pressure drop. The minimum pressure drop occurs at a helix angle between 50 and 60 degrees. Below a helix angle of 50

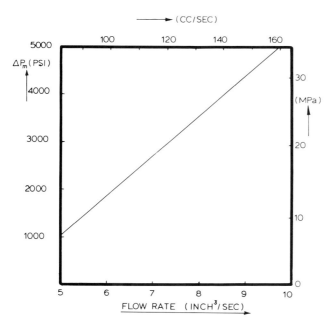

Figure 8–71. Pressure Drop versus Flow Rate

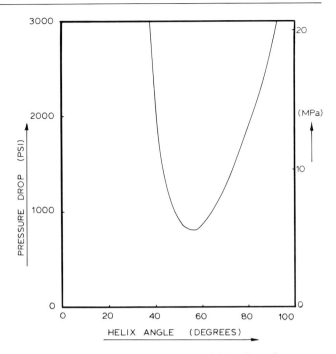

Figure 8-72. Pressure Drop
versus Helix Angle

degrees and above 60 degrees the pressure drop increases quite rapidly. Thus, the proper value of the helix angle is around 50 to 60 degrees.

The effect of the barrier flight width is shown in Figure 8-73.

The pressure drop increases in an approximately proportional fashion with the barrier flight width. This is true if the pressure drop over the clearance is positive ($\Delta P_{cl} > 0$). When the pressure drop over the clearance is made zero ($\Delta P_{cl} = 0$), the width of the barrier flight no longer affects the total pressure drop (see equation 8-146). In all cases, however, the width of the barrier flight will strongly influence the power consumption and the viscous heat generation in the material.

The effect of the barrier clearance is shown in Figure 8-74.

When the clearance is less than about 1/2 mm (0.020 inch), the pressure drop increases quite dramatically. When the clearance is larger than about 3/4 mm (0.030 inch), the effect of the clearance becomes quite small. In fact, when the clearance is larger than 1 mm (0.040 inch), the pressure drop starts to increase because of the reduced drag flow in the inlet channel. It should be noted, however, that changes in the barrier flight width and barrier clearance directly affect the dispersive mixing capability of the mixing section.

Dispersion of agglomerates or gels requires the application of a certain minimum stress to break down the particles. The minimum stress level depends on the nature of the particle as discussed by Martin (54) and Tadmor et al (55). For carbon black, the critical stress level as determined by Martin (54) was found to be around 60 kPa (9 psi). In addition to a minimum stress, there is also a minimum high stress exposure time as discussed by Martin (54). When the duration of high stress is below a minimum exposure time, no dispersion will occur even at very high stress levels. For carbon black, Martin (54) found the minimum exposure time to be about 0.2 seconds. This means that the width of the barrier flight should be large enough so that the residence time of the polymer in the clearance exceeds the minimum exposure time t_{min}. Therefore, the width of the barrier clearance should be:

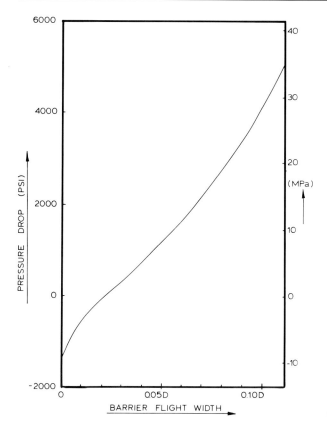

Figure 8-73. Pressure Drop versus Barrier Flight Width

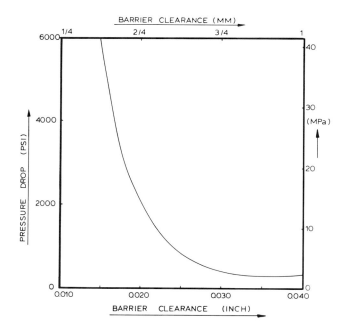

Figure 8-74. Pressure Drop versus Radial Barrier Flight Clearance

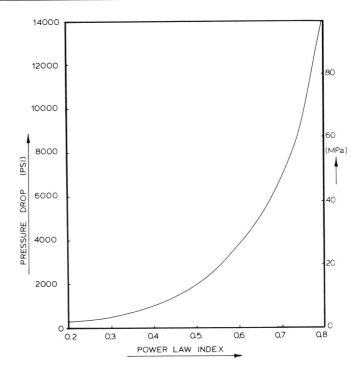

Figure 8–75. Pressure Drop versus Power Law Index

$$w_{cl} > \frac{\pi D N \sin\varphi}{120} \, t_{min} \qquad\qquad (8\text{-}147)$$

where N is expressed in revolutions per minute.

If the critical stress level is τ_{min}, the barrier clearance should be:

$$\delta < \frac{\pi D N}{60} \left(\frac{m}{\tau_{min}} \right)^{\frac{1}{n}} \qquad\qquad (8\text{-}148)$$

Thus, the barrier flight width w_{cl} and the barrier clearance δ have to be designed for both pressure drop and dispersive mixing capacity.

The effect of the degree of non-Newtonian behavior is shown in Figure 8–75.

The effect of the pseudoplasticity is determined by evaluating the local viscosity with the power law equation:

$$\mu = m \, \dot\gamma^{\,n-1} \qquad\qquad (8\text{-}149)$$

where m is the consistency index, $\dot\gamma$ the local shear rate, and n the power law index (see also equation 6–23). If the local shear rate is approximated by the Couette shear rate, the viscosity in the clearance becomes:

$$\mu_{cl} = m\left(\frac{\pi DN}{\delta}\right)^{n-1} \tag{8-150}$$

Similarly, the viscosity in the channel:

$$\mu = m\left(\frac{\pi DN}{H}\right)^{n-1} \tag{8-151}$$

As the power law index increases, the pressure drop increases quite substantially. There-fore, materials with relatively Newtonian flow characteristics can be expected to cause higher pressure drops than strongly non-Newtonian materials. For example, a material like LLDPE will give a much higher pressure drop than regular LDPE of the same melt index because of the larger power law index of LLDPE (4, 5).

The effect of the axial length of the mixing section is shown in Figure 8-76.

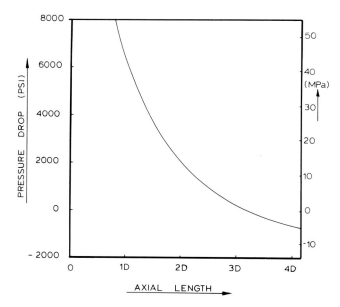

Figure 8-76. Pressure Drop versus Axial Length

The pressure drop reduces substantially when the axial length is increased. Axial lengths of less than 2 D generally create excessive pressure drops. Finally, the effect of the number of inlet channels is shown in Figure 8-77.

The pressure drop initially reduces with the number of inlet channels, but later increases. Thus, there is an optimum number of inlet channels that results in the lowest pressure drop across the mixing section. The optimum number of inlet channels is generally about three or four.

The most important design features for a fluted mixing section can be summarized as fol-lows. The helix angle should be about 50 to 60 degrees, the clearance should not be smaller than about 1/2 mm (0.020 inch), and the axial length should not be less than two diameters.

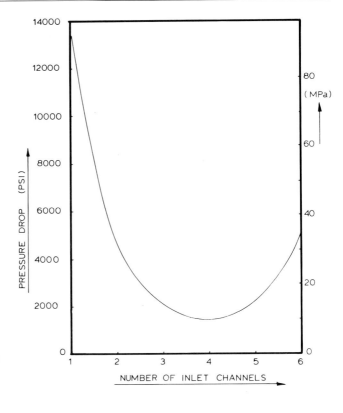

Figure 8-77. Pressure Drop
versus Number of Inlet
Channels

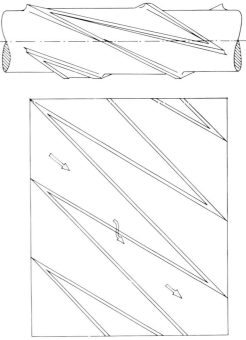

Figure 8-78. Z-Shaped Fluted Mixing Section

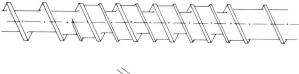

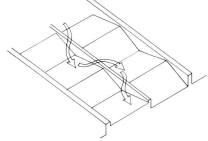

Figure 8-79. Double-Wave Screw

Further, the number of inlet channels should be three or four. The chance of hold-up of material can be reduced by tapering the channel depth. The chance of hold-up can be further reduced by tapering the channel width. This leads to the geometry shown in Figure 8-78.

This geometry has the additional advantage that the pressure drop at the inlet to the mixing section is substantially reduced. This entrance pressure drop has not been taken into account in the analysis. However, it is obvious that the entrance pressure drop in conventional fluted mixing sections can be substantial because the material is forced from the wide screw channel into a number of narrow inlet channels.

Obviously, all barrier-type extruder screws (see Section 8.6.2) impart some degree of dispersive mixing to the polymer because all the polymer has to flow over the barrier flight to leave the extruder. Thus, every polymer element is exposed to a brief but relatively intensive shearing in the barrier clearance. Dispersive mixing also occurs in the double-wave mixing screw, see Figure 8-79.

This screw was developed by Kruder of HPM (39) and is basically an extension of the single channel wave screw (40). Polymer is forced over the center barrier flight by a cyclic variation of the channel depth. When one channel is increasing in depth, the other is reducing. When the first channel reaches its maximum depth, the other channel reaches its minimum depth. Then the first channel starts to reduce in depth and the other channel starts to increase in depth. This process is repeated many times. This screw design gives good mixing performance, but the screw is relatively expensive to manufacture.

8.7.2   DISTRIBUTIVE MIXING ELEMENTS

Distributive mixing is needed where different polymers are blended together with the viscosities reasonably close together. Distributive mixing is easier to achieve than dispersive mixing. Essentially any disruption of the velocity profiles in the screw channel will cause distributive mixing. A common distributive mixing element is the pin mixing section, see Figure 8-80.

The pins cause disturbances in the velocity profile, and thus cause mixing. Many different patterns have been used to place the pins; however, there is little agreement as to what pattern is most effective.

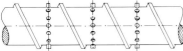

Figure 8-80. Pin Mixing Section

Figure 8-81. Dulmage Mixing Section

Another well known mixing element is the Dulmage mixing section shown in Figure 8-81.

The polymer is divided into many narrow channels combined, divided again, etc. This design was patented almost 30 years ago with Dow Chemical as the assignee (41). A relatively similar mixing section is the Saxton mixing section, shown in Figure 8-82.

The main difference is the orientation of the non-flighted screw sections divided the multi-flighted screw section. This design was patented in 1961 with the assignee being E. I. duPont de Nemours and Company (42). Pineapple-shaped mixing sections are also used for distributive mixing, see Figure 8-83.

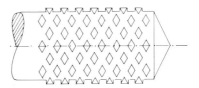

Figure 8-82. Saxton Mixing Section

Figure 8-83. Pineapple Mixing Section

Another technique used to enhance distributive mixing is to put slots in the flight, see Figure 8-84.

A Swedish company, Axon, has patented such a design in many European countries (43). Another distributive mixing section is the cavity transfer mixing (CTM) section shown in Figure 8-85.

Figure 8-84. Slotted Screw Flight

This mixing section was developed by Gale at the RAPRA. It was licensed in the U.S. by David-Standard. It is interesting to note that the cross-cavity mixer concept was described in a patent as early as 1961 (45). The CTM mixer has cavities both in the rotor and barrel housing; the combination of shearing and reorientation appears to give effective distributive mixing. A mixer similar to the CTM mixer was developed earlier by Barmag in Remscheid, West Germany (58). The mixer reportedly performs both dispersive and distributive mixing with the ability to reduce the particle or drop size of additives down to the micrometer (10^{-6} m) range.

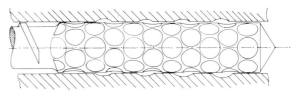

Figure 8-85. Cavity Transfer Mixing Section

It should be remembered, however, that static mixing devices can also be quite effective in distributive mixing capacity, see Section 7.7.2. Thus, if distributive mixing is required, one should certainly consider application of a static mixing device.

8.8   SCALE-UP

One of the first articles on scale-up was written by Carley and McKelvey (59). By analyzing the melt pumping function, they showed that output and power consumption increase by the diameter ratio cubed if channel depth and width are increased in proportion to the diameter ratio and if the screw speed is kept constant. This analysis is only applicable to melt fed extruders where the polymer melt exhibits Newtonian flow characteristics. In plasticating extruders, one has to be concerned with solids conveying, melting, pumping, mixing, and the temperature profiles in the polymer melt. The actual scale-up factors will generally be a compromise between the various functional requirements because different functional requirements often result in conflicting scale-up factors, e.g., mixing and heat transfer.

8.8.1   COMMON SCALE-UP FACTORS

The most commonly used scale-up method maintains constant shear rate by increasing the channel depth proportional to the square root of the diameter ratio and by reducing the screw speed by the square root of the diameter ratio (60). The resulting scaling factors for output, residence time, melting capacity, power consumption, and specific energy consumption are shown in Table 8–2.

The pumping capacity can be checked by using equation 7–218. The drag flow rate can be written as:

$$\dot{V}_d = \left(\frac{4+n}{10}\right) pWH \pi DN \cos \varphi \tag{8–152}$$

If the helix angle φ is constant, $W \propto D$, $H \propto \sqrt{D}$, and $N \propto 1/\sqrt{D}$, the drag flow rate ratio becomes:

$$\frac{\dot{V}_{d\,2}}{\dot{V}_{d\,1}} = \left(\frac{D_2}{D_1}\right)^2 \tag{8–153}$$

The pressure flow rate can be written as:

$$\dot{V}_p = \frac{pWH^3}{4\mu(1+2n)} \frac{\Delta P}{\Delta z} \tag{8–154}$$

The pressure flow rate ratio becomes:

$$\frac{\dot{V}_{p\,2}}{\dot{V}_{p\,1}} = \left(\frac{D_2}{D_1}\right)^{2.5} \frac{\Delta z_1}{\Delta z_2} \tag{8–155}$$

In order for the pressure flow rate to increase by the same rate as the drag flow, the helical length for pressure build-up has to be:

$$\frac{\Delta z_2}{\Delta z_1} = \left(\frac{D_2}{D_1}\right)^{0.5} \tag{8-156}$$

This means that the axial length of the metering section can be increased by the square root of the diameter ratio. Thus, the L/D of the metering section can be reduced by the square root of the diameter ratio.

The melting capacity can be evaluated by examining equation 7-139. The melting rate $\dot{M}_p$ can be written as:

$$\dot{M}_p = \left(\frac{\rho_m k_m v_b \sin\varphi \Delta T_b F_1(K_2)}{\Delta H}\right)^{1/2} \Delta z \sqrt{W_s} \tag{8-157}$$

Thus, the ratio of melting capacity becomes:

$$\frac{\dot{M}_{p2}}{\dot{M}_{p1}} = \left(\frac{v_{b2}}{v_{b1}}\right)^{0.5} \frac{\Delta z_2}{\Delta z_1} \left(\frac{W_{s2}}{W_{s1}}\right)^{0.5} \tag{8-158}$$

With $v_b \propto \sqrt{D}$, $\Delta z_2 \propto D$, and $W_{s2} \propto D$, the ratio becomes:

$$\frac{\dot{M}_{p2}}{\dot{M}_{p1}} = \left(\frac{D_2}{D_1}\right)^{1.75} \tag{8-159}$$

This indicates that the increase in melting rate does not keep up with the increase in output. This problem can be alleviated by increasing the length of the melting section more than a simple proportional increase. In order to keep the increase in melting rate the same as the increase in total throughput, the length of the melting section should increase by $\Delta z_2 \propto D^{1.25}$. Thus, the L/D of the melting section will increase by $L/D_{melt} \propto D^{0.25}$. In order to match the melting capacity and the pumping capacity, the L/D of the melting section has to be increased and the L/D of the pumping section reduced.

The solids conveying rate can be evaluated by using equations 7-46 and 7-48:

$$\dot{M}_s = \rho H p W v_b \left(\cos\varphi - \frac{\sin\varphi}{\tan(\theta+\varphi)}\right) \tag{8-160}$$

If the solids conveying angle is considered to be relatively constant, the solids conveying ratio can be written as:

$$\frac{\dot{M}_{s2}}{\dot{M}_{s1}} = \frac{H_2}{H_1} \frac{W_2}{W_1} \frac{v_{b2}}{v_{b1}} = \left(\frac{D_2}{D_1}\right)^2 \tag{8-161}$$

Thus, the solids conveying rate increases at the same rate as the pumping capacity.

The screw power consumption can be evaluated by considering that the power consumption is roughly determined by the barrel surface area A, the shear stress acting at the barrel surface area τ, and the barrel velocity v_b:

$$Z \propto A \tau v_b \tag{8-162}$$

Considering that $A = \pi D L$ and that the shear stress will be constant because the shear rate is constant as long as the polymer melt viscosity remains constant, the power consumption becomes:

$$Z \propto D^2 L N \tag{8-163}$$

If $L \propto D$ and $N \propto 1/\sqrt{D}$, the power consumption becomes:

$$Z \propto D^{2.5} \tag{8-164}$$

This is not a good situation because the power consumption increases more rapidly than the output, causing an increase in specific energy consumption and thus melt temperature. If the length L is reduced less than proportional to D, then the increase in power consumption can be reduced. In order to match the increase in power consumption to the increase in output, the total length L should increase as:

$$L \propto \sqrt{D} \tag{8-165}$$

This would mean a reduction in the total L/D by $\sqrt{D}$. For the pumping section this is possible; however, it is not possible for the melting section. The other alternative is to reduce the screw speed by more than the square root of the diameter ratio.

TABLE 8-2. COMMON SCALE-UP FACTORS

	Small Extruder	Large Extruder
Diameter	D_1	D_2
Channel width	W_1	$W_2 = W_1 (D_2/D_1)$
Channel depth	H_1	$H_2 = H_1 (D_2/D_1)^{.5}$
Screw speed	N_1	$N_2 = N_1 (D_1/D_2)^{.5}$
Output rate	$\dot{V}_1$	$\dot{V}_2 = \dot{V}_1 (D_2/D_1)^2$
Shear rate	$\dot{\gamma}_1$	$\dot{\gamma}_2 = \dot{\gamma}_1 (D_2/D_1)^0$
Circumferential speed	v_{b1}	$v_{b2} = v_{b1} (D_2/D_1)^{.5}$
Residence time	t_1	$t_2 = t_1 (D_2/D_1)^{.5}$
Plasticating capacity	$\dot{M}_{p1}$	$\dot{M}_{p2} = \dot{M}_{p1} (D_2/D_1)^{1.75}$
Solids conveying rate	$\dot{M}_{s1}$	$\dot{M}_{s2} = \dot{M}_{s1} (D_2/D_1)^2$
Screw power	Z_1	$Z_2 = Z_1 (D_2/D_1)^{2.5}$
Specific energy consumption	$\hat{Z}_1$	$\hat{Z}_2 = \hat{Z}_1 (D_2/D_1)^{.5}$

The effect of the common scale-up factors on extruder performance is presented in tabular form in Table 8-2.

8.8.2   SCALE-UP FOR HEAT TRANSFER

In Section 5.3.3, heat transfer was analyzed in a Newtonian fluid between two plates, one stationary at temperature T_0 and one moving at velocity v and at temperature T_1. When conduction, convection, and dissipation all play a role of importance, the temperature profile is described by equation 5-69. Two dimensionless numbers determine the temperature distribution, the Graetz number and the Brinkman number. If these numbers remain the same in scale-up, the temperature profile in the polymer melt will also remain the same. A constant Graetz number requires that:

$$\frac{vH^2}{\alpha L} = \text{constant} \qquad\qquad (8\text{--}166)$$

It can be assumed that the thermal diffusivity is constant. Considering that $v = \pi DN$ and the L/D ratio is usually constant, equation 8-166 can be written as:

$$NH^2 = \text{constant} \qquad\qquad (8\text{--}167)$$

A constant Brinkman number requires that:

$$\frac{\eta v^2}{k \Delta T} = \text{constant} \qquad\qquad (8\text{--}168)$$

If it can be assumed that the viscosity η, the thermal conductivity k, and the imposed temperature difference ΔT are constant, equation 8-168 becomes:

$$v = \text{constant} \qquad\qquad (8\text{--}169)$$

This means that the circumferential speed ($v = \pi DN$) has to be constant. With equation 8-167 and 8-169 the channel depth and screw speed can be expressed as a function of diameter:

$$N \propto 1/D \qquad\qquad (8\text{--}170a)$$

$$H \propto \sqrt{D} \qquad\qquad (8\text{--}170b)$$

The effect of these scale-up factors on extruder performance is shown in Table 8-3. It can be seen that there is a good match between pumping rate, melting rate, and solids conveying rate. Further, the specific energy consumption remains constant, thus the melt temperature level should be about the same in scale-up. The main disadvantage of this approach is that the output will be considerably lower than with the common scale-up factors, see Table $8-2$. For instance, if $D_1 = 50$ mm and $\dot{M}_1 = 100$ kg/hr, then for $D_2 = 150$ mm, the common scale-up factor will give $\dot{M}_2 = 900$ kg/hr, whereas the scale-up for heat transfer will give $\dot{M}_2 = 520$ kg/hr. In practice, therefore, the screw speed will be increased until the melt temperature almost reaches the maximum acceptable temperature. Another advantage is that the residence time increases faster than with the common scale-up factors.

TABLE 8-3. SCALE-UP FOR HEAT TRANSFER

	Small Extruder	Large Extruder
Diameter	D_1	D_2
Channel width	W_1	$W_2 = W_1 \ (D_2/D_1)$
Channel depth	H_1	$H_2 = H_1 \ (D_2/D_1)^{.5}$
Screw speed	N_1	$N_2 = N_1 \ (D_1/D_2)$
Output rate	$\dot{V}_1$	$\dot{V}_2 = \dot{V}_1 \ (D_2/D_1)^{1.5}$
Shear rate	$\dot{\gamma}_1$	$\dot{\gamma}_2 = \dot{\gamma}_1 \ (D_1/D_2)^{.5}$
Circumferential speed	v_{b1}	$v_{b2} = v_{b1} \ (D_2/D_1)^{0}$
Residence time	t_1	$t_2 = t_1 \ (D_2/D_1)$
Plasticating capacity	$\dot{M}_{p1}$	$\dot{M}_{p2} = \dot{M}_{p1} \ (D_2/D_1)^{1.5}$
Solids conveying rate	$\dot{M}_{S1}$	$\dot{M}_{S2} = \dot{M}_{S1} \ (D_2/D_1)^{1.5}$
Screw power	Z_1	$Z_2 = Z_1 \ (D_2/D_1)^{1.5}$
Specific energy consumption	$\hat{Z}_1$	$\hat{Z}_2 = \hat{Z}_1 \ (D_2/D_1)^{0}$

8.8.3.   SCALE-UP FOR MIXING

If it is assumed that the two most important parameters in mixing are shear rate and residence time, then scale-up rules can be derived that will keep these parameters constant. The shear rate is approximately:

$$\dot{\gamma} \simeq \frac{\pi DN}{H} \qquad (8\text{-}171)$$

The residence time is:

$$t = \frac{L}{\pi DN} \qquad (8\text{-}172)$$

If L/D is constant, then equation 8-172 requires that the screw speed be constant. From equation 8-171, it can be seen that with constant N the ratio of diameter to channel depth also must be constant. Thus, the scale-up factors for mixing become:

$$N = \text{constant} \qquad (8\text{-}173a)$$

$$H \propto D \qquad (8\text{-}173b)$$

The effect of these scale-up factors on extruder performance is shown in Table 8-4. The main problem with this scale-up approach is that the output increases much faster than the melting capacity. This approach, therefore, will not work unless special design changes are made in the melting section of the screw to substantially enhance the melting capacity, see Section 8.2.2, Optimizing for Plasticating. The advantage of this scale-up approach is that very high outputs are obtained and the specific energy consumption remains constant.

TABLE 8-4. SCALE-UP FOR MIXING

	Small Extruder	Large Extruder
Diameter	D_1	D_2
Channel width	W_1	$W_2 = W_1(D_2/D_1)$
Channel depth	H_1	$H_2 = H_1 \ (D_2/D_1)$
Screw speed	N_1	$N_2 = N_1 \ (D_2/D_1)^0$
Output rate	$\dot{V}_1$	$\dot{V}_2 = \dot{V}_1 \ (D_2/D_1)^3$
Shear rate	$\dot{\gamma}_1$	$\dot{\gamma}_2 = \dot{\gamma}_1 \ (D_2/D_1)^0$
Circumferential speed	v_{b2}	$v_{b2} = v_{b1} \ (D_2/D_1)$
Residence time	t_1	$t_2 = t_1 \ (D_2/D_1)^0$
Plasticating capacity	$\dot{M}_{p1}$	$\dot{M}_{p2} = \dot{M}_{p1}(D_2/D_1)^2$
Solids conveying rate	$\dot{M}_{s1}$	$\dot{M}_{s2} = \dot{M}_{s1} \ (D_2/D_1)^3$
Screw power	Z_1	$Z_2 = Z_1 \ (D_2/D_1)^3$
Specific energy consumption	$\hat{Z}_1$	$\hat{Z}_2 = \hat{Z}_1 \ (D_2/D_1)^0$

A comparison of the effect on output of the different scale-up strategies is shown in Table 8-5. If the diameter of the small extruder is 50 mm (2 inch) and the output 100 kg/hr (220 lbs/hr), the output for a 150-mm (6-inch) extruder will be as shown in Table 8-5.

TABLE 8-5. OUTPUT ACCORDING TO VARIOUS SCALE-UP
            RULES

	Output 150-mm Extruder
Common scale-up factors	900 kg/hr (1985 lbs/hr)
Scale-up for heat transfer	520 kg/hr (1145 lbs/hr)
Scale-up for mixing	2700 kg/hr (5952 lbs/hr)

From Table 8-5, it can be seen that scale-up for mixing results in unrealistically high output values. Summarizing, it can be stated that scale-up for heat transfer will result in matched solids conveying, melting, and pumping in addition to constant specific energy consumption. Scale-up for heat transfer, therefore, will result in good extruder performance and melt temperature control. However, outputs according to the scale-up for heat transfer are rather low. Outputs can be increased by increasing the screw speed by more than $N \propto 1/D$, however this will result in an increase in specific energy and insufficient melting capacity. Thus, the melt quality will deteriorate. In practice, one would increase the screw speed to just below the point where the melt quality becomes unacceptable.

REFERENCES – CHAPTER 8

1. R. T. Fenner and J. G. Williams, Polym. Eng. Sci., 11, 474-483 (1971).
2. C. J. Rauwendaal, SPE ANTEC, Chicago, 186-199 (1983).
3. J. Dekker, Polytechnisch Tijdschrift (Dutch), 31, 742-746 (1976).
4. C. J. Rauwendaal, SPE ANTEC, Chicago, 151-154 (1983).
5. C. J. Rauwendaal, Plastics Technology, August, 61-63 (1983).
6. C. J. Rauwendaal, U. S. Patent 4,129,386.
7. C. J. Rauwendaal, SPE ANTEC, New York, 110-113 (1980).
8. B. Miller, Plastics World, March, 34-38 (1982).
9. R. M. Bonner, SPE J., October, 1069-1073 (1963).
10. W. Backhoff, R. von Hooren, and F. Johannaber, Kunststoffe, 6, 307 (1977).
11. J. L. Duda, J. S. Vrentas, S. T. Ju, and H. T. Liu, AIChE J., 28, 279 (1982).
12. D. Anders, in "Entgasen von Kunststoffen", VDI-Verlag, Duesseldorf (1980).
13. J. F. Carley, SPE J., 24, 36-41 (1968).
14. W. H. Willert, paper given at Newark section of the SPE on February 8 (1961).
15. B. H. Maddock and P. P. Matzuk, SPE J., 18, 405-408 (1962).
16. D. Anders, Kunststoffe, 66, 250-257 (1976).
17. H. Werner and J. Curry, SPE ANTEC, Boston, 623-626 (1981).
18. N. J. Brozenick and G. A. Kruder, SPE ANTEC, San Francisco, 176-181 (1974).
19. R. J. Nichols, G. A. Kruder, and R. E. Ridenour, SPE ANTEC, Atlantic City, 361-363 (1976).
20. E. C. Bernhardt, SPE J., 12, 40-57 (1956).
21. B. Franzkoch, in "Entgasen von Kunststoffen", VDI-Verlag, Duesseldorf (1980).
22. B. H. Maddock, SPE ANTEC, San Francisco, 247-251 (1974).
23. Ch. Maillefer, Swiss Patent 363,149.
24. Ch. Maillefer, German Patent 1,207,074.
25. Ch. Maillefer, British Patent 964,428.
26. P. Geyer, U. S. Patent 3,375,549.
27. J. F. Ingen Housz and H. E. H. Meijer, Polym. Eng. Sci., 21, 352-359 (1981).
28. R. Barr, U. S. Patent 3,698,541.
29. C. I. Chung, U. S. Patent 4,000,884.
30. W. H. Willert, European Patent Application 34,505, August 26, 1981.
31. R. F. Dray and D. L. Lawrence, U. S. Patent 3,650,652.
32. J. S. Hsu, U. S. Patent 3,858,856.
33. H. T. Kim, U. S. Patent 3,867,079.
34. H. T. Kim, U. S. Patent 3,897,938.
35. J. F. Ingen Housz, U. S. Patent 4,218,146.
36. J. F. Ingen Housz and H. E. H. Meijer, Polym. Eng. Sci., 21, 1156-1161 (1981).
37. G. LeRoy, U. S. Patent 3,486,192.
38. B. H. Maddock, SPE J., July, 23-29 (1967).
39. G. Kruder, U. S. Patent 4,173,417.
40. G. Kruder, U. S. Patent 3,870,284.
41. F. E. Dulmage, U. S. Patent 2,753,595.
42. R. L. Saxton, U. S. Patent 3,006,029.
43. German Patent 2,026,834.
44. G. M. Gale, SPE ANTEC, Chicago, 109-112 (1983).
45. English Patent 930,339.
46. F. J. Brinkschroeder and F. Johannaber, Kunststoffe, 71, 138-143 (1981).
47. R. B. Gregory and L. F. Street, U. S. Patent 3,411,179.
48. R. B. Gregory, U. S. Patent 3,788,614.
49. R. G. Dray, U. S. Patent 3,788,612.

50. R. A. Worth, Polym. Eng. Sci., 19, 198–202 (1979).
51. Z. Tadmor and I. Klein, Polym. Eng. Sci., 13, 382 (1973).
52. B. Elbirli, J. T. Lindt, S. R. Gottgetreu, and S. M. Baba, SPE ANTEC, Chicago, 104 (1983).
53. C. Rauwendaal, SPE ANTEC, New Orleans, 59–63 (1984).
54. G. Martin, Industrie-Anzeiger, 14, 2651 (1971).
55. I. Manas-Zloczower, A. Nir, and Z. Tadmor, Rubber Chem. Techn., 55, 1250 (1983).
56. P. Heidrich, German Patent 1145 787 (1959).
57. F. K. Lacher, U. S. Patent 3,271,819.
58. M. H. Pahl, "Dispersives Mischen mit dynamischen Mischern", VDI-Verlag, Duessel-dorf, 177–196 (1978).
59. J. F. Carley and J. M. McKelvey, Ind. Eng. Chem., 45, 985 (1953).
60. C. I. Chung, Polym. Eng. Sci., 24, 626–632 (1984).
61. J. C. Miller, Tappi Journal, 67, 64–67, June (1984).
62. M. Kearney and P. Hold, SPE ANTEC, Washington, DC, 17–22 (1985).
63. C. J. Rauwendaal, Polymer Extrusion III Conference, London, 7/1–7/16, September 11–13 (1985)

9   DIE DESIGN

Die design is one aspect of extrusion engineering that has remained more of an art than any other aspect. The obvious reason is that it is quite difficult to determine the optimum flow channel geometry from engineering calculations. Description of the flow of the polymer melt through the die requires knowledge of the viscoelastic behavior of the polymer melt. The polymer melt can no longer be considered a purely viscous fluid because elastic effects in the die region are very important. Unfortunately, there are no simple constitutive equations that adequately describe the flow behavior of polymer melt over a wide range of flow conditions. Thus, a simple die flow analysis is generally very approximate, while more accurate die flow analyses tend to be quite complicated. Many of the more accurate die flow analyses make use of the finite element method (FEM). For more information on die design and the use of FEM in die design, the reader is referred to Michaeli's book on extrusion dies (1). Detailed information on the use of FEM in the analysis of non-Newtonian flow can be found in the book by Crochet, Davies, and Walters (29).

9.1   BASIC CONSIDERATIONS

The objective of an extrusion die is to distribute the polymer melt in the flow channel such that the material exits from the die with a uniform velocity. The actual distribution will be determined by the flow properties of the polymer, the flow channel geometry, the flow rate through the die, and the temperature field in the die. If the flow channel geometry is optimized for one polymer under one set of conditions, a simple change in flow rate or in temperature can make the geometry non-optimum. Except for circular dies, it is essentially impossible to obtain a flow channel geometry that can be used, as such, for a wide range of polymers and for a wide range of operating conditions. For this reason, one generally incorporates adjustment capabilities into the die by which the distribution can be changed externally while the extruder is running. The flow distribution is generally changed in two ways: i) by changing the flow channel geometry by means of choker bars, restrictor bars, valves, etc., and ii) by changing the local die temperature. Such adjustment capabilities complicate the mechanical design of the die but enhance its flexibility and controllability.

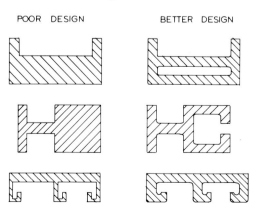

Figure 9-1. Some Examples of Profile Design Guidelines

Some general rules that are useful in die design are:

- no dead spots in the flow channel
- steady increase in velocity along the flow channel
- assembly and disassembly should be easy
- land length about 10 × land clearance
- avoid abrupt changes in flow channel geometry
- use small approach angles

In die design, problems often occur because the product designer has little or no appreciation for the implications of the product design details on the ease or difficulty of extrusion. In many cases, small design changes can drastically improve the extrudability of the product. Some basic guidelines in profile design to minimize extrusion problems are:

- use generous internal and external radii on all corners; the smallest possible radius is about 0.5 mm
- maintain uniform wall thickness (important!)
- avoid very thick walls
- make interior walls thinner than exterior walls for cooling
- minimize the use of hollow sections

Figure 9-1 illustrates applications of these guidelines to a few different profiles.

9.2   FILM AND SHEET DIES

Dies for flat film are essentially the same as dies for sheet extrusion. The difference between sheet and film is primarily the thickness. Webs with a thickness of less than 0.5 mm are generally referred to as film; while webs with a thickness of more than 0.5 mm are generally referred to as sheet. Three distribution channel geometries used in sheet dies are shown in Figure 9-2.

Figure 9-2a shows the T-die. This flow channel geometry is simple and easy to machine. However, the distribution of the polymer melt is not very uniform and the flow channel geometry is not well streamlined. Thus, this die is not suitable for high viscosity polymers with limited thermal stability. The T-die is used in extrusion coating applications. Analyses of the flow in T-die have been made by Weeks (2, 3), Ito (4) and Pearson (5).

The fishtail die is shown in Figure 9-2b. This die results in a more uniform melt distribution than the T-die; however, a completely uniform distribution is still difficult to obtain with this geometry. Analyses of the flow in fishtail dies have been made by Ito (6) and Chejfec (7). Figure 9-2c shows the coathanger die. This is a die geometry commonly used in sheet extrusion. The geometry of the coathanger section can be designed to give a very uniform distribution of the polymer melt. Obviously, the coathanger die is more difficult to machine and, therefore, more expensive than the T-die and fishtail die. Analyses of the flow in coathanger dies have been made by Ito (8, 9), Wortberg (10), Goermar (11, 12), Chung (13), Klein (14), Schoenewald (15), Vergnes (16), Matsubara (17, 18), and probably many more. Goermar used the sinh law (see equation 6-32) as the constitutive equation for the polymer melt. He obtains a remarkably simple expression for the geometry of the coathanger section, see Figure 9-2c. For the radius as a function of distance, Goermar derived the following expression:

$$R(x) = R_0 \left(\frac{x}{b}\right)^{1/3}$$ (9–1a)

and for the land length:

$$L(x) = L_0 \left(\frac{x}{b}\right)^{2/3}$$ (9–1b)

Figure 9–3 shows two commonly used techniques to change the flow channel geometry in sheet dies. The first is the flex lip adjustment. A number of bolts along the width of the die allow local closing of the final land gap. This allows fine adjustments of the extrudate thickness at discrete points. The gap can be adjusted by as much as 1 mm or more if the flex lip is properly designed. The choker bar is not used as often as the flex lip. The choker bar adjustment works in a similar fashion as the flex lip adjustment. The choker bar can be locally deformed by a number of bolts located along the width of the die. The deformation of the choker bar causes a change in the height of the flow channel and, thus, allows an adjustment of the flow distribution in the die. A third adjustment possibility, not shown in Figure 9–3, is the die temperature. Local heating or cooling of certain die sections enhance or restrict flow; this is another means of flow distribution adjustment. Temperature adjustment will be more effective with polymers whose viscosity is quite sensitive to temperature; this includes most amorphous polymers, see also Table 6–1.

Figure 9–3 also shows a feature of the sheet die that is useful when a large thickness range is necessary. By incorporating a removable lower die lip, the die can be used, for instance, for 1-mm sheet extrusion and for 4-mm sheet extrusion by changing only the lower die lip.

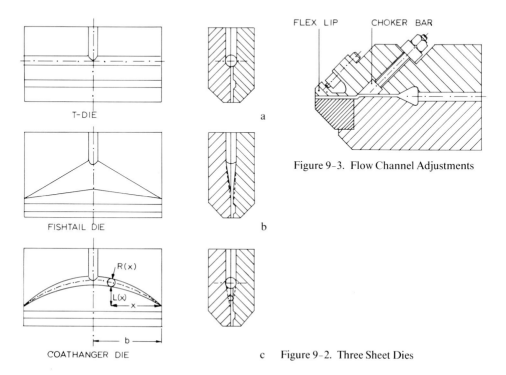

FLEX LIP   CHOKER BAR

T-DIE

FISHTAIL DIE

R(x)
L(x)
x
b
COATHANGER DIE

a

b

c

Figure 9-3. Flow Channel Adjustments

Figure 9-2. Three Sheet Dies

Without the removable die lip, another sheet die would have to be used because the flex lip can only adjust over a limited range. In some automated extrusion lines, the gap of the sheet die is adjusted automatically. This is done with heat expandable die lip bolts (19), first developed by Welex in the early seventies. A similar type of sheet die using thermal bolts is now also offered by Egan Machinery Company and a number of other companies. Another concept is used by Harrel, Inc. Their sheet die has die temperature adjustment capability at various locations along the width of the die. By raising or lowering the temperature automatically, the sheet thickness can be controlled without changing the die lip gap. An interesting automatically adjustable sheet die was developed by Hexco. This sheet die uses a PC servo positioner to hydraulically manipulate each die bolt. This design is said to reduce the response time from minutes, which is typical in thermally adjusted sheet dies, down to about two seconds. However, Hexco went out of business in 1983 and it is not clear whether this sheet extrusion system will still be commercially available.

A major drawback of the conventional coathanger die with the teardrop shape distribution channel is the fact that the distribution changes when the power law index of the material changes. Thus, the distribution will change when a change in polymer is made and also when the output is changed because the power law index is generally somewhat dependent on shear rate. Therefore, flex lips and choker bars are generally used to compensate for the imperfect distribution in the die. A modified coathanger geometry was proposed by Winter and Fritz (30) that eliminates the problem of the power law index dependence of the distribution.

The main feature of the new coathanger geometry is a slit shaped distribution in the shape of a horseshoe channel, see Figure 9-4.

If the ratio of channel width W to channel depth H is larger than about 10, the shape factor becomes independent of the power law index. For a manifold with constant channel width W, the distribution channel geometry is given by (see Figure 4):

$$H(x) = h \sqrt{(b-x)/W} \qquad\qquad (9-2a)$$

and

$$y(x) = 2W \sqrt{(b-x)/W} - 1 \qquad\qquad (9-2b)$$

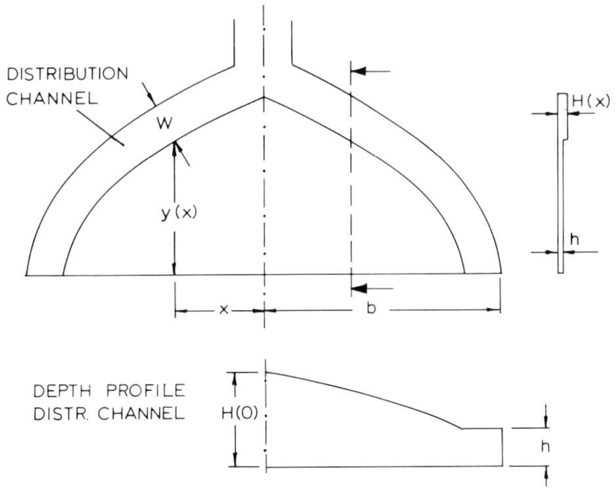

Figure 9-4. Horseshoe Distribution Channel

For a manifold with constant aspect ratio, $W/H = a$, the depth profile is given by:

$$H(x) = h\left(\frac{b-x}{ah}\right)^{1/3} \tag{9-2c}$$

The corresponding contour line is given by:

$$y(x) = \frac{3b}{2A}\left[\left(\frac{1+g(x)}{g^2(x)}\right)^{1/2} - 0.5\ln\left(\frac{\sqrt{1+g(x)}-1}{\sqrt{1+g(x)}+1}\right)\right] + C \tag{9-2d}$$

where:

$$g(x) = \frac{h^2}{F^{2n}H^2(x)-h^2} \tag{9-2e}$$

The integration constant C is determined by setting $H(x) = h$ at $y = 0$. The front factor is defined as:

$$A = \frac{F^{3n}b}{ah} \tag{9-2f}$$

The shape factor F for a rectangular flow channel is given by equations 7-169 and 7-171. If the shape factor is taken as unity, the contour line becomes:

$$y(x) = \frac{3}{2}ah\left[\frac{\sqrt{1-\alpha^2}}{\alpha^2} - \frac{1}{2}\ln\left(\frac{1-\sqrt{1-\alpha^2}}{1+\sqrt{1-\alpha^2}}\right)\right] + h \tag{9-2g}$$

where:

$$\alpha = \frac{h}{H(x)} = \left(\frac{ah}{b-x}\right)^{1/3} \tag{9-2h}$$

The new coathanger geometry can be applied to flat sheet dies as well as annular dies. Winter (30) reports on applications with blow molding dies with a circumferential thickness distribution of the parison between 5 percent and 8 percent. Very good results were also obtained on a flat sheet die of 0.25 meter width and a flat profile die with a width of 2 meters.

9.3   PIPE AND TUBING DIES

The difference between pipe and tubing is mainly determined by size. Small diameter products (less than about 10 mm) are generally referred to as tubing, while large diameter products are generally referred to as pipe. Tubing is often extruded on crosshead dies, see Figure 9-5.

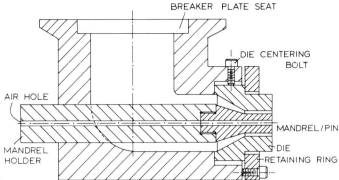

Figure 9-5. Crosshead Die

The direction of the inlet flow is perpendicular to the outlet flow. The polymer melt makes a 90° turn and splits at the same time over the core tube. The polymer melt recombines below the core tube; this is where a weld line will form. After the 90° turn, the polymer melt flows through the annular flow channel where it adopts more or less the shape of the final land region.

Weld lines are usually unavoidable in hollow extruded products. The polymer has to be given sufficient opportunity to "heal" along the knit line. This healing process is essentially a re-entanglement of the polymer molecules. Important parameters in this process are time, temperature, and pressure. Analyses of the healing process have been made by Prager and Tirrell (20) and Wool et al. (21–23). The problem of weld lines also occurs in injection molding, as discussed, for instance, by Malguarnera and Manisali (24). The healing time reduces with temperature but increases with molecular weight. In practical terms, this means that the point of weld line formation has to be a reasonably large distance upstream of the die exit to enable the polymer to heal sufficiently.

The crosshead die is also used for wire coating. In wire coating, a conductor passes through the hollow center of the core tube and becomes coated with polymer melt close to the die exit. The conductor may be a bare conductor or it may already have been coated with one or

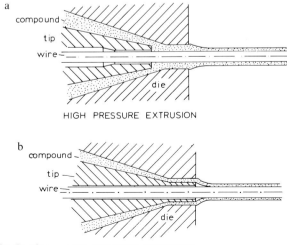

Figure 9-6. High and Low Pressure Wire Coating

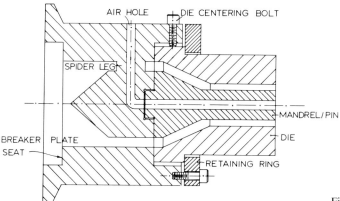

Figure 9-7. In-Line Pipe Die

more layers of polymer. In wire coating, one distinguishes between high pressure extrusion and low pressure extrusion, see Figure 9-6.

In high pressure extrusion, Figure 9-6a, the polymer melt meets the conductor before the die exit. This allows for good contact between the conductor and the polymer. In low pressure extrusion, Figure 9-6b, the polymer melt meets the conductor after the die exit. The polymer is tubed down over the conductor. Low pressure extrusion is used when good contact between the wire and the polymer is not essential, for instance, when a loose jacket needs to be extruded over a coated wire.

In most crosshead dies, the location of the die relative to the tip can be adjusted by means of centering bolts, see Figure 9-5. This allows adjustment of the wall thickness distribution and concentricity. In some extrusion lines with in-line wall thickness measurement, an automatic wall thickness control is obtained by using the signal from the wall thickness probe to automatically adjust the position of the die. In some cases, a slight internal air pressure is applied through the center of the core tube in order to maintain the ID of the tubing or to prevent collapse of the tubing. This is particularly useful in extrusion of tubing with very small ID, 0.1 mm or less. In wire coating, sometimes a vacuum is applied to the center of the core tube to prevent air being dragged along with the conductor; this could cause imperfect contact between the polymer and the conductor.

Pipe is often extruded with in-line pipe dies, see Figure 9-7.

In these dies, the center line of the die is in line with the center line of the extruder. The central torpedo is supported by a number of spider legs, usually three or more. The spider legs are relatively thin and streamlined to minimize the disruption of the velocity profile. Of course, as the polymer recombines after the spider leg, a weld line will form. Thus, the location of the spider support should be far enough from the die exit to enable the polymer to heal. The location of the die is generally adjustable relative to the pin, just as in the crosshead die.

9.3.1 CALIBRATORS

In pipe extrusion, the extrusion die is often followed by a sizing die, where the actual dimensions of the pipe are primarily determined. Such a device is generally referred to as a calibrator. Calibrators are required if the extrudate emerging from the die has insufficient

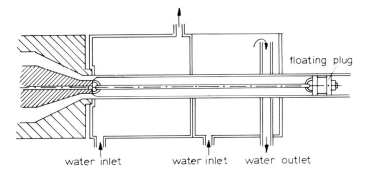

Figure 9-8. Calibrator With Internal Air Pressure

water inlet   water inlet   water outlet

melt strength to maintain the required shape. The calibrator is in close contact with the polymer melt and cools the extrudate. When the extrudate leaves the calibrator, it has sufficient strength to be pulled through a haul-off device, such as a catapuller. In a vacuum calibrator, a vacuum is applied to ensure good contact between the calibrator and the extrudate and to prevent collapse of the extrudate. The use of a vacuum calibrator is generally easier than the use of positive air pressure within the extrudate because a vacuum is easier to maintain at a constant level. Positive internal air pressure tends to vary with the length of the extrudate and it is difficult to maintain when the extrudate has to be cut into discrete lengths. Calibrators are useful when good, accurate shape control is important. When the requirements for shape control are less stringent, the extrudate shape is often maintained by support brackets placed downstream of the extrusion die. In fact, the extrudate shape can be modified substantially by the support brackets. This can be useful because it allows modification of the shape of the extrudate without changing the die, but only by changing the shape of the support brackets.

As discussed by Michaeli (1), there are five types of calibrators:

1. Slide calibrators
2. External calibrators with internal air pressure
3. External vacuum calibrators
4. Internal calibrators
5. Precision profile pultrusion (Technoform process)

Slide calibrators are used for simple and open profiles. The extrudate is more or less in contact with cooled plates and the profile is pulled through the calibrator, causing some amount of drawdown. An example of an external calibrator with internal air pressure is shown in Figure 9-8.

The air pressure is maintained by a sealing mandrel located inside the extrudate downstream of the calibrator. The mandrel is attached to the tip by a cable to fix its position. This

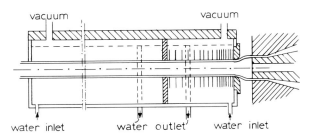

Figure 9-9. Calibrator With External Vacuum

water inlet   water outlet   water inlet

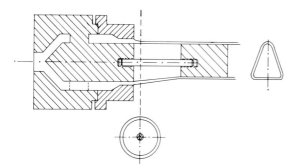

Figure 9-10.   Internal Calibrator

type of calibration is used in larger diameter pipe for PVC (D > 350 mm) and PE (D > 100 mm). An example of an external vacuum calibrator is shown in Figure 9-9.

The profile is pulled through apertures with a reduced pressure applied to this section; the aperture section is followed by a simple water cooling section. External vacuum calibrators are used for hollow profiles and smaller diameter pipe. An example of an internal calibrator is shown in Figure 9-10.

This internal calibrator modifies the annular shape of the extrudate emerging from the die into a more or less triangular shape. This allows shape modification just by the calibrator. However, internal calibration is not used very often. In precision profile pultrusion, a small amount of polymer is allowed to accumulate between the die and the calibrator. The accumulation is controlled by a sensor through adjustment of the haul-off speed. The extrudate is pulled through a short, intensely cooled calibrator followed by a water bath. This type of calibration is referred to as the Technoform process; it was developed by Reifenhauser KG.

9.4   BLOWN FILM DIES

The most common die used in blown film extrusion is the spiral mandrel die. In this die, the polymer is divided into a number of spiraling channels with the depth of the channels reducing in the direction of flow. The popularity of the spiral mandrel die is due to its relatively low pressure requirement and its excellent melt distribution characteristics. Spiral mandrel dies can be used with a wide range of materials over a wide range of operating conditions. A simpler die is the in-line tubing die, see Figure 9-7, but the presence of weld lines makes this die less suitable for use in blown film. Another die that can potentially be used for blown film is the crosshead tubing die, see Figure 9-5. However, the distribution characteristics of this die are generally not good enough for application in blown film extrusion, where wall thicknesses are generally quite small (the typical range is 0.005 mm to 0.25 mm). The crosshead die has the same problem as the in-line tubing die in that it produces weld lines in the polymer. Thus, the spiral mandrel die has become the most common die for blown film extrusion.

A simplified picture of a spiral mandrel die is shown in Figure 9-11.

The incoming polymer melt stream is divided into separate feed ports. Each feed port feeds the polymer into a spiral groove machined into the mandrel. The cross-sectional area of the groove decreases with distance, while the gap between the mandrel and the die increases towards the die exit. This multiplicity of flow channels results in a smearing or layering of

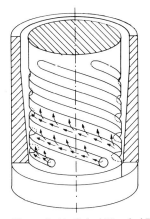

Figure 9-11. Spiral Mandrel Die

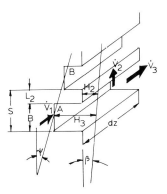

Figure 9-12. Flow in Spiral Section

polymer melt from the various feed ports, yielding a good distribution of the polymer melt exiting from the die. It is obvious that local gap adjustment is not possible as it is with flat sheet dies. As a result, the wall thickness uniformity with spiral mandrel dies is generally not as good as with flat sheet dies. The latter can generally achieve a wall thickness uniformity of about ±5 percent, while the blown film die achieves a wall thickness uniformity of about ±10 percent. For this reason, the die is generally made to rotate to evenly distribute the wall thickness nonuniformities. If this were not done, the final roll of product would show very noticeable variations in diameter.

Analyses of flow in spiral mandrel dies have been made by Ast (25), Wortberg and Schmidtz (26), Proctor (27), Predoehl (28), and others. Proctor (27) used the following assumptions to analyze the flow in the spiral mandrel die:
 - The polymer melt behaves as a power law fluid.
 - The curvature of the mandrel can be neglected.
 - The spiral groove is assumed to have a rectangular cross-section.
 - Symmetrical flow is assumed in the groove and in the clearance.
 - The leakage flow does not affect the flow in the spiral.
 - Constant pressure gradient along the channel.

The incoming flow $\dot{V}_1(z)$ is divided into leakage flow $\dot{V}_2(z)$ and spiral flow $\dot{V}_3(z)$, see Figure 9-12.

$$\dot{V}_1(z) = \dot{V}_2(z) + \dot{V}_3(z) \qquad (9\text{-}3)$$

Leakage flow $\dot{V}_2$ is a pressure-driven flow through a slit. It can be described by using Table 7-1.

$$\dot{V}_2(z) = \frac{dzH_2^2}{2(s+2)} \left[\frac{H_2 \Delta P_2}{2mL_2} \right]^s \qquad (9\text{-}4)$$

Similarly, $\dot{V}_3$ can be written as:

$$\dot{V}_3(z) = \frac{BH_3^2}{2(s+2)} \left[\frac{H_3}{2m} g_z \right]^s \qquad (9\text{-}5)$$

Pressure drop ΔP_2 is related to the down channel pressure gradient g_z by:

$$\Delta P_2 = \frac{Sg_z}{\sin\varphi} \qquad (9\text{--}6)$$

The ratio of $\dot{V}_2$ to $\dot{V}_3$ is:

$$\frac{\dot{V}_2(z)}{\dot{V}_3(z)} = \frac{dz}{B}\left[\frac{H_2(z)}{H_3(z)}\right]^{S+2}\left[\frac{S}{L_2(z)\sin\varphi}\right]^S \qquad (9\text{--}7)$$

The ratio of $\dot{V}_2$ to $\dot{V}_1$ is:

$$\frac{\dot{V}_2(z)}{\dot{V}_1(z)} = \frac{1}{\dfrac{\dot{V}_3(z)}{\dot{V}_2(z)} + 1} \qquad (9\text{--}8)$$

If $\dot{V}_0$ is the inlet flow to the spiral flow channel, $\dot{V}_1(z)$ can be written as:

$$\dot{V}_1(z) = \dot{V}_0 - \int_0^z \frac{\dot{V}_2(z)}{dz}\,dz \qquad (9\text{--}9)$$

The leakage flow from spiral channel B can be expressed as a function of the leakage flow from the previous spiral channel A by:

$$\dot{V}_2(z)_B = \dot{V}_1(z)_B + \dot{V}_2(z)_A - \dot{V}_3(z)_B \qquad (9\text{--}10)$$

The problem is solved by a stepwise progression from the start of the spiral flow channel. The geometrical parameters B, H_2, H_3, L_2, φ, β and S can be determined in such a way as to yield the most uniform flow distribution.

9.5   PROFILE EXTRUSION DIES

Apart from rectangular and annular extrudates, there is a tremendous variety of extruded profiles with other shapes. Profile dies usually describe dies used to produce shapes other than rectangular or annular. Profile extrusion is generally the most difficult type of extrusion. This is, to a large extent, due to the fact that it is very difficult to accurately predict the required geometry of the flow channel that will yield an extruded product of proper shape and dimensions.

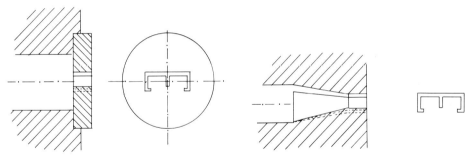

Figure 9–13. Plate Die   Figure 9–14. Streamlined Profile Die

In profile extrusion, there are two extremes in die design. One extreme is the plate die shown in Figure 9–13.

In this design, a plate with the required opening is placed abruptly at the end of the die flow channel with a minimum amount of streamlining. This type of die is simple, easy to make, and easy to modify. However, there is a large dead flow region and degradation is a definite concern with polymers with limited thermal stability. This type of die, therefore, should be used with relatively stable polymers and preferably only for short times.

The other extreme is the highly streamlined profile die shown in Figure 9–14.

In this die, there is a very gradual transition from the geometry of the die inlet channel to the geometry of the die outlet channel. Obviously, this die is more complex, more difficult to manufacture, and more difficult to modify. Good streamlining is important if the polymer is susceptible to degradation. A well-streamlined profile die is more likely to be used for a long extrusion run because the manufacturing cost can be spread over a larger amount of product and because degradation becomes more of a problem in longer runs. Obviously, there are an infinite number of intermediate die designs between the totally non-streamlined plate die and the fully streamlined profile die.

9.6   COEXTRUSION

A relatively new extrusion technique is coextrusion. It is the simultaneous extrusion of two or more polymers through a single die where the polymers are joined together such that they form distinct, well-bonded layers forming a single extrusion product. Coextrusion has been applied in film, sheet, tubing, blown film, wire coating, and other types of profile extrusion.

Advantages of coextrusion are better bonds between layers, reduced materials and processing costs, improved properties, and reduced tendency for pinholes, delamination and air entrapment between the layers. Coextrusion is used in a variety of packaging applications to obtain good moisture resistance, gas barrier properties, reduced costs, etc. A combination of polyethylene/nylon/polyethylene is popular in sterile-packaged disposables. A combination of LDPE/HDPE is used for shrink film and shopping bags to obtain a balance of rigidity and low cost. PS/foamed-PS coextrusion is used in production of egg cartons and meat trays. In sheet extrusion, the combination ABS/polystyrene is used for refrigerator door liners and margarine tubs. The ABS is applied for chemical resistance and the poly-

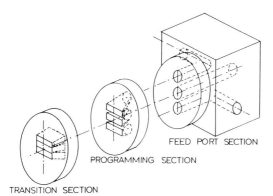

FEED PORT SECTION

PROGRAMMING SECTION

TRANSITION SECTION

Figure 9-15. Schematic of a Feed Block Sheet Die

styrene for economy. Essentially the number of applications is infinite and certainly the number of coextruded products will continue to rise.

There are basically three different techniques for coextrusion. The first employs feed block dies where the various melt streams are combined in a relatively small cross-section before entering the die. The advantage of this system is simplicity and low cost. Existing dies can be used with little or no modification. Disadvantages are that the flow properties of the different polymers have to be quite close to avoid interface distortion. There is no individual thickness control of the various layers, only an overall thickness control. Figure 9-15 shows a schematic of a feed block sheet die.

The second coextrusion technique uses multimanifold internal combining dies. The different melt streams enter the die separately and join just inside the final die orifice. The advantage of this system is that polymers with large differences in flow properties can be combined with minimum interface distortion. Individual thickness control of the different layers is possible; this enables a higher degree of layer uniformity. Disadvantages are complex die design, higher cost, and limited number of layers that can be combined.

Figure 9-16 shows a multimanifold wire coating coextrusion die.

Figure 9-17 shows a multimanifold blown film coextrusion die enabling extrusion of three different polymers.

Figure 9-18 shows a multimanifold sheet die for two-layer coextrusion.

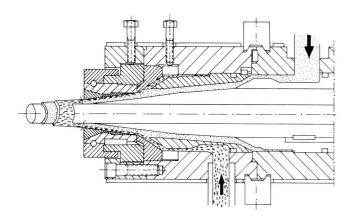

Figure 9-16. Multimanifold Wire Coating Coextrusion Die

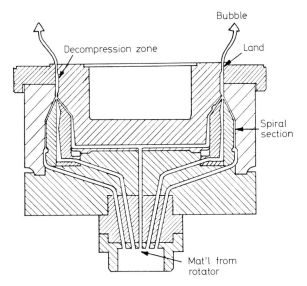

Figure 9-17. Multimanifold Blown
Film Die

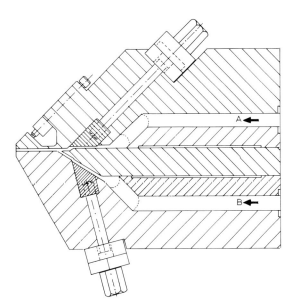

Figure 9-18. Multimanifold Sheet Die

Each individual layer can be adjusted with a choker bar, while the final combined layer thickness can be adjusted with the flex lip. Figure 9-19 shows a triple-layer coextrusion sheet die.

It is clear that coextrusion of more than three layers would be quite difficult in a sheet die. Figure 9-20 shows another design of a triple-layer coextrusion sheet die.

In this design, the layer thickness can be adjusted by die vanes. The internal vane adjustment allows a greater degree of adjustment in overall actual layer thickness; however, it does not enable local thickness adjustment as is possible with the choker bar.

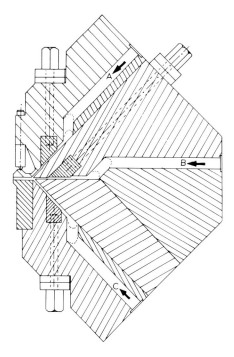

Figure 9-19. Triple-Layer Coextrusion Sheet Die

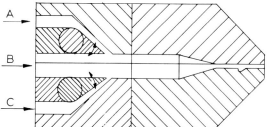

Figure 9-20. Triple-Layer Coextrusion Sheet Die

The third coextrusion technique uses multimanifold external-combining dies, which have completely separate manifolds for the different melt streams as well as distinct orifices through which the streams leave the die separately, joining just beyond the die exit. This technique is also referred to as multiple lip coextrusion. The layers are combined after exiting while still molten and just downstream of the die. For flat film dies, pressure rolls are used to force the layers together. In blown film extrusion, air pressure inside the expanding bubble provides the necessary pressure for combining the layers. This technique is more expensive than the feed block technique; however, gage control of individual layers is more accurate, pinholes are eliminated, and the system is easier to start up.

One interesting benefit that can be obtained with coextrusion is a more uniform temperature distribution in the material. This can be realized by coextruding a thin, low viscosity outer layer over a high viscosity inner layer. The highest shear rate and heat generation normally occurs at the wall. By having a low viscosity material at the location of maximum shear rate, the heat generation is reduced at this point and a more even temperature profile is obtained. This is a useful technique for thermally unstable polymers. As discussed before in Section 7.5.3, this technique can also avoid the occurrence of shark skin and melt fracture type flow instabilities.

REFERENCES – CHAPTER 9

1. W. Michaeli, "Extrusionswerkzeuge fuer Kunststoffe", Carl Hanser Verlag, Munich (1979) - in German.
   W. Michaeli, "Extrusion Dies", Macmillan, New York, N. Y. (1984) – in English.
2. D. J. Weeks, Brit. Plastics, 31, 156–160 (1958).
3. D. J. Weeks, Brit. Plastics, 31, 201–205 (1958).
4. J. M. McKelvey and K. Ito, Polym. Eng. Sci., 11, 258–263 (1971).
5. J. R. A. Pearson, Trans. Plast. Inst., 32, 239–244 (1964).
6. K. Ito, Japan Plastics, 4, 27–30 (1970).
7. M. B. Chejfec, Plast. Massy, 12, 31–33 (1973).
8. K. Ito, Japan Plastics, 2, 35–37 (1968).
9. K. Ito, Japan Plastics, 3, 32–34 (1969).
10. J. Wortberg, Ph. D. thesis, RWTH Aachen, W. Germany (1978). Also in: "Berechnen von Extrudierwerkzeugen", VDI-Verlag GmbH, Duesseldorf, W. Germany (1978).
11. P. Fischer, E. H. Goermar, M. Herner, and U. Kosel, Kunststoffe, 61, 342–355 (1971).
12. E. H. Goermar, Ph. D. thesis, RWTH Aachen, W. Germany (1968).
13. C. I. Chung and D. T. Lohkamp, SPE ANTEC, Atlanta, 363–365 (1975).
14. I. Klein and R. Klein, SPE J., 29, 33–37 (1973).
15. H. Schoenewald, Kunststoffe, 68, 238–243 (1978).
16. B. Vergnes, P. Saillard, and B. Plantamura, Kunststoffe, 70, 750–752 (1980).
17. Y. Matsubara, Polym. Eng. Sci., 19, 169–172 (1979).
18. Y. Matsubara, Polym. Eng. Sci., 20, 716–719 (1980).
19. J. Sneller, Modern Plastics, 48–52 (1983).
20. S. Prager and M. Tirrell, J. Chem. Phys., 75, 5194–5198 (1981).
21. Y. H. Kim and R. P. Wool, P. G. R. 117, University of Illinois, Urbana (1982).
22. R. P. Wool and K. M. O'Connor, J. Polym. Sci., Letters Ed., 20 (1982).
23. R. P. Wool and K. M. O'Connor, J. Appl. Phys., 52 (1981).
24. S. C. Malguarnera and A. Manisali, SPE ANTEC, New York, 124–128 (1980).
25. W. Ast, Kunststoffe, 66, 186–192 (1976).
26. J. Wortberg and K. P. Schmidtz, Kunststoffe, 72, 198–205 (1982).
27. B. Proctor, SPE ANTEC, Washington, D. C., 211–218 (1971).
28. W. Predoehl, "Technologie extrudierter Kunststoffolien", VDI-Verlag GmbH, Duesseldorf, W. Germany (1979).
29. M. J. Crochet, A. R. Davies, and K. Walters, "Numerical Simulation of Non-Newtonian Flow", Elsevier, Amsterdam, The Netherlands (1984).
30. H. H. Winter and H. G. Fritz, SPE ANTEC, New Orleans, 49–52 (1984).

10   TWIN SCREW EXTRUDERS

10.1   INTRODUCTION

The first twin screw extruders for use in polymer processing were developed in the late 1930's in Italy. Roberto Colombo developed the co-rotating twin screw extruder and Carlo Pasquetti developed the counter-rotating twin screw extruder. Initially, twin screw extruders had a number of mechanical problems. The most important limitation was the thrust bearing design. Because of the limited space, it is difficult to design a thrust bearing with good axial and radial load capability. The early thrust bearings were not strong enough to give the twin screw extruders good mechanical reliability. In the late 1960's, special thrust bearings were developed especially for application in twin screw extruders. Since that time, the mechanical reliability of twin screw extruders has been comparable to that of single screw extruders. However, twin screw extruders generally still do not have as high a thrust bearing rating as single screw extruders.

Twin screw extruders have established a solid position in the polymer processing industry. The two main areas of application for twin screw extruders are profile extrusion of thermally sensitive materials (e. g., PVC) and specialty polymer processing operations, such as compounding, devolatilization, chemical reactions, etc. Twin screw extruders used in profile extrusion have a closely fitting flight and channel profile and operate at relatively low screw speeds, in the range of about 10 rpm. These machines offer several advantages over single screw extruders. Better feeding and more positive conveying characteristics allow the machine to process hard-to-feed materials (powders, slippery materials, etc.) and yield short residence times and a narrow residence time distribution (RTD) spectrum. Better mixing and larger heat transfer area allow good control of the stock temperatures. Good control over residence times and stock termperatures obviously are key elements in the profile extrusion of thermally sensitive materials. Twin screw extruders used in profile extrusion are always closely intermeshing and usually counter-rotating, although a few co-rotating twin screw extruders are used.

Specialty polymer processing operations are performed on a number of twin screw extruders with a variety of designs. An overview of the different types of twin screw extruders is shown in Table 2–2, Chapter 2. High speed intermeshing co-rotating extruders are used in compounding, as continuous chemical reactors, and devolatilization devices. These machines operate at high speeds, with maximum screw speeds ranging from 300 to 600 rpm. Non-intermeshing extruders are used for mixing, chemical reactions, and devolatilization. The conveying mechanism in non-intermeshing extruders is considerably different from that in intermeshing extruders; it is closer to the conveying mechanism in a single screw extruder, although there are substantial differences. As a result, non-intermeshing twin screw extruders do not have positive conveying characteristics. However, it should be realized that positive conveying characteristics imply poor axial mixing capability. Thus, if axial mixing is required, the conveying characteristics cannot be positive.

10.2   TWIN VERSUS SINGLE SCREW EXTRUDER

The characteristics of twin screw extruders may be better appreciated by considering the fundamental differences between single and twin screw extruders. One major difference is the type of transport that takes place in the extruder. Material transport in a single screw extruder is a drag-induced type of transport; frictional drag in the solids conveying zone and viscous drag in the melt conveying zone. Therefore, the conveying behavior is to a large extent determined by the frictional properties of the solid material and the viscous properties of the molten material. There are many materials with unfavorable frictional properties that cannot be fed into single screw extruder without getting into severe feeding problems. On the other hand, the transport in an intermeshing twin screw extruder is to some extent a positive displacement type of transport. The degree of positive displacement depends on how well the flight of one screw closes the opposing channel of the other screw. The most positive displacement is obtained in a closely intermeshing, counter-rotating geometry. For instance, a gear pump can be considered to be a counterrotating twin screw extruder with the helix angle of the screw flights being 90 degrees or slightly less than 90 degrees. However, even a gear pump is not a pure positive displacement device because the machine cannot be designed with zero clearances. Thus, leakage flows will reduce the degree of positive conveying that can be achieved in a twin screw extruder.

Another major difference between the single and twin screw extruder is the velocity patterns in the machine. The velocity profiles in single screw extruders are well defined and fairly easy to describe, see Figure 7-28. The situation in twin screw extruders is considerably more complicated. The velocity profiles in twin screw extruders are quite complex and more difficult to describe. A number of workers have analyzed the flow patterns by neglecting the flow in the intermeshing region (1-5). However, the mixing characteristics and the overall behavior of the machine is primarily determined by the leakage flows occurring in the intermeshing region. Thus, results from analyses that do not consider the flow in the intermeshing region have little practical applicability. On the other hand, analyses that attempt to accurately describe the flow in the intermeshing region can easily become very complex (6-7).

The complex flow patterns in twin screw extruders have several advantages, such as good mixing, good heat transfer, large melting capacity, good devolatilization capacity, and good control over stock temperatures. One disadvantage of the complex flow patterns is that they are difficult to describe. The theory of twin screw extruders is not nearly as well developed as the theory of single screw extruders. Therefore, it is difficult to predict the performance of a twin screw extruder based on extruder geometry, polymer properties, and processing conditions. Conversely, it is equally difficult to predict the proper screw geometry when a

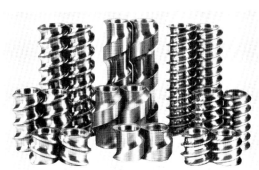

Figure 10-1. Screw Elements for a Co-Rotating Twin Screw Extruder

Figure 10-2. Screw Elements for a Counter-Rotating Twin Screw Extruder

certain performance is required in a particular application. This situation has led to twin screw extruders of modular design. These machines have removable screw and barrel elements. The screw design can be altered by changing the sequence of the screw elements along the shaft. In this way, an almost infinite number of screw geometries can be put together. The modular design, therefore, creates tremendous flexibility and allows careful optimization of screw and barrel geometry to each particular application. A number of modular screw elements for a co-rotating extruder are shown in Figure 10-1.

Modular screw elements for a counter-rotating extruder are shown in Figure 10-2.

10.3   INTERMESHING CO-ROTATING EXTRUDERS

There are two types of intermeshing co-rotating extruders, the low speed extruder and the high speed extruder. The two machines are different in design, in operating characteristics, and in areas of application. The low speed co-rotating twin screw extruder is primarily used in profile extrusion, while the high speed extruder is used in specialty polymer processing operations.

10.3.1 CLOSELY INTERMESHING EXTRUDERS

The low speed extruder has a closely intermeshing screw geometry where the flight profile fits closely into the channel profile, i.e., a conjugated screw profile. A typical screw geometry of the closely intermeshing co-rotating (CICO) twin screw extruder is shown in Figure 10-3.

The conjugated screw profile shown in Figure 10-3 appears to form a good seal between the two screws. However, a cross-section through the intermeshing region, shown in Figure 10-4, reveals the presence of relatively large openings between the channels of the two screws.

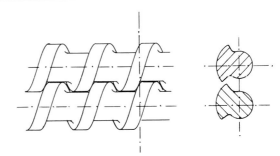

Figure 10-3. Screw Geometry of a
CICO Extruder

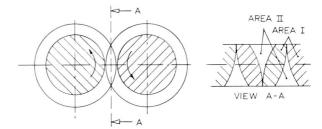

Figure 10-4. Cross-Section
Through the Intermeshing Region in a CICO Extruder

Therefore, the conveying characteristics of the CICO extruder are not as positive as those of
a closely intermeshing counter-rotating extruder (CICT), see also Figures 10–27 and 10–28.

The co-rotating twin screw extruder has a sliding type of intermeshing as shown in Figure
10–5.

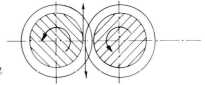

Figure 10-5. Sliding Type of Intermeshing in Co-Rotating
Twin Screw Extruders

The screw velocities in the intermeshing region are in opposite direction. Therefore, material entering the intermeshing region will have little tendency to move through the entire intermeshing region unless the flight flank clearance is quite large; this situation is shown in
Figure 10–6a.

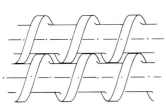

Figure 10-6a. Intermeshing Region With a
Large Flank Clearance

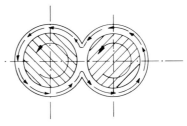

Figure 10-6b. Movement of Material in Fig-
ure-8 Pattern

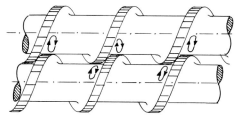

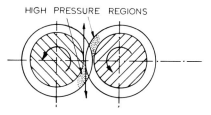

HIGH PRESSURE REGIONS

Figure 10-7. Circulatory Flow at the Passive Flight Flank.

Figure 10-8. High Pressure Regions at the Entrance to the Intermeshing Region

Because of the relatively large open areas between the channels, material entering the intermeshing region will tend to flow into the channel of the adjacent screw. The material will move in a figure-8 pattern, as shown in Figure 10-6b, while at the same time moving in axial direction.

The material close to the passive flight flank cannot flow into the channel of the adjacent screw because it is obstructed by the flight of the adjacent screw. This material, therefore, will undergo a circulatory flow as shown in Figure 10-7.

This material fraction will move forward at velocity v_a:

$$v_a = \pi D N \tan \varphi \qquad\qquad (10\text{--}1)$$

The obstructed material fraction will contribute to the positive conveying characteristics of the extruder. If the obstructed area (area I in Figure 10-4) is large relative to the open area (area II in Figure 10-4), the conveying characteristics will be quite positive. If the open area is large relative to the obstructed area, then the positive conveying characteristics will be considerably reduced, resulting in a wide residence time distribution (RTD) and a more pressure-dependent throughput. CICO extruders have relatively positive conveying characteristics because their screw geometry is such that the open area is small relative to the obstructed area.

The sliding type of intermeshing will result in high pressure regions at the point where the material enters the intermeshing region; this is shown in Figure 10-8.

The pressure build-up occurs primarily because of the reduction in cross-sectional area of the flow channel as the material enters the intermeshing region. Pressure build-up also occurs because of the change in flow direction that occurs as the material enters the intermeshing region. Obviously, the pressure build-up will be most severe when the open area is small compared to the obstructed area, which is the case in CICO twin screw extruders. These high pressure regions will result in lateral forces on the screws trying to push the screws apart. These separating forces will increase with screw speed. Clearly, the separating forces should not be as large as to cause contact between the screws and barrel, since this would result in severe wear. Therefore, CICO extruders have to run at low speed in order to avoid large pressure peaks in the intermeshing region.

10.3.2 SELF-WIPING EXTRUDERS

High speed co-rotating extruders have a closely matching flight profile, as shown in Figure 10-9.

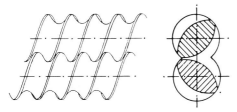

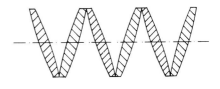

Figure 10-9. Flight Geometry in CSCO Extruders

Figure 10-10. Cross-Section Through the Intermeshing Region in CSCO Extruders

There is considerable openess from one channel to the adjacent channel. This is obvious both from the top view of the screws shown in Figure 10-9 as well as from the cross-section through the intermeshing region, shown in Figure 10-10.

Thus, the open area II is large relative to the obstructed area I. Therefore, there is relatively little tendency for large pressure peaks to form in the intermeshing region. The screws can therefore be designed with relatively small clearances between the two screws; the screws are then closely self-wiping. Twin screw extruders of this design are generally referred to as closely self-wiping co-rotating extruders (CSCO).

Since the tendency to develop large pressure peaks in the intermeshing region is quite small with CSCO extruders, they can run at high speeds, as high as 600 rpm. This is possible by the relatively large open area in the intermeshing region. However, this geometrical characteristic also results in a relatively non-positive conveying characteristic with a corresponding wide RTD and pressure-sensitive throughput. These machines, therefore, are not well suited for direct profile extrusion. A large fraction of the material will follow the figure-8 flow pattern discussed earlier. This fraction in the CSCO extruder will be considerably larger than in the CICO extruder. The progression of the material in one channel is shown in Figure 10-11 for a double-flighted geometry.

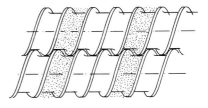

Figure 10-11. Transport in a Double-Flighted CSCO Extruder

Note that the material is displaced an axial distance of three times the pitch when it reenters the screw. Thus, in a double flighted geometry, there are three more or less independent down-channel flows. When the number of parallel flights is p, the number of independent down channel flows n_i is:

$$n_i = 2p-1 \qquad\qquad (10-2)$$

10.3.2.1 GEOMETRY OF SELF-WIPING EXTRUDERS

The flight and channel geometry of self-wiping co-rotating twin screw extruders is determined by the screw diameter, centerline distance, helix angle, and the number of parallel

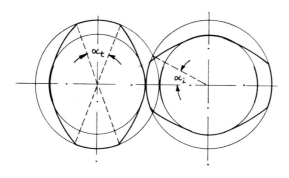

Figure 10-12. Geometry of Self-Wiping Extruders

flights. When these geometrical parameters are selected, the cross-section geometry is fixed and can be determined from kinematic principles as described in detail by Booy (8). If α_t is the tip angle, α_i the angle of intermesh, p the number of parallel flights, and D the screw diameter, then the centerline distance L_c can be determined from:

$$L_c = D \cos\left(\frac{\pi}{2p} - \frac{\alpha_t}{2}\right) \qquad (10-3)$$

Thus, when the diameter, centerline distance, and number of flights are selected, the tip angle and thus the flight width are fixed by expression 10-3. The angle α_t and α_i are angles in the plane perpendicular to the screw axis, see also Figure 10-12.

The angle of intermesh α_i is related to the tip angle by:

$$\alpha_i = \frac{\pi}{2p} - \frac{\alpha_t}{2} \qquad (10-4)$$

Thus, the centerline distance can be expressed simply as:

$$L_c = D \cos\alpha_i \qquad (10-5)$$

When the angle of intermesh is zero, the centerline distance simply equals the screw diameter. The maximum channel depth is given by:

$$H_{max} = D - L_c \qquad (10-6)$$

The channel depth as a function of the circumferential angle θ, see Figure 10-13, can be expressed as:

$$H(\theta) = \frac{D}{2}(1 + \cos\theta) - \left(L_c^2 - \frac{1}{4}D^2\sin^2\theta\right)^{1/2} \qquad (10-7)$$

The axial coordinate l is related to the circumferential angle θ by:

$$l = \frac{D}{2}\theta\tan\varphi \qquad (10-8)$$

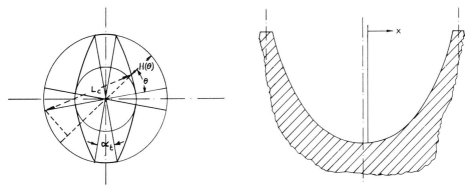

Figure 10-13. Channel Depth as a Function of   Figure 10-14. Cross-Channel Depth Profile
Circumferential Angle

Thus, the channel depth profile as a function of axial distance l is:

$$H(l) = \frac{D}{2}\left[1+\cos\left(\frac{2l}{D}\cot an\varphi\right)\right] - \left[L_C^2 - \frac{1}{4}D^2\sin^2\left(\frac{2l}{D}\cot an\varphi\right)\right]^{1/2} \quad (10-9)$$

The cross-channel coordinate x is related to the circumferential angle θ by:

$$x = \frac{D}{2}\theta\sin\varphi \quad (10-10)$$

Thus, the channel depth as a function of cross-channel distance is

$$H(x) = \frac{D}{2}\left[1+\cos\left(\frac{2x}{D\sin\varphi}\right)\right] - \left[L_C^2 - \frac{1}{4}D^2\sin^2\left(\frac{2x}{D\sin\varphi}\right)\right]^{1/2} \quad (10-11)$$

The resulting geometry in cross-channel direction is shown in Figure 10-14.

When the coordinate x is zero, the channel depth reaches its maximum value, see equation 10-6. The maximum channel depth is maintained over a circumferential angular distance of α_t, which corresponds to a cross-channel distance $0.5D\alpha_t\sin\varphi$.

In CSCO extruders, a major portion of the material entering the intermeshing region in a channel of one screw will transfer to an adjacent channel of the other screw. This is shown in Figure 10-15 where the material transport in one channel is shown in a cross-section perpendicular to the screw axes.

Figure 10-16 shows how the channel area changes as material enters the intermeshing region.

Just before the intermeshing region, the flow channel area is determined by the area between the screw and the barrel, i.e., the screw channel area A_1. In the intermeshing region itself, the flow channel area is determined by the area between the two screws and the barrel. Initially, the flow channel area increases to a maximum value and then it reduces back to area A_1 at the end of the intermeshing region. This action causes a relatively effective

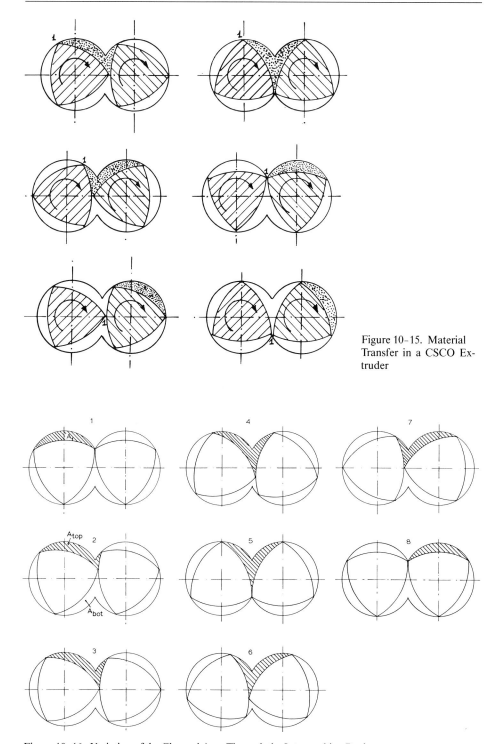

Figure 10-15. Material Transfer in a CSCO Extruder

Figure 10-16. Variation of the Channel Area Through the Intermeshing Region

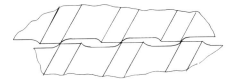

Figure 10-17.Co-Rotating Extruder with Wide Screw Flights.

transfer of material from one screw to the other and vice versa if the flight width is small relative to the width of the channel – this is generally the case in CSCO extruders. As the flight width increases, the interscrew material transfer becomes more restricted, resulting in increased circulatory flow at the entrance to the intermeshing region and increased pressure buildup at this point. In fact, when the flight width becomes sufficiently large, the characteristics of the extruder will change to those of a CICO extruder, see Figure   10-17.

The cross-sectional area of the barrel is:

$$A_b = \frac{1}{2}(\pi - \alpha_i)D^2 + \frac{1}{2}L_c D\sin\alpha_i \qquad (10\text{--}12)$$

The cross-sectional area of one screw is:

$$A_s = p\alpha_i L_c^2 - \frac{1}{2}pL_c D\sin\alpha_i + \frac{1}{2}p\alpha_t\left(L_c^2 + \frac{1}{2}D^2 - L_c D\right) \qquad (10\text{--}13)$$

The open cross-sectional area between barrel and screw is simply:

$$A_o = A_b - 2A_s \qquad (10\text{--}14)$$

This can be written as:

$$A_o = D^2\left[\left(p - \frac{1}{2}\right)\alpha_i + \left(p + \frac{1}{2}\right)\sin\alpha_i\cos\alpha_i - \pi\cos^2\alpha_i + (\pi - 2p\alpha_i)\cos\alpha_i\right]$$

$$(10\text{--}15)$$

Thus, at a certain diameter, the open area is primarily a function of the number of parallel flights p and the intermeshing angle α_i, see Figure 10-18.

The open volume is obtained by simply multiplying the open area with axial screw length L:

$$V_o = A_o L \qquad (10\text{--}16)$$

The surface area of the screw is obtained by multiplying the periphery of the screw with axial length L:

$$A_s^S = \pi L_c L \qquad (10\text{--}17)$$

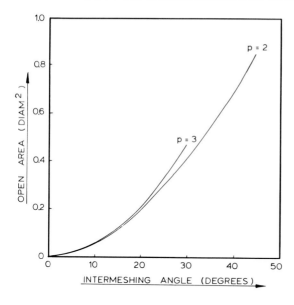

Figure 10-18. Open Area versus p and α_i

The surface area of the barrel is:

$$A_b^S = 2LD(\pi - \alpha_i) \tag{10-18}$$

The channel area formed between the screw and barrel, screw channel area A_1, is:

$$A_1 = \frac{0.25\,\pi D^2 - A_S}{p} \tag{10-19}$$

By using equation 10-13 for A_S, equation 10-19 can be written as:

$$A_1 = D^2\left[\frac{1}{2}\alpha_i - \left(\alpha_i + \frac{1}{2}\alpha_t\right)\cos^2\alpha_i + \left(\frac{1}{2}\sin\alpha_i + \frac{1}{2}\alpha_t\right)\cos\alpha_i\right] \tag{10-20}$$

When the leading tip of a screw channel enters the intermeshing region, the flow channel area increases because of the contribution of the adjacent screw, see Figure 10-16. The flow channel area reaches a maximum, A_{max}, when the leading tip reaches the end of the inter-meshing region. The flow channel area then reduces again and becomes area A_1 when the trailing tip of the screw channel enters the intermeshing region. The open cross-sectional area A_0 is thus formed by $2p-1$ channels, two of which are in the intermeshing region, A_{top} and A_{bot}. When A_{top} is increasing, A_{bot} is decreasing and vice versa. The total area of in-termesh A_{int} is the sum of A_{top} and A_{bot}; this is a constant. Area A_{int} is obtained from:

$$A_{int} = A_{top} + A_{bot} = A_0 - (2p-3)A_1 \tag{10-21}$$

The maximum value of A_{top} and A_{bot} is simply:

$$A_{max} = A_0 - (2p-2)A_1 \qquad\qquad (10-22)$$

If angle θ is the angle from the leading tip to the start of the intermeshing region, see Figure 10-19, then the area A_{top} can be expressed as a function of angle θ.

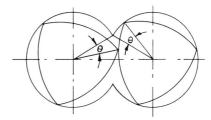

Figure 10-19. Angle θ

This relationship is shown graphically in Figure 10-20 for a geometry with $p=3$ and $\alpha_t=0$.

Area A_{top} increases from $A_1=0.0856D^2$ to reach a maximum of $A_{max}=0.1259D^2$, an increase of 47 percent. At the same time, area A_{bot} reduces from the maximum A_{max} down to A_1. Thus, when area A_{top} is decompressed, area A_{bot} is compressed and vice versa. This action will promote leakage flow through the nip when the degree of fill is high and will cause an alternating dynamic pressure field in area A_{top} and A_{bot}.

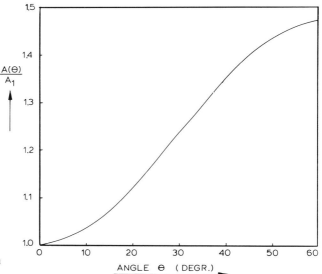

Figure 10-20. A_{top} as Function of Angle θ

10.3.2.2 CONVEYING IN SELF-WIPING EXTRUDERS

The conveying process in single screw extruders is generally analyzed by using the flat plate model, see Chapter 7. A similar analysis can be used in CSCO extruders when both screws are unrolled onto a flat plate as shown in Figure 10–21.

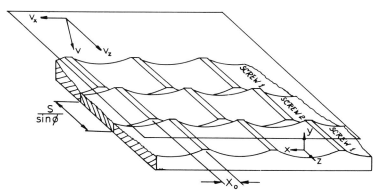

Figure 10–21. Flat Plate Model in CSCO Extruders

The down channel length of each flat screw segment is $S/\sin\varphi$, where S is the pitch of the screw flight and φ the helix angle. The screws are offset in cross-channel direction by a distance X_0, where X_0 is often taken to be equal to the cross-channel flight width. Obviously, this model is a severe simplification of the actual conveying process because it cannot accurately represent the interscrew material transfer in the intermeshing region. However, the advantage of the flat plate model is that one can follow a very similar approach to the one used in single screw extruders.

In most cases, the CSCO extruder will be starve fed. Thus, the output from the extruder is determined by the device feeding the extruder and not by the extruder itself. This means that the screw channels are only partially filled with material over a considerable length of the extruder; only the final length of the extruder will be fully filled if appreciable diehead pressure must be built up. The fully filled length of the machine will increase with diehead pressure. The reason that the fully filled length of the machine is determined by the pressure generating requirement is that pressure can only be generated when the channel is completely filled with material. If the screw channel is only partially filled with material, no pressure can be generated in down channel direction. Local sections of the screw can be fully filled if a restrictive screw element is placed along the screw, such as a reversed flighted screw element or a kneading block.

10.3.2.2.1 PARTIALLY FILLED SCREWS

Following Werner (9), the degree of fill ε is defined as the ratio of filled channel area A_ε to total channel area:

$$\varepsilon = \frac{A_\varepsilon}{A} \tag{10-23}$$

This is shown graphically in Figure 10–22, where the areas are determined perpendicular to the screw flights.

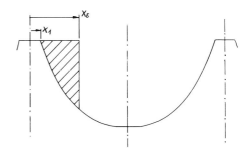

Figure 10–22. Partially Filled Screw Channel

The total cross-channel area A is simply:

$$A = A_1 \sin\varphi \qquad\qquad (10\text{--}24)$$

where A_1 is the channel area in a plane perpendicular to the screw axis and is given by equation 10–20.

Area A_ε is related to the filled cross-channel distance x_ε by:

$$A_\varepsilon = \int_{x_1}^{x_\varepsilon} H(x)\,dx \qquad\qquad (10\text{--}25)$$

where $x_1 = 0.25\alpha_t\, D\sin\varphi$.

A typical relationship between the degree of fill and a normalized cross-channel distance is shown in Figure 10–23.

The velocity profiles can be determined if the following assumptions are made:

1. The fluid is Newtonian
2. The flow is steady and fully developed
3. The flow is isothermal
4. No slip at the wall
5. Body and inertia forces are negligible
6. Channel curvature in down channel direction is negligible

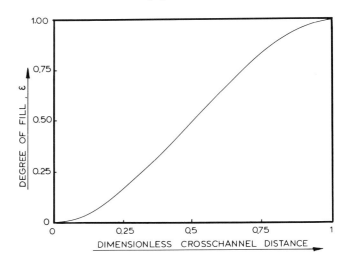

Figure 10–23. Degree of
Fill versus Normalized
Cross-Channel Distance

The equation of motion in down channel direction can be written as:

$$\frac{\partial^2 v_z}{\partial x^2} + \frac{\partial^2 v_z}{\partial y^2} = 0 \tag{10-26}$$

The solution to this equation for a rectangular channel is given in Section 7.4.1.1. However, the channel of a CSCO extruder does not have a rectangular shape but resembles more the segment of a circle. Since the channel depth is a rather lengthy function of the cross-channel distance, see equation 10-11, a simple analytical solution to equation 10-26 is not very likely to be found. Thus, one has to resort to numerical techniques to solve equation 10-26. If it is assumed that:

$$\frac{\partial^2 v_z}{\partial x^2} << \frac{\partial^2 v_z}{\partial y^2} \tag{10-27}$$

then the down channel velocity profile becomes simply:

$$v_z = \frac{y v_{bz}}{H} \tag{10-28}$$

The equation of motion in cross-channel direction can be written as:

$$\frac{\partial^2 v_x}{\partial x^2} + \frac{\partial^2 v_x}{\partial y^2} = \frac{1}{\mu} \frac{\partial P}{\partial x} \tag{10-29}$$

If it is assumed that:

$$\frac{\partial^2 v_x}{\partial x^2} << \frac{\partial^2 v_x}{\partial y^2} \tag{10-30}$$

then the cross-channel velocity profile can be written as:

$$v_x = \frac{y^2 + Hy}{2\mu} \frac{dP}{dx} + \frac{y v_{bx}}{H} + v_{bx} \tag{10-31}$$

where y is measured from the barrel surface in negative direction.

The cross-channel pressure gradient can be determined from the condition that the net flow in x-direction is zero, i.e., leakage flow is neglected. Thus, the cross-channel pressure gradient is:

$$\frac{dP}{dx} = \frac{6 \mu v_{bx}}{H^2} \tag{10-32}$$

and the resulting cross-channel velocity profile becomes:

$$v_x = \frac{3v_{bx}y^2}{H^2} + \frac{4v_{bx}y}{H} + v_{bx} \tag{10-33}$$

If the leakage flow over the flights is neglected and also the forced positive conveying of the obstructed material fraction, see equation 10-1, then the volumetric flow rate can be expressed as:

$$\dot{V} = (2p-1) \int_{-H}^{o} \int_{x_\varepsilon}^{x_1} v_z \, dx \, dy \tag{10-34}$$

If it is assumed that the down channel velocity profile can be approximated with equation 10-28, then the volumetric throughput is approximately:

$$\dot{V} = \frac{1}{2}(2p-1)A_\varepsilon v_{bz} \tag{10-35}$$

By using equation 10-23, the equation can be written as:

$$\dot{V} = \frac{1}{2}(2p-1)\,\varepsilon A\,\pi DN \cos\varphi \tag{10-36}$$

Thus, the throughput according to equation 10-36 is directly proportional to the degree of fill ε and the screw speed N. The power consumption in the screw channel can be determined from:

$$dZ_{ch} = (2p-1)v_{bz}dz \int_{x_1}^{x_\varepsilon} \tau_{yz}(0)\,dx + (2p-1)v_{bx}dz \int_{x_1}^{x_\varepsilon} \tau_{yx}(0)\,dx \tag{10-37}$$

The power consumption in the flight clearance can be written as:

$$dZ_{cl} = \frac{(2p-1)v_b^2 \mu_{cl} D \sin\varphi\, dz}{2\,\delta} \tag{10-38}$$

The total power consumption per unit down channel length is:

$$dZ = dZ_{ch} + dZ_{cl} \tag{10-39}$$

The specific energy consumption (SEC) over axial length L is:

$$SEC = \frac{\int_o^{L/\sin\varphi} dZ}{\rho \dot{V}} \tag{10-40}$$

Another approach to the analysis of the conveying process of partially filled screws was proposed by Booy (29). By assuming that the bank of material at the leading flight edge is approximately symmetrical with respect to the bisectrix, see Figure 10-24, and that the screw flank contacting the bank of material is reasonably flat, the velocity distribution in the bank can be analyzed as follows. The motion of the bank of material can be considered as a superposition of the motions of Figures 10-24b and c.

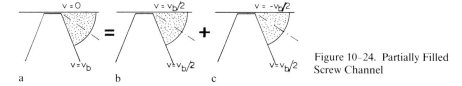

Figure 10-24. Partially Filled Screw Channel

The velocity distribution in Figure 10-24c will be such that no net flow occurs in down channel direction. The velocity distribution in Figure 10-24b will cause a downchannel flow at a uniform velocity of $v_{bz}/2$. The average total velocity of the material in the bank is the resultant of v_{bx} and $v_{bz}/2$, as shown in Figure 10-24. The axial velocity component is:

$$v_a = \frac{1}{2} v_b \sin\varphi \cos\varphi \qquad (10\text{-}41)$$

When the flight clearances are neglected, this approach yields the same expression for throughput as the one derived earlier, see equations 10-35 and 10-36. When the flight clearance is not neglected, it can be assumed that the thin layers smeared out on the screw and barrel surfaces are mostly stagnant and about half the thickness of the clearance. With the expressions for the screw surface area A_s^S, barrel surface area A_b^S, and open area A_0, the degree of fill resulting from non-zero clearances can be expressed as:

$$\varepsilon_1 = \frac{\pi L_c \delta_c + 2D(\pi - \alpha_i)\delta}{2A_0} \qquad (10\text{-}42)$$

where δ_c is the clearance between the screws and δ the clearance between screw and barrel.

Open area A_0 is given by equation 10-15. The rest of the material moves at an axial velocity v_a and corresponds to a degree of fill ε_0. The cross-section A_L through the moving material is:

$$A_L = \varepsilon_0 A_0 \qquad (10\text{-}43)$$

The flow rate is then:

$$\dot{V} = \varepsilon_0 A_0 v_a = \frac{1}{2}\varepsilon_0 A_0 v_b \sin\varphi \cos\varphi \qquad (10\text{-}44)$$

The total degree of fill now becomes:

$$\varepsilon = \varepsilon_0 + \varepsilon_1 \qquad (10\text{-}45)$$

When the extruder is run empty, the remaining degree of fill will be ε_1 when the effect of gravity is negligible.

10.3.2.2.2 FULLY FILLED SCREWS

Booy (29) analyzed the pumping performance of fully filled co-rotating twin screw extruders by distinguishing two flow regimes. One flow regime (I) bounded by both screw and barrel surface and one flow regime (II) bounded primarily by the two screw surfaces in the intermeshing region, see Figure 10-25.

The flow in regime I is analyzed by unwrapping the screws as shown in Figure 10-26.

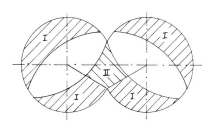

Figure 10−25. Flow Regimes in Twin Screw Extruder

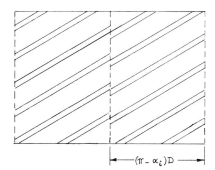

Figure 10−26. Unwrapped Twin Screw Geometry

The drag flow rate is written as:

$$\dot{V}_d = \frac{1}{2}(2p-1)F_d H_{max} W v_{bz} \tag{10-46}$$

The pressure flow rate is written as:

$$\dot{V}_p = \frac{(2p-1)F_p H_{max}^3 W}{12\mu} \frac{\partial P}{\partial z} \tag{10-47}$$

Equations 10-46 and 10-47 can be applied only when the tip and angle α_t is small. Regime II is assumed not to contribute to the pressure generation and is further assumed to move forward at a rate of one pitch per revolution. If the cross-sectional area of regime II is A_a, then the flow rate through this domain is:

$$\dot{V}_a = v_a A_a = A_a v_b \tan\varphi \tag{10-48}$$

The total flow rate then becomes:

$$\dot{V} = \dot{V}_d + \dot{V}_a - \dot{V}_p \tag{10-49}$$

When the width W of the channel is large relative to the depth H_{max} of the channel, the shape factor for drag can be approximated by (29):

$$F_d \simeq \int_{-W/2}^{+W/2} \frac{H(x)dx}{WH_{max}} \qquad (10\text{--}50a)$$

Similarly, the shape factor for drag can be approximated by:

$$F_p \simeq \int_{-W/2}^{+W/2} \frac{H(x)^3dx}{WH_{max}^3} \qquad (10\text{--}50b)$$

Expressions for the limiting shape factors when the width of the channel is small relative to the depth ($W << H_{max}$) are given by Booy (29). However, this type of channel geometry is generally not encountered in commercial twin screw extruder systems.

10.4   INTERMESHING COUNTER-ROTATING EXTRUDERS

A typical screw geometry of a closely intermeshing counter-rotating (CICT) twin screw extruder is shown in Figure 10-27.

A cross-section through the intermeshing region, see Figure 10-28, shows that the openings between the channels of the two screws are quite small.

These openings are considerably smaller than in intermeshing co-rotating extruders, see also Figures 10-3 and 10-4. As a result, CICT extruders can achieve relatively positive conveying characteristics.

The counter-rotating twin screw extruder has a milling type of intermeshing as shown in Figure 10-29.

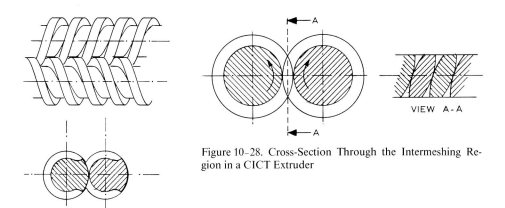

Figure 10-28. Cross-Section Through the Intermeshing Region in a CICT Extruder

VIEW A-A

Figure 10-27. Screw Geometry of a CICT Extruder

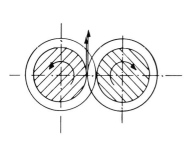

Figure 10-29. Milling Type of Intermeshing in Counter-Rotating Extruders

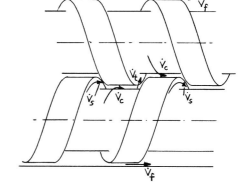

Figure 10-30. Leakage Flows in CICT Extruders

The screw velocities in the intermeshing region are in the same direction. Therefore, the material entering the intermeshing region will have a strong tendency to flow through the intermeshing region. If the clearances between the two screws are rather small, the flow through the intermeshing region will be quite small. This will result in a bank of material accumulating at the entry of the intermeshing region. The material drawn into the nip will exert considerable pressure on the two screws. This can cause deflection of the screws. Therefore, CICT extruders generally run at low speed to avoid excessive pressures developing in the intermeshing region. By designing the screws with larger clearances, the allowable screw speeds can be increased; however, this is at the expense of the positive conveying characteristics. Thus, the maximum allowable screw speed on a CICT extruder is often a good indication of the conveying characteristics of the machine. A low maximum screw speed (about 20 to 40 rpm) indicates a machine with positive conveying characteristics with the most likely area of application being profile extrusion. A high maximum screw speed (about 100 to 200 rpm) indicates a machine with less positive conveying characteristics with likely areas of application being compounding, continuous chemical reactions, and other specialty polymer processing operations.

The theoretical maximum output of a CICT extruder is:

$$\dot{V}_{max} = 2pNV \qquad (10-51)$$

where p is the number of parallel flights, N the screw speed, and V the volume of the C-shaped chamber.

In equation 10-51, it is assumed that the screw channels are fully filled with material and that there is no leakage of material. Equation 10-51 was first proposed by Schenkel (10). It was found that the actual throughput of CICT extruders is usually considerably below $\dot{V}_{max}$. Doboczky (11, 12) and Klenk (13, 14) introduced correction factors to bring the predicted throughput values in line with actual throughput values. However, these correction factors were mostly empirical and, thus, of limited usefulness. Janssen (15) performed a detailed analysis of the leakage flows in counter-rotating extruders. He distinguished four kinds of leakage, see Figure 10-30.

1. Leakage through the clearance between the screw flight and barrel, $\dot{V}_f$.
2. Calender leakage, $\dot{V}_c$, between the root of the screw and the tip of the flight.

3. Interscrew leakage through the gap between the flight flanks (the tetrahedron gap), $\dot{V}_t$, in radial direction.
4. Leakage through the side gap, $\dot{V}_s$, in tangential direction.

The volume of the C-shaped chamber is approximately:

$$V = \frac{\pi DHW_m}{\cos\varphi_m} - \frac{w_m D^2 (2\alpha_i - \sin 2\alpha_i)}{4\cos\varphi_m} \qquad (10\text{-}52)$$

See Figure 10-31.

For screws with straight flight flanks, the flight width is:

$$w(r) = w + 2(R-r)\tan\psi \qquad (10\text{-}53)$$

where ψ is the flight flank angle.

Thus, the mean flight width w_m is:

$$w_m = w + (D-H)\tan\psi \qquad (10\text{-}54)$$

and the mean channel width is:

$$W_m = \frac{\pi(D-H)\sin\varphi_m}{p} - w_m \qquad (10\text{-}55)$$

The axial pressure gradient in the screw channel, using the same assumptions as in Section 10.3.2.2, becomes:

$$\frac{\partial P}{\partial l} = \frac{6\mu v_a}{H^2} \qquad (10\text{-}56)$$

The axial velocity v_a is given by equation 10-1. The derivation of the axial pressure gradient is essentially the same as the derivation of the cross-channel pressure gradient given in equation 10-32. It is assumed that the diehead pressure P_d is built up uniformly along the filled length of the extruder L_f, the drag induced axial channel pressure (equation 10-56) can be superimposed on the linear pressure profile, as shown in Figure 10-32.

The pressure drop ΔP_B over the axial channel width B becomes:

$$\Delta P_B = \frac{6\mu v_a B}{H^2} \qquad (10\text{-}57)$$

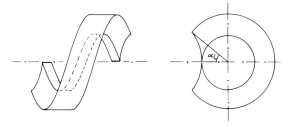

Figure 10-31. Geometry of a C-Shaped Chamber

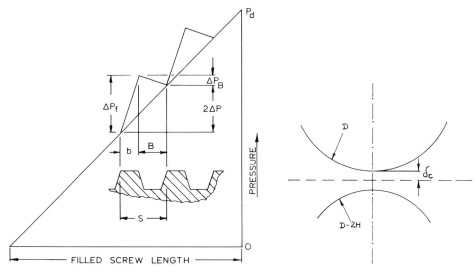

Figure 10-32. Pressure Profile in a CICT Extruder   Figure 10-33. Calender Gap Geometry

The axial drop over a screw flight ΔP_f is:

$$\Delta P_f = 2\Delta P + \Delta P_B \qquad (10-58)$$

where ΔP is the pressure drop per chamber as a result of the diehead pressure.

The leakage flow over the flight of a C-shaped chamber can be written as:

$$\dot{V}_f = (\pi - \alpha_i)D\left[\frac{1}{2} v_a \delta + \frac{\delta^3 \Delta P_f}{12 \mu b}\right] \qquad (10-59)$$

The pressure drop across the calender gap ΔP_c, according to Janssen (15), can be written as:

$$\Delta P_c = \frac{3\pi\mu(D\delta_c)^{1/2}}{4\delta_c^3}\left[\frac{3b\dot{V}_c}{4} - 2\pi N\delta_c(D-H)\right] \qquad (10-60)$$

where δ_c is the calender gap, see Figure 10-33.

Pressure drop ΔP_c must equal the tangential pressure drop over a chamber plus the contribution of the diehead pressure. In order to determine the calendar leakage flow $\dot{V}_c$, the tangential pressure drop over a chamber must be known. If $\dot{V}_{tan}$ is the tangential flow in the C-shaped chamber, the tangential pressure gradient $\partial P/\partial\theta$ after Janssen (15) can be determined from the following relationship:

$$\dot{V}_{tan} = \frac{4v_a B^2}{\pi^2}\sum_{n=1,3,5..}^{\infty}\left\{\frac{1}{n^2}\left[A(n)\left[I_0\left(\frac{\pi n R}{B}\right) - I_0\left(\frac{R-H}{B}\pi n\right)\right]\right]\right.$$

$$\left. - B(n)\left[K_0\left(\frac{\pi n R}{B}\right) - K_0\left(\frac{R-H}{B}\pi n\right)\right] - \frac{2B}{n^2\pi^2\mu v_a}\left(\frac{\partial P}{\partial\theta}\right)\ln\left(\frac{R}{R-H}\right)\right\}$$

$$(10-61)$$

where:

$$A(n) = \frac{C(n)}{F(n)} \left[\left(\frac{R-H}{R} - \frac{2}{n\pi C(n)} \right) K_1\left(\frac{R-H}{B}\pi n \right) - K_1\left(\frac{R\pi n}{B} \right) \right] \qquad (10-62)$$

$$B(n) = \frac{C(n)}{F(n)} \left[I_1\left(\frac{R\pi n}{B} \right) - \left(\frac{R-H}{R} - \frac{2}{\pi n C(n)} \right) I_1\left(\frac{R-H}{B}\pi n \right) \right] \qquad (10-63)$$

$$C(n) = \frac{B^2}{v_a(R-H)n^3\pi^3} \left(-\frac{\partial P}{\partial\theta} \right) \qquad (10-64)$$

$$F(n) = I_1\left(\frac{R-H}{B}\pi n \right) K_1\left(\frac{R\pi n}{B} \right) - I_1\left(\frac{R\pi n}{B} \right) K_1\left(\frac{R-H}{B}\pi n \right) \qquad (10-65)$$

Functions I_0 and I_1 are modified Bessel functions of the first kind and zero and first order; K_0 and K_1 are modified Bessel functions of the second kind and zero and first order. A correction factor is required, which has to be subtracted from the throughput. This correction factor F_c is the channel volume displaced in one revolution multiplied with the rotational speed:

$$F_c = 2\pi N \left[B\left(RH - \frac{1}{2}H^2 \right) + \left(\frac{2}{3}H^3 - H^2R \right)\tan\Psi \right] \qquad (10-66)$$

Equations 10-61 through 10-66 allow the determination of the tangential pressure gradient, which, in turn, allows the determination of the leakage flow through the calender gap.

Leakage through the tetrahedron gap causes interscrew material transfer. In fact, it is the only leakage flow that causes interscrew transfer. The tetrahedron gap increases when the flight flank angle is increased. Janssen, Mulders, and Smith (16) developed an empirical formula for the leakage flow through the tetrahedron gap:

$$\dot{V}_t = \frac{0.0054\Delta P_B R^3}{\mu} \left(\frac{H}{R} \right)^{1.8} \left[\Psi + 2\left(\frac{\delta_f + \sigma_c\tan\Psi}{H} \right)^2 \right] \qquad (10-67)$$

where δ_f is the clearance between the flight flanks:

$$\delta_f = \frac{S}{2p} - b - H\tan\Psi \qquad (10-68)$$

The drag component $\dot{V}_{sd}$ of the side leakage is:

$$\dot{V}_{sd} = \pi N(2R-H)(H-\delta_c)(\delta_f + \delta_c\tan\Psi) \qquad (10-69)$$

The pressure component $\dot{V}_{sp}$ of the side leakage flow is approximately:

$$\dot{V}_{sp} = \frac{\Delta P_c F_\delta^3 (H-\delta_c)^4}{12\mu L\cos^4\Psi} (1 - 0.630F_\delta + 0.052F_\delta^5) \qquad (10-70)$$

where:

$$F_\delta = \frac{\delta_f + \delta_c \tan\Psi}{H - \delta_c} \cos^2\Psi \qquad (10\text{-}71)$$

and

$$L = \frac{2R\sin\alpha_i}{\cos\varphi} \qquad (10\text{-}72)$$

The calender leakage and side leakage can be combined as:

$$\dot{V}_{cs} = K_1\Delta P_c + \dot{V}_2 \qquad (10\text{-}73)$$

The pressure generation in the C-shaped chamber can be determined from:

$$\dot{V}_{tan} = K_{tan}\Delta P_{tan} + \dot{V}_{tan2} \qquad (10\text{-}74)$$

The drag induced pressure generation in the C-shaped chamber plus the pressure rise due to the diehead pressure should equal the pressure drop through the calender gap:

$$\Delta P_{tan} + 2\Delta P = \Delta P_c \qquad (10\text{-}75)$$

where ΔP is the pressure drop per chamber as a result of the diehead pressure.

Since the combined calender and side leakage flow equals the tangential flow in the chamber, this flow can be expressed as:

$$\dot{V}_{cs} = \dot{V}_{tan} = \frac{2K_{tan}K_1}{K_{tan}-K_1}\Delta P + \frac{\dot{V}_{tan2}K_1 - K_{tan}\dot{V}_2}{K_{tan}-K_1} \qquad (10\text{-}76)$$

The total output of the extruder can be determined from:

$$\dot{V} = 2pNV - 2p\dot{V}_{cs} - 2\dot{V}_f - \dot{V}_t \qquad (10\text{-}77)$$

This relationship allows the determination of the throughput pressure relationship for the filled length of a CICT extruder as a function of viscosity and machine geometry. Experimental verification with Newtonian fluids (15) has shown this relationship to be accurate to about 5 to 10 percent. Figure 10-34 shows the dimensionless output $\dot{V}^0$ as a function of flight flank angle when the dimensionless pressure drop $\Delta P^0 = 1E4$.

The dimensionless output is:

$$\dot{V}^0 = \frac{\dot{V}_{actual}}{2pNV} \qquad (10\text{-}78)$$

The dimensionless pressure drop ΔP^0 is:

$$\Delta P^0 = \frac{\Delta P}{\mu N} \qquad\qquad (10\text{–}79)$$

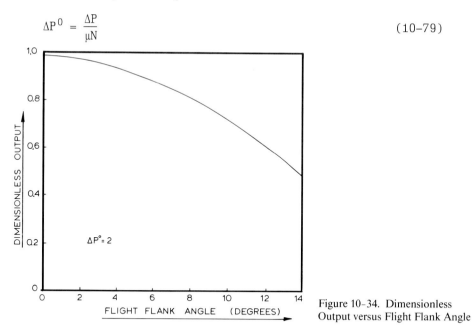

Figure 10-34. Dimensionless Output versus Flight Flank Angle

Figure 10–34 is valid for a CICT extruder with the following dimensions: $p = 1$, $D = 70$ mm, $S = 20$ mm, $H = 10$ mm, $\delta = 0.1$ mm, $\delta_c = 0.2$ mm, and $\delta_f = 0$. Increasing the flight flank angle causes a significant drop in output. Thus, increased flight flank angle strongly reduces the positive conveying characteristics. Figure 10–35 shows the dimensionless output as a function of the calender clearance when $\Delta P^0 = 1E4$ and $\Psi = 6$ degrees; all other dimensions as in figure 10–34.

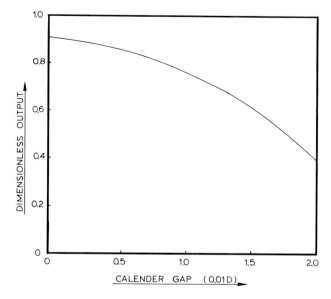

Figure 10-35. Dimensionless Output versus Calender Clearance

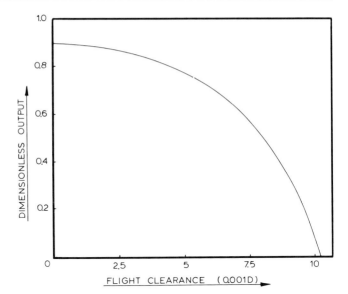

Figure 10-36. Dimension-
less Output versus Radial
Flight Clearance

Clearly the effect of calender clearance is similar to the effect of flight flank angle.

The effect of the radial flight clearance is shown in Figure 10-36 when $\Delta P^0 = 1E4$.

The output starts to drop off rapidly when the flight clearance becomes larger than 0.005 D. The radial flight clearance normally ranges from 0.001D to 0.002 D. Figure 10-37 shows the effect of side clearance when $\Delta P^0 = 1E4$. Increased side clearance causes a strong reduction in output.

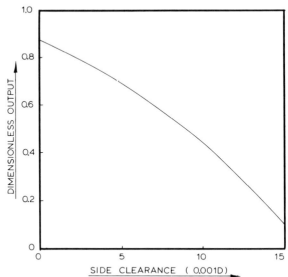

Figure 10-37. Dimensionless Output
versus Side Clearance

The effect of flight pitch is shown in Figure 10-38.

When the pitch is larger than 1/4 D, the dimensionless output is relatively insensitive to the pitch. However, the dimensionless output reduces strongly when the pitch becomes less than 1/4 D.

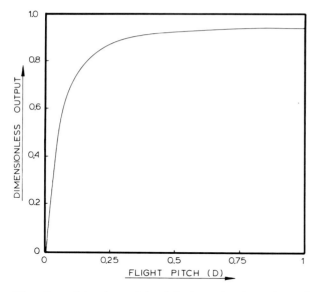

Figure 10-38. Dimensionless Output versus Flight Pitch

The effect of the channel depth is shown in Figure 10-39 for two values of the pressure differential, $\Delta P^0 = 2E4$ and $\Delta P^0 = 5E4$.

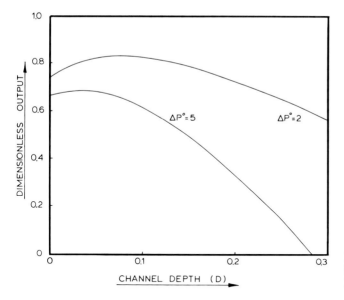

Figure 10-39. Dimensionless Output versus Channel Depth

At each pressure gradient, there is an optimum channel depth for which the output reaches a maximum value. This is similar to the situation in single screw extruders. The optimum channel depth in these examples ranges from about 0.05D to 0.10D. As the pressure gradient increases, the optimum channel depth decreases.

10.5 NON-INTERMESHING TWIN SCREW EXTRUDERS

Non-intermeshing twin screw extruders are double screw machines where the center line distance between the screws is larger than the sum of the radii of the two screws. Commercial examples of non-intermeshing twin screw extruders are counter-rotating (NOCT extruders). The conveying in NOCT extruders is similar to that in a single screw extruder. The main difference is the fact that there is a possibility of exchange of material from one screw to another. If the apex area, see Figure 10-40, were zero, the NOCT extruder would behave as two single screw extruders.

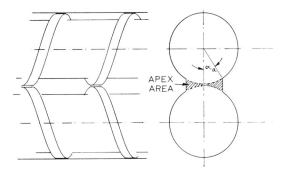

APEX AREA

Figure 10-40. NOCT Extruder Cross-Section

Because of the non-zero apex area, the output of the NOCT extruder will be less than twice the output of a single screw extruder with the same screw diameter.

The NOCT extruder has less positive conveying characteristics than a single screw extruder. As a result, however, it has better backmixing characteristics than a single screw extruder. Therefore, the NOCT extruder is primarily used in blending operations, devolatilization, chemical reactions, etc. The particular conveying characteristics of the NOCT extruder make it less desirable for profile extrusion. In one commercial example of a NOCT extruder, the screws are of different length such that the last section of the extruder has a single screw discharge. This design is shown in Figure 10-41.

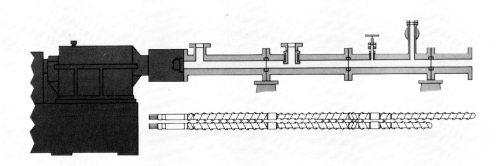

Figure 10-41. NOCT Extruder with Unequal Screw Lengths (Courtesy Welding Engineers, Inc.)

Two advantages of this configuration are improved pumping characteristics and a thrust load on one screw only. The thrust load on the secondary (short) screw is very small. Thus, the thrust bearing design is greatly facilitated. A disadvantage of this construction is a non-symmetrical conveying process with a chance of hangup of material in the transition region.

The first theoretical study of the conveying process in non-intermeshing twin screw extruders was made by Kaplan and Tadmor (17). They simplified the actual geometry, see Figure 10–42, to a flat plate model.

The flat plate model involves three plates, the two outside plates representing the screw surface and the middle plate representing the barrel, see Figure 10–43.

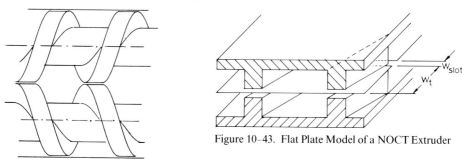

Figure 10–43. Flat Plate Model of a NOCT Extruder

Figure 10–42. Geometry of a NOCT Extruder

The center plate has slots that run perpendicular to the circumferential velocity v_b. The tangential slot width is:

$$W_{slot} = D\alpha_a \qquad (10-80)$$

where α_a is the apex angle, see Figure 10–40.

The tangential distance between the slots is:

$$W_t = D(\pi - \alpha_a) \qquad (10-81)$$

This model yields the following output-pressure relationship for Newtonian fluids:

$$\dot{V} = \frac{1}{2}\,WHv_{bz}F_{DTW} - \frac{WH^3}{12\mu}\left(\frac{\Delta P}{\Delta z}\right)F_{PTW} \qquad (10-82)$$

where:

$$F_{DTW} = \frac{4f}{1+3f} \qquad (10-83)$$

$$F_{PTW} = \frac{4}{1+3f} \qquad (10-84)$$

$$f = \frac{\alpha_a}{\pi} \qquad (10-85)$$

Equation 10–82 gives the output for one screw. Thus, the total output is twice the value from equation 10–82. The factor f is the ratio of uninterrupted barrel circumference to the total barrel circumference. Considering that the drag flow occurs as a result of adherence to the barrel surface, it seems that the drag flow correction factor F_{DTW} overestimates the drag flow considerably. If the barrel circumference is reduced by a factor f, one would expect a proportional reduction in the drag flow rate. Thus, the drag flow correction factor should be approximately:

$$F_{DC} = f \qquad\qquad (10–86)$$

The difference between the two drag flow correction factors is shown in Figure 10–44.

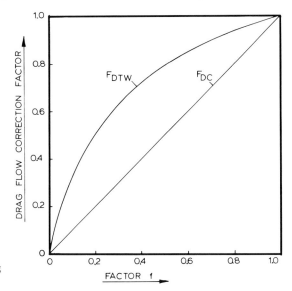

Figure 10–44. Comparison of Two Drag Flow Correction Factors

In the normal range of f, equation 10–83 overestimates the drag flow correction factor by about 10 to 20 percent. Another drawback of equation 10–82 is the fact that the slots in the barrel plate are considered infinitely thin. This corresponds to assuming an essentially zero apex area. Furthermore, the pressure flow correction assumes that pressure leakage because of the open barrel occurs in the crosshatched area shown in Figure 10–45.

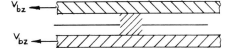

Figure 10–45. Assumed Area of Pressure Induced Leakage

However, a top view, see Figure 10–46, shows that in a matched screw configuration, the screw flights will prevent such leakage from taking place.

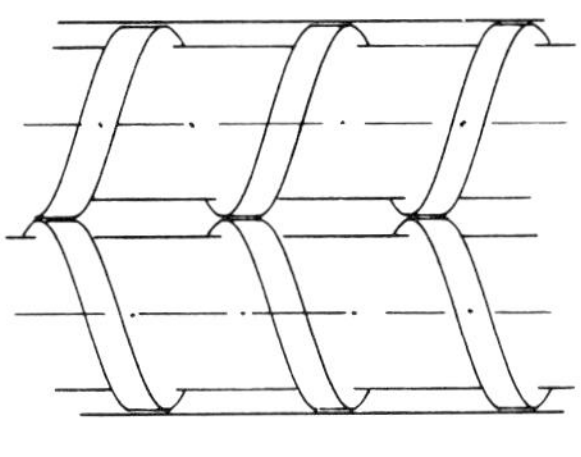

Figure 10–46. Top View of a Matched Screw Geometry

If the slot width is considered infinitely small, pressure leakage can only take place in down channel direction. This would indicate that the pressure flow does not require a correction factor when the screw flights are matched. However, if the screw configuration is staggered, as shown in Figure 10-47, the pressure flow does require a correction term.

In actual NOCT extruders, the apex area has a non-zero value. The effect of the apex area on the pumping performance is quite important, as discussed by Nichols and Yao (18). In actual experiments on a 2-inch NOCT extruder with Dimethylsiloxane, Nichols (19) found less than satisfactory agreement between actual output values and output predictions based on equation 10-82. The actual apex area A_T, shown in Figure 10-40, is:

$$A_T = \frac{1}{2} D^2 \left(2\sin\alpha_a - \sin\alpha_a\cos\alpha_a - \alpha_a + 4\alpha_a \frac{\delta}{D} \right) \qquad (10\text{-}86)$$

Nichols (24) developed a modified output model in cooperation with Lindt. The model is based on a flat three-plate model with the thickness of the center plate having a finite thickness equal to the apex width W_a, see Figure 10-48.

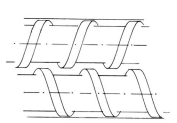

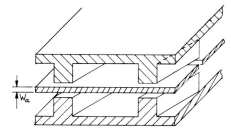

Figure 10-47. Staggered Screw Configuration

Figure 10-48. Modified Flat Plate Model for a NOCT Extruder

This model yields the following output relationship for Newtonian fluids:

$$\dot{V} = \frac{1}{2} WHv_{bz}F_{DCRT} - \frac{WH^3}{12\mu}\left(\frac{\Delta P}{\Delta z}\right) F_{PCRT1} \qquad (10\text{-}87)$$

where:

$$F_{DCRT} = \frac{fF_dF_p(2+W_a/H)^3}{f(2+W_a/H)^3+2F_p(1-f)} \qquad (10\text{-}88)$$

and

$$F_{PCRT1} = \frac{F_p(2+W_a/H)^3}{f(2+W_a/H)^3+2F_p(1-f)} \qquad (10\text{-}89)$$

where F_d and F_p are the shape factors for drag flow and pressure flow as given by equations 7-168 and 7-169, respectively.

When $W_a = 0$, equations 10-88 and 10-89 become the same as equations 10-83 and 10-84, respectively. In addition to the correction factors established from the three-plate model, another correction is introduced to account for leakage in the apex area. The apex area is approximated by a triangle as shown in Figure 10-49.

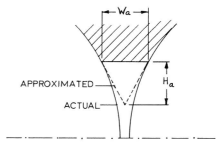

Figure 10-49. Apex Approximation

The pressure flow rate through a triangle is expressed as (see reference 32 of Chapter 1, page 567):

$$\dot{V} = \frac{W_a H_a^3 M_o}{12\,\mu} \left(\frac{\Delta P}{\Delta L}\right)_{f1}$$
(10-90)

where W_a is the apex width and H_a the height of the triangle shown in Figure 10-49, and M_o a shape factor for a triangle.

The pressure gradient over the flight is taken as:

$$\left(\frac{\Delta P}{\Delta L}\right)_{f1} = \left(\frac{\Delta P}{\Delta L}\right) \frac{S}{b}$$
(10-91)

where S is the pitch and b the axial flight width.

With equations 10-87 through 10-91, an additional pressure flow correction term F_{PCRT2} can now be formulated:

$$F_{PCRT2} = \frac{M_o W_a H_a^3 \pi D}{W H^3 w}$$
(10-92)

where M_o is shown graphically in reference 32 of Chapter 1, page 574.

The two pressure flow correction factors are now combined to give:

$$F_{PCRT} = F_{PCRT1} + F_{PCRT2}$$
(10-93)

which yields the final result obtained by Nichols (24):

$$\dot{V} = \frac{1}{2} W H v_{bz} F_{DCRT} - \frac{W H^3}{12\,\mu} \left(\frac{\Delta P}{\Delta z}\right) F_{PCRT}$$
(10-94)

Comparison of predictions from equation 10–94 to experimental results (19) yielded considerably improved agreement as compared to predictions from equation 10–82. A drawback of equation 10–94 is the fact that the derivation of the pressure flow correction term F_{PCRT2} for the apex area is not consistent with the flat three-plate model. Furthermore, the approximated apex area is smaller than the actual apex area as shown in Figure 10–48. The graphical results of the shape factor for triangles shown in reference 1–32, page 574, do not extend beyond a height to base ratio of unity. In reality, the H_a/W_a ratio will usually be larger than unity; thus, the graphical results in reference 1–32 cannot be used. Finally, the actual pressure gradient over the screw flight also contains a drag induced component, which is not taken into account in equations 10–92 through 10–94.

Another approach used to predict the output-pressure characteristics of non-intermeshing extruders is to abandon the three-plate model and follow more closely the analysis used for single screw extruders. The circumference of the barrel is interrupted for a fraction 1-f, see Figure 10–50.

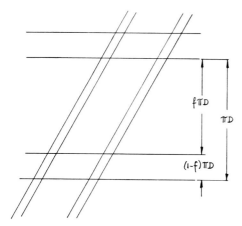

$f\pi D$

πD

$(1-f)\pi D$

Figure 10–50.　Effective Barrel Circumference

Since drag flow occurs as a result of polymer melt adhering to the barrel surface, the reduced barrel circumference will affect the drag flow rate. The drag flow per revolution is found by moving the barrel with respect to the screw over a distance of πD. With a full barrel circumference, this results in a volume per revolution equal to:

$$V_d = \frac{1}{2} BH \pi D \qquad\qquad (10\text{–}95)$$

This results in the familiar drag flow rate equation:

$$\dot{V}_d = \frac{1}{2} WHv_{bz} \qquad\qquad (10\text{–}96)$$

However, when the barrel circumference is reduced by 1-f, the volume dragged forward per revolution will be reduced correspondingly:

$$V_d = \frac{1}{2} BH \pi D - \frac{1}{2}(1-f)BH \pi D \qquad\qquad (10\text{–}97)$$

The actual drag flow rate becomes:

$$\dot{V}_d = \frac{1}{2} fWHv_{bz} \qquad (10-98)$$

The pressure flow in the screw channel outside of the apex area will be the same as it is in a single screw extruder. However, in the apex area there will be an additional pressure induced leakage flow. If the apex area is approximated by an isosceles triangle of width W_a and height H_a, see Figure 10-51, an expression can be derived for leakage in this region.

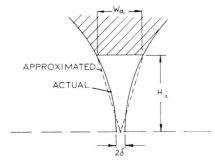

Figure 10-51. Apex Area Approximation

An expression for the pressure flow through an isosceles triangle was derived by Bird et al. (25) by following the variational principle due to von Helmholtz. Other expressions have been proposed by Kozicki et al. (26, 27), who used a simple geometric parameter method to predict the pressure drop-flow rate relationship in flow channels of arbitrary cross-section. Following Bird's approach, the output-pressure relationship for an isosceles triangle can be written as:

$$\dot{V}(\Delta) = \frac{7m^3 H_a^4(27m^2+5)}{90(55m^4+38m^2+3)} \left(\frac{\Delta P}{\mu \Delta L}\right) \qquad (10-99)$$

where:

$$m = \frac{W_a}{2H_a} \qquad (10-99a)$$

For values of W_a less than H_a (m < 0.5), equation 10-99 can be approximated reasonably well by:

$$\dot{V}(\Delta) = 0.017(H_a - 0.61W_a)W_a^3 \left(\frac{\Delta P_f}{\mu b}\right) \qquad (10-100)$$

where:

$$W_a = D(1-\cos\alpha_a) + 2\delta \qquad (10-101)$$

$$H_a = \frac{1}{2} D\sin\alpha_a \qquad (10-102)$$

The difference between expression 10–99 and 10–100 over most of the range is only about 2 or 3 percent, see also Figure 10–52.

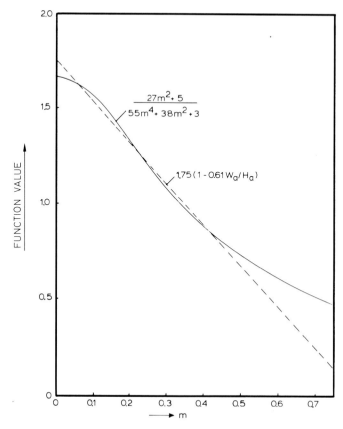

Figure 10–52. Output versus m According to Equations 10–99 and 10–100

Thus, the pressure induced leakage flow through the apex area can be written as:

$$\dot{V}_{12} = 0.017(H_a - 0.61\ W_a)W_a^3\left(\frac{g_a S}{\mu p b}\right) \qquad (10\text{–}103)$$

where g_a is the axial pressure gradient.

In addition to the pressure induced leakage flow through the apex area, there is the usual leakage flow through the flight clearance, see Section 7.4.1.2. equation 7–175b:

$$\dot{V}_{11} = \frac{\pi D\delta^3 \cos\varphi}{12\mu_{cl}w}\left(\frac{\pi Dg_z\cos\varphi}{p} + \frac{6\mu v_{bx}W}{H^2}\right) + \frac{pw\delta^3}{12\mu_{cl}}g_z \qquad (10\text{–}104)$$

Thus, the output per screw for a non-intermeshing twin screw extruder can be written as:

$$\dot{V} = \frac{1}{2}\ fWHv_{bz} - \frac{WH^3g_z}{12\mu} - \dot{V}_{11} - \dot{V}_{12} \qquad (10\text{–}105)$$

This relationship is valid for a matched screw geometry. The apex leakage term $\dot{V}_{12}$ is determined by the apex width cubed, see equation 10-103. Thus, increases in the apex angle α_a will cause very strong increases in the apex leakage flow. Figures 10-53 and 10-54 compare output predictions made with equation 10-105 to the experimental results obtained by Nichols with dimethylsiloxane polymeric fluids (19).

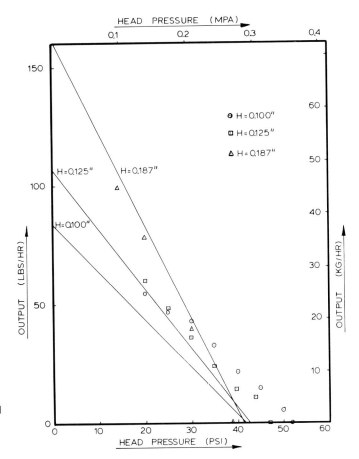

Figure 10-53. Comparison of Output Predictions Using Equation 10-105 to Experimental Data Obtained by Nichols (19); Viscosity 12 Pasec.

It can be seen that reasonable agreement is obtained between predictions and actual data.

The analysis of the conveying characteristics becomes considerably more complicated when the screws are placed in a staggered configuration. In this case, there will be considerably more leakage in the apex region as illustrated in Figure 10-55.

With the matched screw configuration, the tangential pressure profiles in the two screws are symmetrical. Thus, there will be little interscrew material transfer. In the staggered screw configuration, however, the tangential pressure profiles are non-symmetrical because the flights are 180 degrees offset. When the flight of the left screw is approaching the apex area, the pressure on the left side (P_L) of the apex area will be larger than the pressure on the right side (P_R). As a result, material will flow through the apex area from left to right, as shown in Figure 10-55.

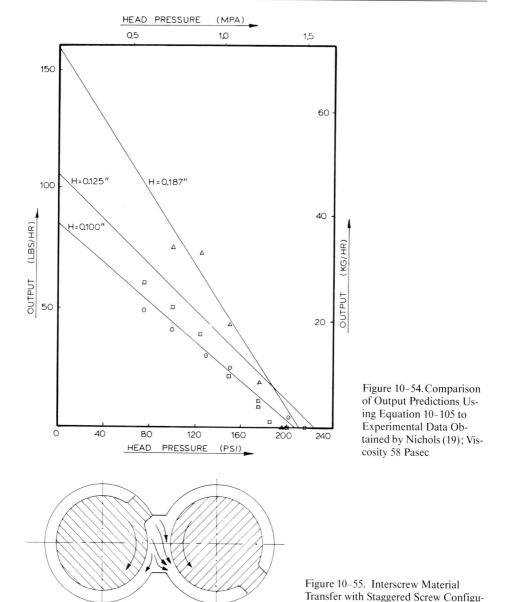

Figure 10-54. Comparison of Output Predictions Using Equation 10-105 to Experimental Data Obtained by Nichols (19); Viscosity 58 Pasec

Figure 10-55. Interscrew Material Transfer with Staggered Screw Configuration

When the flight of the right screw approaches the apex area, P_R will be larger than P_L, causing a flow from right to left. Therefore, in the staggered configuration there will be a significant amount of interscrew material transfer. This material transfer will change direction every half turn of the screw. In the staggered screw geometry, the pressure induced leakage flow through the apex region will be larger because of the larger area. In the staggered configuration, the drag induced leakage flow through the apex region also has to be taken into account because of the non-symmetrical tangential pressure profiles. The apex leakage flow can be written as:

$$\dot{V}_{12} = \dot{V}_{1d} + \dot{V}_{1p} \qquad\qquad (10\text{--}106)$$

where $\dot{V}_{1d}$ is the drag induced leakage flow and $\dot{V}_{1p}$ is the pressure induced leakage flow.

The drag induced leakage flow can be determined from the drag induced pressure gradient g_{d2}. If the effective channel height is H_{eff} and the effective channel width W_{eff}, then the drag induced leakage flow can be written as:

$$\dot{V}_{1d} = \frac{W_{eff}\,H_{eff}^3\,g_{d2}}{12\,\mu} \qquad\qquad (10\text{--}107)$$

In order to determine the drag induced leakage flow, the drag induced pressure gradient needs to be known. Consider the simplified situation shown in figure 10-56.

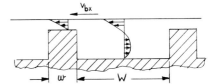

Figure 10-56. Simplified Geometry to Determine the Drag Induced Pressure Gradient

The drag induced pressure gradient in the clearance g_{d2} can be related to the drag induced pressure gradient in the channel g_{d1} by stating that the net cross-channel flow equals the flow through the clearance:

$$\int_0^\delta v_{x2}\,dy = \int_0^{H+\delta} v_{x1}\,dy \qquad\qquad (10\text{--}108)$$

where:

$$v_{x1} = \left(\frac{v_{bx}}{H+\delta} - \frac{g_{d1}(H+\delta)}{2\mu}\right) y + \frac{g_{d1}y^2}{2\mu} \qquad\qquad (10\text{--}109)$$

$$v_{x2} = \left(\frac{v_{bx}}{\delta} - \frac{g_{d2}\,\delta}{2\mu}\right) y + \frac{g_{d2}y^2}{2\mu} \qquad\qquad (10\text{--}110)$$

In the case of pure drag flow, the following relationship must be satisfied:

$$g_{d2}w + g_{d1}W = 0 \qquad\qquad (10\text{--}111)$$

With these equations, the following expression for the drag induced pressure gradient in the clearance is obtained:

$$g_{d2} = \frac{6\,\mu H v_{bx} W}{w(H+\delta)^3 + W\delta^3} \qquad\qquad (10\text{--}112)$$

Considering that the actual clearance value ranges from the flight clearance to the channel depth, the pressure gradient over the screw flight can be approximated by:

$$g_{d2} \approx \frac{6\mu v_{bx}}{\left(H + \frac{1}{2} W_a\right)^2} \qquad (10\text{--}113)$$

The resulting leakage flow can be approximated by:

$$\dot{V}_{ld2} = \frac{\pi D (1-f)(H + 0.5W_a)^3 g_{d2}}{12\mu} \qquad (10\text{--}114)$$

The total drag induced leakage flow in the apex region now becomes:

$$\dot{V}_{ld} = \frac{1}{2}(1-f)\pi D(H + 0.5W_a)v_b \sin\varphi \qquad (10\text{--}115)$$

The pressure induced leakage flow in the apex region can be approximated by:

$$\dot{V}_{lp} = \frac{0.5\pi D(1-f)(H + 0.5W_a)^3 g_a}{12\mu} \qquad (10\text{--}116)$$

The output per screw for the staggered screw geometry can be written as:

$$\dot{V} = \frac{1}{2}fWHv_{bz} - \frac{WH^3 g_z}{12\mu} - \dot{V}_{ll} - \dot{V}_{ld} - \dot{V}_{lp} \qquad (10\text{--}117)$$

Figure 10–57 compares output predictions made with equation 10–117 to the experimental results obtained by Nichols (19).

The agreement between predictions and experimental results is reasonable. Both the drag flow rate and the pressure generating capacity of the staggered screws are lower than the same screws in a matched configuration. However, the mixing capability of the staggered screws will be significantly better as a result of the interscrew material transfer. The mixing process in non-intermeshing twin screw extruders was studied by Howland and Erwin (28) for screws in a matched configuration. They found that the mixing efficiency of the twin screw extruder was markedly better than the single screw extruder. Howland and Erwin did not report on the mixing efficiency of the twin screw extruder with the staggered screw configuration. However, it can be expected that this will be considerably better than the mixing efficiency of screws in a matched configuration.

A different type of NOCT extruder is the Farrel Continuous Mixer (FCM). It is a short twin screw mixer that runs at high speed, up to 1200 rpm for the smallest unit (2 FCM). A schematic picture of an FCM is shown in Figure 10–58.

The length of the screws is about 5 D, where the first 2 D is conventional screw geometry and the last 3 D a sigma type geometry, similar to the Banbury mixer. The actual mixing takes place in the sigma screw section. Since the twin screw mixer does not generate much

pressure, material is dumped into a discharge extruder or a gear pump for pressure generation. A typical residence time in the FCM is about three to five seconds. The mixing action,
therefore, is very intensive because energy is dissipated in the material at a very fast rate.
This type of mixing is particularly useful in applications where dispersive mixing is required. The intensive mixing action is attended with high shear stresses, causing effective
breakdown of gels and agglomerates in the polymer.

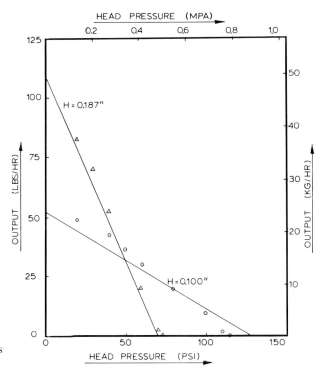

Figure 10-57. Output Predictions Using Equation 10-117 to
Experimental Results by Nichols
(19); Viscosity 58 Pasec

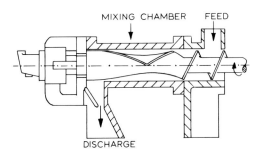

Figure 10-58. Farrel Continuous Mixer

10.6   COAXIAL TWIN SCREW EXTRUDERS

An unusual type of twin screw extruder is the coaxial twin screw (CTS) extruder. The CTS extruder is basically a single screw machine where the main screw is hollow towards the end of the screw. In the hollow portion of the main screw an inner screw is placed to aid in the conveying process in the extruder. The inner screw is generally stationary with a cantilever support against a disk at the end of the barrel. Thus, the I. D. of the main screw forms a rotating barrel for the stationary inner screw. If the flights of the inner screw have an opposite pitch to the flights of the main screw, the material transport in the inner screw will be forward. If the flights of the inner screw have the same pitch as the flights of the main screw, the material transport in the inner screw will be backward.

There are two versions of CTS extruders commercially available. In one version, molten material from the outer screw is transferred to the inner screw where it is pumped towards the die. In this case, the inner screw is used for forward pumping, see Figure 10–59.

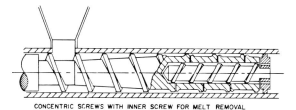

CONCENTRIC SCREWS WITH INNER SCREW FOR MELT REMOVAL

Figure 10–59. Inner Screw for Forward Melt Conveying

If the channel depth of the outer screw is reduced to zero, all the polymer melt is pumped towards the die by the inner screw. This inner melt removal (IMR) screw was invented by Kovacs (20) of Midland-Ross Corporation. It is essentially a very complicated version of a barrier screw and does not offer any obvious advantages over conventional barrier screws.

Another version of a CTS extruder is the solids draining screw (SDS) originally developed by Klein and Tadmor (21) of Scientific Process & Research, Inc. In this design, unmolten polymer drains into the inner screw, see Figure 10–60.

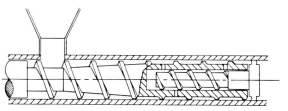

CONCENTRIC SCREWS WITH INNER SCREW FOR SOLIDS RECIRCULATION

Figure 10–60. Solids Draining Screw

Material is transported upstream in the inner screw and plasticated. At the end of the inner screw, the polymer, which is now molten, is pumped back into the channel of the main screw. A modified version of the SDS screw for use in a molding machine was patented on September 22, 1981 (22). Another modification of the SDS screw involves using a barrier type main screw to improve the solids draining process. This barrier SDS screw was patented on June 14, 1983 (23). The SDS screw is claimed to give higher output and lower energy consumption. However, from a functional analysis it is difficult to see why recirculation of a fraction of the polymer flow would increase output or reduce power consumption.

The mechanical design of CTS extruders is considerably more complex than conventional single screw extruders. Since material has to leak through holes in the main screw, there is a chance of plugging and of stagnant flow areas. Also, maintenance and operating procedures of CTS extruders will be considerably more complex than single screw extruders.

10.7   DEVOLATILIZATION IN TWIN SCREW EXTRUDERS

Twin screw extruders are finding increasing use in specialty operations such as reactive processing of polymers and devolatilization. Twin screw extruders are used as continuous chemical reactors for polymerization and polymer modification, e.g. grafting of side groups. Both co-rotating (e.g. 30–32) and counter-rotating (e.g. 33–35) twin screw extruders are used for this purpose, intermeshing as well as non-intermeshing (36). In the extrusion of reacting materials, another degree of difficulty is added to the description of the process because the material properties will change as the reaction progresses along the machine. The theory of extrusion of reacting materials is still in an early stage of its development. However, one aspect of specialty polymer processing operations, namely continuous devolatilization in twin screw extruders, has reached a point where a reasonably accurate description of the process is possible.

Todd (37) proposed an equation to describe devolatilization in co-rotating twin screw extruders based on the penetration theory discussed in Section 5.4 and Section 7.6. The equation contains the Peclet number, see equation 7–248 c, which represents the effect of longitudinal backmixing. The Peclet number must be measured or estimated to predict the devolatilizing performance of an extruder. Todd selected a Peclet number of 40 to correlate predictions to experimental results. A similar approach was followed by Werner (38). A visualization study was made by Han and Han (39), particularly to study foam devolatilization. They found substantial entrainment of the bubbles in a circulatory flow region in a partially filled screw devolatilizer. Collins, Denson and Astarita (40) published an experimental and theoretical study of devolatilization in a co-rotating twin screw extruder. The experimentally determined mass transfer coefficients were about one-third those predicted by the mathematical model. They concluded, therefore, that the effective surface area for mass transfer is substantially less than the sum of the areas of the screws and barrel.

Secor (41) presented an intermeshing model for devolatilization in co-rotating twin screw

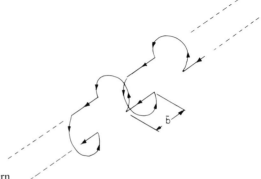

Figure 10-61.   Predominant Fluid Flow Pattern

extruders that incorporates the major characteristics of fluid motion. These characteristics were experimentally observed in a twin screw extruder with a transparent barrel. The observed flow pattern consisted of alternating rotation in tangential direction with the screw and axial forward motion at the entrance to the intermeshing region, see Figure 10-61.

These fluid flow patterns are fully consistent with the flow patterns described in Section 10.3. The model is based on the following assumptions:

1. All the fluid is exposed for a series of intervals, each of duration λ, in which devolatilization occurs.

2. Between exposures, perfect mixing occurs during the forward axial motion in a time that is very small compared with λ.

3. The diffusion coefficient is constant.

4. The fluid layers in the screw channel are effectively of infinite depth.

5. The volumetric flow rate of the fluid is constant.

6. No flow of fluid relative to the underlying flight faces takes place during devolatilization.

7. Nucleation and bubble growth are negligible.
    A material balance on the volatile component can be written as:

$$\dot{V}C_0 = \dot{V}C_1 + \dot{E}_1 \qquad\qquad (10\text{--}118)$$

where

C_0 = concentration of the volatile component in the feed to the extruder in gr/cm^3.

C_1 = average concentration of the volatile component in the liquid at the end of the first exposure in interval in gr/cm^3.

$\dot{V}$ = volumetric liquid flow rate cm^3/sec.

$\dot{E}_1$ = average rate of evaporation in the first exposure interval in gr/sec.

If the surface concentration of the volatile component is maintained at zero, the rate of evaporation is:

$$\dot{E}_1 = A_e C_0 \left(\frac{4D'}{\pi\lambda} \right)^{1/2} \qquad\qquad (10\text{--}119)$$

This equation corresponds to equation 7-242 for devolatilization in single screw extruders. In equation 10-119, the variables are:

A_e = effective area for evaporation in cm^2/sec.
D' = diffusion coefficient in cm^2/sec.
λ = exposure time in sec.

The effective area for evaporation is given by:

$$A_e = f \ A \qquad (10\text{--}120)$$

where A is the total leading face area of a 360° section of a single screw and f is the ratio of channel area outside the intermeshing region to the channel area inside the intermeshing region. Thus, the exposure time λ can be written as:

$$\lambda = \frac{f}{N} \qquad (10\text{--}121)$$

where N is the rotational speed of the screws in rev/sec. With equations 10–120 and 10–121, the rate of evaporation becomes:

$$\dot{E}_1 = AC_0 \left(\frac{4D' \ fN}{\pi} \right)^{1/2} \qquad (10\text{--}122)$$

Substituting equation 10–122 into equation 10–118 gives:

$$\frac{C_1}{C_0} = 1 - \frac{A}{\dot{V}} \left(\frac{4D' \ fN}{\pi} \right)^{1/2} \qquad (10\text{--}123)$$

For the n-th exposure, equation 10–123 becomes:

$$\frac{C_n}{C_{n-1}} = 1 - \frac{A}{\dot{V}} \left(\frac{4D' \ fN}{\pi} \right)^{1/2} \qquad (10\text{--}124)$$

For a sequence of n exposures:

$$\frac{C_n}{C_{n-1}} \frac{C_{n-1}}{C_{n-2}} \cdots \cdots \frac{C_1}{C_0} = \left[1 - \frac{A}{\dot{V}} \left(\frac{4D' \ fN}{\pi} \right)^{1/2} \right]^n \qquad (10\text{--}125)$$

Each exposure is followed by a short axial movement which has been determined to be equal to the mean flight thickness $\bar{b}$. If the total length of the screws is L, the number of exposures is given by:

$$n = \frac{L}{\bar{b}} \qquad \qquad \text{-}(10\text{--}126)$$

Thus, equation 10–125 can be written as:

$$\frac{C_n}{C_0} = \left[1 - \frac{A}{\dot{V}} \left(\frac{4D' \ fN}{\pi} \right)^{1/2} \right]^{L/\bar{b}} \qquad (10\text{--}127)$$

Equation 10–127 is graphically represented in Figure 10–62.

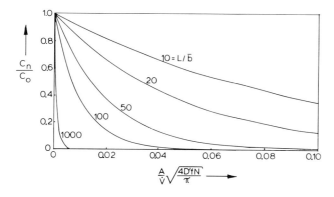

Figure 10–62. Graphical Representation of Equation 10–127

The concentration ratio decreases with increasing diffusivity, screw speed, area, and number of exposures. The concentration ratio increases with increasing flow rate. When the following inequality is fulfilled:

$$\frac{A}{\dot{V}} \left(\frac{4d' fN}{\pi} \right)^{1/2} < 0.02 \qquad\qquad (10\text{--}128)$$

equation 10–127 reduces to:

$$\frac{C_0}{C_n} = \exp \left[\frac{LA}{b\dot{V}} \left(\frac{4D' fN}{\pi} \right)^{1/2} \right] \qquad\qquad (10\text{--}129)$$

with an error of less than one percent.

Experiments were performed on a 20 cm twin screw extruder with a screw length of 27 cm. The liquid was a polybutene and the volatile halocarbon. Figure 10–62 shows the correlation between experimental results and theoretical predictions.

The agreement between theory and experiments is quite good. Even though bubble formation was observed during some of the experiments, the mass transfer rate was not significantly affected. Reasons for this include axial short-circuiting of fluid along the bottom of the figure-eight bore and incomplete mixing at points of transfer between the screws. Another reason is the likelihood of the bubble actually being retained by the liquid phase. This is supported by visual observations by Han and Han (39).

Evaporation from the liquid on the barrel was assumed to be of minor importance compared with evaporation from the liquid in the screw channels. This assumption is supported by experimental work by Biesenberger and Lee (42) who found the contribution of devolatilization through the barrel film to be essentially negligible. This is an important point because if the contribution through the melt film is negligible in a twin screw extruder, it should also be of minor importance in a single screw extruder. One note of caution is that Secor's experiments were performed on a machine with a rather unusual geometry. The screw channel had a very small width to depth ratio (W/H), the actual value being W/H = 0.384, and the screw length was very short (1.36 D). In this case, the area of the melt pool is much larger than the area of the melt film. Thus, in this geometry, the contribution of

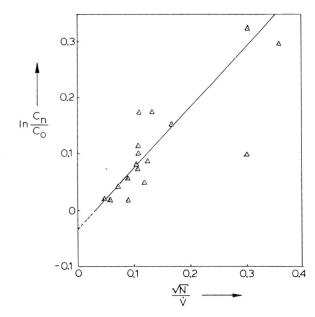

Figure 10-63. Correlation of
Experimental Results (41)

the melt film has to be very small from simple geometric arguments. In most commercial extruders, however, the W/H ratio is at least an order of magnitude higher and the screw
length is also at least an order of magnitude higher. Thus, the Secor model may not give
equally good results with twin screw extruders of more standard geometry.

REFERENCES – CHAPTER 10

1. K. Eise, S. Jakopin, H. Herrmann, U. Burkhardt, and H. Werner, Adv. Plast. Techn.,
   April, 18–39 (1981).
2. K. Burkhardt, H. Herrmann and S. Jakopin, Plast. Compounding, Nov./Dec. 73–78
   (1978).
3. H. Herrmann and U. Burkhardt, Kunststoffe, 11, 753–758 (1978).
4. C. D. Denson and B. K. Hwang, Jr., Polym. Eng. Sci., 20, 965–971 (1980).
5. C. E. Wyman, Polym. Eng. Sci., 15, 606–611 (1975).
6. J. C. Maheshri and C. E. Wyman, Ind. Eng. Chem. Fundam., 18, 226–233 (1979).
7. J. C. Maheshri, Ph. D. thesis, Univ. of New Hampshire, Durham, N. H. (1977).
8. M. L. Booy, Polym. Eng. Sci., 18, 973–984 (1978).
9. H. Werner, Ph. D. thesis, Univ. of Munich, W. Germany (1976).
10. G. Schenkel, "Kunststoff-Extrudertechnik", Carl Hanser Verlag, Munich (1963).
11. Z. Doboczky, Plastverarbeiter, 16, 57–67 (1965).
12. Z. Doboczky, Plastverarbeiter, 16, 395–400 (1965).
13. P. Klenk, Plastverarbeiter, 22, 33–38 (1971).
14. P. Klenk, Plastverarbeiter, 22, 105–109 (1971).
15. L. P. B. M. Janssen, "Twin Screw Extrusion", Elsevier, Amsterdam (1978).
16. L. P. B. M. Janssen, L. P. H. R. M. Mulders, and J. M. Smith, Plastics & Polymers, June,
    93–98 (1974).

17. A. Kaplan and Z. Tadmor, Polym. Eng. Sci., 14, 58-66 (1974).
18. R. J. Nichols and J. Yao, SPE ANTEC, San Francisco, 416-422 (1982).
19. R. J. Nichols, SPE ANTEC, Chicago, 130-133 (1983).
20. L. Kovacs, U. S. Patent 3,689,182.
21. I. Klein and Z. Tadmor, U. S. Patent 3,924,842.
22. R. Klein, Edison, and I. Klein, U. S. Patent 4,290,702.
23. R. Klein, Edison, and I. Klein, U. S. Patent 4,387,997.
24. R. J. Nichols, SPE ANTEC, New Orleans (1984).
25. R. B. Bird, R. C. Armstrong, and O. Hassager, "Dynamics of Polymeric Liquids", Volume 1, Fluid Mechanics, Wiley, New York (1977).
26. W. Kozicki, C. H. Chou, and C. Tiu, Chem. Eng. Sci., 21, 665 (1966).
27. W. Kozicki, C. J. Hsu, and C. Tiu, Chem. Eng. Sci., 22, 487 (1967).
28. C. Howland and L. Erwin, SPE ANTEC, Chicago, 113-116 (1983).
29. M. L. Booy, Polym. Eng. Sci., 20, 1220-1228 (1980).
30. W. A. Mack and R. Herber, Chem. Eng. Prog., 72, Jan., 64-70 (1976).
31. L. Wielgolinski and J. Nangeroni, Adv. Pol. Tech., 3, 99-105 (1984).
32. M. Eyrich, 3 rd Int'l Congress on Reactive Processing of Polymers, Strasbourg, France, Sept., 165-180 (1984).
33. L. P. B. M. Janssen, B. J. Schaart, and J. M. Smith, Polymer Extrusion II Conference, London, England, May, 15.1-15.7 (1982).
34. N. P. Stuber and M. Tirrell, 3 rd Int'l Congress on Reactive Processing of Polymers, Strasbourg, France, Sept., 193-201 (1984).
35. J. A. Speur and L. P. B. M. Janssen, 3 rd Int'l Congress on Reactive Processing of Polymers, Strasbourg, France, Sept., 363-372 (1984).
36. R. J. Nichols and R. K. Senn, Paper Presented at 53 rd Annual Meeting of The Society of Rheology, Louisville, KY, October 15 (1981).
37. D. B. Todd, SPE ANTEC, 472-475 (1974).
38. H. W. Werner, Kunststoffe, 71, 18-26 (1981).
39. H. P. Han and C. D. Han, SPE ANTEC, Washington, DC (1985).
40. G. P. Collins, C. D. Denson and G. Astarita, AIChE J., Aug., 1288-1296 (1985).
41. R. M. Secor, 3 rd Int'l Congress on Reactive Processing of Polymers, Strasbourg, France, Sept., 153-164 (1984).
42. J. A. Biesenberger and S. T. Lee, SPE ANTEC, Washington, DC (1985).

11   TROUBLESHOOTING EXTRUDERS

When an extruder develops a problem, it is very important to be able to diagnose the extruder quickly and accurately in order to minimize downtime or off-quality product. Two important requirements for efficient troubleshooting are good instrumentation and good understanding of the extrusion process. As discussed in Chapter 4, instrumentation is very important in process control, but it is absolutely essential in troubleshooting. Without good instrumentation, troubleshooting is a guessing game at best, no matter how well one understands the entire process. Thus, lack of instrumentation can prove to be very costly if it causes a certain problem to remain unsolved for even a limited length of time. The second requirement for efficient troubleshooting, namely a good understanding of the extrusion process, may be more difficult to meet than the first requirement. It is the author's hope that this book will aid in meeting this second requirement.

In this chapter, the problems that will be primarily focused on are upsets. These are problems that occur in an existing extrusion line for some unknown reason. If the extrusion line has been running fine for a considerable period of time, then it is clear that there must be a solution to the problem. Thus, the objective of troubleshooting is to find the cause of the upset and to eliminate it. On the other hand, when one deals with a development problem, there may not be a solution. In a development problem, one tries to establish a condition that has not been achieved before. If the condition that one is trying to establish is physically impossible, then clearly there is no solution to the problem. From a functional analysis of the process, see Chapter 7, one should be able to determine the bounds of the conditions that can be realized in practice.

11.1   MATERIAL RELATED PROBLEMS

A change in the material can cause a problem in extrusion when it affects one or more polymer properties that determine the extrusion behavior of the material. These properties and the measurement thereof are discussed in Chapter 6. They are the bulk properties, the melt flow properties, and the thermal properties. If a material problem is suspected, one should first examine the quality control (QC) records on incoming material to see if a change in feed stock properties was determined. Unfortunately, often the only QC test on incoming material is a MI (melt index) test, see Section 6.2.6.2. This test is only able to detect a very limited number of material-related extrusion problems. Thus, in many cases, material testing may have to be more extensive than the regular QC testing.

There are a number of problems associated with making measurements on the critical properties of the feed stock. The total number of properties that need to be measured is about ten, with some of the measurements being fairly time consuming. Thus, it may take considerable time to fully characterize the extrusion properties of a material; this does not help when a quick solution is required. Another problem is the fact that some important properties are difficult to measure with a high degree of accuracy and reproducibility. The most notable property in this respect is the external coefficient of friction, see Section 6.1.2. A further problem can be that instruments to measure all the pertinent properties may not be readily available. Not all companies can afford to maintain a fully instrumented laboratory to completely characterize the extrusion characteristics of a certain compound. Finally, even after a material is fully characterized and no significant changes in properties have been found, there is no guarantee that the extrusion problem is not material related because the material sample used for testing may not have been a representative sample. Since most tests are done on samples of about 0.01 kg and most extruders run at a throughput of several hundred to several thousand kg/hr., there is a considerable chance that the test sample is not representative of the entire feed stock.

A practical test for a material related extrusion problem is to extrude some material from an old batch to see if the problem will disappear. If this is indeed the case, then this provides a very strong indication that the problem is material related. For this reason, it is helpful to retain some material of older batches and also to provide a reference for more detailed measurements. If the problem is material related, there are two possible solutions. The easiest solution from an extrusion point of view is to change the material back to the way it was before the problem developed. However, this may not always be possible for other reasons. Thus, if the change in the material is permanent, then the extrusion process will have to be adjusted to accommodate the material change. At this point, the nature of the problem may change from an upset to a development problem. The chance of solving the problem will depend on the nature and the magnitude of change in the material.

11.2   MACHINE RELATED PROBLEMS

In machine related problems, mechanical changes in the extruder cause a change in extrusion behavior. These changes can affect the drive system, the heating and cooling system, the feed system, the forming system, or the actual geometry of screw and barrel. The main components of the drive are the motor, the reducer, and the thrust bearing assembly. Drive problems manifest themselves either in variations in rotational speed and/or the inability to generate the required torque. Problems in the reducer and thrust bearings are often associated with clear audible signals of mechanical failure. If the problem is suspected to be the drive, make sure that the load conditions do not exceed the drive capability. The motor drive system generally consists of a DC motor, a power conversion unit (PCU), and operator controls. A frequent problem with the motor itself is worn brushes; these should be replaced at regular intervals as recommended by the manufacturer. In troubleshooting an extruder drive, one should follow the procedure recommended by the manufacturer of the drive. A typical troubleshooting guide for a DC motor is shown in Table 11-1.

TABLE 11-1. TROUBLESHOOTING GUIDE FOR D.C. MOTOR

Problem	Possible Cause	Action
Motor will not start	Low armature voltage	Check motor nameplate to insure that motor is connected to proper voltage
	Weak field	Check for resistance in shunt field circuit
	Open circuit in armature or field	Check for open circuit
	Short circuit in armature or field	Check for short circuit

TABLE 11-1. TROUBLESHOOTING GUIDE FOR D.C. MOTOR
            (continued)

Problem	Possible Cause	Action
Motor runs too slow	Low armature voltage	Check for resistance in armature circuit
	Overload	Reduce load or use larger motor
	Brushes ahead of neutral	Determine proper neutral position for brush location
Motor runs too fast	High armature voltage	Reduce armature voltage
	Weak field	Check for resistance in shunt field circuit
	Brushes behind neutral	Determine proper neutral position for brush locations
Brushes sparking	Brushes worn	Replace
	Brushes not seated properly	Reseat
	Incorrect brush pressure	Measure brush pressure and correct
	Brushes stuck in holder	Free brushes, make sure brushes are of proper size
	Commutator dirty	Clean
	Commutator rough or eccentric	Resurface commutator
	Brushes off neutral	Determine proper neutral position for brush location
	Short circuit in commutator	Check for shorted commutator, check for metallic particles between commutator segments
	Overload	Reduce load or use larger motor
	Excessive vibration	Check driven machine for balance
Brush chatter	Incorrect brush pressure	Measure and correct
	High mica	Undercut mica
	Incorrect brush size	Replace
Bearings hot	Belt too tight	Reduce belt tension
	Misaligned	Check alignment and correct
	Bent shaft	Straighten shaft
	Bearings damaged	Inspect and replace

The most important component of the feed system in a flood fed extruder is the feed hopper with possible stirrer and/or discharge screw. A mechanical malfunction of this system can be determined by visual inspection. If the feed hopper is equipped with a discharge screw (crammer feeder), constancy of the drive should be checked. For proper functioning of a crammer feeder, the drive of the crammer feeder should have a torque feedback control to ensure constant feeding and to avoid overfeeding.

The heating and cooling system is used to achieve a certain degree of control of the polymer melt temperature. However, stock temperature deviations do not necessarily indicate a heating or cooling problem because heat is added directly to (or subtracted from) the barrel and only indirectly to the polymer. The rate of addition is determined by the temperature in the barrel as measured with a thermocouple (TC). Thus, the rate of heating or cooling is actually determined by a local barrel temperature and the temperature that is controlled is actually a barrel temperature.

The stock temperature is generally controlled by changing the setpoints of the temperature zones along the extruder. However, due to the slow response of the melt temperature to changes in heat input, only very gradual stock temperature changes can be effectively controlled by setpoint changes. Rapid stock temperature fluctuations, cycle time less than about five minutes, can usually not be reduced with a melt temperature control system. Such fluctuations are indicative of conveying instabilities in the extrusion process and can only be effectively reduced by eliminating the cause of the conveying instability.

The heating system can be checked by changing the setting to a much higher temperature, for instance $50°C$ above the regular setting. The heater should turn on a full 100 percent and the measured barrel temperature should start rising in about one to two minutes. If the heater does not turn full on, the barrel temperature measurement is in error or there is a problem in the electronic circuit of the temperature controller. If the heater turns full on but the temperature does not start to rise within two to four minutes, either the barrel temperature measurement is incorrect or there is poor contact between heater and barrel.

The cooling system can be checked by changing the setting to a much lower temperature, for instance $50°C$ below the regular setpoint. The cooling should turn full on and the measured barrel temperature should start to drop in about one to two minutes. If the cooling does not turn full on, the barrel temperature measurement is in error or there is a problem in the circuit of the temperature controller. If the cooling turns full on but the temperature does not start to drop within two to four minutes, either the barrel temperature is incorrect or the cooling device is inoperable. This check-out procedure is summarized in Table 11-2.

TABLE 11-2. HEATING AND COOLING SYSTEM CHECK

Increase setpoint of temperature zone by 50°C:	
– Heater on full blast TC rises in about 2 minutes	– System normal

– Heater on full blast
  TC does not change
	– Poor contact of heater
	  Insufficient heating
	  capacity
	  TC failure

– Heater output does not change	– Heater failure Controller bad
Reduce setpoint of temperature zone by 50°C:	
– Cooling on full blast TC drops in about 2 minutes	– System normal

– Cooling on full blast
  TC does not change
	– TC failure
	  Insufficient cooling
	  capacity

– Cooling output does not change	– Cooling system bad Controller bad

If a substantial amount of cooling is required to maintain the desired stock temperature, this is generally a strong indication of excessive internal heat generation by frictional and viscous dissipation. The most effective way to reduce the internal heat generation is often by changing the screw design, as discussed in Section 8.3

Mechanical changes in the forming system relate to the extrusion die and downstream equipment. These elements can be subjected to simple visual inspection and mechanical changes can thus be easily determined. Changes in the geometry of screw and barrel are often caused by wear. Since wear is a very important phenomenon in extrusion, it will be discussed in detail in the following section.

11.2.1 WEAR PROBLEMS

Wear in extruders generally causes an increase in the clearance between screw flight and barrel. As discussed in Section 8.1.3, wear often occurs towards the end of the compression section. This type of wear is more likely to occur when the screw has a high compression ratio. Wear in the compression section of the screw reduces the melting capacity, see Section 8.2.2, and will lead to temperature non-uniformities and pressure fluctuations. Wear in the metering section of the screw will reduce the pumping capacity; however, the reduction in pumping capacity is generally quite small as long as the wear does not exceed two to three times the design clearance. An increased flight clearance will also reduce the effectiveness of the heat transfer from the barrel to the polymer melt and vice versa; this may contribute to temperature non-uniformities in the polymer melt.

Wear can only be detected by disassembling the extruder and by inspection of the screw and barrel. If the wear is serious enough to affect the extruder performance, it will often be noticeable with the naked eye. However, it is recommended to measure the ID of the barrel and the OD of the screw over the length of the machine. If this is done regularly, then one can determine how fast wear is progressing with time. By extrapolating to the maximum allowable wear, one can determine at what point in time the screw and/or barrel should be replaced or rebuilt. If replacement as a result of wear is necessary after several years of operation, the easiest solution is to simply replace the worn parts. However, if replacement as a result of wear becomes necessary within a short period of time, for instance several months, then simple replacement will not provide an acceptable solution. In short term wear problems, the cost of downtime and replacement parts can easily become unacceptable and the solution has to be found in reducing the actual wear rate instead of simply replacing the worn parts. To reduce the wear rate, one has to understand the wear mechanism(s) in order to determine the most effective way to reduce wear.

11.2.1.1 WEAR MECHANISMS

Five mechanisms of wear can be distinguished:

1. adhesive wear
2. abrasive wear
3. laminar wear
4. surface-fatigue wear
5. corrosive wear

When wear occurs, often more than one mechanism is at work. Adhesive wear occurs with metal-to-metal contact under high stresses. Since the actual contact area is much smaller than the apparent contact area, local welds can form at points of contact. This phenomenon is often referred to as cold welding. The sliding motion causes a rupture in the weld region and small fragments of the weld region are carried away with one member of the sliding system. Usually fragments of the softer material transfer to the harder material. Adhesive junctions are only formed between clean surfaces. The attrition rate depends on the shear strength of the adhesive junctions. Adhesive wear is generally most severe with sliding contact of similar metals. Adhesive wear between similar metals is often referred to as galling. In sliding motion between dissimilar metals, the adhesive junction will contain a spectrum of compositions. Adhesive wear can be significantly reduced when the spectrum of compositions in a junction contains brittle intermetallic compounds that fracture easily. Lubricants are often used to reduce the chance of adhesive wear. When oxide layers form at the interface, this will also reduce adhesive wear because oxides will not bond.

Abrasive wear occurs by a microcutting process. In two-body adhesive wear, the asperities of the harder member penetrate the softer one and remove material as a result of the sliding motion. In three-body abrasive wear, hard particles are embedded in the material of at least one member of the wear system. The hardness ratio has been found to be the most important material characteristic in abrasive wear, although the influence of fracture toughness seems to play a role (1). Krushchov (2) found that the wear resistance of pure metals and annealed steels increases proportionally with hardness. Strain hardening or precipitation hardening does not result in improved abrasive wear resistance since the microcutting process already yields maximum local strain hardening.

Laminar wear occurs when the shear strength in the heterogeneous interfacial layer is higher than the shear strength of the homogeneous portion of the interfacial layer. Laminar wear takes place only at the thin outer layers of the interface. Laminar wear is sustained only if the outer layer of the heterogeneous interface continuously regenerates as an oxidic or other reactive layer. The formation of wear reducing reactive layers can be controlled by additives in the lubricant. Laminar wear, to a certain extent, is a mild form of corrosive wear. A mild corrosive or oxidative action affects only a thin layer of the newly generated metallic surface. When the new surface layer is formed, the reaction will stop.

In surface fatigue wear, there is a separation of microscopic and macroscopic material particles from the surface which is caused by fatigue crazing, cracking, and breakup under specific mechanical, thermal, and chemical load in rolling contact between two surfaces. Fatigue cracking is initiated by alternating thermal or mechanical load. This type of wear can occur even with direct metallic contact. Surface fatigue wear is characterized by considerable induction times and relatively large depth of penetration. A familiar example is the pitting of roller bearings and gears.

In corrosive wear, a chemical reaction attacks at least one of the sliding surfaces. Corrosive wear in extrusion occurs usually in combination with one or more of the other wear mechanisms. The combined chemical and mechanical attack of the sliding surface can cause wear rates far in excess of what would be expected based on their individual contribution.

In extrusion, the most important wear mechanisms are adhesive, abrasive, and corrosive wear.

11.2.1.2 TEST METHODS FOR WEAR

There are basically two ways to test the wear characteristics in the extrusion process. One method is to run the actual machine under normal operating conditions and to measure the progress of wear at regular intervals. This approach is time consuming and expensive, but it does yield accurate and representative results. However, it does not allow a simple analysis of the parameters that influence the wear process.

An interesting technique to do relatively quick wear studies on actual extruders was developed at the IKT in Stuttgart, West Germany by Fritz and coworkers (31). A reference surface of the machine is made radioactive by proton and neutron bombardment to a depth of 30 μm to 80 μm. The impulse rate from the measuring isotope reduces linearly with activation depth. This allows accurate measurement of wear over short time periods. It was found that the wear process can be accurately characterized in about one to three hours. In the particular study mentioned (31), wear was measured in the feed section of a screw; the extruder was equipped with a grooved barrel section. The abrasive filler was titanium dioxide, which was added to the virgin polymer as a masterbatch. When the virgin polymer was in pellet form, considerable wear occurred, while no wear was measured when the virgin polymer was in powder form.

Another method involves testing on model systems. In such a test, a test specimen is subjected to certain load conditions to simulate actual service conditions. Such wear testers allow the tribological relevant loads to be preselected; tribological parameters such as temperature, coefficient of friction, etc., can be measured and recorded continuously. This method allows a relatively quick and inexpensive determination of wear characteristics. However, information obtained from a wear tester can only be transferred to practice if the wear conditions in the model system are essentially the same as those in the real system, i.e., the extruder. Many mistakes have been made in transferring information from a short wear test to actual extruders, simply because of differences in the tribological conditions of the wear process. It is often not realized that the tribological parameters friction and wear are not material properties, but properties of a complex system. Thus the transfer from a model system to an actual extruder has to be made cautiously. Measurements on the actual extruder are required to ensure that results from the model system are also valid for the real system. A good review and analysis of test methods for wear in the polymer processing industry was given by Mennig and Volz (3). Considering that there are many different types of wear in polymer processing, there is, unfortunately, no universal wear tester. Mennig and Volz distinguish four types of testing: metal-liquid wear, metal-solid polymer wear, corrosive wear, and metal-to-metal wear.

As early as 1944, a test device was proposed by Mehdorn (4) to measure metal-liquid wear. This test was to simulate wear conditions in a press used for injection of thermosets. The test geometry is shown in Figure 11-1.

A molten or liquid mass of polymer is forced onto a test specimen, from which it is deflected; the material exits through a small clearance of 0.4 mm. This test device gives relatively quick results. Disadvantages are the complex geometry of the clearance, non-uniform flow

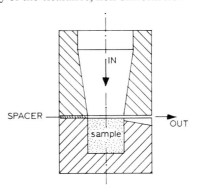

Figure 11-1. Wear Test Device Proposed by Mehdorn

conditions at the specimen, and that increased wear changes the resistance to flow and thus the wear conditions. Another method was developed by Bauer, Eichler, and John (5) in 1967. Figure 11–2 shows the geometry of their test apparatus.

The specimen is a diamond-shaped obstruction in the center of a flow channel. A commercial wear test apparatus based on this geometry is the Tribotest from Brabender OHG in Duisburg, West Germany. This test is often referred to as the "Siemens-Methode" wear test. Eichler and Frank (6) modified this test to make it more suitable for injection molding, see Figure 11–3.

An entirely different test geometry was developed at the DKI (Deutsches Kunststoff Institut). This test utilizes a flat plate geometry as shown in Figure 11–4.

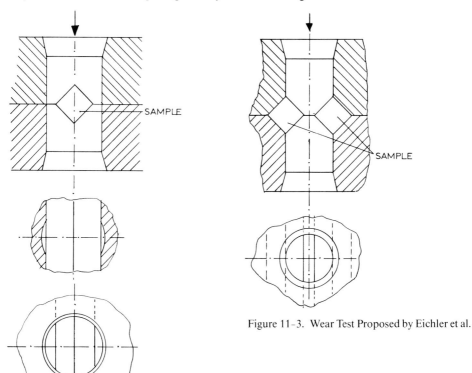

Figure 11–3. Wear Test Proposed by Eichler et al.

Figure 11–2. Wear Test Proposed by Bauer et al.

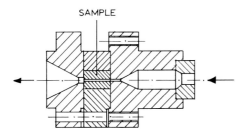

Figure 11–4. DKI Flat Plate Wear Test

The rectangular test gap has a length of 12 mm, a width of 10 mm, and a height that is adjustable from 0.1 mm to 1.0 mm. This geometry has been used for studies with thermoplastics (7) as well as thermosets (8). A modification of the flat plate wear tester is the BASF wear tester, see Figure 11–5.

This test is to simulate the wear process in a molding machine.

Another test apparatus developed at the DKI is the ring method, see Figure 11–6.

This test is to simulate conditions occurring in the annular space between the tip of the screw flight and the extruder barrel (7, 9). Plumb and Glaeser (10) developed a test method for filled elastomers based on a capillary rheometer, see Figure 11–7.

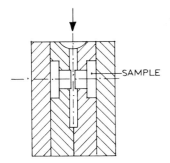

Figure 11–5. BASF Wear Tester

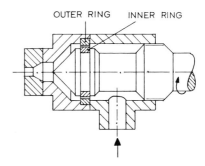

Figure 11–6. DKI Ring Wear Test

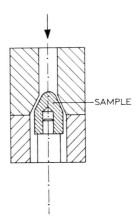

Figure 11–7. Capillary Rheometer Wear Test

This specimen is a cone-shaped torpedo in the flow channel. The flow conditions change in axial direction and with it, the local wear rate. A test developed at Georgia Marble Company, a supplier of calcium carbonate, uses an aluminum breaker plate at the end of a screw extruder to evaluate abrasive wear of mineral fillers. The amount of wear is determined by measuring the weight loss of the breaker plate over a certain run time (25).

Metal-to-solid polymer wear occurs in the solids conveying zone of the extruder. The introduction of the grooved barrel extruder has significantly increased the interest and concern about wear in this portion of the extruder. As discussed in Section 7.2.2.2, grooved barrel sections substantially increase the shear and normal stresses between the polymer solid bed and the metal surfaces. As a result, grooved barrel sections are much more susceptible to wear than smooth barrels. The first systematic study of wear in the solids conveying section of extruders was made by Fritz (11). He used a diamond-shape specimen that protruded into the screw channel as shown in Figure 11–8.

A model system was developed at the DKI in the form of a disk wear tester. This universal disk-tribometer was discussed by Volz (12). The concept of the disk-tribometer is partially based on a modified friction tester developed at Enka Glanzstoff (13). The geometry of the disk-tribometer is schematically shown in Figure 11–9.

The test disk is made up of two annular rings. The outer ring I is used for preconditioning the polymeric sample. The inner ring II contains the metal specimen. The disk-tribometer allows measurement of frictional force, normal force, and wear at contact pressures of up to 150 MPa.

A test for corrosive wear was first proposed by Calloway, Morrison, and Williams (14). They use a vacuum pyrolysis at normal process temperature, see Figure 11–10.

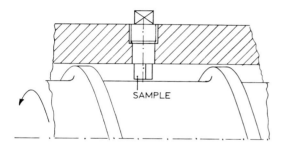

SAMPLE

Figure 11–8. Wear Test in a Grooved Barrel Section

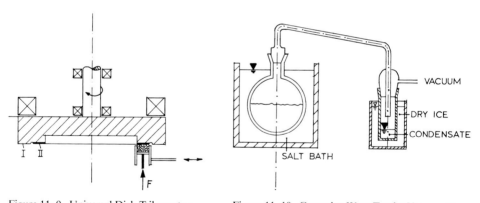

Figure 11–9. Universal Disk-Tribometer

Figure 11–10. Corrosive Wear Test by Vacuum Pyrolysis

A metal specimen is suspended in the extracted volatiles for 24 hours at room temperature. A similar apparatus was used by Mahler (7) and Braun and Maelhammar (15) to test for corrosion at elevated temperatures and pressures. A disadvantage of vacuum pyrolysis is the fact that the corrosion conditions are different from those existing in the actual extrusion process. This drawback is reduced in the test procedure used by Knappe and Mahler (9), see Figure 11–11.

Other tests have been described by Moslé et al. (16) and Maelhammar (17). The latter test is a combination of metal-liquid wear and corrosive wear. The apparatus was modified by

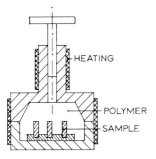

Figure 11–11. Corrosive Wear Test
Proposed by Knappe and Mahler

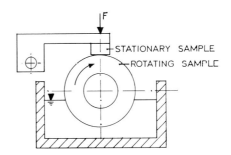

Figure 11–12. Alpha LFW-1 Wear Tester

Volz (18) for thermosets. The volatiles are extracted from the polymer melt, which has been prepared in an injection molding machine and has been sheared through a test gap. Through electrochemical measurements, Volz could prove that significant differences exist in the corrosive action of volatiles separated from injection molded samples of thermosetting polymers.

Tests for metal-to-metal wear can utilize the standard test methods, provided the proper intermediate material can be introduced between the metallic surfaces. Broszeit utilized the cylinder-disk apparatus to study metal-to-metal wear (19). Saltzman, et al. (20–22) used the Alpha LFW-1 test apparatus, see Figure 11–12.

A stationary block is forced against a rotating ring by a dead-weight load. The bottom part of the ring is immersed in a water-oil emulsion. The presence of the water-oil emulsion is a drawback in this test because the actual wear behavior in an extruder with a polymer melt as the intermediate material between screw and barrel is bound to be substantially different from the wear behavior in the LFW-1 test apparatus. No data have been published on metal-to-metal wear with a polymer melt as the intermediate material.

11.2.1.3 CAUSES OF WEAR

In polymer-metal wear, the main causes of wear are abrasive wear and corrosive wear. Abrasive wear is generally due to abrasive components in the polymer matrix. Whether these components are classified as fillers, reinforcements, or additives, they can cause significant wear when the filler is hard and available in significant amounts. Factors affecting the wear are particle hardness, particle size, particle shape, and loading (23). One indication of the abrasive wear ability of a filler is its ranking on the Mohs scale. This scale ranks a material from 1 to 10 according to its ability to scratch another material or to be scratched by another material. A very soft material, such as talc, is ranked at the bottom of the scale, rank 1, and a very hard material, such as diamond, is ranked at the top of the scale, rank 10. The Mohs scale ranking of several fillers is given in Table 11–3.

It should be realized, however, that the filler hardness only partially determines the wear characteristics of a filled compound. The particle size is another important parameter. Generally, the severity of wear reduces with reducing particle size. In glass fiber reinforced compounds, the wear reduces when the fiber length reduces (12). This is believed to be due to greater mobility and reduced kinetic energy of the smaller glass fibers. The particle shape has a very strong influence on the wear characteristics. Experiments done by Mahler (26) with glass fiber reinforced nylon and glass bead reinforced nylon showed the wearing inten-

TABLE 11–3. MOHS SCALE RANKING OF VARIOUS FILLERS

Calcined Kaolin	7
Silica	6.5
Glass	6
Perlite	5.5
Wollastonite	5.5
Mica	3
Calcium Carbonate	3
Kaolin	2
Alumina trihydrate	1
Talc	1

sity of the fiber reinforced compound to be 14 times higher than the compound reinforced with glass spheres. The ability to wear is larger when the particle has sharp corners and a large aspect ratio. The particle shape that best minimizes wear is a spherical shape. Unfortunately, the spherical shape is often undesirable with respect to mechanical properties, electrical properties, etc.

In a study on glass-reinforced polymers, Mahler (7) found that with some polymers, the wear intensity can be much higher than with others as a result of corrosive wear in addition to abrasive wear. He found that the wear intensity of glass fiber reinforced nylon 6,6 against 9 S20K steel was about 13 times higher than reinforced SAN and PC. A similar finding was made by Olmsted (27) in injection molding of glass fiber reinforced nylon 6, where severe wear occurred on both the screw and the barrel. The wear was found to be primarily corrosive-type wear, caused by a silane wetting agent on the fiber. The decomposition temperature of the wetting agent was lower than the process temperature and, hence, degradation occurred with resulting corrosive attack of screw and barrel. However, similar work by Mahler (7) with nylon 6,6 filled with glass fibers coated with an aminosilane coupling agent did not reveal any corrosive wear resulting from the coupling agent.

Surface treatment of fillers can reduce wear. A comparison between coated and uncoated calcium carbonate showed the wear intensity of a rigid PVC compound with the coated filler to be about 3 to 9 times lower than the same compound with the uncoated filler (25). The wear intensity was measured by using the aluminum breaker plate test discussed earlier. When incorporating abrasive fillers, such as glass fibers, it is good practice to add the filler at a location where the polymer is already molten. This allows the melt to coat the filler and reduce the wear intensity. When abrasive fillers are added with solid polymer particles, the abrasive action is much more severe and rapid wear will occur in the solids conveying section of the extruder. This is why glass fiber is generally added in a downstream barrel opening or in a downstream extruder in the case of a tandem extrusion setup.

Corrosive wear also occurs in non-filled polymers; well known examples are fluoropolymers and chlorine containing polymers, such as PVC_2, and PVC. Fluoropolymers have a tendency to form hydrofluoric acid at high temperatures in combination with air and moisture. PVC tends to generate hydrochloric acids at elevated temperatures. The corrosion problem is generally more severe with rigid PVC than with flexible PVC. With polymers like these, the metal parts in contact with the polymer should be made out of a corrosion-resistant metal, such as Hastelloy, 17–4 PH, 15–5 PH, etc.

Corrosion can also occur with hygroscopic polymers, such as ABS, PA, PET, PMMA, etc., when moisture is released under high temperature and pressure, forming high pressure steam. Braun and Maelhammar (15) found that PA 6,6 splits into various corrosive components. Calloway et al. (14) found that the corrosive attack of HIPS is dependent on the chlorine and sulphur content of the carbon blacks. Moslé et al. (16) found that degradation products of ABS can cause corrosive attack in extruders.

Sometimes abrasive components in the compound are foreign objects resulting from contamination or human error. Hard foreign objects, such as wrenches, bolts, knives, etc., can cause severe wear in a very short period of time. Magnetic traps are available to catch metallic objects. However, there is not a simple method to successfully remove all foreign objects from the feed stock. Good housekeeping procedures and conscientious personnel will go a long way in reducing the chance of foreign objects ending up in an extruder. One word of caution is in order for barrier-type screws. If small particles are present in the polymer feed stock that do not melt at operating temperatures, they will get trapped at the end of the barrier section when the particle size is larger than the barrier clearance. This can result in very rapid wear, as the author has been able to verify personally.

Another important cause of wear in extrusion equipment is metal-to-metal wear. Unfortunately, relatively little is known about this type of wear. A number of circumstances can cause metal-to-metal contact between screw and barrel. A number of these have been described in detail in Sections 8.1 and 8.2. In addition, metal-to-metal contact will occur at startup. It can also occur as a result of misalignment, or a warped barrel, or a warped screw. Metal-to-metal contact can occur at the feed opening of an extruder as described by Luelsdorf (28), particularly when the feed opening is offset and when it forms a sharply tapered angle with the circumference of the screw, see also Figure 3-10b. This is discussed in more detail in Section 3.3. The author has also experienced cases where metal-to-metal wear occurred in the feed throat of an extruder as a result of improper feed opening geometry. Radiographic analysis of wear particles indicate (28) that temperatures in the contact zone between screw and barrel may exceed 800° C. This is confirmed by personal observations of a screw subjected to severe metal-to-metal wear with a very noticeable discoloration of the metal in the wear region, see also Section 8.1.3. Discoloration indicates exposure to temperatures over 500° C.

Metal-to-metal wear can also occur in intermeshing twin screw extruders as discussed in Chapter 10. Counter-rotating twin screw extruders are particularly susceptible to metal-to-metal wear. These machines, therefore, generally operate at rather low screw speeds.

Lai Fook and Worth (29) proposed a modified flight geometry to increase the centering force on the screw in order to reduce metal-to-metal contact. The two flight geometries they proposed based on theoretical calculations are shown in Figure 11-13.

Actual measurements of the tangential pressure profile differed from predicted values by a factor of about 10. This indicates that the analysis employed was not entirely realistic; in

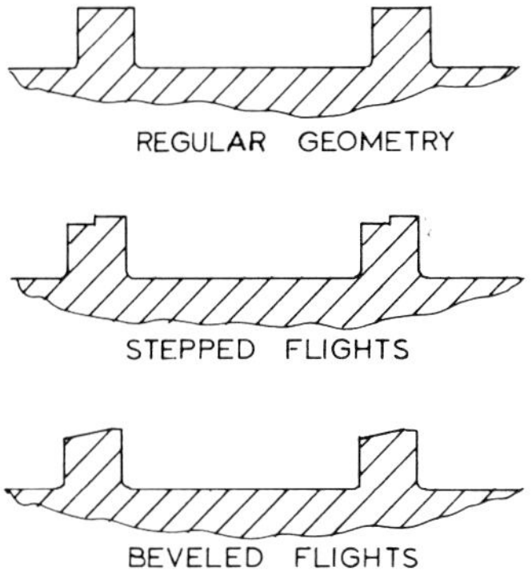

REGULAR   GEOMETRY

STEPPED   FLIGHTS

BEVELED   FLIGHTS

Figure 11-13. Flight Geometry to Reduce Chance of Metal-to-Metal Contact

particular, neglecting side leakage results in large errors in the predicted values. No data was presented on the actual centering force acting on the screw. Winter (30) analyzed the non-isothermal flow of a power law fluid in the flight clearance; solutions were obtained by employing a numerical procedure. Very high temperature increases were found to occur in the polymer melt in the clearance; these temperature changes affect the velocity profile. Since the mass flow rate cannot change, Winter adjusted the pressure gradient along the gap to satisfy the continuity equation. This led to a prediction of large negative pressure gradients at the leading edge of the flight and large positive pressure gradients at the trailing edge of the flight, as shown qualitatively in Figure 11-14.

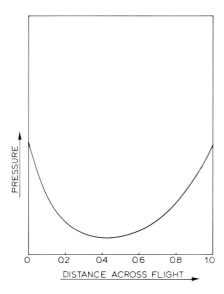

Figure 11-14. Pressure Gradient in the Flight Clearance as Predicted by Winter (30)

Extreme values of the actual pressure occur at about 1/3 from the leading edge and the calculated values range from 5 MPa to 20 MPa.

The predicted pressure profile is obviously a direct result of the assumptions made in the calculations. Winter assumed isothermal conditions at the barrel wall and adiabatic conditions at the flight tip. With stock temperature increases in the order of 100°C and more, it is unlikely that the isothermal boundary condition is valid for the barrel. By the same token, it is unlikely that the adiabatic boundary condition is valid for the flight tip, particularly since the rest of the screw will be at much lower temperature. Unfortunately, it is difficult to measure actual temperature and pressure profiles. Thus, the predicted temperature and pressure profiles have not been compared to experimental results.

Winter postulated that the pressure minimum in the middle of the clearance can cause the screw to be pushed against the barrel by pressure on the other side of the screw. This would only be true if the clearance pressure profile changes along the helical length of the screw flight and if the pressure profiles in the screw channel itself do not play a role of significance. Most likely, the actual situation will be considerably more complex. Following Lai Fook and Worth, Winter recommended beveling the flight tip to create a tapered gap between flight and barrel. Another recommendation was to alter the thermal boundary conditions, for instance by heating the barrel above the temperature of the screw flight. It is interesting to note that the chance of metal-to-metal contact is reduced when the flight clearance and helix angle are increased and when the flight width is decreased, according to Winter's analysis. If this is true, then these measures will have a dual benefit because they will also reduce the power consumption as discussed in Section 8.3.

A disadvantage of the stepped or beveled flight geometry is that when the extruder is running empty or partially empty, the apparent contact area between screw and barrel will be considerably reduced. As a result, the stresses at the actual contact area will be considerably increased and wear, particularly adhesive wear, is more likely to occur. Nonetheless, Volz (12) reports that flight lands with hydrodynamic slide bearing function are used by a number of extruder manufacturers.

From the analysis in Section 8.1, it is clear that metal-to-metal contact between screw and barrel will occur when the extruder is empty. Wear that would result from running an extruder empty would be maximum at the very end of the screw and reduce in upstream direction. In practice, however, the location of maximum wear generally occurs towards the end of the compression section of the screw. This indicates that the screw is supported by the polymer melt at the end of the screw and is subjected to a substantial lateral force in the compression section of the screw. It is unlikely that this lateral force is caused by temperature induced pressure gradients in the flight clearance because the flow process in the flight clearance will not change significantly from the start of melting to the very end of the screw. Therefore, it seems more likely that the lateral force is caused by the conveying process in the screw channel.

As discussed in Section 7.3, in the melting zone of the extruder, there is a continuous deformation of the solid bed. In the compression section of the screw, the solid bed is compressed between the root of the screw and the barrel. Very large pressures can be built up in this screw section, particularly when the compression ratio is high. These rapid pressure changes along the screw can easily cause an imbalance of the lateral forces acting on the screw, as discussed in Section 8.1.3. Thus, it seems that rapid pressure changes along the screw channel are the most likely cause of metal-to-metal contact between screw and barrel. Considering the frequent occurrence of this wear problem, it is surprising how little attention this problem has received as judged from the open technical literature.

11.2.1.4 SOLUTIONS TO WEAR PROBLEMS

The key to finding the best solution to a wear problem is to identify the cause of wear and the wear mechanism. For example, if the screw OD is worn but not the flight flanks and the root diameter, then the problem is probably caused by contact between screw and barrel. In this case, one could put hard facing on the tip of the flight. However, this would not eliminate the cause of the problem, but it may reduce the magnitude of the problem. The actual problem can be eliminated by a change in screw design or by another method that will alter the pressure profile along the screw, e.g., by starve feeding the extruder.

Corrosive wear can usually be identified by a pitted, worn surface. The best solution to corrosive wear is to eliminate the corrosive component from the compound. However, this is often not possible for other reasons. In this case, one has to use corrosion-resistant materials of construction, such as stainless steel, Monel, Hastelloy, etc. In order to select the best material of construction, one should know the chemical species that are causing the corrosive attack. Various metal handbooks contain information about the chemical resistance of many metals against a number of chemical species.

A large number of materials is available for the screw and barrel. As discussed in Section 3.3, most extruder barrels in the U.S. have a liner which is centrifugally cast into the barrel. The barrel liner is made of a wear-resistant material, often boron-stabilized white irons with a Rockwell C hardness of about 65 containing iron chromium boron carbides. Bimetallic barrels provide better wear resistance than nitrided barrels as reported, for instance, by O'Brien (32) and Thursfield (33). The liner material can be formulated to give good abrasion resistance, good corrosion resistance, or a combined abrasion and corrosion resistance. It should be kept in mind that the correct choice of screw material will depend to some extent on the liner material, particularly if metal-to-metal contact takes place. Recommended screw materials for several commercial barrel liners are shown in Table 11-4. The recommendations are based on metal-to-metal wear tests on the Alpha LFW-1 wear test machine. As mentioned in Section 11.2.1.2, this test does not simulate actual conditions in an extruder too well; however, results from more realistic tests are not available in the open literature.

TABLE 11-4. RECOMMENDED SCREW FLIGHT MATERIALS
(Courtesy of Ref. 22)

| | Barrel Liner Material | | |
	Xaloy 101	Xaloy 306	Xaloy 800
Screw Materials	Colmonoy 56	Colmonoy 6	Colmonoy 6
	Colmonoy 6	Colmonoy 56	Colmonoy 56
	Haynes 711	Colmonoy 63	Xaloy 008
	Colmonoy 5	Stellite 1	Nye-Carb
	Colmonoy 63	Nye-Carb	Xaloy 830
	Stellite 1	Stellite 6	Ferro-Tic (Iron)
	Ferro-Tic (Iron)	Stellite 6H (severe wear)	Stellite 6H (severe wear)
	HC-250		
	Colmonoy 84		
	Triballoy T-700		

A number of screw materials are given in Table 3-6. The most common material is 4140 steel. Advantages of 4140 steel are low price, good machinability, and the ability to be used with hard facing and chrome plating. A disadvantage of the 4140 steel is its relatively poor wear resistance, as discussed, for instance, by Hoffmann (34). As a result, 4140 screws are often flame-hardened, plated or hard-faced when used in more demanding applications. Chrome plating is often used on extruder screws. This produces a hard corrosion-resistant layer, up to $70 R_c$ hardness. The layer is usually quite thin, about $25 \mu m$ to $75 \mu m$, and does not form a hermetic seal to most corrosive substances. Thus, the chrome plating does not substantially improve the corrosion resistance of the screw due to its porosity. Chrome plating is quite resistant to abrasive wear, but because of its limited thickness, it does not provide much protection.

Nickel plating is also used quite frequently on extruder screws. It is applied with a thickness similar to chrome plating. The surface hardness of nickel plating tends to be somewhat lower than chrome plating. However, the thickness of the nickel coating is generally more uniform and the coating is much less porous than chrome plating. Therefore, nickel plating usually offers better protection than chrome plating.

A very large number of proprietary plating materials and plating processes are available today. Without exception, the claims made by the supplier of the proprietary plating are quite impressive. Unfortunately, these claims are rarely based on long-term extrusion tests. Thus, one has to be quite cautious in selecting a proprietary plating.

An interesting proprietary plating is the Poly-Ond plating. It is basically an electroless nickel plating impregnated with a fluoropolymer. This yields a moderately hard, corrosion-resistant layer with a low coefficient of friction. The low coefficient of friction makes it attractive to use in injection molds and on extruder screws. Luker from Killion Extruders reported (35) on tests with Poly-Ond plated extruder screws. He reported output increases ranging from 5 percent to 36 percent for a number of different polymers. Another Teflon-impregnated nickel plating is Nedox plating as discussed by Levy (74). A Teflon-impregnated chrome plating was discussed by Trompler (75).

Metal coatings such as chrome plating and nickel plating are generally more effective in reducing wear than hardening of the base material. Various methods are used for hardening, such as flame hardening, nitriding, carburizing, and induction hardening. All of these techniques are basically case-hardening processes with limited depth of penetration and limited improvement in wear characteristics. Further, heat treated steels have reduced hardness and wear resistance at elevated temperatures.

Another technique used for improving the wear resistance of screws and barrels is hard facing. Hard facing materials are generally nickel or cobalt based containing various metal carbides, such as chromium carbide, tungsten carbide, etc. They are applied by welding, spraying or casting with a thickness ranging from 1 mm to 3 mm. Hard facing can offer substantial improvements in wear resistance over heat treated steels. Lucius reports increases in service life by a factor of 2 to 25 by using hard facing on a screw used in extrusion of glass fiber reinforced nylon 6,6 (36). Two common hard facing materials are Stellite (trademark Cabot Corp.) and Colmonoy (trademark Wall Colmonoy Corp.). Colmonoy is a nickel-based alloy containing chromium, iron, boron, and silicon. Stellite is a cobalt-based alloy containing chromium and tungsten. Other hard facing materials are Haynes 711, HC-250, Triballoy T-700, Ferro-Tic HT6 and M6, Nye-Carb, and Xaloy 008, 830, 101, 306, and 800. Metal-to-metal wear tests of these materials were described by McCandles and Maddy (22). Welding techniques for applying these materials are tungsten inert gas (TIG), transferred arc plasma, and oxyacetylene.

Molybdenum-based hard facing alloys have also been used on extruder screws. These materials are relatively soft, about 40 Rockwell C, but have good lubricity. In some instances, wear resistance with the molybdenum hard facing improved about 500 percent over the more common, but harder, hard facing compounds. Molybdenum-based hard facing alloys are used primarily with nitrided barrels, not only in single screw extruders but also in twin screw extruders. The use of molybdenum hard facing in bimetallic barrels often results in rapid wear of the screw. Since most extruders in the U.S. have bimetallic barrels, molybdenum based hard facing for extruder screws has not found widespread use.

In some cases, improvements in wear resistance can be obtained by diffusion coating hard facing alloys. Panzera and Saltzman (37) tested treated and untreated hard facing alloys against carburized SAE 4620 rings using the Alpha Model LFW-1 wear tester. Three case-hardening processes were selected; aluminum diffusion coating, boron diffusion coating and ion nitriding. It was found that cobalt-based hard facing alloys exhibited significant improvement in wear resistance against carburized SAE 4620 steel; however, nickel-based hard facing alloys were unaffected by ion nitriding. Aluminized cobalt based alloys showed improved wear resistance after a porous outer layer was ground off. Boriding reduces the wear resistance of both nickel and cobalt-based hard facing alloys. Removing the porous outer layer improved the wear properties of the borided nickel based alloy.

The effect of work hardening of hard facing alloys was also studied by Panzera and Saltzman. The work hardening was done by shot peening. The cobalt-based alloy hardened to a depth of about 250 µm; the nickel-based alloy did not work harden. It was found that work hardening the cobalt based alloy did not improve its wear resistance against carborized SAE 4620 steel as measured on the LFW-1 wear tests. Another process that has been used to surface-harden extruder screws is chemical vapor deposition (CVD). The process and some of its applications have been described by Bonetti (73). This process has been used to apply a thin layer, approximately 4 micron, of a very hard titanium nitride coating to the screw surface. Hardness values of about 110 Rockwell C can be obtained. Obviously, with such extreme hardness of the screw, one has to be very careful that the barrel material is compatible with the screw material.

11.2.1.5 REBUILDING WORN SCREWS AND BARRELS

In a correctly designed extruder, the majority of the wear should be concentrated on the screw because the screw can be replaced and rebuilt more easily than the barrel. In fact, the rebuilding of extruder screws has become so common that the rebuilding business has become a major segment of the extrusion industry. There are about 70 or more companies in the U.S. involved in rebuilding of extrusion equipment. For a number of these companies, screw rebuilding constitutes the major part of their business. One reason for the popularity of screw rebuilding is the fact that rebuilding is usually considerably less expensive than replacement with a new screw. Rebuilding is usually done with hard facing materials. With the proper choice of hard facing material, the rebuilt screw can be better than the original screw.

Rebuilding barrels is usually considerably more difficult than rebuilding screws. If the barrel wear does not exceed about 0.5 mm, the whole barrel can be honed to a larger diameter and an oversized screw can be placed in the machine. The obvious disadvantage of this procedure is that non-standard barrel and screw dimensions result. Thus, screws from other machines can no longer be used in the non-standard extruder. If barrel wear occurs near the end of the barrel, a sleeve can be placed in the barrel. In most cases, however, the barrel wear is such that replacement of the barrel makes more sense than sleeving or increasing the ID by honing.

11.3   POLYMER DEGRADATION

Polymer degradation is a frequent problem in extrusion. Degradation usually manifests itself as discoloration, loss of volatile components (smoking) or loss of mechanical properties. According to the mode of initiation, the following types of degradation can be distinguished: thermal, chemical, mechanical, radiation, and biological.

Degradation processes are generally quite complex; often more than one type of degradation is operational, e.g., thermooxidative degradation, thermo-mechanical degradation, etc. This situation is quite similar to wear in extruders, where usually more than one wear mechanism is operational at any one time.

11.3.1 TYPES OF DEGRADATION

In extrusion, the first three types of degradation are the most important, i.e., thermal, mechanical, and chemical degradation.

11.3.1.1 THERMAL DEGRADATION

Thermal degradation occurs when a polymer is exposed to an elevated temperature in an inert atmosphere under exclusion of other compounds. The resistance against such degradation depends on the nature and the inherent thermal stability of the polymer backbone. There are three main types of thermal degradation: depolymerizaton, random chain scission, and unzipping of substituent groups.

Depolymerization or unzipping is a reduction in length of the main chain by sequential elimination of monomer units. Polymers that degrade by this mechanism are polymethylmethacrylate, polyformaldehyde, polystyrene, etc. Polystyrene unzips to some extent during degradation, although only about 40 percent is converted to monomer. Random scission occurs in many polyolefins because of their simple carbon chain backbone. Unzipping of substituent groups is an important thermal degradation mechanism since it is the primary breakdown process for polyvinylchloride.

It is often difficult to distinguish between thermal and thermo-chemical degradation because polymers are rarely chemically pure. Impurities and additives can react with the polymeric matrix at sufficiently high temperatures.

11.3.1.2 MECHANICAL DEGRADATION

Mechanical degradation refers to molecular scission induced by the application of mechanical stresses. The stresses can be shear stresses or elongational stresses or a combination of the two. Mechanical degradation of polymers can occur in the solid state, in the molten state, and in solution. An extensive review of the field of mechanically induced reactions in polymers was published by Casale and Porter (38). In an extruder, mechanical stresses are applied mostly to the molten polymer.

Various theoretical approaches have been developed to describe mechanical degradation. One of the earlier studies was made by Frenkel (39) and Kauzmann and Eyring (40). They proposed that linear macromolecules are extended in a shear field in the direction of motion. The strain of the molecules is primarily concentrated at the middle of the chain. No degradation is expected when the degree of polymerization is below a certain critical value. Bueche (41) predicts that entanglements produce preferential tension in the mid-section of macromolecules. Thus, chain scission is more likely to occur in the center of the chain. He also predicts that main chain rupture increases dramatically with increasing molecular weight.

These theoretical considerations suggest that mechanical degradation in polymer melts or solutions is a non-random process, producing new low molecular weight species with molecular weights of one-half, one-fourth, one-eighth, etc. the original molecular weight. Mechanical degradation in polymer melts is essentially always combined with thermal degradation, and possibly chemical degradation, because of the elevated temperature of the melt. When a polymer melt is exposed to intense mechanical deformation, local temperatures can rise substantially above the bulk temperature if the rate of deformation is non-uniform.

Thus, bulk temperature measurements may not properly reflect actual stock temperatures. This is the case in screw extruders where very high local temperature increases can occur between flight and barrel. The same holds true for high intensity internal mixers. In such devices, pure mechanical degradation is unlikely to occur. Therefore, degradation processes in polymer melts involving mechanical stresses tend to be rather complex.

Some workers have reported that degradation at processing conditions is almost exclusively thermal (43, 44), while others conclude that degradation is mainly mechanical (45, 46). Most workers, however, deduce that, though the nature of degradation is basically thermal, there is a distinct reduction in the temperature needed for reaction due to the mechanical energy stored within the polymer chains as a result of the mechanical deformation. This corresponds to a shear-induced change in the potential energy function for thermal bond rupture as proposed by Arisawa and Porter (42). What this means in practice is that the polymer induction time, see Section 6.3.7, determined under quiescent conditions, will be longer than the actual induction time if the polymer is exposed to a mechanical deformation.

Because of the aforementioned complications in mechanical degradation in polymer melts, mechanical degradation can be more easily studied in polymer solution. Casale and Porter (38) have reviewed most work in this area up to 1978. Work in this area published between 1978 and 1984 is summarized in a later publication (77). More recent work by Odell, Keller, and Miles (47) describes an elegant technique to continuously monitor the molecular weight distribution (MWD) of a polymer solution undergoing mechanical deformation. They use a cross-slot device to apply pure elongational flow field to dilute solutions of narrow MWD atactic polystyrene. By measuring birefringence, information was obtained on the MWD of the polymer. They observed repeated breakage of the stretched molecules at their centers, as shown by the MWD before and after mechanical deformation of the polymer, see Figure 11–15.

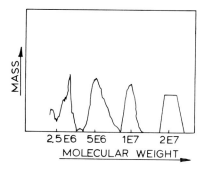

Figure 11–15. MWD After Mechanical Deformation (47)

11.3.1.3 CHEMICAL DEGRADATION

Chemical degradation refers to processes which are induced under the influence of chemicals in contact with a polymer. These chemicals can be acids, bases, solvents, reactive gasses, etc. In many cases, a significant conversion is only observed at elevated temperatures because of high activation energy for these processes.

Two important types of chemical degradation are solvolysis and oxidation. Solvolysis reactions concern the breaking of $C-X$ bonds, where X represents a non-carbon atom. Hydrolysis is an important type of solvolysis; the reaction basically occurs as follows:

$$
\begin{array}{ccccc}
\overset{|}{\underset{|}{-\text{C}-}} & & & \overset{|}{\underset{|}{-\text{C}-}} & \text{HO} \\
\text{X} & + & \text{H}_2\text{O} \longrightarrow & \text{X} & + & \overset{|}{\underset{|}{-\text{C}-}} \\
\overset{|}{\underset{|}{-\text{C}-}} & & & \text{H} &
\end{array}
$$

This type of degradation occurs in polyesters, polyethers, polyamides, polyurethanes and polydialkylsiloxanes. Polymers that tend to absorb water are more likely to undergo hydrolysis. Thus, in the extrusion of, for instance, polyester and polyamide, it is very important that the polymer be properly dried before extrusion. The stability of polymers against solvolytic agents is important in many applications. Some important polymers that have poor stability against acids and bases at room temperature are PVC, PMMA, PA, PC, PETP, PU, PAN, and POM. Polyolefins and fluoropolymers tend to have good stability against these solvolytic agents.

Oxidative degradation is a very common type of degradation in polymers. In extrusion, oxidation occurs at elevated temperatures; thus, the degradation becomes a thermo-oxidative degradation.

Polymer degradation starts with the initiation of free radicals. Free radicals have a high affinity for reacting with oxygen to form unstable peroxy radicals. The new peroxy radicals will abstract neighboring labile hydrogens, producing unstable hydroperoxides and more free radicals that will start the same process again. This results in an autocatalytic process, i.e., one that self-propagates once the process has started. Under continuous initiation, the reaction rate is accelerated, resulting in an exponential increase in conversion with reaction time. The process will stop when a reacting chemical species is depleted or when the propagation is inhibited by reaction products.

The oxidative degradation in polymers is generally combatted with the addition of antioxidants. The purpose of the antioxidant is the interception of radicals or prevention of radical initiation during the various phases of a polymer's life: polymerization, processing, storage, and end use. According to their functionality, antioxidants can be classified as primary or secondary antioxidants. Primary antioxidants or chain terminators interrupt chain reactions by tying up free radicals. They are also referred to as free-radical scavengers. Secondary antioxidants, or preventive antioxidants, destroy hydroperoxides. They are also referred to as peroxide decomposers. Primary antioxidants consist primarily of hindered phenols and aromatic amines. These materials tie up polymeric peroxy radicals through hydrogen donation, forming polymeric hydroperoxide groups and relatively stable antioxidant species. Secondary antioxidants consist of various phosphorous or sulphur containing compounds, particularly phosphites and thioesters. These materials reduce hydroperoxides to inert products, thus preventing the proliferation of alkoxy and hydroxy radicals. Selecting an effective antioxidant package is a key factor to the success of a plastic product. Some of the factors that should be considered in the selection of an antioxidant are toxicity, volatility, color, extractability, odor, compatibility, supply, cost, and performance.

11.3.2 DEGRADATION IN EXTRUSION

Degradation during the extrusion process will often be a combination of thermal, mechanical, and chemical degradation. Factors that are important in determining the rate of degradation are:

i)   Residence time and residence time distribution (RTD).
ii) Stock temperature and distribution of stock temperatures.
iii) Deformation rate and deformation rate distribution.
iv) Presence of solvolytic agents, oxygen, or other degradation promoting agents.
v)   Presence of antioxidants and other stabilizers.

The first three factors are strongly influenced by the machine geometry and by the operating conditions. The presence of solvolytic agents or oxygen can be influenced by operating conditions, e.g., oxygen can be eliminated from the extruder by putting the feed hopper under a nitrogen blanket. The presence of antioxidants and other stabilizers is part of the material selection process. Proper selection of a stabilizer package is very important; however, the details of determining the right stabilizer package are outside the scope of this book.

11.3.2.1 RESIDENCE TIME DISTRIBUTION

Knowledge of the residence time distribution (RTD) of an extruder provides valuable information about the details of the conveying process in the machine. The RTD is directly determined by the velocity profiles in the machine. Thus, if the velocity profiles are known, the RTD can be calculated. Various workers have made theoretical calculations of the RTD in single screw extruders (48–50). Obviously, theoretical calculations of the RTD require knowledge of the velocity profiles in the machine. Thus, the predicted RTD is only as accurate as the velocity profiles that form the basis of the calculations. In single screw extruders, the velocity profiles can be determined reasonably well, although usually a substantial number of simplifying assumptions are made, see Chapter 7. In other screw extruders, e.g., twin screw extruders, calculation of velocity profiles is rather complex and thus prediction of the RTD more difficult.

Experimental determination of the RTD of an extruder yields information about the conveying process in the extruder. This information is useful in a number of areas, not just to analyze the chance of degradation in the machines. The RTD can be used to analyze the mixing process in an extruder. When an extruder is used as a continuous chemical reactor, the RTD provides important information for process design and process analysis. The RTD also provides a good selection criterion, e.g., an extruder used in profile extrusion should have a narrow RTD and short residence time. Experimental studies of RTD in single screw extruders have been reported by Wolf and White (51), Bigg and Middleman (49), Schott and Saleh (55), Rauwendaal (52), Golba (53), and Kemblowski and Sek (54). Experimental studies of RTD in twin screw extruders have been reported by Todd (56), Janssen et al. (57, 58), Rauwendaal (52), Walk (59), Nichols et al. (60).

The RTD is determined by measuring the output response of a change in input. This is referred to as the stimulus response method as discussed by Levenspiel (61) and Himmelblau and Bischoff (62).

The system is disturbed by a stimulus and the response of the system to the stimulus is measured. Two common stimulus response techniques are the step input response and the pulse input response, see Figure 11-16.

Other stimuli that can be used are a random input and a sinusoidal input. The response of a step input is an S-shaped curve, see Figure 11-16a. The response of a pulse input is a bell-shaped curve, see Figure 11-16b. The ideal pulse input is of infinitely short duration; such an input is called a delta function or impulse. The normalized response to a delta function is called the C curve. Thus, the total area under the curve equals unity.

The definition of RTD functions is due to Danckwerts (63). The internal RTD function

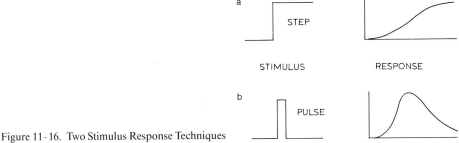

Figure 11-16. Two Stimulus Response Techniques

g(t)dt is defined as the fraction of fluid volume in the system with a residence time between t and t+dt. The external RTD function f(t)dt is defined as the fraction of exiting flow rate with a residence time between t and t+dt. The cumulative internal RTD function G(t) is defined as:

$$G(t) = \int_0^t g(t)dt \qquad (11-1)$$

G(t) represents the fraction of fluid volume in the system with a residence time between o and t. The cumulative external RTD function is defined as:

$$F(t) = \int_{t_0}^t f(t)dt \qquad (11-2)$$

where t_0 is the minimum residence time.

F(t) represents the fraction of exiting flow rate with a residence equal to or shorter than t. For very long times t, both function G and F become equal to unity:

$$G(\infty) = F(\infty) = 1 \qquad (11-3)$$

The mean residence time $\bar{t}$ is given by the following expression:

$$\bar{t} = \int_0^\infty tf(t)dt \qquad (11-4)$$

The mean residence time is determined by the volume of the machine V, the degree of fill X of the machine, and the volumetric flow rate $\dot{V}$:

$$\bar{t} = \frac{XV}{\dot{V}} \qquad (11-5)$$

The relationship between the internal RTD function and the external RTD function is given by:

$$g(t) = \frac{1-F(t)}{\bar{t}} \qquad (11-6)$$

In the flow of a Newtonian fluid through a pipe, the RTD can be calculated rather easily by using the expression for the velocity profile given in Table 7-2. The external RTD function is:

$$f(t) = \frac{2t_0^2}{t^3} \qquad \text{for } t \geq t_0 \qquad (11-7)$$

where the minimum residence time is:

$$t_0 = \frac{4\mu L^2}{\Delta P R^2} \qquad (11-8)$$

Figure 11-17 shows a typical cumulative RTD curve for a single screw extruder as determined experimentally (52).

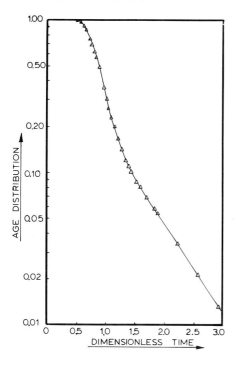

Figure 11-17. RTD of a Single Screw Extruder

The curve is for a 25-mm extruder running at 20 rpm with an output of 2.3 kg/hr. The mean residence time in this example is 5.9 minutes. This type of information is useful because one can easily tell how large a fraction of the material spends how long a time in the machine. For instance, in Figure 11-17, more than 1 percent of the material is exposed to a residence time of three times the mean residence time, i.e., 17.7 minutes! If the induction time of the material at process temperature is less than 17.7 minutes, one can expect more than 1 percent of the material to be degraded.

Figure 11-18 shows several cumulative RTD curves for an intermeshing counterrotating twin screw extruder (52).

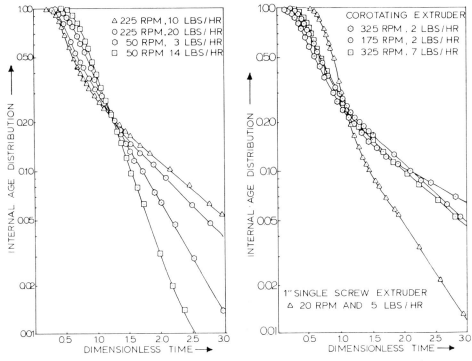

Figure 11-18. RTD of a CICT Extruder

Figure 11-19. RTD of a CSCO Extruder

It can be seen that the shape of the curve changes substantially when the processing conditions are changed. The narrowest RTD is obtained by running the extruder at low speed and high output. Figure 11-19 shows several cumulative RTD curves for an intermeshing corotating twin screw extruder (52).

It is clear that the curves indicate considerable deviations from positive conveying characteristics for the corotating twin screw extruder.

A major advantage of these normalized RTD curves is that conveying characteristics of different extruders can be directly compared. From comparison of Figures 11-17, 11-18, and 11-19, it is clear that the conveying characteristics of the single screw extruder are quite positive compared to the two twin screw extruders. This is partially due to the plug flow of the solid bed in the single screw extruder. The solid bed in a twin screw extruder is not continuous and generally does not extend over a long length of the machine. It should be realized that the RTD is strongly dependent on the screw design and the operating conditions. This point was discussed in some detail by Kemblowski and Sek (54) with regard to single screw extruders and by the author (52) with regard to twin screw extruders.

11.3.2.2 TEMPERATURE DISTRIBUTION

Obviously, the residence time and its distribution only partially determine the chance of degradation in an extruder. The other factors that play an important role are the actual stock temperatures and the strain rates to which the polymer is exposed. The actual stock temperatures and strain rates are closely related. In the extruder, there are two major areas of concern, the screw channel and the flight clearance. Temperature distribution of the polymer melt in the screw channel has been measured by Janssen, Noomen, and Smith (65). Temperature distribution of the polymer right after the end of the screw was measured, for instance, by Anders, Brunner, and Panhaus (66). The temperature variations in the screw channel at the end of the screw are generally less than $5°C$ to $10°C$ and relatively close to the barrel temperature.

The situation in the screw clearance is substantially different from the screw channel. The strain rates in the screw channel are relatively low and the temperature variations are also relatively low. In the screw clearance, however, the strain rates are very high and the stock temperature increase can also be very high. This can be verified by the following simple analysis. The shear rate in the clearance is approximately the Couette shear rate:

$$\dot{\gamma}_{cl} = \frac{\pi DN}{\delta} \qquad (11\text{--}9)$$

The corresponding viscous heat generation per unit volume for a power law fluid is given by equation 5-76:

$$\dot{E}_{cl} = m \left(\frac{\pi DN}{\delta} \right)^{1+n} \qquad (11\text{--}10)$$

If it is assumed that there is no exchange of heat between the polymer melt and the screw and between the polymer melt and the barrel, the average adiabatic temperature rise can be determined from:

$$\overline{\Delta T}_a = \frac{\dot{E}_{cl}\overline{t}_R}{C_p \rho} \qquad (11\text{--}11)$$

where $\overline{t}_R$ is the average residence time of the polymer melt in the flight clearance. The average residence time in the flight clearance is approximately:

$$\overline{t}_R = \frac{2w}{\pi DN\sin\varphi} \qquad (11\text{--}12)$$

Combining equations 11-10, 11-11, and 11-12, the average adiabatic temperature rise in the clearance can be written as:

$$\overline{\Delta T}_a = \frac{2mw(\pi DN)^n}{\rho \sin\varphi C_p \delta^{1+n}} \qquad (11\text{--}13)$$

The average temperature rise is directly proportional to the consistency index m and the tangential flight width $w/\sin\varphi$. The temperature rise is strongly dependent on the radial clearance δ, the power law index of the polymer melt n, and the screw speed N. Figure 11-20 shows the effect of flight clearance δ and the power law index n for a 114-mm

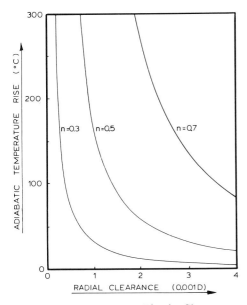

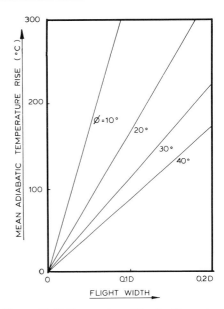

Figure 11-20. Temperature Rise in Clearance versus δ and n

Figure 11-21. Temperature Rise in Clearance versus w and φ

(4.5-inch) extruder running at 100 rpm; the specific heat is 0.8 cal/gr° C, the melt density is 0.9 gr/cc, and the consistency index is 10^4 Pa-sec^n.

It is clear that extremely high temperature rises can occur in the flight clearance. Therefore, the flight clearance is one of the most critical areas with regard to degradation in the extruder. Figure 11-21 shows the effect of perpendicular flight width w and helix angled φ.

It is clear that the temperature rise in the clearance can be substantially reduced by simply reducing the flight width and increasing the flight helix angle. These same measures will also substantially reduce the power consumption of the extruder, as discussed in Section 8.3. Thus, proper design of the screw flight is of great importance when it comes to reducing power consumption and reducing the chance of degradation in the extruder.

Another reason that the flight clearance is so important in degradation processes occurring in the extruder is the fact that, in addition to very high stock temperatures, the polymer melt is exposed to very high strain rates, both elongational and shear. As discussed in Section 11.3.1.2, this causes a flow-induced change in the potential energy function for thermal bond rupture. Thus, the degradation will be more severe than it would be based on just the effect of temperature. Obviously, another important point is to eliminate dead spots in the screw design and in the die design. Hangup of material can be very detrimental and should be avoided if at all possible. For instance, fluted mixing sections with a 90° helix angle should not be used with polymers that have a tendency to degrade because such mixing sections have stagnating regions, as discussed in Section 8.7.1.

The values of the temperature increase in the flight clearance calculated with equation 11-13 are surprisingly high, particularly considering the very short residence time of the polymer in the flight clearance, which is usually in the order of 0.1 second. Obviously, in reality the temperature rise will not be as high as the adiabatic temperature rise because there will be exchange of heat with the screw and with the barrel. The lowest temperature rise will occur in the extreme case that both screw flight surface and barrel surface can be maintained at constant temperature, i.e., isothermal boundary conditions. This situation was analyzed by Meijer, Ingen Housz, and Gorissen (67) with the primary purpose to determine the thermal development length. They assumed that the clearance flow is dominated by drag flow in tangential direction. The thermal development length for the Newtonian case was found to be approximately $0.36 N_{Pe}$, where N_{Pe} is the Peclet number, see equation 5-70. Thus, the length required for thermal development can be written as:

$$L_1 = \frac{0.36 v_b \delta^2}{\alpha} \tag{11-14}$$

where α is the thermal diffusivity.

The thermal development length is directly proportional to the barrel velocity v_b (and thus the screw speed) and to the radial clearance squared. With thermal diffusivity values of about 10^{-7} m^2/sec, the thermal entrance length will be the same order of magnitude as the tangential flight width when the clearance has the normal design value ($\delta \approx 0.001$ D). Thus, the temperature profile at the exit of the flight clearance will be very close to the fully developed temperature profile. The fully developed temperature profile for the isothermal case can be written as (67):

$$T_1 = \frac{\mu v_b^2}{2k} \left(\frac{y}{\delta} - \frac{y^2}{\delta^2} \right) + \frac{(T_b - T_s) y}{\delta} + T_s \tag{11-15}$$

The maximum temperature T_{max} that can develop in the isothermal case is:

$$T_{max} = \frac{\mu v_b^2}{8k} + \frac{1}{2} T_b + \frac{1}{2} T_s \tag{11-16}$$

For the example used earlier (114 mm extruder at 100 rpm) with a thermal conductivity k of 0.25 J/msec° C, the maximum temperature is 26.17° C above the average barrel and screw temperature ($0.5 T_b + 0.5 T_s$). Thus, the maximum temperature rise in the isothermal case is about one order of magnitude below the adiabatic case. In reality, true isothermal conditions cannot be achieved because this would require excessively high heat fluxes at the screw and barrel interface. Thus, the actual maximum stock temperature in the clearance will be somewhere between the adiabatic and the isothermal case.

Winter (30) has performed numerical calculations of the developing temperature profile in the flight clearance for power law fluids. He assumed isothermal conditions at the barrel wall and adiabatic conditions at the screw flight surface. These assumptions are considerably more realistic than the purely adiabatic case or the purely isothermal case, although a better boundary condition would probably be a prescribed maximum heat flux. Winter calculates a typical maximum temperature increase of about 150° C. This value is closer to the maximum temperature rise in the adiabatic case than the maximum temperature rise in the isothermal case. These analyses indicate that the temperature rise in the flight clearance can be quite significant and can play a very important role in degradation in extruders.

11.3.2.3 REDUCING DEGRADATION

The chances of degradation in the extruder can be reduced by making the following changes to the process:

i)   reduce residence time and achieve a narrow RTD
ii) reduce stock temperatures and avoid high peak temperatures
iii) eliminate degradation-promoting substances

The residence time can be reduced by designing the screw for maximum throughput. The procedure for this is discussed in detail in Section 8.2. Low stock temperatures and reduced peak temperatures can be obtained by designing the screw for minimum specific energy consumption; this is discussed in detail in Section 8.3. Further, stagnating regions should be avoided if at all possible. Thus, the design of both the screw and die has to be made as streamlined as possible. If degradation occurs by a thermo-oxidative mechanism, air should be excluded from the extruder. This can be done by putting a nitrogen blanket on the feed hopper, vent port, or at the die, depending on where the air is introduced. If degradation occurs by hydrolysis, moisture has to be excluded from the process. If degradation occurs by a chemical reaction with the metal surfaces of screw and barrel, a non-corrosive material of construction has to be selected for the screw and barrel.

11.4 EXTRUSION INSTABILITIES

Variations in extruder performance is perhaps the most frequent problem encountered in extrusion. One possible reason for the frequent occurrence of instabilities is the fact that they can have a large number of causes, some of which are:

> bulk flow problems in the feed hopper
> solids conveying problems in the extruder
> insufficient melting capacity
> solid bed break – up
> melt temperature non – uniformities in the die
> barrel temperature fluctuations
> screw temperature fluctuations
> variations in the take – up device
> melt fracture/shark skin
> variations in screw speed
> barrel wear/screw wear
> insufficient mixing capacity
> very low diehead pressure
> insufficient pressure generating capacity

Proper instrumentation is vitally important to be able to diagnose a problem quickly and accurately, as discussed in detail in Chapter 4. A prerequisite for stable extrusion is a good extruder drive, good temperature control system, good take-up device, and most importantly, a good screw design. Probably more instabilities result from improper screw design than from any other cause. However, a change in screw design is often only considered as the very last option. The extruder drive should be able to hold the screw speed constant to about 0.1 percent or better; the same holds true for the take-up device. However, this is not always the case on actual extrusion lines. The extruder should be equipped with some type of proportioning temperature control, preferably a PID-type control or better. On-off temperature control is inappropriate for most extrusion operations, as discussed in Section 4.5.

11.4.1 FREQUENCY OF INSTABILITY

Various workers (68, 69) have classified extrusion instabilities based on the time frame in which they occur. The frequency of the instability is often an indication of the cause of the problem. Most earlier workers have distinguished only three or four types of instabilities based on the frequency. However, it is probably more appropriate to distinguish at least five types of instabilities:

1. High frequency instablities occurring faster than the frequency of screw rotation.

2. Screw frequency instabilities occurring at the same rate as the frequency of screw rotation.

3. Low frequency instabilities occurring about 5 to 10 times slower than the frequency of screw rotation.

4. Very slow fluctuations occurring at a frequency of at least several minutes.

5. Random fluctuations.

High frequency instabilities are often associated with die flow instabilities, such as melt fracture, shark skin, or draw resonance, see Section 7.5.3. They can also be caused by drive problems or melt temperature non-uniformities. Screw frequency instabilities occur to a small extent in essentially every extrusion operation. This can be partially due to the fact that the intake of polymer from the feed hopper is interrupted every time the flight passes by the feed opening. This causes a cyclic pressure change that can be noticed on practically every extruder, provided it has an accurate pressure readout. Screw frequency instabilities are generally only minor and do not cause significant problems. One way to reduce the unsteady intake of polymer from the feed hopper is to use a double flighted screw geometry in the feed section; this is discussed in Section 8.4.2. According to Wheeler (68), the screw frequency instability is more likely to occur at very low diehead pressures.

In most cases, the major cause of screw frequency instabilities will be the pressure difference between the leading and trailing edge of the flight in the pump section. This pressure difference is inherent to the conveying process. It occurs even if no pressure is developed in the pump section because this pressure difference is a drag-induced pressure difference. If the flight clearance can be neglected, the cross-channel pressure gradient g_x is given by equation 7-160. In this case, the pressure difference ΔP is:

$$\Delta P = \frac{6\,\mu v_{bx}W}{H^2} \simeq \frac{6\,\mu\pi^2 D^2 N \sin^2 \varphi}{H^2} \tag{11-17}$$

Thus, the pressure difference increases with viscosity, diameter, screw speed and helix angle; the pressure difference reduces with channel depth. When the helix angle increases from 17.5 degrees to 25.0 degrees, the pressure difference will double. Thus, large helix angle screws will exhibit more screw frequency pressure fluctuations than small helix angle screws. The screw frequency pressure fluctuation can be reduced by placing a multi-flighted screw section at the end of the screw, e.g. a pineapple mixing section, see Figure 8-83, or a Dulmage mixing section, see Figure 8-81. It should be noted that these pressure fluctuations will be most severe at the very end of the screw, but they will reduce with increasing distance from the screw because the polymer melt is slightly compressible. Thus, the actual pressure fluctuation will be very much dependent on the location of the pressure transducer. Obviously, the screw frequency pressure fluctuation will be problematic when the value of ΔP is large relative to the actual diehead pressure. This will occur when the diehead pressure is low, as pointed out by Wheeler (68), when the polymer melt viscosity is high, the screw diameter large, the screw speed high, the helix angle or pitch large, or when the channel depth is shallow.

Low frequency instabilities have been associated with solid bed breakup (69, 70). Fenner et al. have attempted to theoretically predict solid bed breakup (69). They proposed that solid bed breakup is due to acceleration of the solid bed in the plasticating zone of the extruder and claimed that no solid bed acceleration occurs in the absence of a melt film between the solid bed and the screw. In practice, formation of a melt film can be avoided by cooling the screw (71). Earlier, Maddock (72) found that screw cooling helped in reducing surging. The most likely reason that screw cooling reduces surging is that it reduces the throughput rate by a substantial amount; about 20 percent in the experiments of Fenner and Edmondson (71). Therefore, the melting process will be completed over a shorter axial distance, reducing the stresses acting on the solid bed. Solid bed breakup is also more likely to occur on screws with a high compression ratio. Fenner et al. (69, 71) found solid bed breakup with high compression ratio screws (3 : 1 and 4 : 1), but did not find solid bed breakup with a low compression ratio screw (2.25 : 1). A low compression ratio screw would seem a better solution than a high compression ratio screw with screw cooling. Another method by which formation of a melt film on the screw surface can be avoided, is by using a barrier screw geometry. Barrier type extruder screws are discussed in detail in Section 8.6.2.

Fluctuations occurring over about 10 to 30 seconds can be caused by temperature fluctuations along the extruder barrel. The temperature fluctuations may not be noticeable from the temperature readouts. This can be due to the slow response of many temperature sensors and because the sensors are often located a considerable distance from the polymer/metal interface. However, if the actual temperature at the interface fluctuates, there will be a corresponding fluctuation in the flow rate. In time-proportioning temperature control systems (see Section 4.5.3.2), power is added or extracted at relatively short intervals, typically about 15 to 20 seconds. These bursts of heating or cooling energy will cause short-term changes in the polymer/metal interface temperature with corresponding variations in throughput rate. The variation in throughput can be as much as 5 to 10 percent. Therefore, from a stability point of view, the true proportioning temperature control is significantly better than the time-proportioning temperature control. These throughput variations caused by wall temperature changes have been described by Gitschner and Lutterbeck (76). They were able to show a very clear correlation between the on and off cycling of barrel cooling and the diehead pressure fluctuations. They also found that the pressure fluctuations correlated very closely with the resulting throughput fluctuations. It should be clear that these temperature-induced throughput fluctuations could be considerably more severe in the case of on-off temperature control, see Sections 4.5.1 and 4.5.4.3.

Very slow fluctuations are often associated with poor temperature control, changes in ambient conditions (room temperature, relative humidity), plant voltage variation, and similar causes. A steady, slow reduction in output is often caused by build-up of contaminants on the screen pack.

Random fluctuations are often associated with irregular feeding. Maddock (72) discussed a case where the extruder performance was very sensitive to the level of fill in the feed hopper. Random fluctuations can also result from a combination of cyclic fluctuations as shown in Figure 11-22.

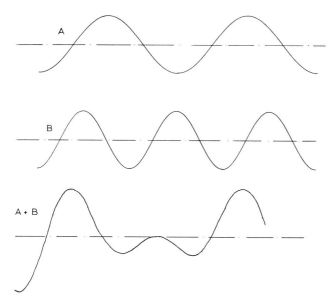

Figure 11–22. Random Fluctuations from a Combination of Cyclic Fluctuations

11.4.2 FUNCTIONAL INSTABILITIES

Another method of classifying instabilities is by the functional zone in which the instability originates. Thus, the following instabilites can be distinguished:

Solids conveying instabilities
Plasticating instabilities
Melt conveying instabilities
Devolatilization instabilities
Mixing instabilities
Die forming instabilities

Solids conveying instabilities have three major causes: flow problems in the feed hopper, internal deformation of the solid bed in the screw channel, and insufficient friction against the barrel surface. Flow problems in the feed hopper can be detected by observing the flow from the feed hopper when it is disconnected from the extruder. Solids conveying problems in the extruder itself are difficult to diagnose. One method that can be used is to teflon-coat the screw. Even though the coating may not last very long, it will substantially reduce the retarding force acting on the solid bed and thus improve solids conveying, see Section 7.2.2 and 8.2.3. If the coating eliminates the instability, this is a strong indication of a solids conveying problem. A more permanent solution is possibly a grooved barrel section or a nickel-plated screw impregnated with a fluorocarbon polymer as discussed in Section 11.2.1.4.

Plasticating problems are likely to occur on screws with a high compression ratio and a short compression section length. Insufficient melting capacity can be diagnosed by preheating the feed stock. If preheating reduces or eliminates the instability, then the problem is most likely insufficient melting capacity. Screw design for improved plasticating performance is discussed in Section 8.2.2.

Most melt conveying or pumping problems are caused by improper design of the metering section. The most common problem is too large a channel depth, the second most common problem is too short a metering section. The problem of too large a channel depth can be counteracted by cooling the metering end of the screw. However, a better solution is to switch to a new screw design with proper dimensions of the metering section. Design of the metering section of the screw is discussed in Section 8.2.1.

Devolatilization instabilities can be caused by plugging of the vent port, variation in the vacuum level, or by variations of the volatile level in the feed stock. Screw design for devolatilization is covered in Section 8.5. Extrusion instabilities are often related to insufficient mixing capacity of the screw. Mixing can be improved a small amount by increasing the diehead pressure. However, this is a relatively ineffective method to improve the mixing capacity of the extruder; while it also increases the chance of degradation. The mixing capacity of a screw can be improved significantly by adding a mixing section; this is discussed in detail in Section 8.7. Die forming instabilities have been discussed in Section 7.5.3.

11.4.3 SOLVING EXTRUSION INSTABILITIES

There are many different causes of extrusion instabilities. Even though the mechanism of the instability is not always clear, the following measures often reduce extrusion instabilities:

    Reduce the screw speed
    Reduce the screw temperature
    Reduce the barrel temperature at delivery end
    Reduce the channel depth in metering section
    Increase the length of compression section
    Increase the rear barrel temperatures
    Increase the diehead pressure

The first approach to the problem is generally adjustment of the temperature profile. If temperature adjustment does not solve the problem, one should check the hardware: thermocouples, controllers, screw, barrel, drive, etc. If the problem is not associated with the hardware, it must be a functional problem and one should determine what functional zone is causing the problem. If the problem cannot be solved by changing the processing conditions, which is, of course, the first choice, then one can generally solve the problem either by making a material change or by making a change in screw or barrel design. In most cases, material changes are not possible. In that case, the problem usually has to be solved by a new screw design, which is discussed in detail in Chapter 8. Another option is to add a gear pump at the end of the extruder; this is discussed in detail in Section 2.2.3. Figure 11–23 attempts to illustrate some of the important interactions that take place in the extrusion process.

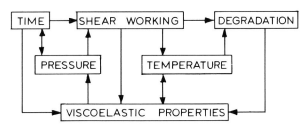

Figure 11–23. Interactions in the Extrusion Process

11.5   AIR ENTRAPMENT

Air entrapment is a rather common problem in extrusion. It is caused by air being dragged in with particulate material from the feed hopper. Under normal conditions, the compression of the solid particulate material in the feed section will force the air out of the solid bed. However, under some circumstances the air cannot escape back to the feed hopper and travels with the polymer until it exits from the die. As the air pockets exit from the extruder, the sudden exposure to a much lower ambient pressure may cause the compressed air bubbles to burst in an explosive manner. However, even without the bursting of the air bubbles, the extrudate is generally rendered inacceptable as a result of the air inclusions.

There are a number of possible solutions to air entrapment. The first approach should be to change the temperature in the solids conveying zone to achieve a more positive compacting of the solid bed. Often a temperature increase of the first barrel section reduces the air entrapment; however, in some cases, a lower temperature causes an improvement. In any case, the temperatures in the solids conveying zone are important parameters in the air entrapment process. It should be realized that both the barrel and screw temperature are important. Thus, if a screw temperature adjustment capability is available, it should definitely be used to reduce the air entrapment problem. The next step is an increase in the diehead pressure to alter the pressure profile along the extruder and to hopefully achieve a more rapid compacting of the solid bed. The diehead pressure can be increased by simply adding screens in front of the breaker plate. Another possible solution is to starve feed the extruder; however, this will probably reduce the output somewhat and requires additional hardware, i.e., an accurate feeding device.

The aforementioned recommended solutions can be implemented rather easily. However, if these measures do not solve the problem, more drastic steps have to be taken. One possibility that needs to be explored is a change in particle size or shape. If this is a reasonable option, it will most likely solve the problem. A rather safe solution is to utilize a vacuum feed hopper system; however, these systems are rather complex and expensive. This point is discussed in more detail in Section 3.4. Another possible solution is to use a grooved barrel section. Pressure development in a grooved barrel section is much more rapid than in a smooth barrel. Thus, a grooved barrel section causes a rapid compacting of the solid bed and, therefore, a much reduced chance of air entrapment. Instead of grooving the barrel, one can opt for reducing the friction on the screw, which would have a similar effect. A coating that might be used for this purpose is described by Luker (35). Air entrapment is also often successfully eliminated by vented extrusion using a multi-stage extruder screw, as discussed in Section 8.5. Increasing the compression ratio of the screw is also likely to reduce air entrapment.

It should be noted that bubbles in the extrudate are not only a sign of air entrapment, but it may also be an indication of moisture, surface agents, volatile species in the polymer itself, or degradation. Thus, before concentrating on solving an apparent air entrapment problem, one should make sure that the problem is indeed caused by air entrapment. In some cases, the pellets contain small air bubbles within the pellet itself. In this case, one of the few possible solutions is vented extrusion; most of the other recommended solutions will not work in this situation.

REFERENCES – CHAPTER 11

1. E. Wagner, Ph. D. thesis, Technische Hochschule Darmstadt, W. Germany (1978).
2. M. M. Krushchov, Ind. Lab (USSR), 20, 372–376 (1962).
3. G. Mennig and P. Volz, Kunststoffe, 70, 385–390 (1980).
4. W. Mehdorn, Kunststoffe, 34, 133–136 (1944).
5. W. Bauer, K. Eichler, and W. John, Kunststoffe, 57, 53–55 (1967).
6. K. Eichler and W. Frank, Ind. Anzeiger, 95, 2033–2035 (1973).
7. W. D. Mahler, Ph. D. thesis, Technische Hochschule Darmstadt, W. Germany (1975).
8. P. Volz, W. D. Mahler, and G. Mennig, Kunststoffe, 66, 428–434 (1976).
9. W. Knappe and W. D. Mahler, Kunststoff-Rundschau, 19, 45–51 (1972).
10. R. S. Plumb and W. A. Glaeser, Wear, 46, 219–229 (1978).
11. H. G. Fritz, Kunststoffe, 65, 176–182 and 258–264 (1975).
12. P. Volz, Kunststoffe, 69, 758–771 (1979).
13. P. Volz, Kunststoffe, 67, 279–283 (1976).
14. G. P. Calloway, E. D. Morrison, and R. F. Williams, Jr., SPE ANTEC, 354–360 (1972).
15. D. Braun and G. Maelhammar, Angew. Makromol. Chem., 69, 157–167 (1978).
16. H. G. Moslé, H. F. Schmidt, and J. Schroeder, Kunststoffe, 67, 220–223 (1977).
17. G. Maelhammar, Ph. D. thesis, Technische Hochschule Darmstadt, W. Germany (1978).
18. P. Volz, Kunststoffe, 69, 259–262 (1979).
19. E. Broszeit, Ph. D. thesis, Technische Hochschule Darmstadt, W. Germany (1972).
20. G. A. Saltzman and J. H. Olson, SPE ANTEC, 173–175 (1974).
21. V. Murer and G. A. Saltzman, Kunststoffe, 66, 219–220 (1976).
22. W. W. McCandles and W. D. Maddy, Plastics Technology, Febr., 89–93 (1981).
23. S. H. Collins, Plastics Compounding, May/June, 113–124 (1982).
24. S. H. Collins, Plastics Compounding, July/Aug., 16–28 (1982).
25. E. L. Moon and R. A. Hunter, "An Abrasion Study of Surface Treated Calcium Carbonate Fillers in Rigid PVC", Technical Bulletin, Georgia Marble Corp., Atlanta, GA.
26. W. D. Mahler, Kunststoffe, 67, 224–226 (1977).
27. B. A. Olmsted, SPE Journal, 26, 42–43 (1970).
28. P. Luelsdorf, VDI-Lehrgang: "Grundlagen der Extrudertechnik", (1975).
29. R. A. Lai Fook and R. A. Worth, SPE ANTEC, Washington, D. C., 450–452 (1978).
30. H. H. Winter, SPE ANTEC, New Orleans, 170–175 (1979).
31. H. Schuele and H. G. Fritz, Kunststoffe, 73, 603–605 (1983).
32. K. O'Brien, Plast. Technology, Febr., 73–74 (1982).
33. G. Thursfield, Modern Plastics, Oct., 94–96 (1975).
34. M. Hoffmann, Plastics Technology, April, 67–72 (1982).
35. K. Luker, Paper presented at the TAPPI Paper Synthetics Conference in Lake Buena Vista, Florida (1983).
36. W. Lucius, Kunststoffe, 63, 433–435 (1973).
37. C. Panzera and G. A. Saltzman, Proceedings of the 2nd International Conference on Wear of Materials, 441–448 (1979).
38. A. Casale and R. S. Porter, "Polymer Stress Reactions", Vol. 1 and 2, Academic Press, New York (1978).
39. Y. I. Frenkel, Acta Physicochim. (USSR), 19, 51 (1944).
40. W. J. Kauzmann and H. Eyring, J. Am. Chem. Soc., 62, 3113 (1940).
41. F. Bueche, "Physical Properties of Polymers", Wiley, New York (1962).
42. K. Arisawa and R. S. Porter, J. Appl. Polym. Sci., 14, 879 (1970).
43. A. Holmstrom, A. Andersson, and E. M. Sorvik, Polym. Eng. Sci., 17, 728–732 (1977).
44. P. W. Springer, R. S. Bradley, and R. E. Lynn, Polym. Eng. Sci., 15, 583–587 (1975).
45. V. L. Folt, Rubber Chem. Technol., 42, 1294 (1969).
46. R. W. Ford, R. A. Scott and R. J. B. Wilson, J. Appl. Polym. Sci., 12, 547 (1968).

47. J. A. Odell, A. Keller, and M. J. Miles, Polymer Comm., 24, 7–10 (1983).
48. G. Pinto and Z. Tadmor, Polym. Eng. Sci., 10, 279–288 (1970).
49. D. Bigg and S. Middleman, Ind. Eng. Chem. Fundam., 13, 66–71 (1974).
50. G. Lidor and Z. Tadmor, Polym. Eng. Sci., 16, 450–462 (1976).
51. D. Wolf and D. H. White, AIChE J., 22, 122–131 (1976).
52. C. Rauwendaal, Polym. Eng. Sci., 21, 1092–1100 (1981).
53. J. C. Golba, SPE ANTEC, New York, 83–87 (1980).
54. Z. Kemblowski and J. Sek, Polym. Eng. Sci., 21, 1194–1202 (1981).
55. N. R. Schott and D. V. Saleh, SPE ANTEC, Washington, D. C., 536–539 (1978).
56. D. B. Todd, Polym. Eng. Sci., 15, 437–443 (1975).
57. L. P. B. M. Janssen, "Twin Screw Extrusion", Elsevier, New York (1978).
58. L. P. B. M. Janssen, R. W. Hollander, M. W. Spoor, and J. M. Smith, AIChE J., 25, 345–351 (1979).
59. C. J. Walk, SPE ANTEC, San Francisco, 423–426 (1982).
60. R. J. Nichols, J. C. Golba, and P. K. Shete, paper presented at the AIChE Annual Meeting, Paper No. 59 F (1983).
61. O. Levenspiel, "Chemical Reaction Engineering", Wiley, New York (1965).
62. D. M. Himmelblau and K. A. Bischoff, "Chemical Process Analysis", Wiley, New York (1966).
63. P. V. Danckwerts, Chem. Eng. Sci., 2, 1 (1953).
64. Z. Tadmor and C. G. Gogos, "Principles of Polymer Processing", Wiley, New York (1979).
65. L. P. B. M. Janssen, G. H. Noomen, and J. M. Smith, Plastics & Polymers, Aug., 135–140 (1975).
66. S. Anders, D. Brunner, and F. Panhaus, Plaste und Kautschuk, 23, 593–598 (1976).
67. H. E. H. Meijer, J. F. Ingen Housz, and W. C. M. Gorissen, Polym. Eng. Sci., 18, 288–292 (1978).
68. N. C. Wheeler, Annual Conv. Wire Ass., Baltimore, MD, October 22–25 (1962).
69. R. T. Fenner, A. P. D. Cox, and D. P. Isherwood, Polymer, 20, 733–736 (1979).
70. Z. Tadmor and I. Klein, "Engineering Principles of Plasticating Extrusion", p. 413, Van Nostrand Reinhold, New York (1970).
71. I. R. Edmondson and R. T. Fenner, Polymer, 16, 49–56 (1975).
72. B. H. Maddock, SPE J., 20, 1277 (1964).
73. R. Bonetti, Metal Progress, June (1981).
74. S. Levy, Plast. Mach. Equip., March (1984).
75. S. Trompler, Kunststoffe, 73, 596 (1983).
76. H. W. Gitschner and J. Lutterbeck, Kunststoffe, 74, 12–14 (1984).
77. R. S. Porter and A. Casale, Polym. Eng. Sci., 25, 129–156 (1985).

CONVERSION CONSTANTS

LENGTH

1 kilometer	km	= 1E3 m (meter)
1 centimeter	cm	= 1E-2 m
1 millimeter	mm	= 1E-3 m
1 micron	μm	= 1E-6 m
1 nanometer	nm	= 1E-9 m
1 Ångstrom	Å	= 1E-10 m
1 inch	in	= 2.54E-2 m
1 milliinch	mil	= 2.54E-5 m
1 foot	ft	= 0.3048 m
1 mile	mile	= 1609 m

VOLUME

1 cubic decimeter	dm^3	= 1E-3 m^3 (cubic meter)
1 cubic centimeter	cc	= 1E-6 m^3
1 liter	l	= 1E-3 m^3
1 deciliter	dl	= 1E-4 m^3
1 milliliter	ml	= 1E-6 m^3
1 cubic inch	in^3	= 1.639E-5 m^3
1 cubic foot	ft^3	= 2.832E-2 m^3
1 barrel	barrel	= 0.159 m^3
1 gallon (US)	gal US	= 3.785E-3 m^3
1 gallon (UK)	gal UK	= 4.546E-3 m^3

MASS

1 ton	t	= 1E3 kg (kilogram)
1 gram	gr	= 1E-3 kg
1 ounce	oz	= 2.83E-2 kg
1 pound	lbm	= 0.4536 kg
1 ton (US)	tn	= 907 kg
1 ton (UK)	ton	= 1016 kg

DENSITY

1 gram per cc	gr/cc	= 1E-3 kg/m^3
1 pound per cu ft	lb/ft^3	= 16.02kg/m^3
1 pound per cu in	lb/in^3	= 2.77E-2 kg/m^3
1 gram per cc	gr/cc	= 3.61E-2 $lbs/inch^3$

FORCE

1 dyne	dyn	= 1E-5 N (Newton)
1 kilogram-force	kgf	= 9.81 N
1 ton-force	tf	= 9810 N
1 pound-force	lbf	= 4.448 N

STRESS

1 megapascal	MPa	= 1E6 Pa (Pascal)
1 dyne per cm^2	dyn/cm^2	= 0.1 Pa
1 Newton per m^2	N/m^2	= 1 Pa
1 Joule per m^3	J/m^3	= 1 Pa
1 atmosphere	atm	= 1.013E5 Pa
1 mm mercury (torr.)	mm Hg	= 133.3 Pa
1 mm water	mm H_2O	= 9.81 Pa
1 bar	bar	= 1E5 Pa
1 pound per $inch^2$	psi	= 6890 Pa
1 kgf per cm^2		= 9.81E4 Pa
1 megapascal	MPa	= 145 psi

VISCOSITY

1 poise	poise	= 0.1 Pas
1 poise	poise	= 1 $dyns/cm^2$
1 pound-sec per in^2	psisec	= 6897 Pas
1 poise	poise	= 14.5E-6 psisec

ENERGY/WORK

1 erg	dyn·cm	= 1E-7 J (Joule)
1 Newton-meter	N·m	= 1 J
1 watt-second	W·s	= 1 J
1 kgf-meter	kgf·m	= 9.81 J
1 foot-pound	ft·lbf	= 1.356 J
1 horsepower-hour	hph	= 2.685E6 J
1 kilowatt-hour	kWh	= 3.60E6 J
1 calorie	cal	= 4.19 J
1 Brit. thermal unit	Btu	= 1055 J
1 inch-pound	in·lbf	= 0.113 J

POWER

1 kilowatt	kW	= 1000 W (Watt)
1 horsepower	hp	= 746 W
1 foot-pound per sec	ftlbf/s	= 1.356 W

SPECIFIC ENERGY

1 calorie per gram	cal/gr	= 4190 J/kg
1 Btu per pound	Btu/lb	= 2326 J/kg
1 hph per pound	hph/lb	= 5.92E6 J/kg
1 kWh per kg	kWh/kg	= 3.60E6 J/kg
1 kWh per kg	kWh/kg	= 0.608 hph/lb

THERMAL CONDUCTIVITY

1 cal/cm·s·°C	= 419 J/ms°K
1 kcal/mh°C	= 1.163 J/ms°K
1 Btu/ft h°F	= 1.73 J/ms°K
1 Btu/in/ft^2 h°F	= 0.144 J/ms°K
1 Btu/ft s °F	= 6230 J/ms°K
1 W/m°K	= 1 J/ms°K

NOMENCLATURE

Lower Case Roman Characters

a = absorptivity

a = acceleration

a_g = gravitational acceleration

a_j = eigenvalues

a_T = shift factor

b = axial flight width

c = constant

c_i = expansion coefficients

e = emissivity

f = coefficient of friction

f = frequency

ff = flow factor

g = gravitational acceleration

g = pressure gradient

h = heat transfer coefficient

h = height

i = integer number

j = integer number

k = stress ratio

k = thermal conductivity

l = l-coordinate

m = consistency index

m = mass

n = power law index

p = number parallel flights

q = heat flux

r = r-coordinate

r = radius

r = striation thickness

s = pseudoplasticity index ($1/n$)

t = time

v = velocity

w = perpendicular flight width

$\hat{w}$ = reduced flight width

x = x-coordinate

y = y-coordinate

z = z-coordinate

Upper Case Roman Characters

A = area

A_l = degree of taper in l-direction

A_z = degree of taper in z-direction

B = axial channel width

C = concentration

C = constant

C_p = specific heat at constant pressure

C_v = specific heat at constant volume

C_B = Boltzmann constant

C_{PL} = Planck constant

C_{SB} = Stefan-Boltzmann constant

D = diameter

D' = diffusion coefficient

E = elastic modulus

E = electric field strength

E = energy

$\dot{E}$ = rate of energy change

$\dot{E}$ = rate of evaporation

F = force

FF = flow function

G = Gibbs free energy (thermodynamics)

G = shear modulus

H = channel depth (screw geometry)

H = enthalpy (thermodynamics)

$\hat{H}$ = specific enthalpy

I = current

I = intensity

I_i = principal invariants

I = moment of inertia

J = diffusional mass flow rate per unit area

J = dimensionless solids conveying rate

J_v = Bessel function of the first kind of order v

K_v = Bessel function of the second kind of order v

L = length

M = mass

M_c = molecular weight between crosslinks

MI = melt index

$\dot{M}$ = mass flow rate

N = screw speed

N_A = Avogadro's number

N_{Bi} = Biot number

N_{Br} = Brinkman number

N_{Fo} = Fourier number

N_{Gz} = Graetz number

N_{Na} = Nahme number

N_{Nu} = Nusselt number

N_{Pe} = Peclet number

N_{Pr} = Prandtl number

N_{Re} = Reynold's number

N_1 = first normal stress difference

N_2 = second normal stress difference

P = pressure

Q = heat

$\dot{Q}$ = heat flow rate

R = gas constant

R = radius

S = entropy

S = pitch

T = temperature

T = torque

T_g = glass transition temperature

T_m = melting point

U = internal energy

V = voltage

V = volume

$\hat{V}$ = specific volume

$\dot{V}$ = volumetric flow rate

W = perpendicular channel width

W = work

X = stage efficiency (devolatilization)

Y = degree of fill

Y = number of complexions

Z = distance in z-direction

Z = power consumption

$\hat{Z}$ = specific power consumption

Lower Case Greek Characters

α = extension ratio

α = hopper half-apex angle

α = thermal diffusivity

α_c = concentration coefficient of diffusion coefficient

α_i = angle of intermesh

α_p = pressure coefficient of viscosity

α_t = tip angle

α_T = temperature coefficient of viscosity

β = cone angle (cone and plate rheometer)

β_e = effective angle of internal friction

β_i = angle of internal friction

β_s = angle of slide

β_w = wall angle of friction

γ = shear strain

$\dot{\gamma}$ = shear rate

δ = flight clearance

δ = loss angle

ε = degree of fill

ε = efficiency

ε = elongational strain

ε = permittivity

$\dot{\varepsilon}$ = elongational strain rate

η = viscosity

θ = dimensionless temperature

θ = solids conveying angle

κ = compressibility

λ = exposure time (degassing)

λ = time constant

λ = wavelength (radiation)

μ = Newtonian viscosity

ν = Poisson's ratio

ξ = dimensionless depth

ρ = density

σ = standard deviation

σ = stress

τ = stress

φ = helix angle

φ = phase angle

φ_i = dynamic angle of internal friction

χ = interaction parameter

ψ = angle between major principle stress and r-axis

ψ = melting rate term defined by equation 8-91 a.

ω = angular frequency

ω = angular velocity

l = length

Upper Case Greek Characters

$\Gamma(p)$ = gamma function

Γ_R = reduced pressure gradient

Δ_{ij} = rate of deformation tensor

Φ = specific fluidity

Ψ = flight flank angle

Ω = melting rate term defined by equation 7-108

AUTHOR INDEX

SUBJECT INDEX

Biography

Chris Rauwendaal earned a post graduate degree in Mechanical Engineering from Delft University of Technology in the Netherlands and a doctorate in Polymer Processing from Twente University of Technology in the Netherlands. He worked 4 years in the man-made fiber industry for American Enka Company, now BASF Fibers, involved in process development in the area of melt spinning and for 12 years at Raychem Corporation of Menlo Park, California, where he was manager of Process Research in Corporate R & D and manager of Extrusion Technology in Corporate Manufacturing. He is now president of Rauwendaal Extrusion Engineering, Inc., a company that offers extrusion engineering services to the plastics industry, including material analysis, extruder screw and design, and training for operating personnel and process engineers.

Chris Rauwendaal has taught for U. C. Berkeley Extension and has given guest lectures at the University of Minnesota at Minneapolis, the University of Massachusetts at Amherst, and the University of Akron, Ohio. Since 1979 he is a regular seminar instructor for the Society of Plastics Engineers, as well as for the Plastics and Rubber Institute in England, teaching seminars on extrusion. He was technical program chairman for the 1987 Antec for the Extrusion Division and was the 1988 Chairman of the Board of Directors for the Extrusion Division of SPE. He twice received the best paper award from the Extrusion Division of the SPE.

He holds several patents in the field of polymer processing and has authored over forty papers on extrusion, a video training course on Extrusion Technology, written and edited a book on mixing titled "Mixing in Polymer Processing", as well as chapters in various handbooks.